ADVANCED ENGINEERING MATHEMATICS

VOLUME ONE

with *Mathematica*® and MATLAB®

REZA MALEK-MADANI

UNITED STATES NAVAL ACADEMY

▲ ADDISON-WESLEY

An imprint of Addison Wesley Longman, Inc.

Reading, Massachusetts • Menlo Park, California • New York • Harlow, England
Don Mills, Ontario • Sydney • Mexico City • Madrid • Amsterdam

Sponsoring Editor: Jennifer Albanese
Assistant Editor: Carolyn Lee
Production Supervisor: Rebecca Malone
Marketing Coordinator: Michael Boezi
Prepress Services Buyer: Caroline Fell
Manufacturing Supervisor: Ralph Mattivello
Cover Designer: Barbara T. Atkinson
Cover Photograph: O. Brown, R. Evans, and M. Carle, University of Miami
                 Rosenstiel School of Marine and Atmospheric Science
Project Management, Composition, and Prepress Services: Techsetters, Inc.

Library of Congress Cataloging-in-Publication Data

Malek-Madani, Reza.
   Advanced engineering mathematics with Mathematica and Matlab /
Reza Malek-Madani.
      p.   cm.
   Includes bibliographical references and index.
   ISBN 0-201-59881-7 (v. 1). -- ISBN 0-201-32549-7 (v. 2)
   1. Engineering mathematics--Data processing.   2. MATLAB.
3. Mathematica (Computer file)   I. Title.
TA345.M34   1997
620'.001'51-- dc21   97-22716
                                                              CIP

1 2 3 4 5 6 7 8 9 10—MA—00999897

*This book is dedicated
to my wife Jo*

to Behzad and Darob

to Ben and Grace

*and to the memory of Mahmoud and Shahnaz
your place is empty*

Contents

Volume Two

Preface

This book is about the basic tools of mathematics as they apply to formulating and solving problems in fluid dynamics, solid mechanics, and electromagnetism. The material covered in this text is based on the notes I have developed during the past several years while teaching two courses on the methods of applied mathematics. In one course, intended for sophomore and junior level students, the three fundamental topics of applied mathematics are presented: linear algebra, differential vector calculus, and partial differential equations. In the other course, whose audience is primarily junior and senior students, the focus is on the basic mathematics needed to understand the concepts of stress and strain in continuum mechanics and to develop the governing equations. This text fits the needs of both courses.

One of the important features of this text is that it incorporates a numerical analysis package and a symbolic manipulator as teaching tools throughout the manuscript. These software packages are fully integrated in every chapter of the text and are not presented as supplements to the *Mathematica* exercises. More precisely, my primary goals in writing this text have been

1. to introduce the basic concepts of applied mathematics in a setting that lends itself to the use of symbolic manipulators and advanced numerical packages.

2. to introduce realistic physical applications that have finer structures than those commonly offered in texts of this nature.

3. to assemble a collection of projects that could be used both as a means for further study of a particular topic of interest as well as a vehicle for producing a writing assignment for students.

I hope that the body of the text will show that these three goals are the motivating force behind the structure of topics and the choice of problems.

Today, more than ever before, the availability of sophisticated computer packages has made it possible for us to bring realistic physical modeling into the classroom at an early stage of our students *Mathematica* education. The computer has give us the ability to spend less time on technicalities and more time on concepts and finer structures, which we used to postpone until more advanced courses.

In the past ten years, the two software packages, *Mathematica* and MATLAB have offered the teachers of applied mathematics an opportunity to rethink the traditional way that the basic topics are introduced to students. Both the packages are extremely powerful tools in teaching and in research because of their underlying structure and logic and, more importantly, because of the ease with which a user interacts with them. Both packages provide the interested student with a sophisticated and interactive calculation power that relieves him of the usual intellectual drawback of tedious computations. Both have excellent graphics.

The examples and exercises in this text are primarily geared toward using these packages. *Mathematica* is represented in a large proportion of the examples and exercises in the text, however, and symbolic manipulator that has this capability (such a MATLAB, *Derive*, *Maple*) could be substituted for *Mathematica*.

Many of the examples and applications that I introduce have their origin in wave propagation and in fluid dynamics. Aside from the fact that the basic concepts of mathematics enter these applications in a natural way, these problems could give the instructor an avenue for describing some of the research issues in applied mathematics today. As an important ingredient for learning to apply the material in this text, in the past I have asked each of my students to start an individual project on a topic in his or her major that is directly related to some of the concepts that have been presented in the course. These projects have ranged from understanding the *Mathematica* definition of the Coriolis force, to the derivation of equations of shallow water waves, to understanding how acoustic waves travel in a block of ice as opposed to a bar made of steel. Each project has had a strong element of computer application. Some of the exercises and the appendices in the text are direct results of such projects.

The text is organized into two volumes: Volume One contains Chapters 1–7, and Volume Two consists of Chapters 8–16. Chapters 1 and 2 introduce the basic syntax of *Mathematica* and MATLAB.

Chapters 3 and 4 review ordinary differential equations. The projects at the end of these chapters include problems that deal with regular perturbation analysis for initial value problems and the shooting method for boundary value problems. These projects are nontrivial examples of programming in *Mathematica* and MATLAB.

Chapter 5 introduces the Laplace transform for ordinary differential equations. This topic, when combined with *Mathematica* and MATLAB, seems to bring home to students why a symbolic algebra system could play an important role in the analysis of problems.

Chapter 6 is an introduction to matrix algebra. The emphasis of this chapter is on making clear the connection between linear transformations and matrices. Special exercises are constructed to emphasize the relation between linear transformations and the stretching and compression of the domain. A section is devoted to the discretizations of a boundary value problem and the role matrix algebra plays in such problems. This section provides another example of the type of projects that the students may pursue further. It also shows the power of *Mathematica* and MATLAB in the implementation of numerical codes.

The matrix theory developed in Chapter 6 is then used in solving linear systems of differential equations in Chapter 7. A treatment of elementary stability theory is also included. In this chapter the student is introduced to standard examples of systems of differential equations in the engineering literature, such as the system that describes the fluid flow past a cylinder and equations that model the Rayleigh–Bénard instability, among others.

Chapter 8 is an introduction to vector and differential calculus. The concepts of gradient, divergence, and curl and their manipulations in *Mathematica* and MATLAB are covered. The student is also introduced to the change of variables from rectangular to polar coordinates and its ramifications for the standard vector operations. The concepts of incompressibility and irrotationality of flows and how they relate to the structure of vector fields are discussed.

Chapter 9 contains two examples of steady state flows, one modeling the flow in a rectangular box (or the flow in a rectangular bay), and the other the flow past a cylinder. At this time the students explore further the tools from vector calculus in the context of a few concrete physical problems. Moreover, the students are introduced to an example of separation of variables.

There are several reasons why introducing the fluid flow problems in this chapter seem appropriate at this juncture of the text. For one, these problems give the student the chance to review the topics of the previous chapters, particularly in context of focused physical problems. Second, this study brings up the important issue in applied mathematics that vector fields are not known explicitly, that the details of their structures are hidden within certain equations whose solutions we seek, that the equations come to us from physics, and that the process of understanding a vector field is an implicit or an inverse problem. Third, both problems lend themselves remarkably well to the type of computations for which the aforementioned software packages are suited well.

In Chapter 10 we return to vector calculus and introduce the integral theorems. Some elements of potential flows are discussed. In two of the projects at the end of this chapter Kelvin's circulation theorem and Strommel's model of the Gulf Stream are studied in detail.

Chapter 11 introduces the basic concepts of complex variables. The emphasis is on the role that analytical functions play in the special fluid flows that possess a pair of potential and stream functions.

In Chapter 12 we take up the rigorous treatment of the Navier–Stokes equations. Their derivations require using the integral theorems of the previous chapters. Some examples of the exact solutions of this system are then studies.

Chapter 13 deals with solving partial differential equations using the characteristics method. I have found this topic invaluable in describing to the students the basic issues in wave propagation. It also lays the foundation for a large class of projects involving waves, whether in a fluid or in a solid medium. The concepts of nonlinearity and shock waves are introduced. In a project at the end of this chapter the problem of the reflection and transmission waves of an incident wave in a medium that changes density is discussed. The characteristics method is used to find formulas for these waves and then *Mathematica* is employed to plot snapshots of the evolution of the solution.

In Chapter 14 the standard results Fourier series are presented. The linkage between this method and the characteristics method of the previous chapter is described. Several examples and exercises demonstrate how *Mathematica* and MATLAB are used to plot solutions of partial differential equations in one dimension and higher-dimensional spaces. The projects for this chapter include the normal mode analysis of the vibration of a heavy string and a vibrating membrane.

Chapter 15 introduces the concept of Fourier transform and its application in obtaining solutions of partial differential equations in unbounded domains.

Chapter 16 concerns the Laplace transform and partial differential equations.

Acknowledgments

I would like to thank the staff of Addison Wesley Longman, Laurie Rosatone, Marianne Lepp, and Jennifer Albanese for all their patience, expertise, and professional effort in seeing this book's production from its inception to its conclusion. My colleagues at the Naval Academy, Sonia Garcia, John Pierce, David Smith, Mario Vieira, and Gabriel Karpouzian provided much support and advice, especially when the book was in its formative stages. My special thanks go to midshipmen Tracey Delk, Chris Irwin, Ryan Eller, Julie Preyer, Julie Wilhelmi, Camille Garrett, Brent Strong, and James Coleman for providing the energy and enthusiasm that is so crucial for a project of this magnitude. I would like to express my heartfelt thanks to the staff of the Computer Aided Design and Interactive Graphics (CADIG), Bob Disque, Don Garner, Lisa Becktold, Carolyn Mayr, and Linda Adlum who were always there and ready to help. The support groups at Wolfram, Inc. and MathWorks, Inc. have my sincerest gratitude for making my life easier by providing such excellent resources.

During the creative stages of this project I was fortunate to have the friendship of Raouf Ali Raouf, a Professor of Mechanical Engineering at the Naval Academy. An energetic, bright, and forward-looking individual, Raouf was a pioneer in recognizing the benefits of applying the new computer technology to teaching and research. His tragic death deprived us all of an invaluable source of courage and energy.

Not unlike the meandering Gulf Stream, this project changed course a few times before steadying itself. Each reconstruction has required an enormous investment of time. To my wife Jo and to our boys Behzad and Darob, who so patiently awaited the completion of this project, I dedicate this book.

Reza Malek-Madani

Chapter 1

An Introduction
to *Mathematica*

1.1 Introduction

This chapter contains a brief introduction to *Mathematica*. The first few sections
will help the reader become familiar with the general syntax of this package. A
glossary at the end of the chapter gives a partial listing of the available commands
with examples of their usage.

One of the most important aspects of any software package is how readily
one is able to access the information from the output of a command and redirect
it as input to another. For example, throughout the text we will make extensive
use of the differential equation solver of *Mathematica* in various applications,
where monitoring the evolution of certain variables is key to our understanding
of a physical model. Typically, we would like to graph some or all of the variables
we obtain from solving a system of differential equations as time varies, or graph
one variable versus another, or integrate and differentiate the output in order
to construct physical quantities that have natural interpretations. This chapter
is an introduction to how one can accomplish such tasks with simple examples
that are similar to the actual circumstances we encounter in the remainder of
the text.

The material presented here is intended primarily to be used as reference for
the exercises in the upcoming chapters. As a result, during the first reading some
of the mathematical language may be unfamiliar. It is hoped that this chapter
will become more useful to the reader as one proceeds with the mathematical
concepts of the future chapters but returns frequently here to review the relevant
syntax.

Using *Mathematica* one is able to

1. determine roots of polynomials, sum series, and evaluate limits of sequences;

2. integrate and differentiate symbolically rather complicated expressions;

3. generate graphics in two and three dimensions;

4. simplify trigonometric and algebraic expressions;

5. solve linear and nonlinear differential equations;

6. determine the Laplace and Fourier transforms of functions,

plus a variety of other mathematical operations. The notation we employ is directed at the usage of *Mathematica* on a SUN workstation. This notation is applicable to other platforms as well, including the Windows version for *Personal Computers* (PCs), after some minor changes that will be clear to the reader.

1.2 A Session in *Mathematica*

There are generally two ways to have *Mathematica* up and running: Either as a stand-alone application (available in DOS and on SUN workstations) or as a Notebook in conjunction with a Windows application such as Windows 3.1 or with xwindows on SUN workstations. The stand-alone version of *Mathematica* is initiated by executing `math.exe` (on PCs) or `math` at the UNIX prompt on a SUN workstation, at which time the package responds with

`In[1]:=`

The program is expecting an input from the user. Commands are entered by typing at the terminal and executed by pressing the **Return** (or **Enter**) key.

In the Notebook version of *Mathematica* one either selects the appropriate icon and executes the application (this is the case on PCs and Apple computers) or enters

`mathematica`

at the system prompt, as is the case on SUN workstations. After some initialization, a window opens in which one is allowed to enter commands. Commands are entered at the keyboard and executed by pressing the **Shift** and **Return** (or **Enter**) keys **simultaneously**. After the first command is executed `In[1]:=` will appear on the screen.

As mentioned earlier, commands are simply entered at the keyboard. For example, to find the roots of the polynomial

$$f(x) = x^2 - 4x + 3,$$

type

`Solve[x^2 - 4 x + 3 == 0,x]`

The program will respond with

`Out[1]=`

`    {{x -> 1}, {x -> 3}}`

that is, 1 and 3 are the roots of the polynomial.

Remark 1.2.1: _Mathematica_ distinguishes between x=y and x==y. In the first expression, $x = y$, x is assigned the value y, while in the second expression _Mathematica_ checks to see if x and y can be equal to each other, that is, it verifies that x and y are compatible objects, and other than that it does not take any action. We will use == to define equations.

Remark 1.2.2: Multiplication in _Mathematica_ can be entered either by leaving a blank space between the operands (for example, a x represents ax) or by putting an asterisk between the terms (that is, a*x). If a is a number, however, _Mathematica_ interprets the combination of a number adjacent to a variable as multiplication. For instance, 2x is understood as 2 times x. This convention does not apply to symbols: the expression ax is interpreted as the variable whose name is ax and not as the product of the variables a and x.

Remark 1.2.3: We will ignore the prompts In[]:= and Out[]:= for the rest of this discussion.

The first important feature of _Mathematica_ is that this program is **case sensitive**; that is, it distinguishes between solve and Solve in the first command that we entered. All functions and programs that are **internally** known to _Mathematica_ must be capitalized. Thus, Solve is a subroutine in _Mathematica_ that is capable of finding roots of polynomials, while solve does not have a special meaning unless the user has defined it previously. Similarly, _Mathematica_ understands that Sin is the usual sine function while sin is a name of a variable that is unknown in this session.

Functions in _Mathematica_ are delimited by [ ] and not by ( ). So Sin[x] is the usual sine function while Sin(x) is the variable Sin multiplied by the variable x. The operator Solve is an example of a function with two arguments; the first is used to define the equation whose solutions we are interested in, and the second is the variable with respect to which the roots should be computed. All of these arguments are grouped by [ ]. To get a better understanding for the usage of arguments in _Mathematica_ let us try the following two examples:

```
b = Solve[x^2 - a x + 3 == 0, x]
```

and

```
c = Solve[x^2 - a x + 3 == 0, a]
```

Do the outputs make sense? In b the symbol x is the variable with respect to which the roots of the polynomial are determined and a is just a parameter, while in c the roles of x and a are reversed.

The above examples show the most important attribute of _Mathematica_: its ability to do symbolic manipulation. In b _Mathematica_ does not need to know the value for a in order to find the roots of the second-order polynomial. This is just one example among many in which *Mathematica* uses its logical power and is able to find the answer to certain questions. There are limitations to this ability, unfortunately. Try finding the roots of the following polynomial on *Mathematica*:

$$x^5 - ax^4 + 3x^3 + 2x^2 + x - 1 = 0.$$

Mathematica's response is its way of saying that it does not know how to find symbolically the roots of fifth-order polynomials. This answer is in agreement with a celebrated theorem in group theory that confirms our inability to write formulas for roots of general polynomials with degrees larger than or equal to five. On the other hand, by simply drawing the graph of this fifth-order polynomial for a fixed a, we recognize that it has some real roots. *Mathematica* is able to find these roots **numerically**, that is, by approximate methods. To accomplish this one needs to furnish a concrete value of the parameter a in the definition of the polynomial. Try

```
NSolve[x^5 - x^4 + 3 x^3  + 2 x^2 + x -1 == 0 ,x]
```

Mathematica finds all five roots of this polynomial:

```
{{x -> -0.533065 - 0.563943 I},

 {x -> -0.533065 + 0.563943 I}, {x -> 0.425529},

 {x -> 0.820301 - 1.7971 I},

 {x -> 0.820301 + 1.7971 I}}
```

NSolve is a program based on a numerical approximation technique that is capable of finding solutions to certain equations; in particular, it is capable of finding roots of polynomials.

In addition to finding roots of polynomials, both Solve and NSolve are capable of finding zeros of general functions as well as solutions to systems of simultaneous algebraic equations. Consider the function $f(x) = \sin \sqrt{x} - \frac{1}{2}$. Its curve shows that f has a zero in the interval $(0, 2)$. We find its value by using

```
Solve[Sin[Sqrt[x]]-1/2==0,x]
```

Mathematica first gives the warning

```
Solve::ifun:
   Warning: Inverse functions are being used by
      Solve, so some solutions may not be found
```

and then continues with the result

```
          2
        Pi
{{x -> ---}}
        36
```

This function, of course, has infinitely many zeros, as its graph shows. We find the solution to the following system of algebraic equations

$$2x - 3y = a, \quad 3x + 2y = b,$$

by entering

```
Solve[{2 x - 3 y == a, 3 x + 2 y == b}, {x, y}]
```

It results in

```
Out[4]=
         1                      1
{{x -> -- (2 a + 3 b), y -> -- (2 b - 3 a)}}
        13                     13
```

Problems

1. Use ? (see the next section for more detail) with `Solve` and `NSolve`, and familiarize yourself with the syntax of these commands.

2. Use `Solve` or `NSolve` to determine the roots or zeros of the following expressions.

   (a) $ax^2 + bx + c$
   (b) $ax^3 + bx^2 + cx + d$
   (c) $x^2 + 1$
   (d) $x^3 + 1$
   (e) $x^{1/3} - x + 1$
   (f) $\sin x - \frac{1}{3}$
   (g) $\sin^2 x - \frac{1}{3}$
   (h) $\sin x^2 - \frac{1}{3}$
   (i) $\sin x - x$. Familiarize yourself with the syntax of the `FindRoot` command by using ?. Then use `FindRoot` in this problem.

3. Determine the solution of the following equations using `Solve`, `NSolve` or `FindRoot`.

   (a) $\tan x - 3x + 1$
   (b) The system of algebraic equations

        i. $3x - 2y = 2, \quad x + y = 7$
        ii. $ax - y = 0, \quad x + ay = 1$
        iii. $x^2 - y^2 = 1, \quad x^2 + y^2 = 4$
        iv. $x^3 - y^3 = 1, \quad x^2 - 3xy + y^2 = 8$
        v. $ax + y + z = 1, \quad x - y + 2z = 0, \quad 2x + 3y - z = 2$
        vi. $3x^2 - 4y^2 + 3z = 1, \quad x + y + z = 0, \quad z^3 - x^2y = -1$

1.3   The Help Command in *Mathematica*

The help command in *Mathematica* is

`?<command>`

For example, to gain information about the internal function `Solve` enter

`?Solve`

Mathematica responds with (this is the response of version 2.2 of *Mathematica* on a SUN platform)

```
Solve[eqns, vars] attempts to solve an equation or
set of equations for the variables vars.
Solve[eqns, vars, elims] attempts to solve the equations for
vars, eliminating the variables elims.
```

In response to

`?Sol*`

Mathematica, which treats the asterisk as a wild character, returns

```
SolutionOf   Solve       SolveAlways  SolveDelayed
```

which is a list of all of the commands that have `Sol` as their beginning characters. We can now obtain more information about each command by using the full name with `?`. The command

`??<string>`

provides even more detailed information about `string` than `?string` does, including the syntax of options that sometimes are not accessed through `?string`. For instance,

`??Solve`

returns

```
Solve[eqns, vars] attempts to solve an equation
or set of equations for the variables vars.
Solve[eqns, vars, elims] attempts to solve the equations for vars,
eliminating the variables elims

Attributes[Solve] = {Protected}

Options[Solve] =
  {InverseFunctions -> Automatic, MakeRules -> False,
    Method -> 3, Mode -> Generic, Sort -> True,
      VerifySolutions -> Automatic,  WorkingPrecision -> Infinity}
```

The double query `??` is particularly helpful when the user becomes more familiar with the inner workings of *Mathematica*, since it provides a list of options that one can manipulate in order to obtain a better and more customized output.

1.4 Factoring and Simplification

Some of the elementary operations that *Mathematica* is capable of performing include expanding, factoring, and simplifying expressions. For instance, the command for expanding

$$(a + b + c)^3$$

is

```
Clear[a, b, c];
Expand[(a+b+c)^3]
```

The first command clears any previous values assigned to a, b, and c. *Mathematica* responds with

```
 3       2         2    3      2                    2
a  + 3 a  b + 3 a b  + b  + 3 a  c + 6 a b c + 3 b  c +
            2       2    3
        3 a c  + 3 b c  + c
```

We can now factor the above expression (assuming that it is stored in Out[4] of the present session) by entering

```
Factor[Out[4]]
```

thereby recovering the original expression. We note that, in place of `Factor[Out[4]]`, we could use

```
Factor[%]
```

to reach the same result if the expression on which `Factor` acts is the latest output of the session.

The operations of `Expand`, `Factor`, and `Simplify` are very useful when we wish to prove the type of identities commonly encountered in elementary algebra. For example, recall the identity

$$(a + b)^2 - (a - b)^2 = 4ab$$

for any two parameters a and b. To verify this identity on *Mathematica* we enter

```
Clear[a,b];
Expand[(a+b)^2 - (a-b)^2]
```

If we repeat the above command but replace `Expand` with `Factor`, we obtain the same result. Now, let us try the previous three commands on

$$a^2 - 2ab + b^2.$$

`Factor` and `Simplify` give us the appropriate alternative expression for $a^2 - 2ab + b^2$, while `Expand` leaves the expression unchanged. Next, we apply these three commands to the expression

$$\frac{1}{x+t} - \frac{1}{x-t}$$

to get a better sense for the range of capabilities of each the above internal functions: `Expand` returns the original expression whereas `Factor` returns

```
    2 t
----------------
(t - x) (t + x)
```

while `Simplify` gives us

```
  2 t
-------
 2    2
t  - x
```

Mathematica is also capable of manipulating trigonometric functions. For example,

`Simplify[Sin[x+y]-Sin[x-y]]`

returns the result we expect from elementary trigonometry. To get a sense for the power of *Mathematica*'s `Simplify` command let us try

`Simplify[((Sin[2 x])^2+(Cos[2 x])^2)^2]`

Does the answer make sense?
    We note that

`Simplify[Sin[x + y]]`

returns the same expression as its output, since `Sin[x + y]` is already in its simplest form. However,

`Simplify[Sin[x +  y] - Sin[x] Cos[y] - Sin[y] Cos[x]]`

returns the value we expect from this identity.

Problems

1. Verify the following identities in *Mathematica*.

   (a) $\sin 2x = 2\sin x \cos x$
   (b) $\cos 2x = 2\cos^2 x - 1$
   (c) $\sin 3x = 4\sin^3 x - 3\sin x$
   (d) $\tan(x + y) = \frac{\tan x + \tan y}{1 - \tan x \tan y}$

2. Use *Mathematica* to discover an identity among the given functions.

   (a) $\cot(x + y)$, $\cot x$, and $\cot y$
   (b) $\sin(4x)$, $\sin x$, and $\cos x$
   (c) $\cos x$ and $\cos \frac{x}{2}$
   (d) $\tan 2x$ and $\cos x$

3. Simplify the following expressions.

   (a) $\cos(x + y) + \cos(x - y)$

   (b) $\cos(x + y) - \cos(x - y)$

   (c) $\cos(x + y) + \sin(x + y)$

   (d) $\cos^2 2x - \sin^2 2x$

   (e) $\cos^2 ax - \sin^2 ax$

   (f) $\cos^3 x - \sin^3 x$ (Try `Factor` followed by `Simplify`.)

1.5 Function Definition

We often need to communicate to *Mathematica* that x in $f(x)$ is a variable and should be viewed as such when we wish to determine the function's zeros, plot it, differentiate, or integrate it. The mechanism for defining x as a variable or a placeholder is provided by the underscore character _. This character is typed immediately after the letter x the first time x is encountered in the definition of f; the underscore character is *not* used in the subsequent referrals to f. For instance, the function

$$f(x) = x \sin^2 x \tag{1.1}$$

is defined by

```
Clear[x, f];
f[x_] = x (Sin[x])^2
```

Once $f(x)$ is defined in *Mathematica* we may refer to it by name in subsequent operations. For example,

```
Solve[f[x]==0,x]
```

requests that *Mathematica* find the roots of (1.1). Its response shows that f has a triple zero at $x = 0$. To see that the argument of f is a dummy variable, let us try

```
Solve[f[t]==0,t]
```

which leads to the same conclusion.

It is also possible to define functions of more than one variable in *Mathematica*. For example, the second-order polynomial $g(x) = x^2 - ax + 3$ is defined by

```
Clear[a,x,g]
g[x_,a_] = x^2 - a x + 3
```

and the operation of finding its roots is accomplished by entering

```
Solve[f[x,a]  == 0, x]
```

Many internal functions in *Mathematica* take as input expressions that involve functions. We have already seen the example of f in (1.1) and the internal function Solve. Another example arises with FindRoot, a variant of the Solve command. This internal function computes roots of functions numerically and is preferable to Solve in cases where the function f is rather complicated and one must resort to approximate methods to seek its roots. Its syntax requires specifying an initial guess for a root. For instance,

FindRoot[f[x], {x, 3}]

determines a root of f by applying a root-finding algorithm to f, based on Newton's method, starting the algorithm at the point $x = 3$. *Mathematica* returns ⏸

{x -> 3.141069642608009}

which is a good (although not a very good) approximation to the exact value π. There are, however, two options available to FindRoot that render the approximate value much closer to the exact value (try ??   FindRoot for the list of options). They are MaxIterations and WorkingPrecision. When we try

FindRoot[f[x], {x, 3}, MaxIterations->50, WorkingPrecision->25]

we end up with

{x -> 3.14159263763142393334862203408029469154`25}

which is accurate to eight digits. Some of the trailing digits in the above answer may differ on your machine. The above calculation was performed on a SUN workstation and PC 486 using version 3.0 of *Mathematica.*

The reader may recall that Newton's method is based on the tangent vector approximation of f, which requires differentiating f. In cases where it is cumbersome to compute the derivative of f, the secant method, which replaces the tangent vector with a chord passing through two points on the graph of f, is preferable. One of the options in FindRoot allows for specifying two starting points, at which time a variant of the secant method is called upon. For example,

FindRoot[f[x] == 0, {x, 3, 4}]

uses function evaluations at $x = 3$ and $x = 4$ to start the approximation algorithm.

Problems

1. Define the following functions in *Mathematica* and evaluate them at the specified points.

   (a) $f(x) = \frac{1-x}{1+x}$; $x = 0, 0.5$, and π

   (b) $f(x) = \log(x + \sqrt{1 - x^2})$; $x = 0, 0.1, 0.2$, and 0.3

2. Consider the function f defined in (1.1). Use `FindRoot` to determine an eight-digit accurate approximation to its root at $x = 0$. Start with an initial guess at $x = 1$.

3. $f(x, t, \omega) = \sin(x - \omega t)$. Use `FindRoot` and find a root of $f(1, t, 4)$ starting with $t = 0$. Repeat with $t = 1$.

4. Use `FindRoot` to find a root of the function f. Experiment with different initial guesses and report on whether the value of the root depends on the initial guess.

   (a) $f(x) = x - \frac{2}{x}$

   (b) $f(x) = x^2 - \frac{2}{x}$

   (c) $f(x) = \frac{x^2 - 1}{x - 2} + x$

   (d) $x \sin x + \cos x$

1.6 Differentiation and Integration

Mathematica is able to differentiate and integrate functions symbolically. The derivative of a function such as $x \sin^2 x$ is found by

```
D[x Sin[x]^2,x]
```

while its integral is determined by entering

```
Integrate[x Sin[x]^2,x]
```

The D and **Integrate** commands of *Mathematica* use the basic properties of the differentiation and integration operators, such as the linearity property, to reduce complicated computations to a series of simpler ones. These properties are then combined with elaborate tables of known derivatives and integrals that allow this software to reach its goal successfully. The power of this software is particularly noticed in the case of integration, where we recall from elementary calculus that we often have to resort to methods such as partial fractions, or special substitutions, or integration by parts to reduce the integrand to a manageable expression. For example, to evaluate the integral

$$\int \frac{1}{1 + x^4} dx$$

using standard tables, we must first note that $1 + x^4$ factors into

$$(1 - \sqrt{2}x + x^2)(1 + \sqrt{2}x + x^2)$$

and apply the method of partial fraction before using the table of integration. On the other hand,

```
Simplify[Integrate[1/(1+x^4),x]]
```

yields

```
(-2*ArcTan[1 - Sqrt[2]*x] + 2*ArcTan[1 + Sqrt[2]*x] -
    Log[-1 + Sqrt[2]*x - x^2] + Log[1 + Sqrt[2]*x + x^2])/
  (4*Sqrt[2])
```

The definite integral

$$\int_0^1 \frac{1}{1+x^4} dx \tag{1.2}$$

is determined in a similar fashion:

```
Simplify[Integrate[1/(1+x^4), {x, 0, 1}]]
```

which results in

```
(-2 ArcTan[1 - Sqrt[2]] + 2 ArcTan[1 + Sqrt[2]] -

   Log[2 - Sqrt[2]] + Log[2 + Sqrt[2]]) / (4 Sqrt[2])
```

To get a decimal approximation to the above value, we apply the N operation to it:

```
N[%]
```

recalling that % stands for the previous output. The new output is

```
0.8669729873399109
```

   Mathematica is also capable of performing numerical integration. The internal function NIntegrate returns an approximate value of the function on which it operates. For example, to evaluate (1.2) numerically we enter

```
NIntegrate[1/(1 + x^4), {x, 0, 1}]
```

in *Mathematica*, which compares well with the result of the numerical value we obtained after evaluating this integral exactly.

   In spite our best effort, it is fair to say the class of functions whose anti-derivative we are able to write down explicitly is rather small. If we just started to list functions at random, we could quickly generate functions whose anti-derivatives are either cumbersome to evaluate or actually impossible to express in terms of elementary functions of calculus. Many functions in mathematical physics fall in the latter category, among them e^{-x^2}, $\sin(x^2)$, and $\frac{1}{\sqrt{1-m\sin^2 x}}$. As a result of efforts of many mathematical analysts in the past few hundred years, properties of such anti-derivatives are tabulated, which are now generally available in most computer algebras, including in *Mathematica*. For example, the function

$$f(x) = \int_0^x e^{-t^2} dt$$

can be accessed as

```
f[x_] = Integrate[Exp[-t^2], {t, 0, x}]
```

Mathematica responds with

```
(Sqrt[Pi]*Erf[x])/2
```

The internal function `Erf[x]` is called the **error function**. It appears prominently in probability theory, among other branches of mathematics. We can now manipulate f just like any other function defined in *Mathematica*. Another example is

$$g(x) = \int_0^x \sin t^2 \, dt.$$

When we enter

```
g[x_] = Integrate[Sin[t^2], {t, 0, x}]
```

we get

```
Sqrt[Pi/2]*FresnelS[Sqrt[2/Pi]*x]
```

where the function `FresnelS` is called the Fresnel Integral, which has applications in the theory of light diffraction. A third example is

$$h(x) = \int_0^1 \frac{1}{\sqrt{1 - x \sin^2 t}} \, dt. \tag{1.3}$$

This time when we try

```
h[x_] = Integrate[1/Sqrt[1 - x Sin[t]^2], {t, 0, 1}]
```

we get

```
EllipticF[1, x]
```

the answer being given in terms of the **Elliptic Function of the First Kind** . We can obtain numerical values and graphical data from this function as before. For example, let us try

```
h[0.1]
```

Mathematica responds with

```
1.014120373414132
```

where a numerical approximation to the integral in h is obtained.

Remark 1.6.1: The function h defined in (1.3) appears naturally in the context of the period of vibration of a nonlinear pendulum.

Let us proceed a bit further with the last example. Since h defines a function in *Mathematica*, we should be able to differentiate it with respect to x. Let $i(x) = h'(x)$, which we evaluate by

```
i[x_] = D[h[x], x]
```

The output is

```
-EllipticE[1, x]/(2*(-1 + x)*x) - EllipticF[1, x]/(2*x) +
  Sin[2]/(4*(-1 + x)*Sqrt[1 - x*Sin[1]^2])
```

To evaluate i at $x = 0.1$ we enter

```
i[0.1]
```

which results in

```
0.1462958973937115
```

Mathematica is capable of determining partial derivatives as well as multiple integrals. Because we will address these topics in detail later in the text, we postpone their treatment in *Mathematica* at this time.

Problems

1. Differentiate the following functions.

    (a) $f(x) = \log \frac{x}{x+1}$
    (b) $f(x) = \sin^3 4x \cos^5 7(x^2 - 2x + 1)$
    (c) $f(x) = x^{x-1}$

2. Combine the differentiation operation with `FindRoot` and determine the extreme points (the maxima and minima) of $f(x) = x \sin^2 x$. Begin by first defining $g(x)$ to be the derivative of this function and then use `FindRoot` to finds its roots.

3. Integrate the following functions.

    (a) $f(x) = \frac{x}{x-1}$
    (b) $f(x) = x \sin x$
    (c) $f(x) = x^2 \sin x$
    (d) $f(x) = x^{10} \sin x$
    (e) $f(x) = e^x \sin x$ (e^x is `Exp[x]` in *Mathematica*)
    (f) $f(x) = \sin^2 x$
    (g) $f(x) = \sin(x^2)$
    (h) $f(t) = t e^{t^2}$
    (i) $f(s) = e^{s^2}$

4. Determine the following indefinite integrals.

(a) $\int_{-\infty}^{\infty} \frac{1}{1+x^2} dx$ (Ans: `Integrate[1/(1+x^2),`
`{x, -Infinity, Infinity})`

(b) $\int_{0}^{\infty} e^{-t^2} dt$

(c) $\int_{0}^{\infty} e^{-at^2} dt$, a is a parameter

(d) $\int_{0}^{\infty} e^{-st} \sin t \, dt$ (This is the Laplace transform of $\sin t$.)

5. Define the function f by

$$f(x) = \int_{0}^{x} \frac{1}{x+t^5} dt.$$

Use *Mathematica* to determine $f'(\frac{1}{2})$. (Ans: -0.0779959)

6. Define the function f by

$$f(x) = \int_{0}^{1} \sin(xt^3) \, dt.$$

Determine $f'(\frac{1}{2})$. (Ans: 0.237662)

7. Verify the Fundamental Theorem of Calculus in *Mathematica*:

$$\int_{a}^{b} f'(t) \, dt = f(b) - f(a), \qquad \frac{d}{dx} \int_{0}^{x} f(t) \, dt = f(x).$$

8. Verify the Leibniz's rule of differentiation in *Mathematica*:

$$\frac{d}{dx} \left(\int_{a(x)}^{b(x)} f(x,t) \, dt \right) = b'(x)f(x,b(x)) - a'(x)f(x,a(x)) + \int_{a(x)}^{b(x)} \frac{\partial f}{\partial x} \, dt.$$

1.7 Two-Dimensional Graphics

To plot the graph of a function whose definition is explicitly known, say $f(x) = \sin(x)$ on the interval $[0, 2\pi]$, we give the command

```
Plot[Sin[x], {x, 0, 2Pi}]
```

Mathematica draws a graph of this function and scales it automatically. There are many options associated with `Plot` that would enable one to customize the graph. Among these options are `AxesLabel`, `PlotLabel`, and `Axes`. For instance,

```
Plot[Sin[x], {x, 0, 2Pi}, AxesOrigin -> {Pi/2,0},
    PlotLabel -> "Graph of Sine", AxesLabel -> {"x","y"}]
```

creates a graph with the proper labels.

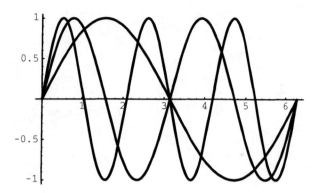

Figure 1.1: The output for `Plot[Evaluate[Table[Sin[n x], {n, 1, 3}]]`, `{x, 0, 2 Pi}]`.

In applications we often need to plot several graphs on the same screen. For instance, as we will see in the later chapters, the set of functions $f(x) = \sin nx$ represents a set of basic functions in terms of which we expand Fourier series of a large class of functions. Each $\sin nx$, with n a positive integer, represents a mode of oscillation with period $\frac{2\pi}{n}$. By graphing these functions one sees how the different modes compare to one another. There are several ways to draw the graphs of $\sin nx$, with $n = 1$, 2 and 3 on the same screen. One way is to draw the graphs of each mode separately, and then to combine them by using `Show`:

```
a1 = Plot[Sin[x], {x, 0, 2Pi}];
a2 = Plot[Sin[2 x], {x, 0, 2Pi}];
a3 = Plot[Sin[3 x], {x, 0, 2Pi}];
Show[a1, a2, a3]
```

Each one of the first three lines draws a separate graph. The last line combines the three graphs on a new screen.

A second way of getting the same result is to use the following syntax:

```
Plot[Evaluate[Table[Sin[n x], {n, 1, 3}]], {x, 0, 2Pi}]
```

Figure 1.1 shows the output.

To obtain a hard copy of any graph we produced in *Mathematica*, we need to take into account the specific features of the software and the platform on which it is installed. For example, to get a hard copy of the above graph in a Notebook session, we must first highlight the cell containing the graph and then select `Print` from the `File Menu`. On the other hand, when the stand-alone version of *Mathematica* is accessed on a SUN workstation, the command `PSPrint` will do the job. For example, assuming that the above graph is stored in Out[24], try

```
PSPrint[Out[24]]
```

or, simply,

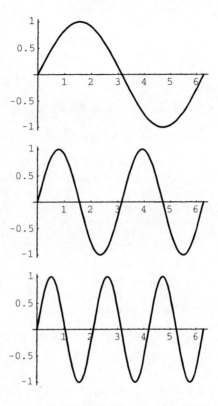

Figure 1.2: The output of the Show[GraphicsArray[...]] command.

```
PSPrint[%]
```

if the graph is the latest output.

Another way to draw the graphs of $\sin nx$ is by using GraphicsArray:

```
Show[GraphicsArray[{{a1}, {a2}, {a3}}]]
```

Now, the above three graphs are plotted separately but on the same screen (cf. Figure 1.2). This form of graphics is useful for displaying wave motion.

1.7.1   Curves in the Plane

Curves are the geometric manifestations of particle motions in a domain. Typically in this text curves represent motions of fluid particles in force fields and therefore, are parametrized by time. The position of the particle at time t is represented by a vector $\mathbf{r}(t)$ whose endpoint denotes the location of the particle at t. For example,

$$\mathbf{r}(t) = \langle 2\sin t, 2\cos t \rangle$$

defines a curve in the plane R^2, where the x and y components of each point are $2\sin t$ and $2\cos t$, respectively. Since $x^2 + y^2 = 4$, this curve is a circle of radius 2. To draw its graph, we apply `ParametricPlot` to **r**:

```
a=ParametricPlot[{2 Sin[t], 2 Cos[t]}, {t, 0, 2Pi}]
```

One of the options of `Show` is `AspectRatio`. Hence,

```
Show[a, AspectRatio -> Automatic]
```

displays a circle with the 1:1 aspect ratio. We have the option of combining the above two commands into one as follows:

```
ParametricPlot[{2 Sin[t], 2 Cos[t]}, {t, 0, 2Pi},
    AspectRatio -> Automatic]
```

`ParametricPlot` has several options, among them `PlotPoints` and `PlotStyle`. The `PlotPoints` option is particularly useful when the curve's domain is highly oscillatory because this option designates the number of points at which the parametrization **r** is to be evaluated. The default value of `PlotPoints` is 25. Compare the outputs of

```
ParametricPlot[{t, 1/(t+1)+Sin[30 t]}, {t, 0, 2Pi}]
```

and

```
ParametricPlot[{t, 1/(t+1)+Sin[30 t]}, {t, 0, 2Pi},
    PlotPoints->50]
```

We will come back to parametrization and `ParametricPlot` in more detail when we review curves, and the concepts of velocity and acceleration, in the context of vector calculus.

Problems

1. Plot the graphs of the following functions.

    (a) $f(x) = \sin(5x)$. What is the period of this function; that is, what is the smallest value of $T > 0$ for which $f(x + T) = f(x)$?

    (b) $g(x) = \sin(2x) + 3\sin(3x)$. What is the period of this function?

    (c) $h(x) = \sin x + \sin\sqrt{2}x$. Draw the graph of this function on the intervals $(0, 5)$, $(0, 10)$, and $(0, 50)$. Do these graphs give any indication as to whether h is periodic or not? Why?

2. Let $f(x) = x^n$. Draw the graph of f for $n = 0, 1, 2, 3, ..., 10$ for $x \in (0, 2)$ on the same screen.

3. Consider the function $f(x) = e^{-x^2}$ with $x \in (-5, 5)$.

    (a) Plot the graph of this function. Use the option `PlotRange -> All` with `Plot` to get the entire range of the function.

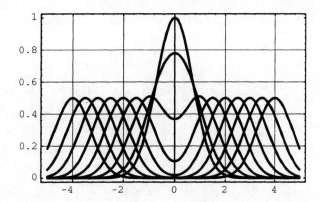

Figure 1.3: A graphics output using the `Frame` and `GridLines` options.

(b) Compare this graph to the graphs of the functions $g(x) = f(x + 4)$ and $h(x) = f(\frac{x}{0.5})$ over the same domain. Is there a scale change among these three graphs?

(c) Draw on the same screen the graph of

$$\frac{1}{2}(f(x - 2t) + f(x + 2t)) \tag{1.4}$$

for t ranging between 0 and 2 at increments of 0.25. Do this part by first defining (1.4) as a new function $g(x, t)$ and then using

```
a=Plot[Evaluate[Table[g[x, t], {t, 0, 2, 0.25}]],
       {x, -5, 5}]
```

to draw all of the graphs on the same screen. Experiment with the options `GridLines` and `Frame` until you obtain a figure similar to Figure 1.3. Is there a scale change in these graphs?

Expression (1.4) is the formula for the way a wave travels in a string of a guitar when the string is plucked in the shape of $f(\frac{x}{0.5})$ at time zero. Notice that the original "disturbance" (that is, $f(\frac{x}{0.5})$) is broken into two parts, each having amplitude half as much as the original wave and propagating to the left and to the right. Determine the speed of propagation; that is, find how long it takes for the maximum at $x = 0$ at $t = 0$ to reach $x = 1$ and $x = -1$.

4. Plot the graphs of the following curves.

(a) $\mathbf{r}(t) = \langle \sin^2 t, \cos^2 t \rangle$; $t \in (0, 2\pi)$

(b) $\mathbf{r}(t) = \langle \sin^5 t, \cos^5 t \rangle$; $t \in (0, 2\pi)$

(c) $\mathbf{r}(t) = \langle \sin^3 t, \cos^2 t \rangle$; $t \in (0, 2\pi)$

(d) $\mathbf{r}(t) = \langle \sin^3 t, \cos 2t \rangle$; $t \in (0, 2\pi)$

(e) $\mathbf{r}(t) = \langle \sin^5 t, \cos(1 + 2t) \rangle$; $t \in (0, 2\pi)$

1.8 Three-Dimensional Graphics

The syntax for plotting graphs of surfaces in three dimensions is very similar to two-dimensional graphics. For instance,

```
Plot3D[Sin[x]Cos[y], {x, 0, Pi}, {y, 0, Pi}]
```

produces the surface of the function f defined by $f(x,y) = \sin x \cos y$ in the domain $(0, \pi) \times (0, \pi)$. Similar to `Plot`, `Plot3D` has several options, which we examine in the context of the graph of a "sombrero": the mathematical equation that describes this surface is

$$f(x,y) = \frac{\sin r}{r}, \quad \text{where } r = \sqrt{x^2 + y^2}.$$

First, we define the polar radius r:

```
Clear[r, x, y];
r=Sqrt[x^2+y^2]
```

and then plot the function f by

```
Clear[a];
a=Plot3D[Sin[r]/r, {x, -10, 10}, {y, -10, 10}]
```

We now take two steps to improve the picture on the screen. First, we use `PlotRange` to force *Mathematica* to show us the entire range of the plot:

```
Show[a, PlotRange -> All]
```

The second step is to use `PlotPoints` with `Plot3D` to require *Mathematica* to use more points on the x- and y-axes in its plotting routine:

```
Plot3D[Sin[r]/r, {x, -10, 10}, {y, -10, 10},
    PlotRange -> All, PlotPoints -> 30]
```

Figure 1.4 contains the output.

The `ParametricPlot3D` is the analogue of `ParametricPlotParametricPlot` for curves whose range is in the three-dimensional space R^3. For example, the graph of a helix whose parametrization is

$$\mathbf{r}(t) = \langle \sin 3t, \cos 3t, t \rangle,$$

with $t \in (0, 2\pi)$, is obtained from

```
ParametricPlot3D[{Sin[3 t], Cos[3 t], t}, {t, 0, 2 Pi}]
```

One can set the aspect ratio of the graph to any desired value using `AspectRatio` with either `ParametricPlot3D` or `Show`.

We can also plot the graphs of surfaces with `ParametricPlot3D`. The parametrization of a surface, by definition, requires two independent parameters. For example, the surface whose equation is given by $z = x^2 + y^2$ within the domain $(x, y) \in (-2, 2) \times (-2, 2)$ can also be viewed as the set of points $(x, y, x^2 + y^2)$. Here, the two independent parameters are x and y, each of which takes on values

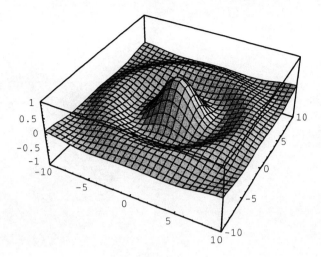

Figure 1.4: The graph of the "sombrero."

in the interval $(-2, 2)$. `ParametricPlot3D`'s syntax for displaying this surface is

```
ParametricPlot3D[{x, y, x^2 + y^2}, {x, -2, 2}, {y, -2, 2}]
```

The surfaces we have considered thus far have the property that one could write down an explicit formula that relates one of the coordinates of the set of points on the surface to the remaining two. For example, the set of points (x, y, x^2+y^2), with $(x, y) \in (a, b) \times (c, d)$, can be expressed by the relation $z = x^2 + y^2$. The function $f(x, y) = x^2 + y^2$ is the argument we pass to `Plot3D` in order to draw the graph of this set of points. We now consider examples of surfaces that cannot be readily expressed as $z = f(x, y)$. Many familiar geometric surfaces—among them cylinders, spheres, and tori—are examples of such surfaces. Let us begin with the example of a sphere of radius 1, whose equation is given by

$$x^2 + y^2 + z^2 = 1. \tag{1.5}$$

It is easy to see that (1.5) is equivalent to

$$z = \pm\sqrt{1 - x^2 - y^2}. \tag{1.6}$$

We can now use the two functions in (1.6) with `Plot3D` and combine the resulting surfaces with `Show`. One complication arises because of the domain in (1.6). When we try

```
Plot3D[Sqrt[1-x^2-y^2], {x, -1, 1}, {y, -1, 1}]
```

Mathematica complains about the complex numbers associated with certain values in the domain (such as $x = y = 1$) but still produces a relatively reasonable graph of part of the sphere.

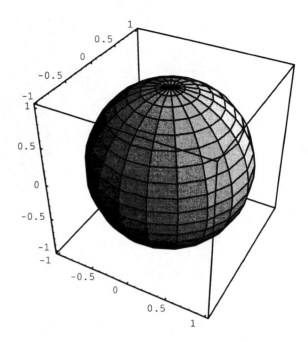

Figure 1.5: The graph of the sphere of radius 1.

Equations (1.6) describe the surface of the sphere in rectangular coordinates, a coordinate system that is not natural or convenient for plotting the sphere. Instead, we look for the description of this surface in spherical coordinates. In this new coordinate system a point whose rectangular coordinates are x, y, and z is represented by (r, u, v) where

$$x = r \sin u \cos v, \quad y = r \sin u \sin v, \quad z = r \cos u, \tag{1.7}$$

where u and v are the standard spherical angles. After substituting (1.7) into (1.6), we note that the equation for the sphere takes the simple form

$$r = 1. \tag{1.8}$$

Equations (1.7) and (1.8), when combined with `ParametricPlot3D`, produce the plot of the unit sphere:

```
ParametricPlot3D[{Sin[u]Cos[v], Sin[u]Sin[v], Cos[u]},
                    {u, 0, Pi}, {v, 0, 2Pi}]
```

The output of the above line is shown in Figure 1.5.

Many of the familiar shapes and surfaces that we study in mathematics are already programmed in *Mathematica* and are available as built-in functions. To access them we must first enter

```
<< Graphics'Shapes'
```

and read the special library of shapes into our session of *Mathematica*. To get, for example, the graph of a torus, we enter the command

```
Show[Graphics3D[Torus[ ]]]
```

while

```
Show[Graphics3D[MoebiusStrip[2,1,80]]]
```

draws the surface of the Moebius Strip with inner radius 1 and the outer radius 2 using 160 polygons.

    Mathematica is capable of rendering animation. In many applications it is possible, and often desirable, to produce a sequence of graphs and put them in motion. For example, consider the snapshots of a "vibrating string" defined by

$$u(x, t) = \sin x \cos t,$$

with $x \in (0, \pi)$. For each fixed t, the graph of the function $u(\cdot, t)$ is a snapshot of the string. Let us first generate a sequence of these snapshots:

```
Clear[u, x, t];
u[x_, t_] = Sin[x] Cos[t];
Table[Plot[u[x, t], {x, 0, Pi},
     PlotRange -> {-1, 1}], {t, 0, 2 Pi, 0.25}]
```

The `PlotRange` option plots all the snapshots over the same vertical range. The period of vibration is 2π (the period of $\cos t$); hence, we let t vary from 0 to 2π, at increments of 0.25. When using *Mathematica*'s Notebook version in Windows 3.1 or on a SUN workstation, we animate the above sequence of graphs by selecting first the cell that contains all of the graphs; next we choose `Cell` from the top menu, from which we select `Animate Selected Graphics`.

    The animation of a sequence of surfaces is accomplished in a similar manner. Consider, for example, the function u defined by

$$u(x, y, t) = \sin 3\pi x \sin 2\pi y \cos \pi t, \qquad (x, y) \in (0, 1) \times (0, 1).$$

Such a function often represents the displacement of an elastic membrane whose boundaries (that is, $x = 0, 1$ and $y = 0, 1$) are constrained not to move. To render an animation of u, first we generate a sequence of its snapshots as follows:

```
Table[Plot3D[Sin[3 Pi x] Sin[2 Pi y] Cos [Pi t], {x, 0, 1},
         {y, 0, 1}, PlotRange -> {-1, 1}], {t, 0, 1, 0.05}]
```

The output is animated as before, by first selecting the cell that contains all of the individual snapshots and then selecting `Animate Selected Graphics` from `Cell`.

Problems

1. Draw the graph of the following curves in the specified domain.

    (a) $\mathbf{r}(t) = \langle t, t, t \rangle$; $t \in (0, 1)$

    (b) $\mathbf{r}(t) = \langle \frac{t}{12}, \frac{t}{4}, \frac{1}{2+\sin t} \rangle$; $-2\pi < t < 2\pi$

    (c) $\mathbf{r}(t) = \langle e^{-\frac{t}{4}} \sin 3t, e^{-\frac{t}{4}} \cos 3t, \frac{t}{12} \rangle$; $t \in (0, 4\pi)$ (use the PlotPoints option of ParametricPlot3D to get a graph with better resolution)

    (d) $\mathbf{r}(t) = \langle \sin t, \cos t, \frac{1}{\sqrt{t^2+1}} \rangle$; $t \in (0, 2\pi)$

    (e) $\mathbf{r}(t) = \langle \sinh \frac{t}{6}, \sin(4t), \cosh \frac{t}{6} \rangle$; $0 < t < 4\pi$

    (f) $\mathbf{r}(t) = \langle t + \sin 3t, \frac{1}{t^2+1} \rangle$; $t \in (-2\pi, 2\pi)$

2. Draw the graph of each of the following curves.

    (a) A circle of radius 2 centered at the origin and located in the xy-plane

    (b) A circle of radius 2 centered at the origin and located in the $z = 1$ plane

    (c) The ellipse located in the xy-plane centered at the origin with major and minor axes of 3 and 2, respectively

    (d) The curve of intersection of $x^2 + y^2 = 1$ and $z = x$

3. Draw the graph of the following surfaces.

    (a) $z = x^2 + y^2$, with $(x, y) \in (-3, 3) \times (-3, 3)$

    (b) $z = \sqrt{x^2 + y^2}$, with $-3 < x < 3$ and $-3 < y < 3$

    (c) $z = 3x^2 + 4y^2$, in $(-3, 3) \times (-3, 3)$

    (d) $z = \sin(x^2 + y^2)$, with $x \in (-\pi, \pi)$ and $y \in (-\pi, \pi)$

    (e) $z = \sin(\sqrt{x^2 + y^2})$, in $(-\pi, \pi) \times (-\pi, \pi)$

    (f) $z = \sin(x^2 + y^2) \cos(x)$, in $(-\pi, \pi) \times (-\pi, \pi)$

    (g) $z = \sin(x^2 + y^2) \cos(y)$, in $(-\pi, \pi) \times (-\pi, \pi)$

    (h) $z = \frac{\sin \sqrt{x^2+y^2}}{\sqrt{x^2+y^2}}$, in $(-\pi, \pi) \times (-\pi, \pi)$

4. Plot the graph of a unit sphere centered at $(1, -1, 0)$.

5. Plot the graph of a sphere of radius 2 centered at $(-1, 1, 0)$. Combine this graph with the sphere in the previous problem on the same screen.

6. Draw the graph of a cylinder of height 2 and radius 3. (Hint: A cylinder of height h and radius a may be parametrized as $\mathbf{r}(u, v) = \langle a \cos u, a \sin u, v \rangle$, where $0 < u < 2\pi$ and $0 < v < h$.)

7. Draw the graph of a "dumbbell," that is, two spheres connected to a cylinder.

8. Draw the graph of the surface given by

$$z^2 + 9\sqrt{x^2 + y^2} = 9$$

for $z > 0$.

9. Animate the sequence of curves or surfaces. In each problem t represents time. Find the period of vibration in each case and generate a sequence of snapshots of the curve or surface over an entire period.

   (a) $u(x, t) = \sin 3x \cos t;\ x \in (0, \pi)$

   (b) $u(x, t) = 2 \sin 3x \cos t - 3 \sin x \cos 2t;\ x \in (0, \pi)$

   (c) $u(x, y, t) = \sin 3\pi x \sin 2\pi y \cos \pi t;\ (x, y) \in (0, 1) \times (0, 1)$

   (d) The graph of a sequence of spheres whose radii at time t are described by $\cos 2\pi t$

1.9 Solving Differential Equations

Mathematica has two internal functions, `DSolve` and `NDSolve`, that are capable of solving special classes of ordinary differential equations. `DSolve` is primarily used to find the exact solution to first-order (nonlinear) equations or linear equations with constant coefficient equations. Here are some examples. Consider the initial value problem

$$v' + v^2 = 0, \quad v(0) = 1.$$

To find a solution to this equation, we enter

```
Clear[a, v, t];
a=DSolve[{v'[t] + v[t]^2 == 0, v[0] == 1}, v[t], t]
```

The reason for the label `a` will become clear shortly. The output is

```
            1
{{v[t] -> -----}}
          1 + t
```

This output is interpreted as a replacement rule, that is, a rule that replaces `v[t]` with

```
  1
-----
1 + t
```

whenever `a` is called upon. For example,

```
v[t] /. a/. t -> 2
```

results in

```
    1
  {-}
    3
```

while

```
Plot[v[t] /. a, {t, 0, 3}]
```

plots the graph of v. If more than one solution is stored in a, Plot displays the graph of all solutions.

It is often convenient to define a function v using a:

```
v[t_] = v[t] /. a[[1]]
```

we can now manipulate the function v just like any other function. Assuming v represents the velocity of a particle of unit mass, we determine its kinetic energy during the time interval $t \in (0, 3)$ by

```
1/2*Integrate[v[t]^2, {t, 0, 3}]
```

Similarly, the acceleration of the particle is obtained by differentiating v[t] once

```
D[v[t], t]
```

Remark 1.9.1: The inclusion of the independent variable t in v[t] is not optional. For example,

```
DSolve[{v' + v^2 == 0, v[0] ==1}, v, t]
```

does not lead to the correct solution of this equation. Also, the operand == cannot be replaced by =. Here $v' + v^2 = 0$ is an equation and not an assignment; hence, it is necessary to use ==.

Next, let us consider the differential equation

$$mv' + kv = -mg, \quad v(0) = 0.$$

Here v represents the velocity of an object of mass m falling under the action of gravity and being resisted by a linear frictional force. From

```
Clear[b, v, t];
b=DSolve[{m v'[t] + k v[t] == - m g, v[0] == 0}, v[t], t]
```

we receive

```
                         (k t)/m
              -(g m) + E          g m
{{v[t] -> -(---------------------)}}
                  (k t)/m
                E        k
```

Assuming k and m are positive, it is clear from the above expression that the terminal (limiting) velocity of the object is $-\frac{mg}{k}$. With $m = 70$, $k = 10$, and

$g = 9.8$, we define

```
v[t_] = v[t] /. b[[1]] /. {m -> 70, k -> 10, g -> 9.8}
```

and determine v's terminal velocity by

```
Limit[v[t], t -> Infinity]
```

Mathematica returns

```
-68.59999999999999
```

DSolve is an effective tool for solving linear higher order differential equations with constant coefficients. Consider the initial value problem

$$x'' + 3x' + 2x = 0, \quad x(0) = 1, x'(0) = 0.$$

The exact solution of this problem is found by

```
Clear[x, t];
DSolve[{x''[t] + 3 x'[t] + 2 x[t] == 0,
     x[0] == 1,x'[0] == 0}, x[t], t]
```

whose output is

```
                 t
        -1 + 2 E
{{x[t] -> ---------}}
            2 t
           E
```

The command DSolve works equally well with systems of equations. Consider the initial value problem

$$x' = 2x + 3y, \quad y' = x, \quad x(0) = -2, \quad y(0) = 2.$$

We input

```
b=DSolve[{x'[t] == 2 x[t] + 3 y[t], y'[t] ==  x[t],
       x[0] == -2, y[0] == 2}, {x[t], y[t]}, t]
```

into *Mathematica* and obtain the output

```
        -2            2
{{x[t] -> --, y[t] -> --}}
          t           t
         E           E
```

In the applications where systems of equations appear, the solution pair $(x(t), y(t))$ is often the parametrization of the path of a fluid particle. To get the graph of this path we enter

```
ParametricPlot[{x[t], y[t]}/. b[[1]], {t, 0, 3}]
```

Mathematica first gives the warning

```
ParametricPlot::ppcom:
   Function {x[t], y[t]} /. b[[1]]
      cannot be compiled; plotting will proceed with
      the uncompiled function.
```

and then proceeds to graph the particle path. To work with compiled functions, first we apply the `Evaluate` command to the solution pair `x[t]`, `y[t]` and then plot the result

```
ParametricPlot[Evaluate[{x[t], y[t]} /. b], {t, 0, 3}]
```

Alternatively, we define a function `f` as the solution pair `x[t]`, `y[t]` and then plot `f`:

```
Clear[f];
f[t_] = {x[t], y[t]} /. b[[1]]
ParametricPlot[Evaluate[f[t]], {t, 0, 3}]
```

The second command in *Mathematica* that is capable of determining solutions of differential equations is `NDSolve`. This program uses a numerical algorithm (based on the standard Runge–Kutta scheme) and solves linear as well as nonlinear systems of differential equations. Consider the forced nonlinear pendulum equation

$$x'' + 0.1x' + \sin x = 0.02 \cos t, \tag{1.9}$$

with initial conditions

$$x(0) = 0, \quad x'(0) = 1.$$

We have the option of giving (1.9) as a second-order equation to *Mathematica* or as a first-order system. In the first case the syntax is

```
Clear[c, x];
c=NDSolve[{x''[t] + 0.1 x'[t] + Sin[x[t]] == 0.02 Cos[t],
     x[0] == 0, x'[0] == 1}, x, {t, 0, 5}]
```

Mathematica responds with

```
{{x -> InterpolatingFunction[{0., 5.}, <>]}}
```

which states that it has successfully obtained an approximate solution to the above differential equation and has interpolated a curve through the data points (t_i, x_i), where $t_i \in (0, 5)$ are the discretized values chosen by the Runge–Kutta algorithm. We now define a function `x` by

```
x[t_] = x[t] /. c[[1]]
```

To get the graph of the solution x of (1.9), we enter

```
Plot[x[t], {t, 0, 5}]
```

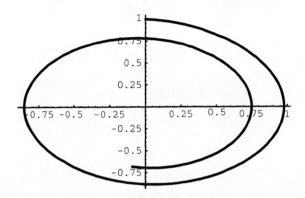

Figure 1.6: The phase plane diagram of the solution to $x'' + 0.1x + \sin x = 0.02 \cos t$.

Its value at a point such as $t = 2.15$ is evaluated by

```
x[2.15]
```

which results in 0.8674379209871004. The pair $(x(t), x'(t))$ is plotted by

```
ParametricPlot[Evaluate[{x[t], x'[t]}], {t, 0, 5}]
```

The last statement brings up the point of solving the second-order differential equation (1.9) as a first-order system so that information about x and x' is available simultaneously. To reduce this equation to a first-order system, let $x' = y$, form which we deduce that $y' = -0.1y - \sin x + 0.02 \cos t$. Thus, (1.9) is equivalent to

$$x' = y, \quad y' = -0.1y - \sin x + 0.02 \cos t, \quad x(0) = 0, \quad y(0) = 1. \qquad (1.10)$$

We are now able to determine the approximate solution of (1.9) and plot its particle path in the following manner:

```
sol = NDSolve[{x'[t] == y[t],
             y'[t] == - 0.1 y[t] - Sin[x[t]] + 0.02 Cos[t],
             x[0] == 0, y[0] == 1},
             {x, y}, {t, 0, 10}];
ParametricPlot[Evaluate[{x[t], y[t]} /. sol], {t, 0, 10}]
```

We first receive the output

```
{{x -> InterpolatingFunction[{0., 10.}, <>],

 y -> InterpolatingFunction[{0., 10.}, <>]}}
```

The particle path is then graphed, as shown in Figure 1.6.

NDSolve has several important options that help considerably with solving nonlinear differential equations accurately. One of these options is MaxSteps.

This parameter sets the number of steps that the Runge–Kutta algorithm is allowed to take in discretizing the domain $t \in (0, T)$. If T is too large, then the value of MaxSteps (whose default value is 500) must be increased.

Problems

1. Use DSolve to solve the following differential equations.

   (a) $x' + 3x = 0$

   (b) $x' + tx = 3e^{-t^2}, \quad x(1) = 2$

   (c) $x' + x^3 = 0$

   (d) $y'' + y = 0, \quad y(0) = 0, y'(0) = 1$

   (e) $x''' + x' + x = 0$

   (f) $y'' + y = 0, \quad y(0) = 1, y(1) = -1$

   (g) $x'' + x = \sin 2t$

   (h) $y'' + y = \sin t$

2. Use NDSolve and draw the trajectories of the following differential systems. Unless otherwise specified, in each case choose a suitable final value for t.

   (a) $x'' + x = 0, \quad x(0) = 0, x'(0) = 1$

   (b) $y'' + 0.1y' + \sin y = 0, \quad y(0) = 0, y'(0) = 1$

   (c) $y'' + 0.1y' + \sin y = 0, \quad y(0) = 0, y'(0) = 3$

   (d) $x'' + 0.1x + \sin x = 0.02 \cos t, \quad x(0) = 0, x'(0) = 1, t \in (0, 100)$

3. Solve each system of differential equations for two sets of initial data $(x(0), y(0)) = (1, 1); \quad (x(0), y(0)) = (2, 2)$. Plot the solutions on the same screen.

   (a) $x' = y, \quad y' = -x$

   (b) $x' = y, \quad y' = -x - 0.1y$

   (c) $x' = y, \quad y' = -x - y$

   (d) $x' = y, \quad y' = -x - y^2$

4. Consider the system

$$x' = y, \quad y' = -\frac{1}{(1 + \epsilon x)^2},$$

   with initial data $x(0) = 0$, $y(0) = 1$, and $t \in (0, 3)$. Solve this system with $\epsilon = 0, 0.1, 0.2$, and 1. Plot the graphs of the solutions $(x_\epsilon(t), y_\epsilon(t))$ on the same screen. What seems to be the effect of ϵ on the solutions?

1.10 Vectors, Matrices, and Lists

In *Mathematica* vectors and matrices are entered as lists. For example, the vector $a = \langle -2, 1, 3 \rangle$ is entered as

```
a = {-2, 2, 3}
```

Similarly, the matrix

$$B = \begin{bmatrix} 1 & -1 & 2 \\ 0 & 1 & 0 \\ -1 & 5 & 1 \end{bmatrix}$$

is entered as

```
B = {{1, -1, 2}, {0, 1, 0}, {-1, 5, 1}}
```

We can write B in matrix form by invoking the `MatrixForm` command. Thus,

```
MatrixForm[B]
```

gives

```
1    -1   2

0    1    0

-1   5    1
```

Elements of a list are accessed by putting the subscript of the element between the delimiters `[[...]]`. For example, the first entry of a is `a[[1]]` while the (1,2)-entry of B is `B[[1,2]]`. Also, `B[[1]]` returns

```
{1, -1, 2}
```

which is the first row of `B`.

The length of a list is the number of elements in the list. For example, the vector `a` and the matrix `B` both have lengths equal to 3 as can be checked from `Length[a]` and `Length[B]`.

The standard matrix multiplication is carried out in *Mathematica* by placing a period (`.`) between the matrices. Thus, to compute the product $A_1 A_2$ of the matrices

$$A_1 = \begin{bmatrix} 2 & -1 \\ -3 & 4 \\ 1 & -1 \end{bmatrix}, \quad A_2 = \begin{bmatrix} 3 & 0 & 2 \\ 1 & 1 & 1 \end{bmatrix}$$

we enter

```
A1 = {{2, -1}, {-3, 4}, {1, -1}};
A2 = {{3, 0, 2}, {1, 1, 1}};
prod1 = A1 . A2;
```

Now

```
MatrixForm[prod1]
```

results in the 3×3 matrix

5   -1   3

-5   4   -2

2   -1   1

Similarly, to determine the product $A_2 A_1$ we enter

```
prod2 = A2 . A1;
MatrixForm[prod2]
```

which results in the 2×2 matrix

8   -5

0   2

Mathematica will return an error message if the dimensions of the matrices being multiplied are not compatible. For example,

```
Clear[a,b];
a={{1,2}, {3,4}};
b={1,2,3};
a . b
```

returns

```
Dot::dotsh:
   Tensors {{1, 2}, {3, 4}} and {1, 2, 3}
      have incompatible shapes.
```

Matrix and vector multiplication are carried out in the same manner. If c is the column vector

$$c = \begin{bmatrix} 1 \\ 3 \\ 9 \end{bmatrix},$$

then, after defining it as

```
Clear[c];
c = {1, 3 , 9}
```

we compute the product of B (defined earlier) and c by

```
B . c
```

which results in

{16, 3, 23}

Also the product of the row vector c^T with B is determined by

c . B

whose output is

{-8, 47, 11}

We note that the vector c could also have been defined as a 3×1 matrix:

```
Clear[c];
c={{1}, {3}, {9}};
```

Now B . c returns

{{16}, {3}, {23}}

while c . B returns the error message

```
Dot::dotsh:
   Tensors {{1}, {3}, {9}} and
     {{1, -1, 2}, {0, 1, 0}, {-1, 5, 1}}
       have incompatible shapes.
```

On the other hand, Transpose[c] . B returns

{{-8, 47, 11}}

There is a second way of multiplying lists in *Mathematica*, using the ∗ operand. This operation between two lists A and B of equal lengths returns a list whose entries are the product of individual entries of A and B. For example,

```
Clear[A, B];
A = {-1, 2}; B = {{1, 2,3 }, {3}};
A * B
```

returns

{{-1, -2, -3}, {6}}

Here both A and B have length 2. On the other hand, the product

```
Clear[A, B];
A = {{-1, 2}}; B = {{1, 2,3 }, {3}};
A * B
```

returns

```
Thread::tdlen:
   Objects of unequal length in
     {{-1, 2}} {{1, 2, 3}, {3}} cannot be combined.
```

The `Append` command appends information to the end of a list. For example,

```
B = {{1, 2, 3}, {-1, 0, 1}, {3, 4, 5}};
c = Append[B, {1, 1, 1}]
```

returns the 4×3 matrix with B as its first three rows and $\langle 1, 1, 1 \rangle$ as its fourth row. This command is particularly useful when, in the course of a calculation, new entries are being computed and this information must be added to a variable and stored for later processing.

The commands `Det`, `Inverse`, `Eigenvalues`, and `Eigenvectors` operate on lists, when applicable, with the standard mathematical results that their names suggest. For example, consider the matrix f

$$f(a) = \begin{bmatrix} 1 & a & 1 \\ -2 & 0 & a \\ -3a & 1 & a \end{bmatrix}.$$

Define this matrix in *Mathematica* by

```
f[a_] = {{1, a, 1}, {-2, 0, a}, {-3a, 1, a}}
```

Now

```
Det[f[a]]
```

returns

```
            2       3
-2 - a + 2 a   - 3 a
```

while

```
Inverse[f[a]]
```

leads to

```
                                           2
              a                       1 - a
 {{-(-------------------),  --------------------,
                 2       3              2       3
      -2 - a + 2 a   - 3 a   -2 - a + 2 a   - 3 a


              2
              a
 -------------------},
              2       3
  -2 - a + 2 a   - 3 a

                2
```

```
       2 a - 3 a                        4 a
  {--------------------,  --------------------,
               2       3              2       3
  -2 - a + 2 a  - 3 a   -2 - a + 2 a  - 3 a


         -2 - a
  --------------------},
               2       3
  -2 - a + 2 a  - 3 a
```

```
                                               2
            2                          -1 - 3 a
  {-(--------------------),  --------------------,
               2       3              2       3
    -2 - a + 2 a  - 3 a     -2 - a + 2 a  - 3 a


         2 a
  --------------------}}
               2       3
  -2 - a + 2 a  - 3 a
```

Similarly, `Eigenvalues[f[1]]` and `Eigenvectors[f[2]]` return the appropriate outputs, respectively. The command `Eigensystem` combines the outputs of `Eigenvalues` and `Eigenvectors`.

We often need to plot a set of ordered pairs of numbers. The command `ListPlot` is the appropriate tool for this task. For example, consider the following four ordered pairs:

$$(1, 0.1), (2, 0.2), (-1, 0.3), (-2, 0.4).$$

To plot them, first we define a list that contains the four pairs and then apply `ListPlot` to it:

```
c = {{1, 0.1}, {2, 0.2}, {-1, 0.3}, {-2, 0.4}};
ListPlot[c, PlotJoined -> True]
```

The output is shown in Figure 1.7.

The above discussion touches on only a small portion of what is available in *Mathematica* concerning lists, matrices, and linear algebra. The reader is encouraged to consult [1] and [2] for more detail on this subject.

1.10.1 Lists and Differential Equations

In certain problems where a large number of unknowns are present, we often end up writing down a system of ordinary differential equations that contains the information about the coupling between the variables. A typical example in engineering arises in the context of a set of masses and springs connected to

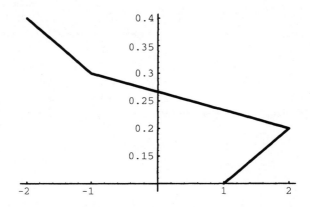

Figure 1.7: The output of `ListPlot`.

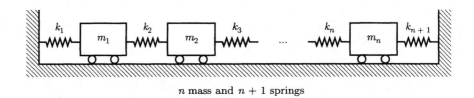

n mass and $n + 1$ springs

Figure 1.8: A mass-spring system.

each other in a series. The objective is to determine the oscillation of all masses in terms of the spring constants, the initial data, and any external forcing that may be present. We now give a simple example of such a system and outline how one proceeds to find the solution. We will use the concept of lists and present an algorithm that is nearly independent of the number of masses and springs present in the system.

Consider the mass–spring system in Figure 1.8.

The governing system of differential equations is (see Chapter 7 in regards to the derivation of this system of equations)

$$
\begin{aligned}
m_1 x_1'' &= -k_1 x_1 + k_2 (x_2 - x_1) \\
m_2 x_2'' &= -k_2 (x_2 - x_1) + k_3 (x_3 - x_2) \\
m_3 x_3'' &= -k_3 (x_3 - x_2) + k_4 (x_4 - x_3) \\
\dots &= \\
\dots &= \\
m_n x_n'' &= -k_n (x_n - x_{n-1}) + k_{n+1} x_n.
\end{aligned}
$$

In matrix notation, the above system is equivalent to

$$M\mathbf{x}'' = A\mathbf{x} \tag{1.11}$$

where

$$M = \begin{bmatrix} m_1 & 0 & 0 & \ldots & 0 \\ 0 & m_2 & 0 & \ldots & 0 \\ \ldots & \ldots & \ldots & \ldots & \ldots \\ \ldots & \ldots & \ldots & \ldots & \ldots \\ 0 & \ldots & \ldots & 0 & m_n \end{bmatrix}$$

and

$$A = \begin{bmatrix} -(k_1 + k_2) & k_2 & 0 & \ldots & 0 \\ k_2 & -(k_2 + k_3) & k_3 & \ldots & 0 \\ 0 & k_3 & -(k_3 + k_4) & \ldots & 0 \\ \ldots & \ldots & \ldots & \ldots & \ldots \\ 0 & 0 & \ldots & -(k_{n-1} + k_n) & k_n \\ 0 & 0 & \ldots & k_n & -(k_n + k_{n+1}) \end{bmatrix}.$$

System (1.11) is supplemented with the initial data

$$\mathbf{x}(0) = \mathbf{x}_0, \quad \mathbf{x}'(0) = \mathbf{x}_1. \tag{1.12}$$

Let us consider the concrete example of three unit masses attached to four identical springs with spring constants 10. The matrix M reduces to the identity matrix and A becomes

$$A = \begin{bmatrix} -20 & 10 & 0 \\ 10 & -20 & 10 \\ 0 & 10 & -20 \end{bmatrix}. \tag{1.13}$$

We consider the initial data

$$\mathbf{x}_0 = \begin{bmatrix} 0 \\ 0 \\ 1 \end{bmatrix}, \quad \mathbf{x}_1 = \begin{bmatrix} 1 \\ -1 \\ 2 \end{bmatrix}. \tag{1.14}$$

Our goal is solve (1.13)-(1.14) in *Mathematica* using NDSolve. We begin a new session of this program and introduce the number of variables num by

```
num = 3
```

and a list representing the solutions $x_i(t)$ by

```
vars = Table[x[i][t], {i, num}]
```

Next, we define the matrix A as

```
A = {{-20, 10, 0}, {10, -20, 10}, {0, 10, -20}}
```

We refer to the right-hand side of (1.11) by **rightside** and generate it by

```
rightside = A . vars
```

Next, we define the set of equations in (1.11) as the list

```
eqns = Table[x[i]''[t] == rightside[[i]], {i, num}]
```

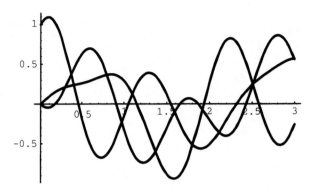

Figure 1.9: The output of the mass-spring system of differential equations.

The initial data (1.14) is stored in `initdata` directly:

```
initdata = {x[1][0] == 0, x[2][0] == 0, x[3][0] == 1,
            x[1]'[0] == 1, x[2]'[0] == -1, x[3]'[0] == 2}
```

Since `NDSolve` requires the differential equations and the initial data as one list, we use `Join` on `eqns` and `initdata`:

```
eqnsANDinit = Join[eqns, initdata]
```

We are now ready to apply `NDSolve`:

```
sol = NDSolve[eqnsANDinit, vars, {t, 0, 3}]
```

Mathematica responds with

```
{{x[1][t] -> InterpolatingFunction[{0., 3.}, <>][t],

   x[2][t] -> InterpolatingFunction[{0., 3.}, <>][t],

   x[3][t] -> InterpolatingFunction[{0., 3.}, <>][t]}}
```

To plot the three displacements on the same screen, we enter

```
Plot[Evaluate[Table[x[i][t] /. sol, {i, num}]], {t, 0, 3}]
```

Figure 1.9 shows the output. We have taken special care in implementing the various parts of (1.13)–(1.14) in such a way that only a few changes need be made in the above procedure when a different set of parameters is supplied to this system. This is particularly useful when *Mathematica* is being used in its Notebook setting, where the cut-and-paste mode offers a handy tool for introducing or altering parameters or definitions. Later in this chapter we will discuss how to write a program in *Mathematica* to reduce the number of changes one needs to make in order to analyze a problem such as (1.11)–(1.12) when the physical parameters are allowed to vary.

Problems

1. Let A and B be defined by

$$A = \begin{bmatrix} -1 & 1 \\ 1 & 3 \end{bmatrix}, \quad B = \begin{bmatrix} -3 & 4 \\ 7 & 5 \end{bmatrix}.$$

   Compute $A + B$, $A - B$, AB, BA, $6A$, and $3A + 2B$.

2. Consider the matrix

$$A = \begin{bmatrix} a & b \\ b & c \end{bmatrix},$$

   where a, b, and c are constants. Find all values of these parameters for which the determinant of A vanishes.

3. Let A be the matrix

$$A = \begin{bmatrix} a & -a & b \\ -a & b & a \\ b & a & 2a \end{bmatrix}.$$

   Find all values of a and b for which the determinant of A vanishes.

4. Let A be the matrix

$$A = \begin{bmatrix} a & b \\ b & a \end{bmatrix}.$$

   (a) Compute A^5 and A^{10} (use `MatrixPower`).

   (b) Use `Exp` and `MatrixExp` with A. Why are the results different?

5. Consider the matrix

$$A = \begin{bmatrix} -1 & 1 \\ 1 & 3 \end{bmatrix}.$$

   Let $s(n)$ be the nth partial sum of the Taylor series expansion of e^A, that is,

$$s(n) = I + \sum_{i=1}^{n} \frac{1}{i!} A^i,$$

   where I is the 2×2 identity matrix. Find $s(2)$, $s(5)$, and $s(10)$. Compare these matrices to the output of `MatrixExp[A]`.

6. Start a session in *Mathematica* and generate Figure 1.9.

7. Generate the analogue of Figure 1.9 in the following settings. All masses are assumed to be one.

   (a) $k_1 = k_2 = 10$, $k_3 = 20$; $x_1(0) = 0, x_2(0) = -1, x_3(0) = 1, x_1'(0) = 0, x_2'(0) = 0, x_3'(0) = 0$; $t \in (0, 3)$

   (b) $k_1 = k_2 = k_3 = k_4 = 10$; $x_1(0) = 0, x_2(0) = 0, x_3(0) = 0, x_4(0) = 1, x_1'(0) = 0, x_2'(0) = 0, x_3'(0) = 0, x_4'(0) = 0$; $t \in (0, 5)$

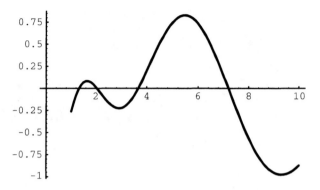

Figure 1.10: The graph of $x_1(3, k)$.

8. Consider the mass–spring system in Figure 1.8 containing five equal masses with $m_i = 5$, $i = 1, ..., 5$, and six springs with spring constants $k_i = 3$, $i = 1, ..., 6$. While all other masses are kept stationary, we perturb this system from its equilibrium by displacing the third mass one unit to its right and releasing it. Find the displacement of all of the masses for $t \in (0, 10)$.

9. Consider the mass–spring system in Figure 1.8 consisting of three unequal masses $m_1 = 1$, $m_2 = 3$, and $m_3 = 5$. Four identical springs with spring constants $k = 20$ are attached to these masses. Assuming that the initial disturbance in this system is caused by displacing the first mass one unit to its left and then releasing it, generate the analogue of Figure 1.9 for $t \in (0, 10)$.

10. Consider the mass–spring system consisting of two identical masses $m_1 = m_2 = 1$, attached to three identical springs with spring constant k. Let $x_1(0) = 1$, $x_2(0) = 0$, and $x_1'(0) = x_2'(0) = 0$. Let $x_i(t, k)$ denote the displacement of the ith mass at time t. Use NDSolve to find the value for $x_1(3, k)$ as k takes on values from 1 to 10 at increments of 0.1. Let S be the list containing the ordered pairs $(k, x_1(3, k))$. Use ListPlot and graph S. Compare your graph with Figure 1.10. (Hint: One way to get Figure 1.10 is to make the following changes to the process discussed in Section 1.10.1. Besides the necessary modifications to prepare for the above two-dimensional system, define the matrix A in terms of k, that is

```
A = {{-2 k, k}, {k, -2k}};
```

Next, enhance the definition of sol to allow for the dependence on the spring constant:

```
sol[j_] := NDSolve[eqnsANDinit/. k -> j, vars, {t, 0, 3}];
```

(the significance of using := as opposed to = is discussed in the next section). Finally, generate the points for the graph by

```
points = Table[{k, First[x[1][t] /. sol[k]/. t->3]},
        {k,1,10,0.1}];
```

)

1.11 The =, :=, ==, ->, /. Operators

The operation = has the same standard definition in *Mathematica* as in many other computer languages. The expression

```
a = b
```

means that the value stored in b replaces the value stored in a and becomes the current value of that symbol. In particular, = should not be used when defining an equation, as we have seen in the discussion of Solve and DSolve, among other internal functions of *Mathematica*. Also, the expression

```
t = t + h
```

does not imply that $h = 0$; it simply states that the old value of t is added to the current of h, and the result becomes the current (new) value of t. This expression is the common way of updating a variable in a recursive process, as we will see in the next section.

The operator := denotes delayed evaluation and is most useful in definitions where we are not yet ready to use an assignment. Compare the output of

```
t = 3;
f[t_] = Sin[t];
f[4]
```

with the output of

```
t = 3;
g[t_] := Sin[t];
g[4]
```

In the first case we see that Sin[3] is the output, since the value of t is passed to the argument of f as soon as its definition is executed. In the second case, where the output is Sin[4], the function Sin[t] is assigned to g, but nothing is executed until the next time g is called upon.

We have already encountered == on many occasions. This operation performs the task of testing for equality and defining equations. For instance,

```
x == y
```

tests to see whether x and y are equal to each other. Try

```
Clear[x,y];
x = 4; y = 5; x == y
```

The operation == is also used in defining equations as in

```
DSolve[x''[t] + x'[t] == 0, x[t], t]
```

The operations /. and -> are used together for the purpose of replacement by the rule(s) defined by ->. For example,

```
Clear[x];
x^2 - 4*x + 3 /. x -> -2
```

results in the evaluation of the quadratic polynomial at $x = 2$. Similarly,

```
Integrate[f[t], {t, 0, 1}] /. f[t_] -> Exp[t]
```

computes $\int_0^1 e^t \, dt$. It is also possible to use -> with multiple rule assignments:

```
Clear[a,x,b,y];
a x^2 + b y^2 /. {a->1, b->-3, x->2, y->-1}
```

or

```
Clear[a,x,b,y,c,d,e,f];
Solve[{a x + b y == c, d x + e y == f/. a -> 1/. c -> -1}]
```

Problems

1. Determine the outcome of the following statements.

    (a) `t == 3`

    (b) `t = 3; f[t_] = Sin[t] Cos[t]; f[Pi]`

    (c) `t = 3; f[t_] := Sin[t] Cos[t]; f[Pi]`

    (d) `Sin[a t + b] /.  a -> Pi /.  b -> 0 /.  t -> 1/2`

    (e) `Solve[x^2 - a == 0 /.  a -> 3, x]`

    (f) `Solve[x^2 - a == 0, x ] /.  a -> 3`

    (g) `D[Sin[t] /.  t -> 3, t]`

    (h) `D[sin[t], t] /.  t -> 3`

2. Find the syntax errors in the following statements.

    (a) `Dsolve[x^2 - 4 = 0, x]`

    (b) `Integrate[f[t] = t^2, {t, 0, 1}]`

    (c) `D[Sin[t], t /.  t -> 3]`

    (d) `DSolve[x'[t] + x[t] == -1, x[0] = 2, x[t], t]`

1.12 Loops and the Do Command

In many numerical applications we need to perform an operation repeatedly while a few parameters may change with each iteration. The Do command in *Mathematica* is the right tool for such a task. As a first example, consider the sum

$$S = \sum_{i=1}^{100} \frac{1}{i^2}.$$

One can find an approximate value for S using Do via

```
S = 1.;
Do[S = S + 1/i^2, {i, 2, 100}];
S
```

Mathematica returns `1.63498`. (Try the last program with `S = 1` replacing the first line. How are the outputs different?) We also get the same result from Sum:

```
Sum[N[1/i^2], {i, 1, 100}]
```

A different context in which Do is useful is in carrying out the type of iterations that arise naturally in the discretization of differential equations. A simple example of this type of application appears in the numerical scheme that produces the Euler approximation of the solution of a first-order differential equation. Consider the first-order differential equation

$$\frac{dy}{dx} = f(x, y), \quad y(x_0) = y_0.$$

The Euler approximation of the solution $y(x)$ seeks a sequence (x_n, y_n) that satisfies the difference equation

$$y_{n+1} = y_n + hf(x_n, y_n),$$

with $x_{n+1} = x_n + h$, where h is a fixed small positive number. The following program shows how to generate this sequence and plot the approximate solution using ListPlot. This program is written for $f(x, y) = -y + \sin x$, $x_0 = 0$, $y_0 = 1$, $h = 0.01$, and $n = 10$.

```
Clear[f, x, y];
f[x_, y_] = -y + Sin[x];
x0 = 0; y0 = 1;
h = 0.01;
n = 10;
x = x0; y = y0;
output = {{x, y}};
```

```
Do[y = y + h*f[x, y]; x = x + h;
   output = Append[output, {x, y}], {i, 1, n}];
ListPlot[output, PlotJoined -> True]
```

The function of Do in the above program is to perform y = y + h*f[x, y] repeatedly, while updating the value of x and appending the result of x and y to output.

Do is also quite useful in finding fixed points of functions: A fixed point of a function f is a point x such that $f(x) = x$. For example, $x = 0$ is a fixed point of the sine function. One way to approximate a fixed point of a function is to generate the sequence

$$x_0 = \text{initial guess}, \quad x_1 = f(x_0), \quad x_2 = f(x_1),$$

and in general

$$x_n = f(x_{n-1}).$$

If $\{x_n\}$ converges to a, and if f is a continuous function, then a is a fixed point of f. This algorithm is rather easy to implement using Do (the following program is written for $f(x) = 2\sin x$, with $x_0 = \frac{1}{2}$):

```
Clear[f, x];
f[x_] = 2 Sin[x];
x = 0.5;
Do[y = N[f[x]]; Print[{x, y}]; x = y, {i, 20}]
```

Its output is

```
{0.5, 0.958851}
{0.958851, 1.63706}
{1.63706, 1.99561}
{1.99561, 1.82223}
{1.82223, 1.93711}
{1.93711, 1.86731}
{1.86731, 1.91272}
{1.91272, 1.88422}
{1.88422, 1.90257}
{1.90257, 1.89093}
{1.89093, 1.89838}
{1.89838, 1.89364}
{1.89364, 1.89667}
{1.89667, 1.89474}
{1.89474, 1.89597}
{1.89597, 1.89519}
{1.89519, 1.89569}
{1.89569, 1.89537}
{1.89537, 1.89557}
{1.89557, 1.89544}
```

Problems

1. Use **Do** and sum the following series.

   (a) $\sum_{i=0}^{20} i$

   (b) $\sum_{i=1}^{10} \frac{1}{i^2}$

   (c) $\sum_{i=1}^{100} \frac{1}{i^2}$. First sum the series using exact arithmetic and then using floating point arithmetic (that is, use the decimal representation of $\frac{1}{i^2}$).

   (d) $\sum_{i=1}^{1000} \frac{1}{i}$. What is the exact value of the sum? Find its forty decimal point approximation (use **N[number, 40]**). Compare this value with the value of the sum when the decimal representation of $\frac{1}{i}$ is used.

2. Find the fixed points of the following functions. Experiment with the number of iterations to get a sense of whether the iteration process converges or not.

   (a) $f(x) = \sin 2x$; $x_0 = \frac{1}{2}$

   (b) $f(x) = \sin 2x$; $x_0 = \frac{3}{2}$

   (c) $f(x) = -\sin 2x$; $x_0 = \frac{1}{2}$

   (d) $f(x) = \sqrt{x} + 1$; $x_0 = 1$

   (e) $f(x) = \sin \sqrt{x} + 1$; $x_0 = 1$

   (f) $f(x) = \frac{1}{x^2+1}$; $x_0 = 1$. Compare the result with the output of

   ```
   Clear[x];
   FindRoot[1/(x^2+1) == x, {x, 1}]
   ```

   (g) $f(x) = \ln 2x$; $x_0 = \frac{1}{2}$

3. Use **Do** to determine the first ten iterations of

$$f_1(x) = \sin x, \quad f_i(x) = \int_0^x e^{-t} f_{i-1}(t)\, dt.$$

1.13 Examples of Programming in *Mathematica*

On several occasions in this chapter we have encountered situations where the process of answering questions in a problem set requires repeated execution of a series of instructions in *Mathematica* in which only a few parameters change. With the Notebook interface of this software we could reduce the amount of typing by cutting and pasting input lines already present in a session. We now introduce an alternative method that is quite efficient and works especially well with projects that require a large number of input statements.

A useful feature of *Mathematica* is that it allows one to input lines of commands from an external file. Using this feature and combining a series of internal functions, we can construct new functions that are specifically customized for certain objectives. We give an example of such a "program" in the context of differential equations. Its task is to solve a system of differential equations and plot the solution to an initial value problem.

Let us consider the system of differential equations

$$\frac{dx}{dt} = f(x, y, t), \quad \frac{dy}{dt} = g(x, y, t) \tag{1.15}$$

subject to the initial data

$$x(0) = x0, \quad y(0) = y0, \tag{1.16}$$

where

$$f(x, y, t) = x - y + \sin t, \quad g(x, y, t) = x + y + \cos t \tag{1.17}$$

and

$$x0 = 0.1, \quad y0 = 1.2. \tag{1.18}$$

We wish to plot the solution of this system over the interval $(0, 3)$. The following lines are saved in a file called ode.m:

```
Clear[x,y,tfinal,sol];
f[x_, y_, t_] = x - y + Sin[t];
g[x_, y_, t_] = x + y + Cos[t];
x0 = 0.1; y0 = 1.2; tfinal = 3;
sol = NDSolve[{x'[t] == f[x[t], y[t], t],
             y'[t] == g[x[t], y[t], t],
             x[0] == x0, y[0] == y0},
             {x, y}, {t, 0, tfinal}];
output = ParametricPlot[Evaluate[{x[t], y[t]} /. sol],
     {t, 0, tfinal}]
```

Remark 1.13.1: When using a word processing software (such as *Word-Perfect*) to create files for use in *Mathematica*, it is a good habit to save the files as text-only.

Then, after initiating *Mathematica*, we input ode.m by entering

```
<<ode.m
```

Clearly, if we intend to solve a different set of differential equations we only need to alter the lines in ode.m that define f and g and input the new ode.m program to *Mathematica*.

An alternative to re-editing the file ode.m for each set of differential equations is to construct a function from the block of instructions in ode.m, whose input would be the functions f and g (and other parameters, if we wish), and whose output is the graph of the solution through the initial data. The **Block** command is the appropriate tool for this task. Additionally, we are often interested in the phase plane diagram of the system (1.15), that is, a graph that contains plots of solutions of this system through several typical initial data. We will use the internal function `Flatten` for this purpose. This function acts on lists and flattens out nested lists. For example,

```
Flatten[{{1,2}, {3,4}}]
```

returns

```
{1, 2, 3, 4}
```

Also, `Flatten[list, n]` flattens `list` to level n. With these remarks in mind, the following is the listing of the program `odesolver.m`. It takes the parameters

```
f, g, data, tfinal
```

and delivers the phase diagram of (1.15) as output:

```
odesolver[f_, g_, data_, tfinal_] := Block[{x, y},
    solution=Table[NDSolve[{x'[t]==f[x[t],y[t],t],
                    y'[t]==g[x[t],y[t],t],
                    x[0]==data[[i]][[1]],
                    y[0]==data[[i]][[2]]},
                    {x,y},{t,0,tfinal}],
        {i,1,Length[data]}];
    solution=Flatten[solution,1];
    OutPut=ParametricPlot[Evaluate[{x[t],y[t]}/.solution],
        {t,0,tfinal}];
    Print["The graph is stored in OutPut"];
]
```

To illustrate the utility of `odesolver.m`, we now plot the graphs of approximate solutions to the system

$$x' = y, \quad y' = -y + \sin x,$$

which pass through the following three initial points

$$(1,1), \quad (2,2), \quad \text{and } (3,3).$$

Start a new session of *Mathematica* and enter

```
<< odesolver.m
```

At the prompt, enter the right-hand side of the system of differential equations

```
f[x_, y_, t_] = y; g[x_, y_, t_] = -y - Sin[x];
```

and the initial data and `tfinal`

```
tfinal = 10; data = {{1,1}, {2,2}, {3,3}};
```

Next, enter

```
odesolver[f, g, data, tfinal]
```

Mathematica now goes to work, plots the graph of the three trajectories, and returns

```
The graph is stored in OutPut
```

(see Figure 1.11). The symbol OutPut now contains all of the information about the phase plane. We could apply Show, PSPrint, or AspectRatio to OutPut, if we wish. It is also possible to make the entire process of entering the functions f and g and the initial data an interactive process. The Input command in *Mathematica* is quite useful in these circumstances. The following program, which we call myode.m, has the interactive feature built into it.

```
Block[{x,y,f,g,data},
    f[x_, y_, t_] = Input["This program integrates
        x' = f(x,y,t), y' = g(x,y,t)
    \n
    Enter f (for example x + y + Sin[t])
    \n"];
    g[x_,y_,t_]=Input["\n
    Input g (for example x - y - Cos[t])
    \n"];
    data=Input["\n
    Enter the initial data (for example {{0, 1}, {-1, 2}}
    \n"];
    tfinal=Input["\n
    Enter the length of integration in t (for example, 3)
    \n"];
    sol=Table[NDSolve[{x'[t]==f[x[t],y[t],t],
                    y'[t]==g[x[t],y[t],t],
                    x[0]==data[[i]][[1]],
                    y[0]==data[[i]][[2]]},
                    {x,y},{t,0,tfinal}],
        {i,1,Length[data]}];
    solution=Flatten[sol,1];
    OutPut=ParametricPlot[Evaluate[{x[t],y[t]}/.solution],
        {t,0,tfinal}];
    Print["The graph is stored in OutPut"];
]
```

Let us apply myode.m to

$$x' = \frac{\partial H}{\partial y}, \quad y' = -\frac{\partial H}{\partial x}. \tag{1.19}$$

Such a system is often called a Hamiltonian system of differential equations, and H is called a hamiltonian or a stream function of the system. Such systems are prevalent in fluid dynamics, where the ordered pair $(x(t), y(t))$ represents the position of a fluid particle at time t. Among them is the system of equations of

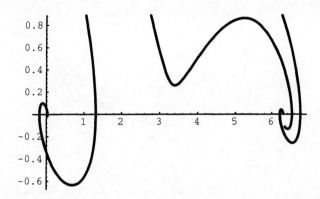

Figure 1.11: The output of **odesolver.m** for the system $x' = y, y' = -y - \sin x$.

motions of particles of fluid that pass a cylinder, where H is explicitly known to be

$$H = y - \frac{y}{x^2 + y^2}.$$

We now use **myode.m** and plot the particle paths of the fluid particles that are initially located at

$$(-3, 0.1i), \quad i = 1, ..., 5.$$

Start a new session of *Mathematica*. First, define H by

```
H = y - y/(x^2 + y^2)
```

Next, call up **myode.m**:

```
<<myode.m
```

When prompted for **f** and **g** enter

```
D[H, y]
```

and

```
- D[H, x]
```

respectively, Next, enter the initial data

```
Table[{-3, 0.1 i}, {i, 1, 5}]
```

Finally, choose **tfinal** to be 5. Figure 1.12 shows the output that is stored in **OutPut**.

Problems

1. Use an editor and create the files **ode.m**, **odesolver.m**, and **myode.m**. Study the logic of each program carefully. Run these programs separately in *Mathematica*, and generate Figures 1.11 and 1.12.

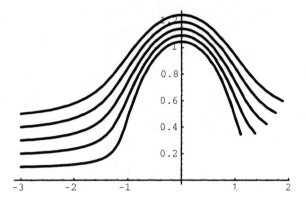

Figure 1.12: The output of myode.m for the flow past cylinder.

2. Use ode.m to plot the trajectories of the following systems of differential equations.

(a) $x' = y, \quad y' = -x; \; x(0) = 2, \; y(0) = -3; \; 0 < t < 4$

(b) $x' = 2x - y, \quad y' = x + y; \; x(0) = -1, \; y(0) = 1; \; 0 < t < 3$

(c) $x' = \dfrac{y}{\sqrt{x^2+y^2}}, \quad y' = -\dfrac{x}{\sqrt{x^2+y^2}}; \; x(0) = -2, \; y(0) = 2; \; 0 < t < 4$

(d) $x' = e^{-t}y, \quad y' = -x + e^{-3t}; \; x(0) = 1, \; y(0) = -1; \; 0 < t < 1.$

3. Execute odesolver.m or myode.m to generate the phase plane diagram for each of the following systems. In each experiment with the initial data and the value of tfinal.

(a) $x' = y, \; y' = -0.1y - \sin y$

(b) $x' = y, \; y' = -y - y^3$

(c) $x' = x + y(1 - x^2 - y^2), \; y' = y - x(1 - x^2 - y^2)$

1.14 Glossary of Useful Commands

1. **PSPrint**

   This command is used primarily in the stand-alone version of *Mathematica*. The entry PSPrint[Out[x]], where x marks the output you wish to print, creates a postscript hardcopy of the graphics created and stored in Out[x]. In the Notebook interface version of *Mathematica*, the Print command in the menu performs this task.

2. **Table**

   The Table command creates a list of objects. Its syntax is

   ```
   Table[expr, {i, n}, {j, m}]
   ```

where a list of copies of `expr` are generated as i and j vary from 1 to n and m, respectively. For example,

```
Table[i*j, {i, 1, 3}, {j, 1, 4}]
```

returns

```
{{1, 2, 3, 4}, {2, 4, 6, 8}, {3, 6, 9, 12}}
```

3. **Plot, Plot3D, ParametricPlot**, and **ParametricPlot3D**

The commands `Plot`, `Plot3D`, `ParametricPlot`, and `ParametricPlot3D` draw graphs of various two- and three-dimensional representations of functions. For example,

```
Plot[Sin[x], {x, 0, 2 Pi}]
```

draws the graph of $f(x) = \sin x$ over the interval $(0, 2\pi)$, while

```
ParametricPlot[{t, Sin[t]}, {t, 0, 2 Pi}]
```

accomplishes the same task using a typical parametrization of the same curve. The commands `Plot3D` and `ParametricPlot3D` have similar syntax:

```
Plot3D[Sin[x] Cos[y], {x, 0, 6 Pi}, {y, 0, 6 Pi}];
```

and

```
ParametricPlot3D[{x, y, Sin[x] Cos[y]},
       {x, 0, 6 Pi}, {y, 0, 6 Pi}]
```

yield the same results.

4. **Solve** and **FindRoot**

`Solve` and `FindRoot` find solutions of equations. Typical examples are

```
Solve[x^2 + 2 x + a == 0, x]
```

and

```
FindRoot[x^2 + 2 x + 1 == 0, {x, 3}]
```

5. **DSolve** and **NDSolve**

`DSolve` and `NDSolve` find solutions to ordinary differential equations. Their syntax follows the pattern

```
DSolve[equations, dependent variables, independent variable]
```

and

```
NDSolve[{equations, initial data}, dependent variables,
        independent variable]
```

For example,

```
DSolve[{x''[t] + 2 x[t] == 0, x[0] == 1, x'[0] == 2}, x[t], t]
```

and

```
NDSolve[{x''[t] + 2 Sin[x[t]] == 0, x[0] == 1, x'[0] == 2},
        x[t], {t, 0, 3}]
```

6. **LaplaceTransform** and **InverseLaplaceTransform**

The package that enables *Mathematica* to compute the Laplace transform of a function is `Calculus'LaplaceTransform'`. It should be entered in *Mathematica* at the beginning of a session as

```
<<Calculus'LaplaceTransform'
```

Typical commands for computing the Laplace transform and inverse transforms of functions are

```
LaplaceTransform[t Sin[t], t, s]
```

and

```
InverseLaplaceTransform[(s-3)/(s^2 + 2s +9), s, t]
```

References
[1] Wolfram, Stephen, *Mathematica*, 3rd Edition, Addison-Wesley Publishing Co., Reading, MA, 1996.

[2] Blachman, Nancy, *Mathematica: A Practical Approach*, Prentice-Hall, Englewood Cliffs, NJ, 1992.

Chapter 2

An Introduction to *MATLAB*

2.1 Introduction

MATLAB is a powerful computer language that is available for PCs (IBM compatibles and Apple computers) as well as on UNIX workstations. This software provides an environment for numerical computation as well as graphical display of outputs. In addition to the main program, the producers of *MATLAB* have made accessible a variety of subpackages (called toolboxes) that concentrate on specific areas of science and engineering. One of these packages is the *Symbolic Math Toolbox*, which enhances the overall performance of *MATLAB* by providing it access to a large number of internal functions of the symbolic algebra system *Maple*. This toolbox, which must be purchased separately, makes *MATLAB* a useful, convenient, and powerful tool for learning as well as research. We assume throughout this text that the reader has access to both *MATLAB* and its *Symbolic Math Toolbox*.

The notation we employ here is intended for *MATLAB* Version 5.1 and for its use on UNIX workstations. Minor changes to this notation may be necessary for other platforms.

Using *MATLAB* one is able to

1. find roots of polynomials, sum series, and determine limits of sequences;

2. symbolically integrate and differentiate rather complicated expressions;

3. plot graphics in two and three dimensions;

4. simplify trigonometric and algebraic expressions;

5. solve linear and nonlinear differential equations;

6. determine the Laplace transform of functions,

plus a variety of other mathematical operations. In this chapter we will review the basic commands and internal functions of this software as they pertain to the topics and concepts of this text. Reference [1] is an excellent companion text to this chapter.

2.2 A Session in *MATLAB*

We start a session of *MATLAB for Windows* on a PC by double clicking on the *MATLAB* icon. On a UNIX workstation we enter

```
matlab
```

at the machine's prompt to start a session of this package. In either platform a new window opens and, after a few introductory statements, the line

```
>>
```

appears, after which the cursor will be blinking. The program is expecting an input. For example, to find the roots of the polynomial

$$f(x) = x^2 - 4x + 3,$$

we enter

```
solve('x^2 - 4*x + 3 = 0','x')
```

The program returns

```
ans =

[1]
[3]
```

that is, 1 and 3 are the solutions of the equation $x^2 - 4x + 3 = 0$.

The above calculation uses the capabilities of the *Symbolic Math Toolbox*. The fact that x and x^2 - 4*x + 3 = 0 are within quotes signifies that these expressions are to be treated as symbols. Alternatively, we may define x as a symbol by

```
x = sym('x')
```

or by

```
sym x;
```

Then

```
solve('x^2 - 4*x + 3 = 0',x)
```

will return 1 and 3 as roots of the polynomial.

It is important to note that the asterisk in the preceding command is necessary to signify multiplication and, if it is omitted (either 4 x or 4x), will lead to a syntax error. Blank spaces between operands indicating addition or multiplication do not cause any problems. For example, in place of 4*x we could write 4 * x or 4* x.

In the previous expression single quotes are used to group **symbolic** expressions; the expressions $x^2 - 4x + 3 = 0$ and x are being treated as abstract symbols.

The string = 0 in solve is optional. Hence,

```
solve('x^2 - 4*x + 3','x')
```

leads to the same conclusion.

The first important feature of *MATLAB* one should note is that this program is **case sensitive**, that is, it distinguishes between solve and Solve in the prior input statement. All functions and programs that are **internally** known to *MATLAB* must be in lower case. Thus, solve is a subroutine in *MATLAB* that is capable of finding solutions of equations, while Solve or sOlve do not have special meanings unless defined by the user. Similarly, *MATLAB* understands that sin is the usual sine function while Sin or sIn is a name of a variable that, unless defined earlier, will be unknown in this session.

Internal functions in *MATLAB* are delimited by ( ): sin(x) and asin(x) are the usual sine and inverse sine functions. Operators are delimited by (,) as well. The operator solve used here is an example of a function that has two arguments; the first is used to define the expression whose roots we seek and the second is the variable with respect to which the root is computed. To gain a better understanding for the usage of symbols in *MATLAB*'s *Symbolic Math Toolbox*, let us try the following two examples:

```
solve('x^2 - a*x -1= 0', 'x')
```

in which x is the variable with respect to which the roots of the polynomial $x^2 - ax - 1$ are being determined, returns

```
ans =

[1/2*a+1/2*(a^2+4)^(1/2)]
[1/2*a-1/2*(a^2+4)^(1/2)]
```

while

```
solve('x^2 - a*x - 1 = 0', 'a')
```

returns

```
(x^2-1)/x
```

since now a is the independent variable.

These examples show one of the most important attributes of _MATLAB_'s _Symbolic Math Toolbox_, namely, its ability to perform symbolic manipulations. There is, however, a limitation to this capability. Consider, for instance, the fifth-order polynomial

$$f(x) = x^5 - ax^4 + 3x^3 + 2x^2 + x - 1.$$

To find its roots, we enter

```
solve('x^5 - a*x^4 + 3*x^3 + 2*x^2 + x -1 = 0','x')
```

in _MATLAB_. The response is

```
??? Error using ==> solve
Error, (in allvalues/rootseq) cannot
       evaluate with symbolic coefficients
```

which is _MATLAB_'s way of stating that it does not know how to find roots of this particular fifth-order polynomial symbolically. On the other hand, by drawing the graph of f for a fixed a, we recognize that it has some real roots. _MATLAB_ is able to find these roots **numerically** using approximate methods. We must first assign a fixed value to the parameter a. For example, let $a = 1$, so that the roots of f satisfy the equation

$$x^5 - x^4 + 3x^3 + 2x^2 + x - 1 = 0,$$

which has coefficients

$$1 \quad -1 \quad 3 \quad 2 \quad 1 \quad -1.$$

The internal function `roots` takes for its input these coefficients and computes an approximation to all five roots;

```
roots([1 -1 3 2 1 -1])
```

results in

```
ans =

   0.8203 + 1.7971i
   0.8203 - 1.7971i
  -0.5331 + 0.5639i
  -0.5331 - 0.5639i
   0.4255
```

Problems

1. Execute `help` (see the next section for more detail) with `solve` and `roots` and familiarize yourself with the range of applications of these functions.

2. Use `solve` or `roots` to find the roots of the following functions.

(a) $ax^2 + bx + c$

(b) $ax^3 + bx^2 + cx + d$

(c) $x^2 + 1$

(d) $x^3 + 1$

(e) $x^{1/3} - x + 1$ (Use `numeric(ans)` to get an approximate value for the roots.)

(f) $\sin x - \frac{1}{3}$

(g) $\sin^2 x - \frac{1}{3}$

(h) $\sin x^2 - \frac{1}{3}$

(i) $\sin x - x$

3. Determine the solution of the following equations using `solve`.

(a) $\tan x - 3x + 1$

(b) The system of algebraic equations

    i. $3x - 2y = 2, \quad x + y = 7$

    ii. $ax - y = 0, \quad x + ay = 1$

    iii. $x^2 - y^2 = 1, \quad x^2 + y^2 = 4$ (Hint: Start with `[x, y] = solve('x^2 - y^2 =1', 'x^2 + y^2 = 4')` and then ask for approximate values for x and y by `numeric(x)` and `numeric(y)`.)

    iv. $x^3 - y^3 = 1, \quad x^2 - 3xy + y^2 = 8$

    v. $ax + y + z = 1, \quad x - y + 2z = 0, \quad 2x + 3y - z = 2$

    vi. $3x^2 - 4y^2 + 3z = 1, \quad x + y + z = 0, \quad z^3 - x^2 y = -1$

2.3   The `help` Command in *MATLAB*

The help command in *MATLAB* is

```
help <string>
```

where `<string>` is an internal function. For instance,

```
help solve
```

returns (only a portion of the response is printed)

```
SOLVE  Symbolic solution of algebraic equations.
    SOLVE('eqn1','eqn2',...,'eqnN')
    SOLVE('eqn1','eqn2',...,'eqnN','var1,var2,...,varN')
    SOLVE('eqn1','eqn2',...,'eqnN','var1','var2',...'varN')

The eqns are symbolic expressions or strings specifying
equations. The vars are symbolic variables or strings
specifying the unknown variables.  SOLVE seeks zeros of the
expressions or solutions of the equations.  If not specified,
```

the unknowns in the system are determined by FINDSYM. If no
analytical solution is found and the number of equations equals
the number of dependent variables, a numeric solution is
attempted.

The statement goes on with describing other features of `solve`. The command
help by itself returns (the output is slightly different for *MATLAB* in UNIX
workstations, and depends on which optional toolboxes are installed)

HELP topics:

```
matlab/general       -  General purpose commands.
matlab/ops           -  Operators and special characters.
matlab/lang          -  Programming language constructs.
matlab/elmat         -  Elementary matrices and matrix
                          manipulation.
matlab/elfun         -  Elementary math functions.
matlab/specfun       -  Specialized math functions.
matlab/matfun        -  Matrix functions - numerical
                          linear algebra.
matlab/datafun       -  Data analysis and Fourier transforms.
matlab/polyfun       -  Interpolation and polynomials.
matlab/funfun        -  Function functions and ODE solvers.
matlab/sparfun       -  Sparse matrices.
matlab/graph2d       -  Two dimensional graphs.
matlab/graph3d       -  Three dimensional graphs.
matlab/specgraph     -  Specialized graphs.
matlab/graphics      -  Handle Graphics.
matlab/uitools       -  Graphical user interface tools.
matlab/strfun        -  Character strings.
matlab/iofun         -  File input/output.
matlab/timefun       -  Time and dates.
matlab/datatypes     -  Data types and structures.
matlab/demos         -  Examples and demonstrations.
simulink/simulink    -  Simulink
simulink/blocks      -  Simulink block library.
simulink/simdemos    -  Simulink demonstrations and samples.
simulink/dee         -  Differential Equation Editor
toolbox/control      -  Control System Toolbox.
control/obsolete     -  (No table of contents file)
toolbox/local        -  Preferences.
nnet/nnet            -  Neural Network Toolbox.
nnet/nndemos         -  Neural Network Demonstrations
                          and Applications.
toolbox/optim        -  Optimization Toolbox.
toolbox/signal       -  Signal Processing Toolbox.
toolbox/splines      -  Splines Toolbox.
stateflow/stateflow  -  Stateflow
```

```
stateflow/sfdemos    -  (No table of contents file)
toolbox/stats        -  Statistics Toolbox.
toolbox/symbolic     -  Symbolic Math Toolbox.
toolbox/tour         -  MATLAB Tour
```

```
For more help on directory/topic, type "help topic".
```

which is a list of some of the most important operations *MATLAB* performs, from which more detailed information is extracted by applying the `help` operation to the specific string in the left column in the preceding list: For example

```
help matlab\demos
```

provides information about available demos in *MATLAB*.

2.4 Factoring and Simplification

Some of the elementary operations that *MATLAB*'s *Symbolic Math Toolbox* is capable of performing include expanding, factoring, and simplifying expressions. We recall that we may declare a quantity `a` as a symbol to *MATLAB* by

```
a = sym('a')
```

Now, the sequence of commands for expanding

$$(a + b + c)^3$$

is

```
a=sym('a'); b=sym('b'); c=sym('c');
expand((a+b+c)^3)
```

which results in

```
ans =
```

```
a^3+3*a^2*b+3*a^2*c+3*a*b^2+6*a*b*c+3*a*c^2+b^3+
    3*b^2*c+3*b*c^2+c^3
```

Alternatively, we could define the symbolic variable `A` by

```
A = sym('(a+b+c)^3');
```

and then apply `factor` to `A`:

```
factor(A)
```

Remark 2.1: The punctuation ; at the end of a line prevents the output of that statement from being printed on the screen. This is an extremely handy tool since, in a typical computation, a large list of intermediate input/output expressions are generated, whose exact form is often of little interest to us. Having said that, removing the semicolon from the end of a line is a very useful device when debugging a program that does not seem to be doing what one expects.

We can factor an expression such as

$$1 - x - y + xy$$

as

```
x=sym('x'); y=sym('y');
factor(1-x-y+x*y)
```

MATLAB returns

```
ans =

(-1+y)*(-1+x)
```

The operations of **expand** and **factor** (and **simplify**) are quite useful when we wish to prove the type of identities we learned in elementary algebra. For example, the identity

$$(a + b)^2 - (a - b)^2 = 4ab$$

is valid for any two values a and b. To verify this identity in *MATLAB* we enter

```
a=sym('a'); b=sym('b');
expand((a+b)^2 - (a-b)^2)
```

and receive the response

```
ans =

4*a*b
```

If we repeat this input but replace **expand** with **factor**, we will reach the same conclusion. Next, let us try these commands on

$$a^2 - 2ab + b^2.$$

factor gives us the appropriate alternative expression for $a^2 - 2ab + b^2$ while **expand** leaves the expression unchanged, as expected. The result of applying the **factor**, **expand**, and **simplify** commands on the expression $\frac{1}{x+t} - \frac{1}{x-t}$ is as follows:

```
x=sym('x'); t=sym('t');
factor(1/(x+t)-1/(x-t))
```

yields

```
ans =

-2*t/(x+t)/(x-t)
```

while

```
expand(1/(x+t)-1/(x-t))
```

returns the same expression unaltered. The function `simplify`, on the other hand, returns the same expression as `factor` does.

A fourth command that is quite helpful in simplifying expressions is `simple`. When we apply this command to $a^2 - 2ab + b^2$ by entering

```
simple(a^2 - 2*a*b +b^2)
```

we receive the output

```
simplify:

a^2-2*a*b+b^2


radsimp:

a^2-2*a*b+b^2


combine(trig):

a^2-2*a*b+b^2


factor:

(a-b)^2


expand:

a^2-2*a*b+b^2


convert(exp):

a^2-2*a*b+b^2


convert(sincos):

a^2-2*a*b+b^2


convert(tan):

a^2-2*a*b+b^2
```

```
collect(b):
```

```
a^2-2*a*b+b^2
```

```
ans =
```

```
(a-b)^2
```

(Use `help` on `simple` to gain a sense of the range of applicability of this command.)

MATLAB is also capable of manipulating trigonometric functions. For example,

```
x=sym('x'); y=sym('y');
expand(sin(x+y)-sin(x-y))
```

returns the familiar result. Note that

```
expand(sin(x + y))
```

returns the alternative expression of the well-known identity while `simplify(sin(x+y))` returns the input string as its output.

Problems

1. Verify the following identities in *MATLAB*.

   (a) $\sin 2x = 2 \sin x \cos x$
   (b) $\cos 2x = 2 \cos^2 x - 1$
   (c) $\sin 3x = 4 \sin^3 x - 3 \sin x$
   (d) $\tan(x + y) = \frac{\tan x + \tan y}{1 - \tan x \tan y}$

2. Use *MATLAB* to discover an identity among the given functions.

   (a) $\cot(x + y)$, $\cot x$, and $\cot y$
   (b) $\sin(4x)$, $\sin x$, and $\cos x$
   (c) $\cos x$ and $\cos \frac{x}{2}$
   (d) $\tan 2x$ and $\cos x$

3. Simplify the following expressions.

   (a) $\cos(x + y) + \cos(x - y)$
   (b) $\cos(x + y) - \cos(x - y)$
   (c) $\cos(x + y) + \sin(x + y)$
   (d) $\cos^2 2x - \sin^2 2x$
   (e) $\cos^2 ax - \sin^2 ax$
   (f) $\cos^3 x - \sin^3 x$

2.5 Function Definition

There are two ways to define functions in *MATLAB*. One way is to use the string mode; for instance, let f be defined by

```
f = '1 - a*x^2'
```

The symbols a and x are both variables in f. The command subs acts on a function such as f and assigns values to a or x. For example,

```
subs(f,2,'a')
```

returns

```
ans =
```

```
1-2*x^2
```

while

```
subs(f,2,'x')
```

evaluates f at $x = 2$ and returns 1-4*a.

A second way to declare a function is to create an M-file and declare a and x as independent variables in f. An M-file is an ASCII file consisting of a sequence of standard *MATLAB* statements that are executed by this program once the file is entered to it. Such an M-file must be created by an editor and saved in a directory path accessible by *MATLAB*. M-files are particularly useful when one needs to execute a long list of commands. Here is an M-file, called fcn.m, that defines the above f:

```
function y=fcn(a,x);
y=1-a*x^2;
```

Now, assuming that the M-file fcn.m is located in the main directory of *MAT-LAB*, we can access and evaluate it as follows:

```
fcn(2,3)
```

MATLAB returns -17. It is important to note that a and x are **numerical** entities and an expression such as

```
fcn(a,3)
```

leads to a syntax error.

Both forms of defining functions in *MATLAB* are extremely useful in practice, as we will see in numerous examples in later chapters.

We often need to perform algebraic operations on functions that are defined symbollically. For example, given the two functions $f(x) = \sin 2x$ and $g(x) =$

$\cos 3x$, we define them to *MATLAB* first by

```
sym x;
f = sin(2*x); g= cos(3*x);
```

Now, to add these functions, enter

```
f+g
```

which returns the desired result. Similarly, try `f*g`, `f^3`, `f/g`. Any of the previous results may be evaluated at a particular x by using `subs`. For example,

```
numeric(subs(f/g,x,2.1))
```

yields

```
ans =

   -1.7607
```

Problems

1. Define the following functions in *MATLAB*.

    (a) $f(x) = \frac{1-x}{1+x}$. Evaluate f at $x = 0, 0.5$, and π.

    (b) $f(x) = \log(x + \sqrt{1 - x^2})$. Evaluate f at $x = 0, 0.1, 0.2$, and 0.3.

    (c) $f(x, t, \omega) = \sin(x - \omega t)$. Use `solve` to find a root of $f(1, t, 4)$.

2. Define each function in *MATLAB*. Use `solve` to find a zero of f.

    (a) $f(x) = x - \frac{2}{x}$ (Hint: Define f by `f='x - 2/x'`. Then find its zeros by `solve(f, 'x')`.)

    (b) $f(x) = x^2 - \frac{2}{x}$

    (c) $f(x) = \frac{x^2-1}{x-2} + x$

    (d) $x \sin x + \cos x$

2.6 Vectors and Matrices

Vectors and matrices are the natural building blocks of *MATLAB*. It is fair to say that *MATLAB* is one of the most popular software packages in use today precisely because of the efficient way that it manipulates vectors and matrices. The standard operations of linear algebra are all available in *MATLAB* and have been packaged in such a way that accessing them is simple and productive. Certain other matrix operations are available in *MATLAB* that at first sight may not seem as natural operations, but after some experience a user appreciates their enormous utility, especially when it comes to the numerical approximation of differential equations. We will see several examples of these operations in the text.

In *MATLAB*, vectors are entered as matrices. For example, the vector $a = \langle -2, 2, 3 \rangle$ is entered as the 1×3 matrix

```
a = [-2 2 3]
```

MATLAB responds with

```
a =

    -2    2    3
```

The transpose of this matrix is the 3×1 matrix

```
b = [-2; 2; 3]
```

which can also be derived from a using the operation ': So

```
b = a'
```

yields

```
b =

    -2
     2
     3
```

Similarly,

```
B=[1 -1 2;0  1  0; -1  5 1]
```

generates the matrix

$$B = \begin{bmatrix} 1 & -1 & 2 \\ 0 & 1 & 0 \\ -1 & 5 & 1 \end{bmatrix}$$

Elements of a vector (matrix) are accessed by putting the address of the element between the delimiters (,). For example, the first entry of a is `a(1,1)` while the (1,2)-entry of B is `B(1,2)`.

We can access a row or a column of a matrix by using : in place of the column or row in the preceding notation. For instance,

```
B(:, 1)
```

yields

```
ans =

     1
     0
    -1
```

the first column of B. Similarly,

```
B(:,2:3)
```

yields

```
ans =

    -1     2
     1     0
     5     1
```

a matrix consisting of the second and third columns of B.

Matrix addition and subtraction are straightforward. Matrix multiplication follows the standard rules from linear algebra: Two matrices a and b can be multiplied, resulting in a matrix c, if a and b are compatible (i.e., if a has as many columns as b has rows). Matrix multiplication is denoted by ∗. So, the two matrices a and B, defined previously, are compatible and their product $c = aB$ is determined in *MATLAB* by

```
c = a*B
```

which results in the 1×3 matrix

```
c =

    -5    19    -1
```

The product Ba is not permissible, since the columns of B and the rows of a are not equal. When we try

```
B*a
```

we receive the message

```
??? Error using ==> *
Inner matrix dimensions must agree.
```

A second way to multiply matrices is by using .∗. Given two matrices a and b with the same dimension (i.e., a and b have the same number of rows and columns), the product a.∗ b is a matrix c whose ijth entry is the product of the ijth entries of a and b. For example, if $a = \langle -1, 2, 3 \rangle$ and $b = \langle 3, 2, 7 \rangle$, then a.∗b is the matrix $c = \langle -3, 4, 21 \rangle$, as shown by

```
a=[-1 2 3]; b=[3 2 7];
a.*b
```

which produces

```
ans =

    -3    4    21
```

Note that this operation allows us to **square** a vector. For example,

`a.*a`

results in

`ans =`

    1   4   9

By the same token, the operation `.^` raises a vector (matrix) to a power where the resulting vector (matrix) has the same dimension as the original vector (matrix) but every entry of which is the corresponding entry of the original vector (matrix) raised to the appropriate power. For example,

`a.^5`

yields

`ans =`

    -1   32   243

We recall that many elementary functions in calculus have power series expansions. For example, the sine function is equivalent to the Taylor (Maclaurin) series

$$\sin x = x - \frac{x^3}{3!} + \frac{x^5}{5!} - \frac{x^7}{7!} + \cdots$$

To evaluate this power series at a specific value of x is tantamount to having the ability to compute x^n, for arbitrary n. Since we have the capability of raising a *vector* to a power n in *MATLAB*, it is then reasonable to expect that we could extend the capability of function evaluation to vectors. For example,

`sin(a)`

returns the vector

`ans =`

   -0.8415   0.9093   0.1411

whose components correspond to $\sin(-1)$, $\sin(2)$, and $\sin(3)$. Similarly, we can compute

`exp(a); log(a); sqrt(a);`

It is important to note that `exp(a)` is **not** generally equal to the standard e^a of linear algebra, when a is a matrix with dimension higher than one. The exponential of a square matrix A is defined by

$$e^A = \sum_{n=0}^{\infty} \frac{A^n}{n!},$$

also exists in *MATLAB*, but it is denoted by

`expm(A)`

In the spirit of multiplying and exponentiating vectors, *MATLAB* also divides vectors. The operation is denoted by `./` and operates on vectors and matrices of equal dimensions. For example, let x be the following vector of equally distributed points on the interval $(0, 1)$:

`x=0:0.01:1;`

In x we have a vector of length 101 (try `length(x)` or `size(x)`). Here is how one plots the function $f(x) = \frac{x}{1+x}$ using the preceding x:

`plot(x, x./(1+x));`

(Try `plot(x, x/(1+x))` instead. What is *MATLAB*'s response?) It is important to note that the 1 in

`1+x`

is interpreted as a vector of length 101, each of whose entries is 1.

The commands `det`, `inv`, and `eig` operate on a square matrix and determine the determinant, inverse, and eigenvalues of that matrix, respectively. We will elaborate on these operations in detail when we encounter their definitions later in the text.

Problems

1. Let A and B be defined by

$$A = \begin{bmatrix} -1 & 1 \\ 1 & 3 \end{bmatrix}, \quad B = \begin{bmatrix} -3 & 4 \\ 7 & 5 \end{bmatrix}.$$

   Compute $A + B$, $A - B$, AB, BA, $6A$, and $3A + 2B$.

2. Consider the matrix

$$A = \begin{bmatrix} a & b \\ b & c \end{bmatrix},$$

   where a, b, and c are constants. Find all values of these parameters for which the determinant of A vanishes.

3. Let A be the matrix

$$A = \begin{bmatrix} a & -a & b \\ -a & b & a \\ b & a & 2a \end{bmatrix}.$$

   Find all values of a and b for which the determinant of A vanishes.

4. Let A be the matrix

$$A = \begin{bmatrix} a & b \\ b & a \end{bmatrix},$$

   Compute A^5 and A^{10}.

5. Plot the graph of each of the following functions by first generating a vector x of discretized points in the domain with $\Delta x = 0.01$, where Δx is the length of a typical subinterval (if the domain is (a, b), then define x = a:0.01:b;).

    (a) $f(x) = \sin x$; $(0, 2\pi)$

    (b) $f(x) = \sin 3x$; $(0, 2\pi)$

    (c) $f(x) = \sin 5x - \sin 4x$; $(0, 2\pi)$

    (d) $f(x) = e^{-x} \sin x$; $(0, 5)$

    (e) $f(x) = e^{-x} \sin 10x$; $(0, 10)$

    (f) $f(x) = \ln(1 + |\sin x|)$; $(0, 2\pi)$

    (g) $f(x) = \dfrac{1}{\sqrt{1 + \sin^2 x}}$; $(0, 2\pi)$

2.7   Differentiation and Integration

MATLAB and its *Symbolic Math Toolbox* are able to differentiate and integrate functions symbolically. To find the derivative of a function such as $f(x) = x \sin^2 x$, we enter

```
diff('x*sin(x)^2','x')
```

which results in

```
ans =
```

```
sin(x)^2+2*x*sin(x)*cos(x)
```

To integrate f, we give the command

```
int('x*sin(x)^2','x')
```

and obtain

```
ans =
```

```
x*(-1/2*cos(x)*sin(x)+1/2*x)+1/4*sin(x)^2-1/4*x^2
```

The *MATLAB* function `pretty` generates an output that is somewhat familiar to the eye. Its syntax is

```
pretty(ans)
```

which applies to the most recent output and returns

```
                                      2           2
    x (- 1/2 cos(x) sin(x) + 1/2 x) + 1/4 sin(x)  - 1/4 x
```

Similarly,

```
pretty(f)
```

rewrites the output stored in f.

Next, let us evaluate the integral $\int_1^2 \frac{1}{1+x^4}\,dx$ via

```
int('1/(1+x^4)', 'x', 1, 2)
```

We receive the output

```
ans =
```

```
1/8*2^(1/2)*log((2*2^(1/2)+5)/
   (-2*2^(1/2)+5))+1/4*2^(1/2)*atan(2*2^(1/2)+1)+1/4*2^(1/2)*
     atan(2*2^(1/2)-1)-1/8*2^(1/2)*
          log((-2-2^(1/2))/(2^(1/2)-2))-1/8*2^(1/2)*pi
```

which, after applying **pretty** to it, yields

```
      1/2              1/2            1/2              1/2
1/8 2     log(5 + 2 2    ) - 1/8 2     log(5 - 2 2    )


           1/2            1/2              1/2            1/2
    + 1/4 2     atan(2 2     + 1) + 1/4 2     atan(2 2     - 1)


         1/2            1/2            1/2            1/2
   - 1/8 2     log(2 + 2    ) + 1/8 2     log(2 - 2    )
           1/2
     - 1/8 2     pi
```

To evaluate this integral, *MATLAB* first determined the anti-derivative of the integrand and then evaluated the result at the two endpoints. To get a decimal approximation to the value of the integral, enter

```
numeric(ans)
```

and obtain

```
ans =
```

```
    0.2032
```

MATLAB is capable of performing symbolic integration of a rather large class of functions. For example, try integrating the function $\sin(x^2)$ on the interval $(0, 1)$ via

```
int('sin(x^2)', 'x', 0, 1)
```

The response

```
ans =
```

```
1/2*FresnelS(2^(1/2)/pi^(1/2))*2^(1/2)*pi^(1/2)
```

is the exact answer in terms of Fresnel integrals. We can now get a numerical approximation to the exact value by applying the `numeric` function to `ans`, which results in

`0.3103`

MATLAB is also capable of performing integrals numerically. For example,

$$\int_0^1 \sin(x^3)dx$$

is outside of its scope of symbolic integration, so when we try

`int('sin(x^3)', 'x', 0, 1)`

it responds with

`Warning: Explicit integral could not be found.`

`ans =`

`int(sin(x^3),x = 0 .. 1)`

We can now obtain a numerical approximation to this integral, however, by typing `numeric(ans)` at the prompt. *MATLAB* responds with

`ans =`

`0.2338`

Problems

1. Differentiate the following functions.

    (a) $f(x) = \log \frac{x}{x+1}$

    (b) $f(x) = \sin^3 4x \cos^5 7(x^2 - 2x + 1)$

    (c) $f(x) = x^{x-1}$

2. Determine the extreme points (the maxima and minima) of $f(x) = x \sin^2 x$.

3. Integrate the following functions.

    (a) $f(x) = \frac{x}{x-1}$

    (b) $f(x) = x \sin x$

    (c) $f(x) = x^2 \sin x$

    (d) $f(x) = x^{10} \sin x$

    (e) $f(x) = e^x \sin x$

    (f) $f(x) = \sin^2 x$

(g) $f(x) = \sin(x^2)$

(h) $f(t) = te^{t^2}$

(i) $f(s) = e^{s^2}$

4. Do the following indefinite integrals.

   (a) $\int_{-\infty}^{\infty} \frac{1}{1+x^2} dx$

   (b) $\int_0^{\infty} e^{-t^2} dt$

   (c) $\int_0^{\infty} e^{-at^2} dt$ (Here a is a parameter.)

   (d) $\int_0^{\infty} e^{-st} \sin t \, dt$ (This is the Laplace Transform of $\sin t$.)

5. Define the function f by

$$f(x) = \int_0^x \frac{1}{x + t^5} \, dt.$$

   Use *MATLAB* to determine $f'(\frac{1}{2})$. (Ans: -0.0779959)

6. Verify the Fundamental Theorem of Calculus in *MATLAB*:

$$\int_a^b f'(t) \, dt = f(b) - f(a), \qquad \frac{d}{dx} \int_0^x f(t) \, dt = f(x).$$

7. Define the function f by

$$f(x) = \int_0^1 \sin(xt^3) \, dt.$$

   Determine $f'(\frac{1}{2})$. (Ans: 0.237662)

8. Verify the Leibniz's rule of differentiation in *MATLAB*:

$$\frac{d}{dx} \left(\int_{a(x)}^{b(x)} f(x,t) \, dt \right) = f(x, b(x)) - f(x, a(x)) + \int_{a(x)}^{b(x)} \frac{\partial f}{\partial x} \, dt.$$

2.8 Two-Dimensional Graphics

To plot the graph of a function whose formula is known, say $f(x) = \sin(x)$ on the interval $(0, 2\pi)$, we give the command

```
ezplot('sin(x)', [0, 2*pi])
```

MATLAB draws a graph of this function and scales it automatically. A second way of generating the graph of this function is to use the **plot** command. First, we must discretize the domain of f. For example, to plot f in the interval $(0, 2\pi)$, we discretize the interval into equidistant subintervals of length 0.01 and use the resulting vector as input in `plot` as follows:

```
x=0:0.01:2*pi;
plot(x,sin(x));
```

The command `title` puts a title on a graph:

```
title('This is the sine function')
```

Multiple graphs can be displayed on the same screen by invoking the `hold on` command:

```
hold on
ezplot('exp(-x)*sin(x)');
ezplot('exp(-x)*cos(x)');
axis([0 3 -2 2]);
```

The last command manually sets the scale for the axes.

The `hold on` command allows one to plot several graphs on the same screen. It is often desirable, however, to plot several graphs on the same screen but on different axes. This is particularly important when the functions, such as displacement and velocity, have different dimensions. *MATLAB*'s `subplot` command is ideal for such a setting. The syntax is

```
subplot(mnp)
```

which subdivides the screen into an $m \times n$ partition and plots the pth graph in the mn subdivision. For example, here is how we can plot the functions $f(x) = e^{-x} \sin x$ and $g(t) = 3e^{-t} \cos t$, with f in the top half of the screen and g in the bottom half:

```
subplot(211);
x=0:0.01:3;
plot(x,exp(-x).*sin(x));
title('exp(-x)*sin(x)');
subplot(212)
t=0:0.05:5;
plot(t,3*exp(-t).*cos(t));
title('3*exp(-t)*sin(t)');
```

(Note the usage of `.*` in the above expressions.) Figure 2.1 shows the output of this code.

We can get a hardcopy of any graph we produce in *MATLAB* by giving the command

```
print
```

after generating the desired graph. Use `help print` to review the many options *MATLAB* provides with printing.

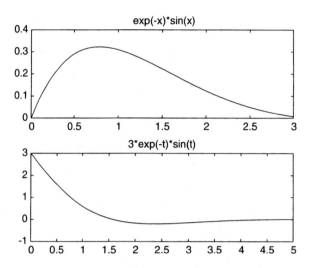

Figure 2.1: The output of the subplot command.

plot is also capable of plotting the graphs of parametrized curves. For instance, to draw the graph of the curve whose parametrization is given by

$$\mathbf{r}(t) = \langle \sin^3 t, \cos^3 t \rangle, \qquad t \in (0, 2\pi),$$

we proceed as

```
t=(0:0.01:2*pi);
plot(sin(t).^3, cos(t).^3)
```

Problems

1. Plot the graph of the following functions.

   (a) $f(x) = \sin(5x)$. What is the period of this function, that is, what is the smallest value of $T > 0$ for which $f(x + T) = f(x)$?

   (b) $g(x) = \sin(2x) + 3\sin(3x)$. What is the period of this function?

   (c) $h(x) = \sin x + \sin \sqrt{2}x$. Draw the graph of this function on the intervals $(0, 5)$, $(0, 10)$, and $(0, 50)$. Do these graphs give any indication as to whether h is periodic or not? Why?

2. Let $f(x) = x^n$. Draw the graph of f for $n = 0, 1, 2, 3, \ldots, 10$ for $x \in (0, 2)$ on the same screen.

3. Consider the function $f(x) = e^{-x^2}$ with $x \in (-10, 10)$.

(a) Plot the graph of this function.

(b) Compare this graph to the graphs of the functions $g(x) = f(x + 4)$ and $h(x) = f(\frac{x}{0.5})$ over the same domain. Is there a scale change among these three graphs?

(c) Draw on the same screen the graph of

$$\frac{1}{2} \left(f\left(\frac{x - \pi t}{0.5} \right) + f\left(\frac{x + \pi t}{0.5} \right) \right),$$

for t ranging between 0 and 4 at increments of 0.5.

4. Plot the graphs of the following curves.

(a) $\mathbf{r}(t) = \langle \sin^2 t, \cos^2 t \rangle, \qquad t \in (0, 2\pi)$

(b) $\mathbf{r}(t) = \langle \sin^5 t, \cos^5 t \rangle, \qquad t \in (0, 2\pi)$

(c) $\mathbf{r}(t) = \langle \sin^3 t, \cos^2 t \rangle, \qquad t \in (0, 2\pi)$

(d) $\mathbf{r}(t) = \langle \sin^3 t, \cos 2t \rangle, \qquad t \in (0, 2\pi)$

(e) $\mathbf{r}(t) = \langle \sin^5 t, \cos(1 + 2t) \rangle, \qquad t \in (0, 2\pi)$

2.9 Three-Dimensional Plotting

The syntax for plotting graphs of curves and surfaces in three dimensions is somewhat similar to plotting in two dimensions. The command `plot3` renders curves in R^3. For example, let $\mathbf{x} = \langle \sin t, \cos t, t \rangle$ be the parametrization of the standard helix. To draw the graph of this curve, first we need to specify a vector of t values

```
t=0:0.01:6*pi;
```

and then invoke `plot3`:

```
plot3(sin(t), cos(t), t)
```

Figure 2.2 shows the output.

To do three-dimensional graphing, we must first create a discrete two-dimensional domain. The command `meshgrid(x,y)` creates a two-dimensional grid from the vectors x and y. For example, let $f(x, y) = \sin \sqrt{x^2 + y^2}$ be the function whose graph we want to plot. To draw the graph of f in the rectangular domain $(x, y) \in (-2\pi, 2\pi) \times (-3\pi, 3\pi)$, first we create two matrices X and Y, each appropriately related to the above domain (**don't forget to put a semicolon at the end of this line!**):

```
[X, Y] = meshgrid(-pi:0.1:pi,-pi:0.1:pi);
```

The command `mesh`, when combined with matrices X and Y by

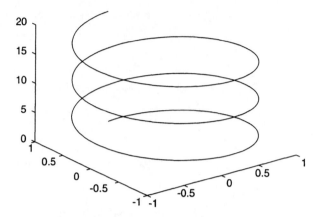

Figure 2.2: The output of plot3 for the helix $\mathbf{x} = \langle \sin t, \cos t, t \rangle$.

```
mesh(sin(sqrt(X.*X+Y.*Y)))
```

renders Figure 2.3. There are many other internal functions in *MATLAB* that aid in visualizing curves and surfaces, among them contour, contour3, pcolor, surf, surfl, shading, and colormap. Use help on these commands for more information.

Problems

1. Draw the graphs of the following curves. In each case choose a range of the parameter.

   (a) $\mathbf{r}(t) = \langle t, t, t \rangle$

   (b) $\mathbf{r}(t) = \langle t^2, t, t^2 \rangle$

   (c) $\mathbf{r}(t) = \langle \sin(t^2), \cos(t^2), t \rangle$

   (d) $\mathbf{r}(t) = \langle \sinh t, t, \cosh t \rangle$

   (e) $\mathbf{r}(t) = \langle t, \frac{1}{t}, 0 \rangle$

2. Draw the graphs of each of the following curves.

   (a) A circle of radius two centered at the origin and located in the xy-plane

   (b) A circle of radius two centered at the origin and located in the $z = 1$ plane

   (c) The ellipse located in the xy-plane centered at the origin with major and minor axes of 3 and 2, respectively

   (d) The curve of intersection of $x^2 + y^2 = 1$ and $z = x$

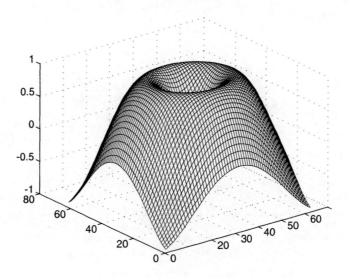

Figure 2.3: An output generated by `mesh`.

3. Draw the graphs of each of the following surfaces.

(a) $z = x^2 + y^2$ on the domain $(-3, 3) \times (-3, 3)$

(b) $z = \sqrt{x^2 + y^2}$ on the domain $(-3, 3) \times (-3, 3)$

(c) $z = 3x^2 + 4y^2$ on the domain $(-3, 3) \times (-3, 3)$

(d) $z = \sin(x^2 + y^2)$ on the domain $(-\pi, \pi) \times (-\pi, \pi)$

(e) $z = \sin(\sqrt{x^2 + y^2})$ on the domain $(-\pi, \pi) \times (-\pi, \pi)$

(f) $z = \sin(x^2 + y^2)\cos(x)$ on the domain $(-\pi, \pi) \times (-\pi, \pi)$

(g) $z = \sin(x^2 + y^2)\cos(y)$ on the domain $(-\pi, \pi) \times (-\pi, \pi)$

(h) $z = \dfrac{\sin\sqrt{x^2+y^2}}{\sqrt{x^2+y^2}}$ on the domain $(-\pi, \pi) \times (-\pi, \pi)$

2.10 Solving Differential Equations

MATLAB has two commands, `dsolve` and `ode45`, each of which is capable of solving special classes of ordinary differential equations. The command `dsolve` is primarily used for finding the exact solution of linear constant coefficient equations. Consider the following example. Let $x(t)$ be the solution to

$$x''(t) + 2x(t) = 0, \tag{2.1}$$

subject to the initial data

$$x(0) = 0, \quad x'(0) = 1. \tag{2.2}$$

The exact solution to this initial value problem is

$$x(t) = \frac{\sin \sqrt{2}t}{\sqrt{2}}. \tag{2.3}$$

To get this solution in *MATLAB*, we give the command

```
dsolve('D2x + 2*x = 0', 'x(0) = 0, Dx(0) = 1')
```

whose output is

```
ans =
```

```
1/2*sin(2^(1/2)*t)*2^(1/2)
```

The function `dsolve` is also capable of finding solutions to boundary value problems. Consider the same differential equation in (2.1) but with boundary conditions

$$x(0) = 0, \qquad x(3) = 1.$$

The exact solution of this problem is

$$x(t) = \frac{1}{\sin(3\sqrt{2})} \sin(\sqrt{2}t),$$

which is the same solution we obtain by running `dsolve`:

```
dsolve('D2x + 2*x = 0', 'x(0) = 0, x(3) = 1')
```

(How would you draw the graph of the solution?)

In some cases *MATLAB*'s `dsolve` is capable of determining the analytic solution of a differential with *nonconstant* coefficients. For instance, one can get the analytic solution to the Cauchy–Euler equation

$$t^2 x'' - 2tx' + 3x = 0,$$

by entering the statement

```
dsolve('t^2*D2x-2*t*Dx+3*x=0')
```

The second function in *MATLAB* capable of solving differential equations is `ode45`. This command uses a numerical algorithm (based on the standard Runge–Kutta scheme) and solves linear as well as nonlinear systems of differential equations. Consider an example in which we let $x(t)$ be a solution to the forced nonlinear pendulum equation

$$x'' + 0.1x' + \sin x = 0.02 \cos t, \quad x(0) = 0, \quad x'(0) = 1. \tag{2.4}$$

This equation, being nonlinear, is not a candidate for `dsolve`. To give this equation to `ode45` and plot its trajectory in the phase plane, we need to create an M-file that contains the equations. First, we convert (2.4) to a first-order

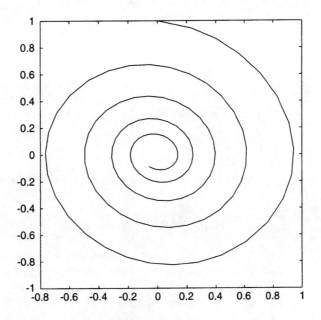

Figure 2.4: The phase plane diagram of the solution to $x'' + 0.1x + \sin x = 0.02 \cos t$.

system:

$$x' = y, \quad y' = -0.1y - \sin x + 0.02 \cos t, \qquad x(0) = 0, \quad y(0) = 1. \qquad (2.5)$$

Then, we use an editor and create the M-file `pendulum.m` containing the lines

```
function yprime=pendulum(t,y);
yprime=[y(2); -0.1*y(2)-sin(y(1))-0.02*cos(t)];
```

Next, at *MATLAB*'s prompt, we enter the line

```
[t,y] = ode45('pendulum',0,3,[0;1]);
```

To get the phase portrait of this particular solution, we need to plot the first column in y versus its second column:

```
plot(y(:,1), y(:,2))
```

(see Figure 2.4).

Problems

1. Use `dsolve` and solve the following differential equations.

   (a) $y'' + y = 0$
   (b) $x''' + x' + x = 0$

(c) $y'' + y = 0$, $y(0) = 1, y'(0) = -1$

(d) $x'' + x = \sin 2t$

(e) $y'' + y = \sin t$

2. Use `ode45` and draw the trajectories of the following differential systems. In each case supply a final value of t.

(a) $x'' + x = 0$, $x(0) = 0, x'(0) = 1$

(b) $y'' + 0.1y' + \sin y = 0$, $y(0) = 0, y'(0) = 1$

(c) $y'' + 0.1y' + \sin y = 0$, $y(0) = 0, y'(0) = 3$

2.11 Loops and Iterations

In many numerical applications we need to perform an operation repeatedly while a few parameters may change with each iteration. The `for ... end` command in *MATLAB* is designed to do just that. For example, consider the sequence of Fibonacci numbers, 1, 1, 2, 3, 5, 8, ..., where the nth term of the sequence x_n is the sum of the two previous terms of the sequence, that is,

$$x_n = x_{n-1} + x_{n-2}, \quad n = 3, 4, \ldots,$$

and $x_1 = x_2 = 1$. One generates the first ten terms of this sequence by

```
x1=1; x2=1;
for i=1:8
    x=x1+x2
    x1=x2; x2=x;
end
```

The first line of the program is self-explanatory and initializes the first two terms of the Fibonacci sequence. The second and the last lines define the "loop" or the "iteration" process, while the lines in between execute the necessary information to generate the subsequent entries of the sequence. When the `for ... end` process begins, the **counter** i has value 1. The line x = x1+x2 executes the formula that defines the Fibonacci numbers, and since this line is not punctuated by a semicolon, the output is printed to the screen. The next line prepares for the next step of the iteration by "updating" the values of x1 and x2: First the current value of x2 is stored in the x1 variable, and next, the current value of x is stored in the x2 variable. At this point one cycle of the iteration is completed and we are ready to carry out the next iteration. Next, i takes on the value 2 and *MATLAB* proceeds to the line x=x1+x2, and so on.

We often need to store the output of an iteration process in a vector for later use. For example, suppose that we wish to save the first 10 elements of a Fibonacci sequence in a vector called `fibonacci`. Two new lines must be added to the preceding program. The first is

```
fibonacci=[1 1];
```

which is inserted before the for ... statement to initialize the fibonacci vector with the first two elements of the sequence. And the second is

```
fibonacci=[fibonacci x];
```

after x has been determined. The effect of this line is to *append* to fibonacci the value of x at each step of the iteration. So now the program reads

```
x1=1; x2=1;
fibonacci=[1 1];
for i=1:8
    x=x1+x2
    fibonacci=[fibonacci x];
    x1=x2; x2=x;
end
```

Assuming these lines are saved in the M-file fib.m, the command

```
fib
```

will now print the content of fibonacci to the screen.

The previous program shows how to access the first n entries of the Fibonacci sequence. There are also two *conditional* commands in *MATLAB*, while ... end and if ... else ... end, that are very helpful in controlling how much information one wishes to store or print. Consider the following variation of the prior program:

```
x1=1; x2=1;
fibnumber=[1 1];
x = 0;
while x < 100
    x=x1+x2
    fibnumber=[fibnumber x];
    x1=x2; x2=x;
end
```

Now, the iteration process continues through the first time the element of the sequence reaches 100 and then terminates. An alternative way of achieving the same goal is

```
x1=1; x2=1;
fibnumber=[1 1];
x=0;
for j=1:1000
    if x < 100
        x=x1+x2
        fibnumber=[fibnumber x];
        x1=x2; x2=x;
    elseif x >= 100
```

```
            break
        end
    end
```

The next example shows another usage of the iteration process in the context of a simple numerical scheme that produces the Euler approximation of the solution of a first-order differential equation. Consider the first-order differential equation $\frac{dx}{dt} = f(t, x), x(t_0) = t_0$. The Euler approximation of the solution $x(t)$ seeks a sequence (t_n, x_n) of the form

$$x_{n+1} = x_n + hf(t_n, x_n), \tag{2.6}$$

where $t_{n+1} = t_n + h$ and h is a fixed small positive number. The following program shows how to generate the sequence of pairs (t_n, x_n) and use the plot command and plot the approximate solution. This program is written for $f(t, x) = -x + \sin t$, $t_0 = 0$, $x_0 = 1$; $h = 0.01$, and $n = 100$.

We first create an M-file, called yprime.m, that defines the function $f(t, x)$:

```
function yp=yprime(t,x);
yp=-x + sin(t);
```

Next, we write the program that carries out the iteration in (2.6):

```
n = 100; h = 0.01; t = 0; x = 1;
output = [0 1];
for i=1:n
    y = x + h*yprime(t,x);
    t=t+h;
    output=[output;t y];
    x=y;
end
plot(output(:,1),output(:,2))
```

The output of this program is shown in Figure 2.5.

Another use of the iteration process falls within the context of graphing functions whose definitions are suited either to symbolic manipulation or numerical approximation. As an example consider the function f defined by

$$f(t) = \int_0^t e^{-x} \tan(x) \, dx. \tag{2.7}$$

This integrand does not yield easily to the *Symbolic Math Toolbox* or *Maple*'s integration routines. The command

```
f = int('exp(-x)*tan(x)','x',0,'t')
```

results in

```
int(exp(-x)*tan(x),x = 0 .. t)
```

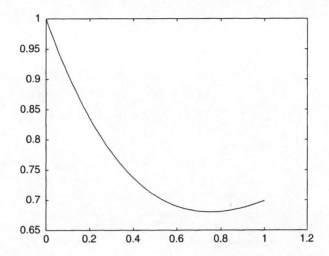

Figure 2.5: The output of the Euler method for the differential equation $x' = -x + \sin t, x(0) = 1$.

Even though it seems that *MATLAB* did not succeed in carrying out the integration, the function f is well defined to this program and the previous answer is still quite useful. Try

```
numeric(subs(f,1,'t'))
```

MATLAB returns 0.3158 for the value of f at $t = 1$. So, if we are interested in graphing f, on the interval $(1, 5)$, say, we could evaluate f at a set of points in that interval and use `plot` to get the desired result:

```
for i=1:101
    a=[a numeric(subs(f,0.05*(i-1),'t'));
end
plot(a)
```

Unfortunately this is a time-consuming solution to graphing (2.7). The difficulty lies in the amount of time spent between the *MATLAB* and *Maple* kernels as the loop is executed. In particular, the vector efficiency, which is the hallmark of *MATLAB*, is absent in the way we have introduced the prior iteration.

Problems

1. Let f be defined by (2.7).

   (a) Try `ezplot(f)`. What happens?

   (b) Try

   ```
   t = 0:0.05:5;
   a=eval(vectorize(f));
   ```

   What is *MATLAB*'s response?

2. Let f be defined by

$$f(t) = \int_0^t \sin(x^2)\, dx. \tag{2.8}$$

Try drawing the graph of this function using `ezplot` and the process described previously. Is either one successful?

References

[1] *The Student Edition of MATLAB*, Prentice-Hall, Englewood Cliffs, NJ, 1995.

[2] Marcus, Marvin, *Matrices and Matlab*, Prentice-Hall, Englewood Cliffs, NJ, 1993.

[3] Etter, D. M., *Engineering Problem Solving with MATLAB*, Prentice-Hall, Englewood Cliffs, NJ, 1993.

Chapter 3

An Introduction to Ordinary Differential Equations

3.1 Introduction

Differential equations are mathematical objects that commonly arise in the description of many problems in science and engineering. In this chapter we will study the elementary tools that one employs to find solutions of these equations. Before going on with a formal discussion, it is worth mentioning that there are many excellent texts on ordinary differential equations that the reader can consult to supplement the material presented in this chapter. Three such texts are [1], [2], and [3].

Many of the details about the topics we discuss here, in addition to a variety of applications on which we may not focus with quite the same degree of emphasis, can be found in the above texts. The main source of applications on which we will concentrate comes to us from the field of fluid dynamics, a branch of science that has enjoyed a rich history, during which it has made fundamental contributions to the evolution of applied mathematics. What is new in this text is the use of available software that will aid us with some of the tedious computations encountered in solving differential equations as well as the ease with which we will be able to display our results, especially in the context of problems involving fluid flows.

Differential equations involve the derivatives of an unknown function y. These equations often come about as a consequence of the laws of physics that relate the *rate of change* of a physical quantity, such as the velocity of a moving object, to the forces present in the physical problem. By a **solution** of a differential equation we mean a function $y = f(t)$, defined on an interval (a, b), and sufficiently differentiable so that when $y = f(t)$ is substituted in the differential equation, an identity in t is achieved.

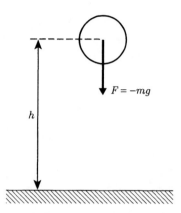

Figure 3.1: In the absence of friction, a falling object satisfies the differential equation $my'' = -mg$.

Example 3.1.1

One of the simplest physical problems in which a differential equation appears is in the case of falling bodies. Suppose that an object of mass m falls, under the influence of gravity, from a height h. Newton's second law, that the rate of change of the linear momentum of a moving body (that is, $my'(t)$) must be balanced by the resultant of all the forces acting on the body, is the governing law in this physical setting. Here $y(t)$ is the distance of the falling body from a reference point such as ground level. If we ignore all forces acting on the body except for its weight, we arrive at the differential equation (see Figure 3.1)

$$my'' = -mg, \tag{3.1}$$

where $y(t)$ is the height of the object (with the positive direction pointing upward). The parameter g in (3.1) is the constant of acceleration of gravity.

Equation (3.1) is easy to solve. After cancelling m from both sides and integrating twice with respect to t, we end up with

$$y(t) = -\frac{1}{2}gt^2 + c_1 t + c_2. \tag{3.2}$$

The parameters c_1 and c_2 in (3.2) are constants of integration. No matter what values these constants take $y(t)$ will satisfy (3.1), as can be verified by direct substitution. Thus, we have found the **general** solution to (3.1). We will use the constants c_1 and c_2 to fit the initial configuration of the falling object to (3.2). The object started from height h, so

$$y(0) = h.$$

Evaluate (3.2) at $t = 0$ and set the lefthand side equal to h. This results in $c_2 = h$. Let us also assume that the object started its motion from rest, so that

$$y'(0) = 0.$$

Differentiate (3.2) once with respect to t and evaluate the result at $t = 0$, which yields $c_1 = 0$. So the unique solution of (3.1) that satisfies the **initial conditions** $y(0) = h, y'(0) = 0$, is

$$y(t) = -gt^2 + h. \tag{3.3}$$

In developing this model of a falling object we ignored all forces except for the gravitational force. As a result, the motion of this object is independent of its mass (m does not appear in (3.2)). According to this model, if a feather and a rock are released at time zero from height h, they reach the ground level at the same time and have the same velocity at the time of impact. Also note that the object's velocity constantly increases, which is primarily due to ignoring the force of air resistance.

This simple and elementary example embodies several steps one often undertakes in applied mathematics to develop a mathematical model of a physical problem.

1. We used a physical law (Newton's second law) to relate the unknowns of a problem to known quantities.

2. We used certain mathematical techniques and determined (possibly all) functions that are solutions of the differential equation. In this example, which unfortunately is not typical because of its simplicity, two direct integrations led to all the solutions of (3.1).

3. We used specific information (such as the initial conditions of the falling body) about the physical problem to choose a unique function that satisfies some additional constraints.

4. Finally, we interpreted the physical attributes of the solution to understand the extent of the applicability of our model.

Example 3.1.2

We know from experience that air resistance is among the physical parameters that play an important role in how an object falls to the ground. In this example we add air resistance to the physical setting considered in Example 3.1.1 and study its effect on the behavior of the solutions of the resulting differential equation.

The force due to drag on a falling object is proportional to how fast the object is moving. Therefore, in addition to the force of gravity (whose magnitude is mg), we have a force given by $-ky'$, with $k > 0$, that acts on the object. The negative sign is used here to emphasize that the drag force acts in the direction opposite to gravity. The proportionality constant k must be determined experimentally.

Equation (3.1) must now be modified to include $-ky'$

$$my'' = -mg - ky'. \tag{3.4}$$

This equation already demonstrates the main challenge we face in solving differential equations, that is, direct integration is often not very useful. In fact, if we follow the path of Example 3.1.1 and integrate (3.4), we get

$$my'(t) = -mgt - ky(t) + c_1. \tag{3.5}$$

Notice that a second integration does not give us any new information; the left-hand side leads to $my(t)$ but the right-hand side involves the *unknown* function $y(t)$, whose anti-derivative *with respect to t* we cannot determine. Therefore, the simple procedure of Example 3.1.1 fails here and we need to resort to other ideas. We will see later that the function

$$y(t) = c_1 e^{-\frac{kt}{m}} - \frac{mg}{k} t + c_2 \tag{3.6}$$

satisfies this equation (which can be verified by direct substitution of (3.6) into (3.4)).

It is remarkable how differently the solutions to (3.1) and (3.5) behave. Except for the fact that both have two arbitrary constants c_1 and c_2, the two solutions are not similar. For example, note that (3.6) depends on the mass m of the object, which is not the case in (3.3). Also, the velocity of the object modeled by (3.4) does not grow indefinitely. In fact, as a direct effect of adding air resistance to the model, the velocity $y'(t)$ in (3.6) reaches a limit (called the **terminal velocity**) given by

$$\lim_{t\to\infty} y'(t) = -\frac{mg}{k}. \tag{3.7}$$

How realistic is the model now that we have added air resistance to it? Does the formula we used to model air resistance encompass all physical situations of interest? For example, is the model equally reasonable in describing the state of a falling feather as well as a heavy rock? Do we now have a reasonable model to describe the descent of a parachutist? These are the types of questions that a researcher in search of a mathematical model of a natural phenomenon would ask of the underlying differential equations. It is fair to say that one should not expect that a single differential equation would have a sufficient amount of information in it to replicate all facets and features of a physical setting. The knowledge we draw from a mathematical model is often even more limited in that a differential equation may only apply for a finite range of a parameter involved in the problem. For example, it may be reasonable, by comparing with experimental data, to use (3.4) to model the descent of a parachutist weighing between 70 and 110 kilograms, but a model with y'^2 in place of y' in (3.4) should be used to address the descent of a parachute with a heavy machinary hanging from it. With all of these limitations taken into account, differential equations still present the best effort man has put forth in his attempt to understand the evolution of many physical phenomena in his environment.

Example 3.1.3
One of the tasks we face repeatedly in this text is to verify that a certain function is a solution of a differential equation. For example, we noted that (3.6) is a solution to (3.4). To prove this analytically (i.e., by hand), one needs to

substitute (3.6) directly into (3.4). Alternatively, we can use *Mathematica* or *MATLAB* to verify this fact. To accomplish this in *Mathematica* we first define (3.6) as

```
y = c1 Exp[- k t/m] - m g/k t + c2;
```

Next, we evaluate y according to (3.4)

```
D[m y, {t, 2}] + k D[y, t] + m g;
Simplify[%]
```

The result is zero, as expected.

The procedure in *MATLAB* is similar. First, we define (3.6) symbolically in *MATLAB*

```
syms y c1 k t m g c2
y = c1*exp(-k*t/m)-m*g/k*t+c2
```

Next, we evaluate (3.4)

```
a=m*diff(y,2,t)+k*diff(y,t)+m*g
simplify(a)
```

MATLAB returns zero.

Example 3.1.4

Newton's law of cooling is another physical law whose application leads to differential equations. This law, in its simplest form, states that the time rate of change of the temperature of a body is proportional to the difference between the temperature of the body and the temperature of the medium surrounding it. If we denote by $T(t)$ the temperature of a body at time t, then Newton's law of cooling gives us the differential equation

$$\frac{dT}{dt} = k(E - T), \tag{3.8}$$

where k is the proportionality constant and E (which may depend explicitly on t) is the temperature of the environment. Given E and k, our task is to determine a formula for $T(t)$ from (3.8) once the temperature at some "initial" time is known.

Example 3.1.5

Hooke's law of mechanical deformation states that the restoring force in a stretched spring is proportional to the amount of stretching. The proportionality constant is called the **spring constant** or the **modulus of elasticity** of the spring. This principle enters in a natural way in the equation of motion of a mass m suspended from a spring with constant k (see Figure 3.2). If we let $y(t)$ stand for the deviation of the mass from its equilibrium, Hooke's law and Newton's second law of motion combine to produce the differential equation

$$my'' + ky = 0. \tag{3.9}$$

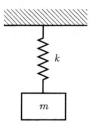

Figure 3.2: In the absence of friction, a mass-spring system satisfies the differential equation $my'' + ky = 0$.

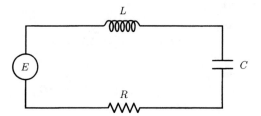

Figure 3.3: The current in an RLC circuit satisfies $LI'' + RI' + \frac{1}{C}I = E'(t)$.

We will see later that the function y defined by

$$y(t) = c_1 \sin \sqrt{\frac{k}{m}}t + c_2 \cos \sqrt{\frac{k}{m}}t \tag{3.10}$$

is the general solution for the vibration of the mass, where c_1 and c_2 are arbitrary constants that must be determined from the initial position and velocity of the mass.

Example 3.1.6

Kirchhoff's law, a fundamental principle in the theory of electrical circuits, states that the sum of voltage drops across an electrical circuit consisting of a resistor, an inductor, and a capacitor—known as an RLC circuit, for short—must be balanced by the total amount of voltage supplied to the circuit (see Figure 3.3.) In a typical RLC circuit, consisting of a resistor with resistance R ohms, an inductor with inductance L henry, and a capacitor with capacitance C farad, the current $I(t)$ in the circuit satisfies the differential equation

$$LI'' + RI' + \frac{1}{C}I = E'(t), \tag{3.11}$$

where $E(t)$ is a known input *electromotive force* (emf) that may model a battery or an alternating voltage. Much like the case of the previous example, our task is to determine a formula for $I(t)$ from (3.11) once L, R, C, and $E(t)$ are supplied.

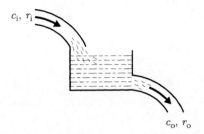

c_i, r_i

c_o, r_o

Figure 3.4: A single-tank mixture problem.

Example 3.1.7 (A Mixture Problem)
Consider a tank containing a mixture of salt (a solute) and water (a solvent). The tank has two openings through which fluid enters and leaves (see Figure 3.4). Let $y(t)$ be the amount of salt in the tank at time t, c_i (in grams per liter, say) the *concentration* of salt in the inflow, and r_i (in liters per minute, say) the *rate of inflow*. Then the amount of salt added to the tank per unit time is

$$r_i c_i.$$

Similarly, let r_o and c_o be the *rate of outflow* and the concentration of salt in the well-mixed outgoing flow. With $V(t)$ representing the volume of the fluid in the tank at any time t, we have

$$c_o = \frac{y(t)}{V(t)}.$$

Therefore, since the rate of change of salt in the tank must be equal to the net gain (or loss) of salt per unit time due to inflow and outflow, the function y must satisfy

$$\frac{dy}{dt} = r_i c_i - \frac{y}{V(t)}. \tag{3.12}$$

This expression is a differential equation for the amount of salt in the tank at any instant t.

As a concrete example, consider a tank with a capacity of 100 liters containing 40 liters of fresh water. At time zero a solution containing 10 grams of salt per liter is pumped into the tank at a rate of 3 liters per minute. At the same time, from a different outlet, the well-mixed solution in the tank is removed at a rate of 2.9 liters per minute. Here

$$r_i = 3, \quad c_i = 10, \quad r_o = 2.9.$$

The volume of the fluid in the tank increases according to ($r_i - r_o = 0.1$)

$$V(t) = 40 + 0.1t.$$

Hence, $y(t)$, the amount of salt in the tank, varies according to the differential equation

$$y' = 30 - \frac{y}{40 + 0.1t}.$$

This equation is supplemented by the initial condition

$$y(0) = 0.$$

Example 3.1.8

The flow of a fluid, such as water, about an obstacle, such as a rock or an island, is one of the fundamental motions we often study in fluid dynamics. In this example we present the differential equations that govern the special motions (called irrotational) of a class of fluids (called incompressible) about a cylindrical obstacle that has radius one and infinite extent otherwise. Let $(x(t), y(t))$ denote the position of a fluid particle in a plane perpendicular to this cylinder. It can be shown that the velocity of each particle (given by $\langle \frac{dx}{dt}, \frac{dy}{dt} \rangle$) is related to its position according to the differential equations

$$x' = 1 + \frac{y^2 - x^2}{(x^2 + y^2)^2}, \qquad y' = -\frac{2\,x\,y}{(x^2 + y^2)^2}. \tag{3.13}$$

In general, it is difficult to find explicit solutions of (3.13), but as we will see later, both *Mathematica* and *MATLAB* give us truly valuable information about the motion of such fluids from (3.13).

The differential equations in this example differ from those cited in all other examples in that in Example 3.1.8 there are two unknowns (namely, x and y) whose dynamics depend on each other. We will refer to such a simultaneous set of equations as a **system** of differential equations.

In all of the preceding examples the unknown function depends on a single independent variable. In Example 3.1.1 the height of the falling object is a function of time only, while the position of a moving fluid particle in Example 3.1.8 depends explicitly on time alone. Equations whose unknown(s) depends on a single variable are called **ordinary differential equations** (ODEs). In the next few examples we consider physical settings in which the unknown function(s) may depend on more than one independent variable. Such equations are called **partial differential equations** (PDEs).

Example 3.1.9

In Example 3.1.4 we developed a differential equation that monitored how a body exchanged heat with its environment. This equation was derived based on Newton's law of cooling under the simplifying assumption that the temperature distribution in the body is *homogeneous*; that is, at every instant of time the temperature of the body is constant throughout the body. A more realistic model would take into account the dissipation of heat in the body and the effect of the boundary (environment) data on the temperature of the material in the interior of the domain.

Consider a heat-conducting rod of length l located on the interval $(0, l)$ of the x-axis. Let ρ and C be the density and the specific heat of the rod, respectively.

Let $u(x,t)$ be the temperature of the rod at point x and at time t. It can be shown that the evolution of this function is governed by the PDE (heretofore referred to as the **heat equation**)

$$\rho C \frac{\partial u}{\partial t} = \lambda \frac{\partial^2 u}{\partial x^2}, \tag{3.14}$$

where λ is the thermal conductivity of the material. In certain experiments we are able to control the temperature at the boundary of the rod. For example, if the ends of the rod are kept in ice, then

$$u(0,t) = u(l,t) = 0. \tag{3.15}$$

It is not difficult to verify that

$$u(x,t) = e^{-\frac{\pi^2 c^2 t}{l^2}} \sin \frac{\pi x}{l} \tag{3.16}$$

is a particular solution that satisfies both (3.14) and (3.15). Here $c^2 = \frac{\lambda}{\rho C}$. The mathematical problem of determining all the solutions, or in other words, the general solution, of (3.14), (3.15) is the subject of several later chapters of this text (see also the exercises at the end of this section).

It is an easy task to verify in *Mathematica* or *MATLAB* that (3.16) is a solution of $u_t = c^2 u_{xx}$. We first define the function in (3.16) in *Mathematica*, say, and then ask it to differentiate (and simplify) the expression $u_t - c^2 u_{xx}$ as follows:

```
u = Exp[- Pi^2 c^2 t/l^2] Sin[Pi x/l];
a = D[u, t] - c^2*D[u, {x, 2}];
Simplify[a]
```

The same task is carried out in *MATLAB* by entering the following two lines:

```
syms u t c x l;
u='exp(-pi^2*c^2*t/l^2)*sin(pi*x/l)';
diff(u,t)-c^2*diff(u,2,x)
```

Example 3.1.10
The **wave equation** is another fundamental equation of mathematical physics. The simplest form of this equation models the vibrations of an elastic string by the PDE (see Figure 3.5)

$$\rho \frac{\partial^2 u}{\partial t^2} = c^2 \frac{\partial^2 u}{\partial x^2}, \tag{3.17}$$

where $u(x,t)$ represents the vertical displacement of an elastic string away from its equilibrium. Once we know the initial displacement and velocity of the string, the goal is to determine the displacement $u(x,t)$ for all time $t > 0$ from (3.17).

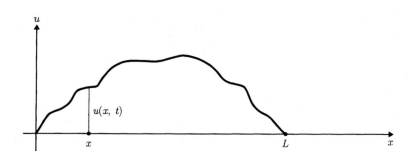

Figure 3.5: The displacement $u(x,t)$ of an elastic string satisfies the wave equation $u_{tt} = c^2 u_{xx}$.

We considered here just a few examples of differential equations that come up naturally in mathematical physics. We will see numerous other occasions in this text where a differential equation will be the underlying mathematical tool that gives a quantifiable description of the physical setting. The exercises at the end of this section introduce a few more of these systems.

Problems

1. State whether each differential equation is an ordinary or partial differential equation. Verify that the given function is a solution. Use *Mathematica* or *MATLAB* to check the calculations.

   (a) $y' + 2y = 0$, $y(x) = 3e^{-2x}$

   (b) $x''' = t^2 + 1$, $x(t) = \frac{1}{60}t^5 + \frac{1}{6}t^3 + \frac{A}{2}t^2 + Bt + C$

   (c) $y'' + 4y = 0$, $y(x) = -7\cos 2x$

   (d) $y'' + 4y' + 4y = 0$, $y(t) = ate^{-2t}$

   (e) $z'' + 4z' + 5z = 0$, $z(t) = ae^{-2t}\sin t$

   (f) $u_t = u_{xx}$, $u(x,t) = e^{-a^2 t}\sin ax$

   (g) $u_t = 9u_{xx}$, $u(x,t) = e^{-9a^2 t}\cos ax$

2. Show that the given function is a solution, and use either *Mathematica* or *MATLAB* to draw its graph.

   (a) $y'' + 3y' + 2y = 0$, $y(x) = e^{-2x}$

   (b) $y^{(iv)} + y = 0$, $y(x) = e^{\frac{\sqrt{2}}{2}x}\sin\frac{\sqrt{2}}{2}x$. ($y^{(iv)}$ means the fourth derivative of y.)

   (c) $x'' + 3x = \sin t$, $x(t) = \frac{1}{2}\sin t$

   (d) $w' + w^2 = 0$, $w(t) = \frac{1}{1+t}$

   (e) $x'' + x = \sin t$, $x(t) = -\frac{1}{2}t\cos t$

(f) $z'' + 4z' + 4z = e^{-2t}$, $z(t) = \frac{t^2}{2}e^{2t}$

(g) $4x^2 y'' + y = 0$, $y(x) = \sqrt{x}\ln x$

(h) $u_{tt} = 4u_{xx}$, $u(x,t) = \cos 2t \sin x$. Plot the graphs of $u(x,t)$ for $x \in (0, \pi)$ and $t = 0, 0.1, 0.2, \ldots, 0.5$ on the same screen.

(i) $u_t + 2u_x = 0$, $u(x,t) = \sin(x - 2t)$. Plot the graphs of $u(x,t)$ for $x \in (-5, 5)$ and $t = 0, 0.1, 0.2, \ldots, 0.5$ on the same screen.

3. Show that the given function is a general solution of the differential equation. Determine the parameters c so that the particular side conditions (constraints) are satisfied.

   (a) $x' + 6x = 0$, $x(t) = ce^{-6t}$, and $x(1) = -1$

   (b) $x'' + 16x = 0$, $x(t) = c_1 \sin 4t + c_2 \cos 4t$, and $x(0) = 0, x'(0) = 1$

   (c) $x'' + 4x = 0$, $x(t) = c_1 \sin 2t + c_2 \cos 2t$, and $x(0) = 0, x'(1) = 1$

   (d) $y' + 2y^2 = 0$, $y(x) = \frac{1}{2x+c}$, and $y(0) = 1$. Is it possible to find a solution to this equation with $y(0) = 0$? Explain.

   (e) $y'' + 3y' + 2y = \sin x$, $y(x) = c_1 e^{-2x} + c_2 e^{-x} + \frac{1}{10}\sin x - \frac{3}{10}\cos x$, and $y(0) = 0, y'(0) = 1$

4. Use *Mathematica* or *MATLAB* to plot the graphs of the following functions as the parameter c varies. In each case choose a representative domain for the independent variable.

   (a) $y(x) = ce^x$, with c ranging from -1 to 1 at increments of 0.1

   (b) $y(x) = \frac{1}{x+c}$, with c ranging from -1 to 1 at increments of 0.1

   (c) $x(t) = c\sin 2t$, with c ranging from -3 to 3 at increments of 0.5

   (d) $u(x) = \sin(x - c)$, with c ranging from -1 to 1 at increments of 0.1

   (e) $y(x) = ce^{-x^2}$, with c ranging from -1 to 1 at increments of 0.1

   (f) $y(x) = ce^{-(x-c)^2}$, with c ranging from -1 to 1 at increments of 0.1

5. Recall that a differential equation such as $y' + y = \sin t$ is an identity in the independent variable t, that is, the equation holds true for every value of t in some interval of the domain of the solution. This fact allows us to evaluate the equation at any particular value of t and to differentiate it as many times as the coefficients allow us in order to get information about the higher derivatives of the solution $y(t)$. In each of the following initial value problems determine the designated derivative of the solution directly from the differential equation.

   (a) $y' + y = 0$, with $y(0) = 1$; determine $y'(0)$ and $y''(0)$

   (b) $y'' + y = 0$, with $y(0) = 1$ and $y'(0) = 0$; determine $y''(0)$ and $y'''(0)$

   (c) $x' + tx = \sin t$, with $x(0) = -2$; determine $x'(0)$ and $x''(0)$

   (d) $tx' + x = 0$, $x(1) = -1$; determine $x'''(1)$

(e) $y'' + 2y' + y = \frac{1}{x} \ln x$, with $y(1) = 1$ and $y'(1) = 2$; determine $y''(1)$

(f) $y'' + 2 \sin t\, y' + y = 0$, with $y(0) = 0$ and $y'(0) = 0$; determine $y''(0)$ and $y'''(0)$

(g) $y'' + ay = t \cos t$, with $y(\pi) = -1$ and $y'(\pi) = -2$ and a a parameter; determine y'' and y''' at $t = \pi$

6. A body of mass 4 kg is located 200 meters above ground level. At time zero it is released and allowed to descend under the action of gravity ($g = 9.8$ meters per second). Assuming that air resistance is negligible, how long does it take for the body to reach the ground? What is its velocity at the time of impact?

7. Do Problem 6 with air resistance taken into account (see Example 3.1.2); suppose that the coefficient of drag k in Example 3.1.2 is 0.2 kg/sec. How long does it take for the body to reach the ground? What is its velocity at the time of impact? Use *Mathematica* or *MATLAB* to draw on the same screen the graphs of y as a function of t for the two cases in this problem and Problem 6.

8. Newton's law of cooling states that the rate of change of temperature of a body is proportional to the difference between the temperature of the body and its environment. Suppose that the temperature of the environment is $100°$. Write down a differential equation for $u(t)$, the temperature of the body.

9. Under some idealized conditions, the rate of change of pressure p in the atmosphere is proportional to the pressure itself. Express this statement in terms of a differential equation.

10. Imagine a tank containing 50 cubic meters of fresh water. At time zero a solution of salty water with a concentration of 2 grams of salt per cubic meters is added to the tank at a constant rate of 1 cubic meter per second. At the same time the solution of fresh and salt water is immediately mixed well and then removed through a hole at the bottom of the tank at the same rate as the inflow rate (see Figure 3.4 and Example 3.1.7). Let $y(t)$ be the amount of salt in the tank at any time $t > 0$. Write down a differential equation for y. What is the initial condition for y?

11. Consider the two-tank cascade system in Figure 3.6. Tanks A and B contain 50 and 40 gallons of brine, respectively, each brine initially containing 23 pounds of salt. Fresh water is added to tank A at a rate of 3 gallons per minute. At the same time, the well-mixed brine is removed from each tank at the rate of 3 gallons per minute. Let $x(t)$ and $y(t)$ be the amounts of salt in tanks A and B, respectively. Show that the pair (x, y) satisfies the following system of differential equations:

$$\frac{dx}{dt} = -\frac{3}{50}x, \quad \frac{dy}{dt} = \frac{3}{50}x - \frac{3}{40}y. \qquad (3.18)$$

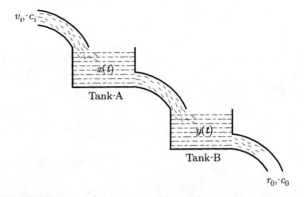

Figure 3.6: A two-tank cascade mixture problem.

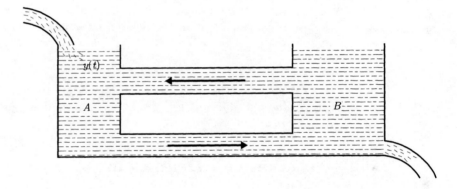

Figure 3.7: Two interconnected tank mixture problem.

What are the initial values for x and y? What do you expect the asymptotic value of x will be as t approaches infinity? What about y's asymptotic value?

12. Consider the two interconnected tank system in Figure 3.7. Tanks A and B contain 200 and 310 liters of brine, respectively. Tank A initially contains 6 kilograms of salt. Tank B initially contains 7 kilograms of salt. Fresh water is added to tank A at a rate of 17 liters per minute. At the same time, the well-mixed brine flows from tank A to tank B at the rate of 23 liters per minute, while brine from tank B flows into tank A at the rate of 6 liters per minute. Finally, through a second outlet, the brine in tank B is removed at the rate of 17 liters per minute. Let $x(t)$ and $y(t)$ be the amounts of salt in tanks A and B, respectively. Show that the pair (x, y) satisfies the system of differential equations

$$\frac{dx}{dt} = -\frac{23}{200}x + \frac{6}{310}y, \quad \frac{dy}{dt} = \frac{23}{200}x - \frac{6}{310}y. \tag{3.19}$$

What are the initial values for x and y?

13. Consider a container in the shape of a cone with height H and radius R. Suppose that a fluid is poured into this container at a rate of c gallons per minute. Let $h(t)$ be the height of the fluid in the container at any time t. Show that h satisfies the differential equation

$$\frac{dh}{dt} = \frac{cH^2}{\pi R^2 (H - h(t))^2}, \quad h(0) = 0. \tag{3.20}$$

Use *MATLAB* or *Mathematica* to show that

$$h(t) = H \left(1 - (1 - \frac{3ct}{\pi R^2 H})^{\frac{1}{3}} \right) \tag{3.21}$$

is a solution of (3.20). Find how long it takes to fill the container with the fluid.

14. The rate of growth or decline of a population of a species is often proportional to its present size, the size of its predator, and the availability of resources. Examine the differential equation

$$\frac{dx}{dt} = \mu x (1 - x),$$

where $x(t)$ is the size of the population at time t, and explain whether the right-hand side is a reasonable formulation of the constraints on the growth of $x(t)$.

3.2   Notation and Definitions

We saw in the previous section that a differential equation is a relation between a function and its derivatives. Typically, these relations arise due to the fundamental laws of physics. Newton's second law, which relates the acceleration of a moving body to the resultant force on that body, is one of these fundamental laws whose mathematical manifestation is often a differential equation. Another is the second law of thermodynamics, which relates the rate of change of temperature in a body with the rate at which temperature dissipates throughout the body. And so is Kirchhoff's law in electricity, which describes the relationship between the rate of change of current and the voltage drop across a circuit.

We also noted that a differential equation is called ordinary if the unknown function in the equation is a function of a single variable. Otherwise the equation is called a PDE. In this section we introduce a few more definitions and some new terminology for ODEs.

A **differential operator** L is an operation that involves a function and its derivatives. Such an operation takes sets of functions for its domain and range.

Example 3.2.1

Let L be defined by

$$L[x] = x'' + 2(x')^2. \tag{3.22}$$

This operator is an example of a differential operator. Given a twice differentiable function $x(t)$, L differentiates it twice with respect to t and adds the result to twice the square of its first derivative. For instance,

$$L[\sin t] = -\sin t + 2\cos^2 t. \tag{3.23}$$

Similarly,

$$L[t^3] = 6t + 18t^4. \tag{3.24}$$

The **domain** of a differential operator L is the set of all functions to which L can apply. In Example 3.2.1 the domain of L is the set of all twice differentiable functions. The **range** of L is the set of all functions one obtains as the result of applying L to the elements in its domain. Unless otherwise stated we consider differential operators that satisfy

$$L[0] = 0.$$

Definition 3.2.1 (Linear Operators)

A differential operator L is called **linear** *if for any two functions x_1 and x_2 in its domain and for any two scalars a_1 and a_2 belonging the set of real numbers R we have*

$$L[a_1 x_1 + a_2 x_2] = a_1 L[x_1] + a_2 L[x_2]. \tag{3.25}$$

A differential operator that is not linear is called **nonlinear**.

Example 3.2.2

Let L be defined by

$$L[x] = x'' + 2x' + 3x. \tag{3.26}$$

This differential operator is linear. To see this, let x_1 and x_2 be two differentiable functions. Then

$$\begin{aligned} L[a_1 x_1 + a_2 x_2] \quad &= (a_1 x_1 + a_2 x_2)'' + 2(a_1 x_1 + a_2 x_2)' + 3(a_1 x_1 + a_2 x_2) \\ &= a_1(x_1'' + 2x_1' + 3x_1) + a_2(x_2'' + 2x_2' + 3x_2) \\ &= a_1 L[x_1] + a_2 L[x_2]. \end{aligned} \tag{3.27}$$

Hence, (3.25) is satisfied. Similarly, $L[x] = x''' + (\sin t)x'$ is linear because

$$\begin{aligned} L[a_1 x_1 + a_2 x_2] \quad &= (a_1 x_1 + a_2 x_2)''' + (\sin t)(a_1 x_1 + a_2 x_2)' \\ &= a_1(x_1''' + (\sin t)x_1') + a_2(x_2''' + (\sin t)x_2') \\ &= a_1 L[x_1] + a_2 L[x_2]. \end{aligned} \tag{3.28}$$

On the other hand, the differential operator

$$L[x] = x'' + \sin x \tag{3.29}$$

is nonlinear because

$$L[a_1x_1(t) + a_2x_2(t)] = (a_1x_1(t) + a_2x_2(t))'' + \sin(a_1x_1(t) + a_2x_2(t))$$
$$\neq a_1 L[x_1] + a_2 L[x_2], \tag{3.30}$$

since $\sin(x + y) \neq \sin x + \sin y$. Similarly, the operator $L[x] = x'' + (x')^2$ is nonlinear.

As the preceding examples suggest, a differential operator is linear if the variable x and its derivatives enter linearly in the formula for L. Therefore, the operators $L_1[x] = x''^2$ or $L_2[x] = x' + \frac{1}{x}$ are nonlinear, while

$$L[x] = a_n(t)x^{(n)} + a_{n-1}(t)x^{(n-1)} + \cdots + a_1(t)x' + a_0(t)x,$$

where the a_is are known functions of t, is linear. Here $x^{(i)}$ stands for the ith derivative of x.

It is rather easy to verify with a symbolic manipulator whether a differential operator is linear or not, as the next example shows.

Example 3.2.3

Going back to the definition of linearity in (3.25), we see that in order to show an operator L is linear, we must establish the validity of the relation

$$L[a_1x_1(t) + a_2x_2(t)] - a_1 L[x_1(t)] - a_2 L[x_2(t)] = 0$$

for all pairs (x_1, x_2) in the domain of L and for all $a_i \in R$. To verify the linearity of an operator in *Mathematica* when L is the first operator in Example 3.2.2 one proceeds as follows:

```
L[x_] := D[x, {t, 2}] + 2 D[x, t] + 3 x;
output = L[a1 x1[t] + a2 x2[t]] - a1 L[x1[t]] - a2 L[x2[t]];
Simplify[output]
```

Remark 3.2.1: It is important to use the assignment operator := in place of the equality operator = when we define L in *Mathematica*. With :=, only an assignment of the right-hand side to the left-hand side is carried out and no differentiation of x is performed until the next time L is called upon, at which time x1 and x2 are functions of t. Had we used = in place of := in the first line of the program, the output would be zero, since x in that line would be considered a parameter and not a function of t.

Next, let us show that $L[x] = x'' + 2x' + 3x$ is a linear operator in *MATLAB* (see Chapter 2 for a review of subs and expand):

```
syms x t a1 a2
operator = diff('x(t)',2,t)+2*diff('x(t)',t)+'3*x(t)';
```

```
leftside = expand(subs(operator,'a1*x1(t)+a2*x2(t)','x(t)'));
part1 = subs(operator,'x1(t)','x(t)');
part2 = subs(operator,'x2(t)','x(t)');
result = leftside-a1*part1-a2*part2
simplify(result)
```

The output is zero.

Remark 3.2.2: It would be worthwhile to use an editor and input this program into a file since only its second line must be altered when we consider a different differential operator.

Definition 3.2.2

An ordinary differential equation is a relation of the form

$$L[x] = f \tag{3.31}$$

for some differential operator and a known function f. When the differential operator is linear and $f \equiv 0$, the differential equation (3.31) is called **homogeneous**. *When L is linear but f is not identically zero, $L[x] = f$ is called a* **nonhomogeneous** *differential equation.*

As we saw in the previous section, a function x is called a **solution** of the differential equation (3.31) if it satisfies $L[x] = f$ for every t in a suitable interval in R.

Example 3.2.4

The differential equation $x'' + x = 0$ is a homogeneous differential equation, while $x'' + x = \sin t$ is nonhomogeneous.

Definition 3.2.3 (Order of Differential Operators)

The **order** *of a differential operator L is the order of the highest derivative in L. The order of a differential equation is the order of the differential operator that defines it.*

Example 3.2.4

The order of $L[x] = 3x'' + (x')^4$ is 2, the order of the derivative in x''. By the same token, the order of the differential equation $x'' + (x')^4 = \sin t + 4\cos t$ is 2.

All of these examples involve differential operators whose domain consist of functions of a single variable. The next example examines the definition of linearity of an operator whose domain may contain functions of several variables.

Example 3.2.5

Consider the differential operator

$$L[u] = \frac{\partial^2 u}{\partial x^2} + \frac{\partial^2 u}{\partial y^2}.$$

To see that L is linear, let $u_1 = u_1(x, y)$ and $u_2 = u_2(x, y)$ be two twice differentiable functions in x and y. Let a_1 and a_2 be two arbitrary scalars. Since the operation of second partial differentiation is linear, we have

$$\frac{\partial^2}{\partial x^2}(a_1 u_1(x, y) + a_2 u_2(x, y)) = a_1 \frac{\partial^2 u_1}{\partial x^2} + a_2 \frac{\partial^2 u_2}{\partial y^2}.$$

A similar expression holds when differentiation with respect to x is replaced by y. It is then easy to see that (3.25) holds.

Problems

1. Use Definition 3.2.1 and verify which one of the following differential operators is linear. Using the analogue of the program in Example 3.2.3 check each answer in *Mathematica* or *MATLAB*.

   (a) $L[x] = x' + 2x$

   (b) $L[x] = (x')^2 + 2x$

   (c) $L[x] = x'' + x$

   (d) $L[x] = x''' + 3x'' - 2x' + x$

   (e) $L[u] = \frac{\partial^2 u}{\partial x^2} + \frac{\partial^2 u}{\partial y^2} + \sin(xy)\frac{\partial u}{\partial x}$

   (f) $L[u] = \frac{\partial u}{\partial x} + \frac{\partial u}{\partial y} + u^2$

2. Verify whether the following differential operators are linear or not. Modify the *Mathematica* or the *MATLAB* program in Example 3.2.3 to allow for operators with coefficients that depend on t.

   (a) $L[x] = tx'' + 2x' + 3x$

   (b) $L[x] = x'' + x + \sin t$

   (c) $L[x] = (\sin t)x'' + 3tx' + (\cos t)x$

   (d) $L[x] = x'' + \sin x$

   (e) $L[x] = x^{(iv)} + 3x''' - (\tan t)x'' - x$

3. Verify whether the following functions are solutions of the corresponding differential equations. Check the answers in *Mathematica* or *MATLAB*.

   (a) $x(t) = \sin t$ and $x'' + x = 0$

   (b) $x(t) = \sin t$ and $(\cos t)x' + x^2 + 1 = 0$

   (c) $x(t) = \sin t$ and $x^{(iv)} - x = 0$

   (d) $x(t) = e^{-2t}$ and $x''' + 2x'' = 0$

4. Choose c so that $x(t) = e^{ct}$ is a solution of the following differential equations.

    (a) $x'' + 9x = 0$

    (b) $x''' + 3x'' + x = 0$ (Hint: Use *Mathematica*'s `FindRoot` command.)

5. Let $L[x] = 0$ be a linear differential equation. Suppose that $x_1(t)$ and $x_2(t)$ are solutions of this equation. Prove that $c_1 x_1 + c_2 x_2$ is also a solution of this equation, where c_1 and c_2 are arbitrary constants.

6. Consider the differential equation $y' + y^2 = 0$. First verify that $y_1(t) = \frac{1}{1+t}$ is a solution of this equation. Next, show that the function $y_2 = 2y_1$ is not a solution of this equation. Why doesn't this fact contradict the conclusion of the previous problem?

7. Consider the differential operators

    (a) $L[x] = x'^a$ and prove that L is a linear operator if and only if $a = 1$.

    (b) $L[x] = x' + t^b x^a$ and find all the values of a and b for which L is linear.

8. Let x_1 and x_2 be two solutions of the nonhomogeneous differential equation $L[x] = f$, where L is a linear operator. Show that the function $y = x_2 - x_1$ is a solution of the homogeneous differential equation $L[x] = 0$.

9. Consider the nonhomogeneous differential equation $L[x] = f$, where L is a linear operator. Let x_p be a solution of this equation. Let x_c be any solution of $L[x] = 0$. Show that $y = x_c + x_p$ is a solution of $L[x] = f$.

3.3 Direction Fields and Existence and Uniqueness

Most of the differential equations one encounters in mathematical physics are extremely difficult to solve analytically. We often resort to approximate methods to gain an idea as to how the solutions of differential equations behave when the independent variable evolves. It then becomes necessary to know that the solutions to these equations exist and are unique for the duration of the approximation. In this section we present a general existence and uniqueness theorem that will serve this purpose for many of the nonlinear differential equations that arise in the applications we will study later. We will also start using some of the internal functions of *Mathematica* and *MATLAB* that are capable of determining exact and approximate solutions of ODEs. Specifically, we will use the functions `DSolve` and `NDSolve` of *Mathematica* and `dsolve` and `ode45` of *MATLAB*. At this juncture the reader should think of these internal functions as "black boxes" into which one inputs a sequence of information and out of which comes the solution of a differential equation. In later sections of this chapter we will learn in detail how to generate these solutions as well; based on the knowledge one gains from the analytic techniques developed there, the structure of these "black boxes" will become considerably more clear.

Before giving the technical statement of the main theorem, we give a geometric interpretation of what a differential equation of the form

$$x' = f(t, x) \tag{3.32}$$

represents. A solution of this equation is a function x whose first derivative with respect to t equals the right-hand side of (3.32). By a **solution curve** we the mean the graph of x in the tx-plane. If we knew x explicitly as a function of t, then $x'(t)$ would give us the slope of the tangent line to the solution curve at t. Alternatively, the right-hand side of (3.32) gives us the same information, without any explicit knowledge of $x(t)$. This geometric point of view is quite valuable in many problems because it gives us a qualitative sense of how the solution curves behave without having explicit formulas for them.

For the sake of concreteness, let us consider the differential equation

$$x' = x^2 + t. \tag{3.33}$$

The solution curve that satisfies the condition $x(1) = 2$ must have slope 5 at $(x, t) = (2, 1)$ since, by (3.33),

$$x'(1) = 2^2 + 1 = 5.$$

Even though we do not have an explicit formula for the exact solution curve that passes through $(2, 1)$, we do know that the tangent line to this curve at $(2, 1)$ is the line

$$x = 5(t - 1) + 2.$$

For values of t near 1, the tangent line is a reasonable approximation of how the solution curve will look.

The idea that a tangent line to a solution curve is an approximation to that curve suggests obtaining a qualitative picture of solutions of a differential equation such as (3.33) by drawing the graph of the tangent lines at various ordered pairs (x, t). The picture we produce this way is referred to as the **direction field** corresponding to the differential equation. Figure 3.8 shows the direction field for the differential equation (3.33). This figure is produced by executing the following commands in *Mathematica*:

```
<<Graphics'PlotField'
DirectionField=PlotVectorField[{1, x^2 + t}, {t, 0, 3},
       {x, -2, 2}, PlotPoints->20, ScaleFunction -> (1&),
       axes -> True]
```

The `ScaleFunction -> (1&)` option in this statement produces *unit* vectors at each point of application.

Figure 3.8 already gives some indication as to how the solutions of (3.33) behave when $t \in (0, 3)$. Figure 3.9 shows the graphs of a few of the solutions of this equation superimposed on Figure 3.8. These solutions were obtained using the `NDSolve` internal function of *Mathematica*. For example, one gets the solution curve to (3.33) that satisfies the initial condition $x(0) = 1$ by entering

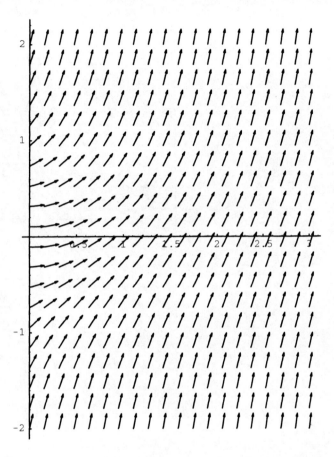

Figure 3.8: Direction field for the equation $x' = x^2 + t$, in *Mathematica*.

```
sol=NDSolve[{x'[t] == t + x[t]^2, x[0] == 1}, x, {t, 0, 0.9}];
graph=Plot[x[t] /. sol, {t, 0, 0.9}]
```

The output of `graph` and `DirectionField` can now be combined using the `Show` command. The resulting output will show the direction field with one solution curve superimposed on it. Other solution curves can be added to the output by changing the initial data in `sol` (see Figure 3.9).

A project at the end of this chapter contains the *MATLAB* M-file `field.m` that, together with the M-file `xprime.m`, produces the direction field of a first-order differential equation. The reader may use a word processor and create these files in the main *MATLAB* directory. Figure 3.10 was produced by executing the following sequence of commands in *MATLAB*:

```
field
[t,x]=ode45('xprime',0,0.9,1);
plot(t,x)
```

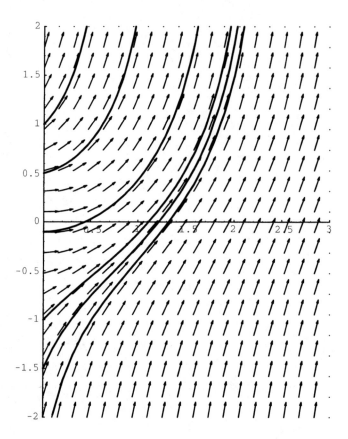

Figure 3.9: Direction field and solution curves for the equation $x' = x^2 + t$, in *Mathematica*.

The first command inputs the contents of the file `field.m` to *MATLAB*, whose output is the direction field of (3.33). The second line uses the internal function `ode45` to determine an approximate solution curve to this equation with initial condition $x(0) = 1$ and $t \in (0, 0.9)$. The last line plots the solution curve on top of the direction field. Figure 3.10 is produced by repeating the last two lines with different initial conditions and t intervals.

An **isocline** of a first-order differential equation $x' = f(t, x)$ is the set of points (t, x) at which the slope x' remains constant, that is, the set of points such that

$$f(t, x) = c,$$

for some constant c. For a given constant these points generally constitute smooth curves whose graphs provide the locus of points along which the solution curves have the same slope. For example, returning to (3.33), the isocline corresponding to $c = 1$ is the parabola $x^2 + t = 1$; any solution that intersects this curve in the tx-plane must do so with slope 1. Similarly, the isocline corresponding to $c = 2$ is given by parabola $x^2 + t = 2$. Figure 3.11 shows some of the isoclines and solution curves of this equation.

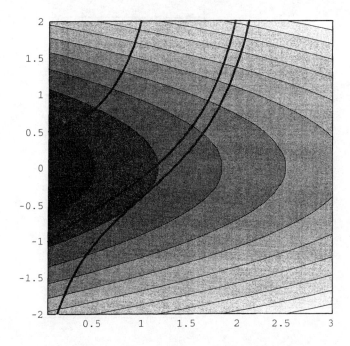

Figure 3.10: Direction field and solution curves for the equation $x' = x^2 + t$, in *MATLAB*.

The definition of an isocline requires that any point (t, x) on that curve satisfy $f(t, x) = c$. In calculus we call such curves **contours** or **level curves** of the function f. Both *Mathematica* and *MATLAB* are capable of drawing contours of a function. The following example shows how one produces a figure such as Figure 3.11 in *Mathematica*.

Example 3.3.1

To draw the isoclines of (3.33) (or equivelently the contours of $x^2 + t$) in *Mathematica*, we use `ContourPlot` together with some of its options. The following is the sequence of commands that led to Figure 3.11 (`NDSolve` was used to generate the solution curves):

```
f = x^2 + t;
isoclines = ContourPlot[f, {t, 0, 3}, {x, -2, 2},
                PlotPoints -> 30, ContourShading -> False];
solution[x0_] := NDSolve[{x'[t] == x[t]^2 + t, x[0] == x0},
                x, {t, 0, 2}];
specificSolution = solution[-1.5];
solutionCurve = Plot[x[t] /. specificSolution, {t, 0, 2}];
finalGraph = Show[isoclines, solutionCurve];
```

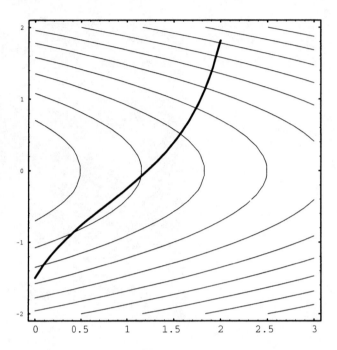

Figure 3.11: Isoclines and solution curves for the equation $x' = x^2 + t$.

The program used to draw Figure 3.11 in *MATLAB* is

```
[t, x] = meshgrid(0:0.05:3,-2:0.05:2);
z = t + x.*x;
isoclines = contour(t,x,z,20);
axis([0 3 -2 2]);
hold on
[tt, xx] = ode45('diffeqn', [0 2], -1.5);
plot(tt, xx)
```

The M-file `diffeqn.m` is defined separately as

```
function xprime = diffeqn(tt,xx);
xprime = tt+ xx.*xx;
```

Remark 3.3.1: The above program superimposes one solution curve on the graph of the isoclines, the output of `isoclines`. One can get the graphs of other solution curves added to this figure by rerunning the last two lines of the main program with other initial conditions; for example,

```
[tt, xx] = ode45('diffeqn', [0 3], -1);
plot(tt, xx))
```

plots the graph of the solution with initial condition $x_0 = -1$ over the interval $t \in (0, 3)$.

This geometric interpretation is quite useful for gaining an understanding of how the solutions of first-order equations behave. A similar idea has also proven very useful for systems of two first-order equations, especially when the dependent variables are the components of the velocity of a fluid flow. In the next example we consider the flow past the cylinder given by (3.34), where $x(t)$ and $y(t)$ are the horizontal and vertical components of the position at time t of a fluid particle whose path takes it past a cylinder of radius one.

Example 3.3.2

Consider the velocity field $(v_1, v_2) = (\frac{dx}{dt}, \frac{dy}{dt})$, where the position of each particle is related to its velocity through the system of first-order equations

$$x' = 1 + \frac{y^2 - x^2}{(x^2 + y^2)^2}, \quad y' = -\frac{2xy}{(x^2 + y^2)^2}. \tag{3.34}$$

We can get a sense of how the fluid particles negotiate their way past the cylinder by drawing some of the typical velocity vectors in the xy-plane. The `PlotVectorField` in *Mathematica* will accomplish this task:

```
<<Graphics`PlotField`
DirectionField = PlotVectorField[{1 + (y^2 - x^2)/(x^2 + y^2)^2,
                         - 2 x y/(x^2 + y^2)^2},
                        {x, -2.1, 2}, {y, -2.01, 2},
                        ScaleFunction->(1&)];
circle = ParametricPlot[{Cos[t], Sin[t]}, {t, 0, 2Pi}];
finalGraph = Show[circle, DirectionField, AspectRatio ->
    Automatic];
```

Figure 3.12 shows the output of this program. The first two lines draw the velocity field in the plane, while the third and fourth lines superimpose the cylinder (circle of a radius one, in the two-dimensional cross section) on the velocity field. Since the velocity field has a singularity at $(x, y) = (0, 0)$, we chose the domain $(-2.1, 2) \times (-2.01, 2)$, which is not symmetric about the origin, with the hope that the discretization used by `PlotVectorField` will miss this point.

Remark 3.3.2: The physical domain in this example is the region outside of the circle of radius one.

Direction fields and isoclines provide some insight about solutions of differential equations. The main theorem of this section, however, guarantees that a unique solution to an initial value problem actually exists, at least in some interval of the independent variable near the values specified in the initial data. Let $x(t)$ be a solution to the nth order differential equation

$$\frac{d^n x(t)}{dt^n} = f(t, x(t), \frac{dx(t)}{dt}, \frac{d^2 x(t)}{dt^2}, \dots, \frac{d^{n-1} x(t)}{dt^{n-1}}) \tag{3.35}$$

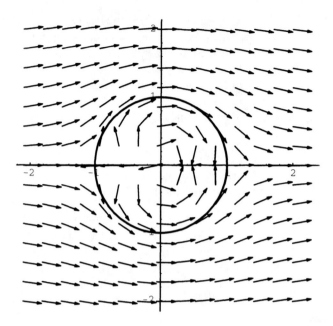

Figure 3.12: Velocity field of the flow past a cylinder.

subject to the initial data

$$x(t_0) = x_0, \frac{dx(t)}{dt}\Big|_{t=t_0} = x_1, \ldots, \frac{d^{n-1}x(t)}{dt^{n-1}}\Big|_{t=t_0} = x_{n-1}, \qquad (3.36)$$

where $\mathbf{x}_0 \equiv (x_0, x_1, \ldots, x_{n-1})$ is a given point in R^n. Let $x(t, t_0, \mathbf{x}_0)$ denote the solution to (3.35), (3.36), if it exists. The following existence and uniqueness theorem gives sufficient conditions on f to guarantee a solution to this problem for an interval about t_0.

Theorem 3.3.1 (Existence and Uniqueness)

Consider the initial value problem (3.35), (3.36). Let $f(t, x, p_1, p_2, \ldots, p_{n-1})$ be a continuous function in each variable for $t_0 - \delta_0 < t < t_0 + \delta_0$, $x_0 - \epsilon_0 < x < x_0 + \epsilon_0$, $x_1 - \epsilon_1 < p_1 < x_1 + \epsilon_1, \ldots$, $x_{n-1} - \epsilon_{n-1} < p_{n-1} < x_{n-1} < +\epsilon_{n+1}$. Also, suppose that f is continuously differentiable in the variables x, $p_1, \ldots, p_{n-1}$ in the respective intervals above. Then (3.35), (3.36) has a unique solution $x(t, t_0, x_0, x_1, \ldots, x_{n-1})$ for $t_0 - \delta_1 < t < t_0 + \delta_1$, where δ_1 depends on $\delta_0, \epsilon_0, \epsilon_1, \ldots$, and ϵ_{n-1}. Furthermore, the function x depends continuously on the variables $t, x_0, x_1, \ldots$, and x_{n-1}.

The proof of this theorem can be found in many advanced texts on ODEs (for example, see J. K. Hale's *Ordinary Differential Equations*, John Wiley and Sons, 1974). Although we will not present the proof in this text, we will utilize the main idea in its argument, that of approximating the **fixed point of a map**, in a project at the end of this chapter.

Even though the statement of the above theorem seems quite technical, its usefulness for the most part is in the ease with which we are able to verify each hypothesis. Intuitively the theorem states that we could expect a unique solution to an initial value problem for a given (possibly) nonlinear differential equation if the nonlinearity behaves reasonably smoothly near the initial condition. We point out, however, that the above theorem guarantees the existence of a unique solution to (3.35), (3.36) for *some* interval of t containing t_0. This interval could possibly be smaller than the original interval $(t_0 - \delta_0, t_0 + \delta_0)$, as some of following examples show.

Example 3.3.3
Consider the differential equation

$$x'' + x = 0, \quad x(0) = 1, x'(0) = 0. \tag{3.37}$$

In the notation of (3.35), $n = 2$, $t_0 = 0$, $x_0 = 1$, $x_1 = 0$, and $f(t, x, p) = -x$. The function f satisfies the hypotheses of Theorem 3.3.1 with all ϵs and δ_0 equal infinity. Hence, there is a unique solution to (3.37) in $(-\delta_1, \delta_1)$ for some $\delta_1 > 0$. As the reader can verify, $x(t) = \cos t$ is this unique solution. The domain of this solution is $(-\infty, \infty)$, although this fact could not be deduced from Theorem 3.3.1.

Example 3.3.4
Consider the differential equation

$$x' = x^2, \quad x(0) = 1. \tag{3.38}$$

In this example $n = 1$, $t_0 = 0$, $x_0 = 1$, and $f(t, x) = x^2$. As in Example 3.3.3, f satisfies the hypotheses of Theorem 3.3.1 for all t and x in R. Therefore, there is a unique solution to this problem for $t \in (-\delta_1, \delta_1)$ for some $\delta_1 > 0$. The reader can check that

$$x(t) = \frac{1}{1 - t} \tag{3.39}$$

is this unique solution. However, even though the function f in this example is everywhere continuously differentiable, the solution x "blows up" at $t = 1$, so that $\delta_1 = 1$. This solution exists for all t in $(-\infty, 0)$ but cannot be extended to $t \geq 1$.

Example 3.3.5
Consider the differential equation

$$x' = \sqrt{x}, \quad x(0) = 0. \tag{3.40}$$

In this example $n = 1$, $t_0 = 0$, $x_0 = 0$, and $f(t, x) = \sqrt{x}$. The function f is continuous in t but is not continuously differentiable in x (since $\frac{\partial f}{\partial x} = \frac{1}{2\sqrt{x}}$, which becomes unbounded at $x = x_0 = 0$). Thus, the hypothesis of Theorem 3.3.1 does not hold and we cannot guarantee the existence of a unique solution to (3.40). In fact, it can easily be checked that (3.40) has at least the two distinct solutions

$$x(t) \equiv 0 \quad \text{and} \quad x(t) = \frac{t^2}{4}. \tag{3.41}$$

Problems

1. Consider the differential equation in (3.33). Use either *Mathematica* or *MATLAB* to generate either Figure 3.9 or Figure 3.10.

2. Use either *Mathematica* or *MATLAB* to generate the graph of the velocity field associated with the flow past the cylinder.

3. In each of the following problems draw the direction field. Choose some representative initial conditions, solve the initial value problems in *Mathematica* or *MATLAB*, and superimpose the graphs of the solution curves on the direction fields.

   (a) $x' = x \sin t$

   (b) $y' = y + \sin x$

   (c) $y' = x + \sin y$

   (d) $y' = \sin y$

   (e) $z' = e^{-z}$

   (f) $x' = \sqrt{t + x^2}$, where $t \geq 0$

   (g) $x' = x^2 + t^2$

   (h) $x' = -x^3 + t^2$

   (i) $y' = \ln(t + y^2)$, where $t > 0$

4. In each of the following problems draw the isoclines together with a few of the solution curves with initial conditions of your own choosing.

   (a) $x' = x + t$

   (b) $x' = x - t$

   (c) $x' = t x$

   (d) $y' = 2y + \sin x$

   (e) $y' = 2x^2 + 3y^2$

5. Draw the velocity field (see Example 3.3.2) in each of the following problems.

   (a) $\frac{dx}{dt} = x, \qquad \frac{dy}{dt} = y$

   (b) $\frac{dx}{dt} = 0, \qquad \frac{dy}{dt} = y$

   (c) $\frac{dx}{dt} = 0, \qquad \frac{dy}{dt} = x$

(d) $\frac{dx}{dt} = y, \qquad \frac{dy}{dt} = x$

(e) $\frac{dx}{dt} = -y, \qquad \frac{dy}{dt} = x$

(f) $\frac{dx}{dt} = -\frac{y}{\sqrt{x^2+y^2}}, \qquad \frac{dy}{dt} = \frac{x}{\sqrt{x^2+y^2}}$

(g) $\frac{dx}{dt} = \sin y, \qquad \frac{dy}{dt} = \cos x$

(h) $\frac{dx}{dt} = e^{-y}, \qquad \frac{dy}{dt} = x$

6. Determine whether the hypotheses of the existence and uniqueness theorem hold for the following differential equations.

(a) $x'' + \sin x = 0, \quad x(0) = 0, x'(0) = 1$

(b) $tx'' + \sin x = 0, \quad x(1) = 0, x'(1) = 1$

(c) $tx'' + \sin x = 0, \quad x(0) = 0, x'(0) = 1$

(d) $xx'' + \sin x = 0, \quad x(0) = 0, x'(0) = 1$

(e) $y' + t^2 - y^2 = 0, \quad y(1) = y'(1) = 0$

(f) $y' = \sqrt{y}, \quad y(0) = 0$

(g) $y' = \sqrt{y}, \quad y(0) = 1$

3.4 Separable Differential Equations

In the next few sections we will study several classes of differential equations for which we are able to construct exact solutions. These classes include a) **separable**, b) **linear first-order**, and c) **exact** equations. In this section we introduce the class of separable equations and present a method for obtaining a closed-form solution.

A first-order differential equation of the form

$$\frac{dy}{dx} = H(x,y) \tag{3.42}$$

is called **separable** if H is separated in x and y, that is,

$$H(x,y) = f(x)g(y). \tag{3.43}$$

Because of the special form of H, we can separate the dependence of y and x from each other in (3.42) and write formally

$$\frac{dy}{g(y)} = f(x)dx. \tag{3.44}$$

Integrate (3.44) once to get

$$\int \frac{dy}{g(y)} = \int f(x)dx + c. \tag{3.45}$$

Equation (3.45) is the solution to (3.42) up to an integration. Let $F(x)$ and $G(y)$ denote the indefinite integrals on the right-hand and left-hand sides of (3.45).

The general solution y of (3.42) then satisfies the relation

$$G(y) = F(x) + c. \tag{3.46}$$

We have thus found the general solution of (3.42) implicitly. The constant c is determined once a constraint such as

$$y(x_0) = y_0 \tag{3.47}$$

is given. After substituting (3.47) in the general solution, we have

$$c = G(y_0) - F(x_0). \tag{3.48}$$

Substitute (3.48) in (3.46) to get the following *implicit* representation of the solution to (3.42), (3.47):

$$G(y) = F(x) + G(y_0) - F(x_0). \tag{3.49}$$

Example 3.4.1
Consider the differential equation

$$\frac{dy}{dx} = -2xy, \quad y(0) = 2. \tag{3.50}$$

This equation is separable and can be written as

$$\frac{1}{y}dy = -2xdx. \tag{3.51}$$

Integrate both sides of (3.51):

$$\ln|y| = -x^2 + c. \tag{3.52}$$

We need to use the initial condition $y(0) = 2$ to determine c. First, note that $y(0) > 0$, which allows us to replace the absolute value of y by y itself. Next, evaluate this equation at $(x, y) = (0, 2)$ to get

$$\ln 2 = c \tag{3.53}$$

so that, after exponentiating (3.52) and using this value of c, we arrive at the solution

$$y(x) = 2e^{-x^2}. \tag{3.54}$$

Equation (3.52) is the general solution of the differential equation, while (3.54) is the particular solution to the initial value problem (3.50).

Example 3.4.2
Consider the differential equation

$$y' = \frac{1 - x}{1 + y^5}. \tag{3.55}$$

This equation is also separable and can be written in the form

$$(1 + y^5)dy = (1 - x)dx. \tag{3.56}$$

After integrating both sides we obtain

$$y + \frac{1}{6}y^6 = x - \frac{1}{2}x^2 + c \tag{3.57}$$

as the general solution of the differential equation. Unlike the case of the previous example, however, it is difficult to invert the function y in (3.57) and get an *explicit* solution of the equation. Instead, we will be content with the *implicit* solution given by (3.57), which carries enough information about the function $y(x)$ to be useful for many applications.

Example 3.4.3

The fact that the solution of a separable differential equation can be obtained via integration makes these equations a good candidate for analysis with symbolic manipulators. For instance, the following program in *Mathematica* shows how one employs the internal functions `Integrate`, `Solve`, and `Plot` to plot the solution to a separable initial value problem. The logic of this program mimics closely the procedure we outlined at the beginning of this section, which led to (3.46). The lines

```
Clear[x, y];
x0 = 0; y0 = 2; f[x_] = -x; g[y_] = 2 y;
```

clear any values x and y may contain and define the various input parameters and functions. Next, we define F and G:

```
F[x_] = Integrate[f[x], x];  G[y_] = Integrate[1/g[y], y];
```

The general solution of the differential equation is determined from an inversion of (3.46):

```
gensol = Solve[G[y] == F[x] + c, y];
```

Next, we compute c from the initial datum $y(x_0) = y_0$:

```
const = Solve[y == y0 /. gensol /. x->x0, c]
```

Mathematica returns

```
{{c -> Log[-Sqrt[2]]}, {c -> Log[Sqrt[2]]}}
```

The first of these constants is complex. The second is real. Thus,

```
sol[x_] = y /. gensol /. const[[2]];
```

Finally, we plot the solution via

```
Plot[sol[x], {x, -2, 2}]
```

Remark 3.4.1: There are several interesting points about the above program that are worth noting. First, `Solve` is used to generate the general (implicit) solution of the differential equation. Next, the same command is used to solve for c (note the double use of `/.`). Alternatively, we could have computed c from (3.48). Finally, the information is collected in `sol[x]`.

The `DSolve` command in *Mathematica*, an internal function of this software, is capable of determining the exact solution of our problem as well. Try

```
a = DSolve[{y'[x] == - 2 x y[x], y[0] == 2}, y[x], x];
Plot[y[x] /. a, {x, -2, 2}]
```

which leads to the same result.

Implementing the separability procedure in *MATLAB* is quite similar. Here is the program that handles the initial value problem in Example 3.4.1:

```
syms F G x y c
f = '-x'; g = '1/(2*y)';
x0 = 0; y0 = 2;
F = int(f); G = int(g);
sol = solve(G-F-c,y); % General Solution
c = solve(subs(sol,x0,'x')-y0,c); % Particular Solution
y = subs(sol,c,'c');
ezplot(y,[-2,2])
```

We note that, unlike *Mathematica*, *MATLAB*'s `solve` returns only the real solution to

$$e^{2c-x_0^2} = y_0.$$

The `dsolve` command in *MATLAB*, an internal function to this software, also reaches the same conclusion:

```
y=dsolve('Dy=-2*x*y','y(0)=2','x')
ezplot(y,[-2,2])
```

Example 3.4.4

In Example 3.4.2 we considered a differential equation that was separable but were able to determine its general solution only implicitly. How does *Mathematica* or *MATLAB* deal with such a case? Let us consider (3.55) with initial condition $y(0) = 1$, whose solution, following (3.57), is

$$y + \frac{1}{6}y^6 = x - \frac{1}{2}x^2 + \frac{7}{6}. \tag{3.58}$$

There are two ways to get *Mathematica* to draw the graph of y as a function of x. One way is to use the `ImplicitPlot` command and draw the graph of (3.58) (it is often desirable to start a new session of *Mathematica* whenever a package, such as `ImplicitPlot`, is used):

```
<<Graphics'ImplicitPlot'
H[x_, y_] = y + 1/6 y^6 - x + 1/2 x^2 - 7/6;
a = ImplicitPlot[H[x, y] == 0, {x, 0, 3}, {y, -1, 3}]
```

A second way is to use the `NDSolve` command (in place of `DSolve`) to generate an approximate solution to (3.57):

```
b = NDSolve[{y'[x] == (1-x)/(1+y[x]^5), y[0] == 1}, y[x],
     {x, 0, 3}];
c = Plot[y[x] /. b, {x, 0, 3}]
```

(What happens when you try to solve this equation in the interval $(0, 4)$?) We will discuss the properties of `NDSolve` in more detail in later sections.

It is interesting to combine the outputs of the latter two programs (by entering `Show[a, c]`) and see that, at least as far as naked eye can discern, there is little difference between these graphs.

The internal function `ode45` in *MATLAB* finds approximate solutions of ODEs. To apply it to the initial value problem (3.55) with $y(0) = 1$, first we create an M-file called `diffeqn1.m` that contains the definition of the right-hand side of (3.55):

```
function yprime = diffeqn1(x,y);
yprime = (1-x)/(1+y^5);
```

and then use this M-file as input to `ode45` as follows:

```
[x, y] = ode45('diffeqn1', 0, 3, 1);
```

Finally, we can plot the graph of $y(x)$ by

```
plot(x, y)
```

It is possible that some information about the solutions of a separable equation may be lost by the method we presented. This happens when the function $g(y)$ is zero at a point y_1, say. Then, one of the solutions of (3.42) is the constant function $y(x) \equiv y_1$, which may not be recovered by our method. For example, consider the differential equation

$$y' = y^2. \tag{3.59}$$

After separating variables as before and integrating the resulting expressions, we end up with

$$y(x) = \frac{1}{c - x} \tag{3.60}$$

as the set of solutions. Note that $y = 0$ is a root of $g(y) = y^2$. Thus, the function $y(x) \equiv 0$ is a solution of (3.59) but can only be recovered from (3.60) if we let c approach infinity. Another example comes to us from the differential equation

$$y' = \sqrt{y}. \tag{3.61}$$

After separating variables, we arrive at the solutions

$$y(x) = (\frac{x}{2} + c)^2. \tag{3.62}$$

Again, $y(x) \equiv 0$ satisfies (3.61). This time there is no value of c, not even infinity, that can be used in (3.62) to recover this solution.

Problems

1. Alter the programs in Example 3.4.3 and use (3.48) to determine the solution to (3.50) in *Mathematica* or *MATLAB*.

2. Modify the programs in Example 3.4.3 so that the final output is only the implicit solution (3.49).

3. Verify that the following equations are separable and find their general solutions. For each problem use the programs in Example 3.4.3 or Problem 2 to check each answer.

   (a) $y' = 2y$

   (b) $y' = 2xy$

   (c) $(1 + x)y' = y^2$

   (d) $y' = \sqrt{xy}$

   (e) $y'' + y' = 0$ (Hint: Integrate this equation once before applying the method.)

   (f) $y' = 1 - y^2$. Are $y(t) = \pm 1$ solutions? Are they recovered by this method?

   (g) $y' = y \tan x$

   (h) $xz' = \frac{(1-x)z}{x(1+z)}$

   (i) $e^t x' = tx$

   (j) $x' = \csc(x)$

   (k) $w' = t \sec(w)$

   (l) $y' = 1 + x^2 + y^2 + x^2 y^2$

   (m) $tx' + x + x^2 = 0$

   (n) $y' + y + 1 = 0$

   (o) $y'^2 - y - 1 = 0$

   (p) $z' + az = 0$, where a is a constant. For what values of a do all solutions go to zero when t approaches infinity?

   (q) $v' + ctv = 0$, where c is a positive constant. Determine $\lim_{t \to \infty} v(t)$.

   (r) $v' + 3v = 4$. Determine $\lim_{t \to \infty} v(t)$. Does the value of the limit depend on the initial value of v? Could one have obtained the limiting value directly from the differential equation?

4. In each of the following problems find the solution to the initial value problem. Check each answer with *Mathematica* or *MATLAB* using the programs in Example 3.4.3 or by using `DSolve` or `dsolve`.

   (a) $y' + xy = 0$, $y(0) = 3$

   (b) $xy' + y = 0$, $y(1) = -1$

   (c) $x' + tx = 1$, $x(0) = 0$

   (d) $y' + x^2 y = 0$, $y(0) = 1$

   (e) $\cos t\, y' + y^2 = 0$, $y(0) = 1$. Try this problem with $y(0) = 0$. What should the solution be? What do `DSolve` or `dsolve` give instead?

   (f) $x' + ax = 0$, $x(0) = 2$, where a is a constant. Plot the graphs of the solutions $x(t, a)$ as a varies from -5 to 5 at increments of 0.1 and $t \in (-2, 2)$.

   (g) $v' + av = 1$, $v(0) = 0$, and a a constant. Plot the graphs of $v(t, a)$ for $a \in (-0.2, 0.2)$ at increments of 0.1 and $t \in (0, 10)$. How does the behavior of v depend on a as $t \to \infty$?

5. (**Newton's law of cooling**) Newton's law of cooling states that, under certain ideal conditions, the time rate of change of the temperature of a body is proportional to the difference between the temperature of the body at time t and the temperature of its environmemnt. Consider a can of soda that is kept in a refrigerator at a constant temperature of $40°$ fahrenheit. At time zero the can is removed from the refrigerator into an environment of $70°$ F. Assuming that Newton's law of cooling applies to this problem and that the proportionality constant is $0.1°$ fahrenheit per minute, how long does it take for the temperature of the soda to reach $65°$? Determine how long it takes for the temperature to reach $69°$? How about $70°$?

6. What happens to the temperature profile of the body in the above problem if Newton's law of cooling is altered to allow for second-order effects? Namely, suppose that the temperature $T(t)$ satisfies

$$\frac{dT}{dt} = k_1(70 - T) + k_2(70 - T)^2, \qquad T(0) = 40. \qquad (3.63)$$

Let $k_1 = 0.1$ and $k_2 = 0.01$. How long does it take for the temperature in the soda can to reach 65?

Remark 3.4.2: The `DSolve` function in *Mathematica* and the `dsolve` function in *MATLAB* are not very helpful with the differential equation in this problem. Both functions look for an explicit solution of this initial value problem and fail to determine it. The differential equation (3.63) is separable, so one can use the method described in Example 3.4.4 to get the graph of an approximate solution. `NDSolve` in *Mathematica* and its counterpart `ode45` in *MATLAB* are also very helpful in getting to the answer. One way of handling this problem in *Mathematica* is

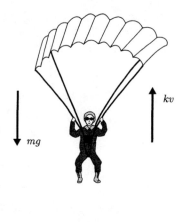

Figure 3.13: Vertical motion with air resistance.

```
a = NDSolve[{T'[t] == 0.1*(70-T[t]) + 0.01*(70-T[t])^2,
             T[0] == 40}, T, {t, 0, 9}];
f[t_] := First[T[t] /. a];
FindRoot[f[t] == 65, {t, 3}]
```

7. (**Air resistance and falling objects**) Consider the vertical motion of a falling object of mass m as indicated in Figure 3.13 (positive direction is pointing downwards). Assume the only forces acting on the body are the gravitational force and a frictional force that is linearly proportional to the velocity of the body. Use Newton's second law to show that the velocity v at any time t satisfies the differential equation

$$m\frac{dv}{dt} = mg - kv, \tag{3.64}$$

where k is the proportionality constant in the force due to air resistance. Find the solution to this problem using the method described in this section. Check the answer in either *Mathematica* or *MATLAB*. Also, determine the steady-state behavior of a typical solution (i.e., what happens to $v(t)$ as t approaches infinity).

8. Consider a mass of 70 kg falling from a height of 5000 meters under the action of gravity and subject to the linear drag force of the previous problem. Suppose that $k = 10$. Solve (3.64) with the initial condition $v(0) = 0$, and show that the velocity field satisfies $v(t) = 68.6(e^{-\frac{1}{7}t} - 1)$. Use this $v(t)$ to determine the function $y(t)$, the height of the mass at time t (note that $y'(t) = v(t)$). Use the `GraphicsArray` function in *Mathematica* or the `subplot` function in *MATLAB* and plot the graphs of $v(t)$ and $y(t)$ on the

same screen with $t \in (0, 100)$. How does one determine the time at which this mass reaches ground level? What is its velocity at impact?

9. Show that when we consider the problem of an object of mass m falling to the ground and model air resistance as a force that depends quadratically on the velocity, the velocity v satisfies the differential equation

$$mv' = mg - kv^2, \quad v(0) = 0. \tag{3.65}$$

Show that (3.65) is separable and its solution is given by

$$v(t) = \sqrt{\frac{gm}{k}} \tanh\left(\sqrt{\frac{gk}{m}} t\right).$$

Check the answer against what one obtains from DSolve in *Mathematica* or dsolve in *MATLAB*. As in the previous problem, consider the special case of a mass of 70 kg falling from a height of 5000 meters. Let $k = 10$. Determine the velocity profile with these parameter values. Use this $v(t)$ and determine the distance-to-the-ground function $y(t)$. Graph $v(t)$ and $y(t)$ on the same screen with $t \in (0, 100)$. Find the time at which the object reaches ground level. Draw the graphs of $v(t)$ for the two models (linear and quadratic) of air resistance on the same screen. In which case does the object fall faster?

10. (**Torricelli's Law**) In this problem we develop a separable differential equation that, under ideal conditions, models the flow of a fluid through a hole or an orifice at the bottom of a container.

    (a) First, use Newton's second law (mass × acceleration = force) to show that the velocity v of an object descending from a standing position and falling under the action of gravity alone (ignoring air resistance) is given by
    $$v = -\sqrt{2g(H - h(t))}, \tag{3.66}$$
    where g is the constant of acceleration of gravity, H is the object's original height, and h is its height at time t.

    (b) Let C be a water tank of a certain shape that has a hole with area s at its bottom. Let $x(t)$ be the depth of the water in the tank at any time t. Torricelli's law assumes that water empties out of the tank with the velocity of an object falling freely under the action of gravity; that is, using (3.66), the velocity of water leaving the tank is given by
    $$v = -\sqrt{2gx}. \tag{3.67}$$
    Show that the rate at which water leaves the hole is given by
    $$-s\sqrt{2gx}. \tag{3.68}$$
    Recalling that g has dimensions of length per time squared (e.g., g's units are cm/s^2), show that the expression in (3.68) has dimensions of volume per unit time.

(c) Let $S(x)$ be the cross-sectional area of the tank at depth x. Show that the rate at which water leaves the tank is equal to

$$S(x)\frac{dx}{dt}. \tag{3.69}$$

(Hint: The volume V of the tank equals $\int_0^x S(\tau)d\tau$. Compute $\frac{dV}{dt}$.) Combine (3.68) and (3.69) and arrive at the differential equation

$$S(x)\frac{dx}{dt} = -s\sqrt{2gx}. \tag{3.70}$$

Equation (3.70) is first order and separable.

(d) Suppose that C is a circular cylinder of height 10 meters and base diameter of 3 meters. At the bottom (base) of the cylinder there is a circular opening of diameter 0.1 meters, which is kept closed while the tank is being filled with water. After the tank is full, the hole at the bottom is opened and the water is allowed to empty out. How long does it take before the tank is half full ($g = 9.8\,\text{m/s}^2$)?

(e) Consider a hemispherical tank of radius 3 meters with a circular opening of diameter 0.01 meters at its bottom. How long does it take for the full tank to be half empty? (Hint: $S(x) = \pi(9 - (3 - x)^2)$.)

(f) Consider the tank C that is a surface of revolution of height 2 meters generated by the parabola $x = z^2$. The hole at the bottom of the tank has area 0.01 m^2. How long does it take for the tank to drain?

11. Consider a hemispherical tank of radius 7 meters. Let s be the area of a hole at the bottom of the tank. Suppose that the tank becomes half empty in 4 hours. Follow the analysis of the previous problem and find s.

12. (**Homogeneous equations**) A homogeneous first-order differential equation is one that has the form

$$x' = F\left(\frac{x}{t}\right). \tag{3.71}$$

(a) Show that any homogeneous differential equation can be written in the form

$$t\frac{dv}{dt} = F(v) - v, \tag{3.72}$$

where v is given by the substitution $v = \frac{x}{t}$. Note that (3.72) is separable.

(b) Show that each of the following equations is homogeneous and find its solution using the above substitution.

   i. $x' = 1 + \left(\frac{x}{t}\right)^2$
   ii. $(x - y)\,y' = x + y$
   iii. $z^3\,z' = t^3 + t^2 z$
   iv. $x' = \frac{1}{t}(x - \sqrt{t^2 - x^2})$

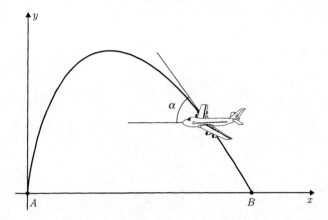

Figure 3.14: An aircraft headed for the airport located at the origin.

(c) (**Flight paths**) An airliner leaves Airport A for Airport B 750 kilometers away. The aircraft directs its controls and radar toward its
destination after takeoff and travels with constant speed of 600 km/h.
A strong wind with a speed of 80 km/h blows in the direction perpendicular to the path between the airports. Find the path that the
aircraft takes in the air by performing the following steps (see Figure
3.14):

i. Choose a coordinate system so that the airports are located on
the x-axis with Airport A at the origin $(0,0)$ and Airport B at
$(750, 0)$. Assume that the wind is blowing in the y direction and
its velocity is given by $\mathbf{v} = \langle 0, 80 \rangle$. Let $(x(t), y(t))$ be the path
of the aircraft. Then $\langle \frac{dx}{dt}, \frac{dy}{dt} \rangle$ is its velocity. Show that x and y
satisfy the system of differential equations

$$x'(t) = -\frac{600x}{\sqrt{x^2 + y^2}}, \quad y'(t) = -\frac{600y}{\sqrt{x^2 + y^2}} + 80. \qquad (3.73)$$

(Hint: Note that $\frac{dx}{dt} = -600\cos\alpha$, where α is the angle between
the flight direction and the x-axis. A similar expression holds for
y. How is α related to x and y?)

ii. Divide the preceding equations and show that

$$\frac{dy}{dx} = \frac{y}{x} - \frac{2}{15}\sqrt{1 + \frac{y^2}{x^2}}. \qquad (3.74)$$

Note that (3.74) is homogeneous. Find its general solution. Use
DSolve in *Mathematica* or dsolve in *MATLAB* and obtain the
general solution. (Hint: With the substitution $xv(x) = y(x)$,
Mathematica's response is

$$\{\{v[x] \to \operatorname{Sinh}[\frac{15\,C[1] - 2\operatorname{Log}[x]}{15}]\}\}$$

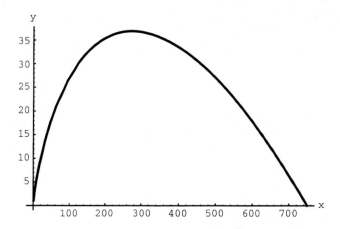

Figure 3.15: The trajectory of an aircraft taking off from $(750, 0)$ and landing at $(0, 0)$.

iii. Show that the initial condition for this problem is

$$v(750) = 0. \tag{3.75}$$

Solve (3.74) and (3.75), draw the graph of $y(x)$, and compare with Figure 3.15.

iv. Determine the time of travel. (Hint: Note that the total time of travel is $T = \int_{750}^{0} \frac{dt}{dx} \, dx$. Use the function $y(x)$ obtained in the previous step in (3.73a).)

(d) The graph in Figure 3.15 is the output of the following *Mathematica* program:

```
a= DSolve[x v'[x] == -2/15*Sqrt[1 + v[x]^2], v[x], x];
b= v[x] /. a /. x -> 750;
c=Solve[b == 0, C[1]];
y[x_] = x*v[x] /.a/. c;
Plot[y[x], {x, 0.001, 750}, AxesLabel -> {"x", "y"}]
```

What happens to the angle α as the aircraft approaches its destination?

(e) Alter the appropriate parameters in this program to find the path of an aircraft when the two airports are 1000 kilometers away, the wind blows 110 km/h perpendicular to the path, and the speed of the aircraft is 800 km/h.

13. A fighter aircraft (which we label A) has spotted an enemy aircraft (labeled B) 100 miles away due west. Aircraft A locks its radar on B and aims to intercept it. Suppose that B travels at a constant speed of 500 m/h in a straight line due north. Aircraft A travels at speed of 600 m/h.

(a) Find the trajectory of aircraft A.

(b) Find how long it takes for A to intercept B.

14. Find a solution to the following differential equations using the suggested substitutions.

   (a) $x' = x + t$ (Hint: Let $v = x + y$ and note that v satisfies $v' = 1 + v$, which is separable.)

   (b) $x' = (x + t + 1)^2$. Try the substitution $v = x + y + 1$.

   (c) $t^2 x' + 2t\, x = 1 + t^2$. Let $v = t^2 x$ and find an equation for v.

   (d) Find the general solution to $f(t)x' + f'(t)x = g(t)$. Try $v = f(t)x$.

3.5 Linear First-Order Equations

A linear first-order differential equation is an equation of the form

$$x'(t) + p(t)x(t) = q(t), \quad x(t_0) = x_0, \tag{3.76}$$

where p and q are smooth functions of t at t_0. These differential equations form a class of equations whose solutions we are able to determine explicitly as a function of t. The existence and uniqueness of solutions to the initial value problem in (3.76) are guaranteed by Theorem 3.3.1.

We begin first with the **homogeneous** case of (3.76), namely, the case when $q \equiv 0$. With its right-hand side vanishing, (3.76) reduces to

$$\frac{dx}{dt} = -p(t)x \tag{3.77}$$

which is separable, and therefore falls in the class of equations we discussed previously. We write (3.77) in the equivalent form

$$\frac{dx}{x} = -p(t)dt \tag{3.78}$$

and integrate both sides to get

$$\ln|x(t)| + c = -\int p(s)\, ds, \tag{3.79}$$

where c is the constant of integration. The initial datum in (3.76) will now determine c and the value of $|x|$. Without loss of generality, we assume that $x_0 > 0$ (what is the solution to (3.77) if $x_0 = 0$?) As a consequence of Theorem 3.3.1, the solution x must remain positive for all t. For the sake of concreteness, we let t_0 be the lower limit of integration in (3.79), so

$$\ln x(t) + c = -\int_{t_0}^{t} p(s)\, ds.$$

Substitute $x(t_0) = x_0$ in this equation and solve for c. This results in

$$c = -\ln x_0.$$

The solution $x(t)$ is now known implicitly through the expression

$$\ln x(t) - \ln x_0 = -\int_{t_0}^{t} p(s)\, ds. \tag{3.80}$$

After exponentiating both sides of (3.80), we find the following explicit formula for x

$$x(t) = x_0 e^{-\int_{t_0}^{t} p(s) ds}. \tag{3.81}$$

Formula (3.81) is the starting point for solving (3.76) when q is not identically zero. The function

$$P(t) = e^{\int_{t_0}^{t} p(s) ds} \tag{3.82}$$

is called the **integrating factor** for (3.76) (note that $P(t_0) = 1$). The role of P is to render (3.76) an exact derivative. To see this, multiply both sides of (3.76) by $P(t)$:

$$P(t)x' + p(t)P(t)x = P(t)q(t). \tag{3.83}$$

Keeping in mind the definition of P in (3.82), we note that the left-hand side of (3.83) is the exact derivative of $P(t)x(t)$. Consequently, (3.83) can be written in the form

$$\frac{d}{dt}(P(t)x(t)) = P(t)q(t). \tag{3.84}$$

After integrating (3.84) once and using the initial condition $x(t_0) = x_0$ to determine the constant of integration, we have

$$P(t)x(t) = x_0 + \int_{t_0}^{t} P(u)q(u)du. \tag{3.85}$$

Substituting the expression for P in (3.85), we obtain the general solution to the original differential equation:

$$x(t) = x_0 e^{-\int_{t_0}^{t} p(s)ds} + \int_{t_0}^{t} e^{\int_{t}^{u} p(s)ds} q(u)du. \tag{3.86}$$

We summarize this result in the following theorem.

Theorem 3.5.1 (Solution to Linear First-Order Equations)
Consider the initial value problem (3.76). The function $P(t)$ defined by

$$P(t) = e^{\int_{t_0}^{t} p(s)ds},$$

*is the **integrating factor** for this equation. In terms of this function the solution to (3.76) is given by*

$$x(t) = \frac{x_0}{P(t)} + \frac{1}{P(t)} \int_{t_0}^{t} P(u)q(u)du. \tag{3.87}$$

Example 3.5.1

Consider the differential equation

$$3t\, x'(t) + 2x(t) = t, \quad x(1) = 1. \tag{3.88}$$

In order to apply the above procedure to (3.88), we must first put this equation in the form (3.76). Divide (3.88) by $3t$

$$x'(t) + \frac{2}{3t}x(t) = \frac{1}{3}, \tag{3.89}$$

so $p(t) = \frac{2}{3t}$ and $q(t) = \frac{1}{3}$. The integrating factor P is then defined by

$$P(t) = e^{\int p(s)ds} = e^{\int \frac{2}{3t}dt} = t^{\frac{2}{3}}. \tag{3.90}$$

Multiply both sides of (3.89) (*not* (3.88)) by (3.90). The left-hand side of the resulting equation is the exact derivative of $P(t)x(t) = t^{\frac{2}{3}}x(t)$, as expected. Therefore, (3.89) is equivalent to

$$\frac{d}{dt}(t^{\frac{2}{3}}x(t)) = \frac{t^{\frac{2}{3}}}{3}. \tag{3.91}$$

Integrating both sides of (3.91) and using the initial datum $x(1) = 1$ yield

$$x(t) = \frac{1}{5}t + \frac{4}{5}t^{-\frac{2}{3}}. \tag{3.92}$$

The method of integrating factors is well suited for *Mathematica* and *MATLAB* because this method reduces the computations for finding the solution to (3.76) to two integrations. For many functions p and q these integrals can be explicitly computed, and for those for which the antiderivatives in (3.87) do not exist the numerical integration capabilities of *Mathematica* or *MATLAB* allow one to get an approximate solution to the problem.

Example 3.5.2

We present two programs in *Mathematica* and *MATLAB* that determine the solution to (3.88) by following the pattern of the logic that led to the conclusion in Theorem 3.5.1. First the *Mathematica* program:

```
t0 = 1; x0 = 1;
p[t_] = 2/(3*t); q[t_] = 1/3;
P[t_] = Exp[Integrate[p[s], {s, t0, t}]];
x[t_] = 1/P[t]*(x0 + Integrate[P[u]*q[u], {u, t0, t}])
```

The syntax for the *MATLAB* program that leads to the solution of (3.88) is

```
syms t u p s
t0 = 1; x0= 1;
p = 2/(3*s); q = 1/3;
P=exp(int(p, s, t0, u));
```

```
intfac=subs(P,t,u);
interim=int(P*q,u,t0,t);
sol=x0/intfac+1/intfac*interim
```

We remark that P and intfac, although the same functions, are represented by different independent variables, u and t, respectively, as required by the formula in (3.87).

Example 3.5.3
Consider the differential equation

$$x' + 2t\,x = \sin t, \quad x(0) = 1. \tag{3.93}$$

Here $p(t) = 2t$ and $q(t) = \sin t$. Following (3.82), the integrating factor P is

$$P(t) = e^{t^2}. \tag{3.94}$$

After multiplying (3.93) by P, integrating the result once, and using the initial condition, we arrive at

$$x(t) = e^{-t^2} + \int_0^t e^{u^2 - t^2} \sin u\, du \tag{3.95}$$

as the solution to (3.93). The formula we have in (3.95) is the closest we get to an analytic representation of the solution because it is a difficult task to evaluate the integral in (3.95) in terms of elementary functions of calculus. This expression, on the other hand, is quite useful from an approximation point of view. Both *Mathematica* and *MATLAB* are capable of approximating the integral in (3.95) numerically at any t and consequently, graphing the solution. We need to make small modifications to the programs in Example 3.5.2 in order to be able to plot the solution curve in the tx-plane. First, we take up the case in *Mathematica*. The last line of the first program in Example 3.5.2 should now read as follows:

```
x[t_] := 1/P[t]*(x0 + NIntegrate[P[u]*q[u], {u, t0, t}])
```

Note the two changes in this line. First, that := is used in place of = so that the right-hand side of x[t] is not evaluated until a value is specified for t. Second, we use NIntegrate in place of Integrate, the former being a numerical integration routine that readily handles integrands of the type used in (3.95). We can now plot the solution curve by invoking the Plot command:

```
Plot[x[t], {t, 0, 6}]
```

Figure 3.16 shows the output.

When we apply the *MATLAB* program described in Example 3.5.2 to (3.93) we get the response

```
sol =

    1/exp(t^2)+1/exp(t^2)*int(exp(u^2)*sin(u), u = .. t)
```

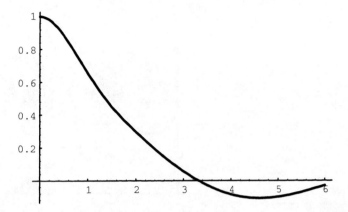

Figure 3.16: The graph of the solution to $x' + 2t\,x = \sin t, x(0) = 1$.

which is *MATLAB*'s (actually in this case, *Maple*'s) way of stating that it cannot determine an antiderivative for $e^{u^2}\sin u$. To see how to invoke *MATLAB*'s numerical integration capabilities, first try

```
trial = subs(sol, 0.4, 't')
```

MATLAB returns the same expression as that stored in `sol` except with 2/5 replacing `t`. We now use `numeric` on `trial`:

```
numeric(trial)
```

MATLAB returns

```
ans =

    0.9251
```

We have thus determined an (excellent) approximation to $x(0.4)$. To get a graph of the solution curve, we need to generate a vector of t and x values:

```
for i=1:60
    a=[a numeric(subs(sol,0.1*i,'t'))];
end
```

and then plot the contents of the vector `a` using `plot`:

```
t=0.1:0.1:6;
plot(t,a)
```

Example 3.5.4
Consider the first-order equation

$$x' + e^{-t^2}x = \sin t, \quad x(0) = 1. \tag{3.96}$$

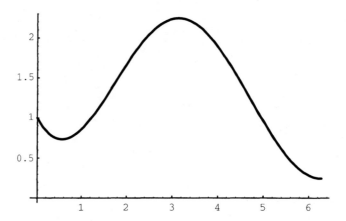

Figure 3.17: The graph of the solution to $x' + \exp(-t^2)x = \sin t, x(0) = 1$.

The integrating factor for this equation is

$$P(t) = \exp\left(\int_{t_0}^{t} \exp(-s^2)\, ds\right). \tag{3.97}$$

This antiderivative cannot be written in terms of the elementary functions of calculus, although it does exist since the function e^{-s^2} is continuous and well-behaved for $t \in R$. The antiderivative of $\exp(-s^2)$ is often called the *error function*, is denoted by erf(t), and defined by

$$\mathrm{erf}(t) = \frac{2}{\sqrt{\pi}} \int_{0}^{t} e^{-s^2}\, ds. \tag{3.98}$$

Mathematica and *MATLAB* both recognize this function as well. Hence, defining P in *Mathematica* or *MATLAB* will not cause any trouble. However, integrating the exponential of this function against $\sin t$ will. We, therefore, pursue the same strategy as in Example 3.5.3 and resort to numerical integration of the integrands. Figure 3.17 shows the output.

Example 3.5.5 (Electrical Circuits)

One of the important applications of first-order differential equations has to do with electrical circuits. Consider an electrical circuit consisting of a resistor with resistance R ohms, an inductor with inductance L henry, and applied voltage of $A \sin \omega t$, where A is the amplitude of the electrical force and ω its frequency, all put in a series (cf. Figure 3.18). Kirchhoff's law states that the voltage drop across all of these devices must add up to zero algebraically. A resistor's usage of voltage is proportional to the current $I(t)$ that passes through it at time t, or

$$V_1 = RI.$$

An inductor's usage of voltage is proportional to the rate at which the current

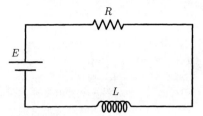

Figure 3.18: An RL circuit.

passes through it:

$$V_2 = L\frac{dI}{dt}.$$

Applying Kirchhoff's law to V_1, V_2, and the applied force, we arrive at the following linear differential equation for the current $I(t)$:

$$L\frac{dI}{dt} + RI = A\sin\omega t. \tag{3.99}$$

Divide this equation by L and note that the integrating factor for this equation is $p(t) = e^{\frac{R}{L}t}$. Multiply (3.99) by p and integrate the result once with respect to t to get

$$I(t) = I(0)e^{-\frac{R}{L}t} + \int_0^t e^{\frac{R}{L}(s-t)}\sin(\omega s)\,ds. \tag{3.100}$$

As the reader may recall, it takes two integration by parts to evaluate the latter integral. More conveniently, either use *Mathematica*'s `Integrate` or *MATLAB*'s `int` to evaluate (3.100). We end up with the following expression for the solution $I(t)$:

$$I(t) = e^{-\frac{R}{L}t}\left(I(0) + \frac{L^2\omega}{L^2\omega^2 + R^2}\right) + \frac{L}{L^2\omega^2 + R^2}\left(R\,\sin(\omega t) - L\omega\,\cos(\omega t)\right). \tag{3.101}$$

Note that (3.101) has two parts to it. The first part, which has information from the initial current in the circuit, decays to zero because of the exponential term. This part is called the **transient** part of the solution. The second part, which consists of periodic trigonometric functions, persists for all time and is called the **steady–state** part of the solution. The steady state describes a balance between the current supplied by the electromotive force applied to the circuit and the amount of current that is used to run the resistor and the inductor.

Example 3.5.6 (Mixing and Concentrations)

Another application of first-order equations concerns inflow-outflow problems and mixing of solutes in fluids. Consider a basin, say the Chesapeake bay, with a source of inflow of fluids, such as the Potomac and the Sesquehenna rivers, and an outlet, the Atlantic Ocean in this case, where the basin's fluid can exit. It is often of interest to understand the effect of the inflow-outflow mechanism on the concentration of vital solutes in the basin. These solutes may be salt, phosphate, agricultural, as well as human wastes, among other possibilities. Here we give a simple and idealized model of the transport of a solute based on differential equations.

Let $x(t)$ be the amount of a solute in the basin at time t. The rate at which this function changes is related to how much fluid enters and leaves the basin and how much of the solute is present in the inflow and the outflow fluids. We have (see Example 3.1.7)

$$\frac{dx}{dt} = r_{in}\, c_{in} - r_{out}\, c_{out}, \tag{3.102}$$

where r_{in} is the rate of the inflow of the fluid into the basin and c_{in} is the amount of the solute of interest in the incoming flow. The terms r_{out} and c_{out} have similar interpretations. We assume the fluid inside the basin is fully mixed. The amount of the solute in the basin at any time t, and hence the amount that leaves the basin, is

$$c_{out} = \frac{x}{V},$$

where V is the volume of the basin. Substitute the latter in (3.102):

$$\frac{dx}{dt} = r_{in}\, c_{in} - r_{out}\, \frac{x}{V}. \tag{3.103}$$

This equation is a linear first-order equation in x, which can be solved by the method described in this section once values for V, r_{in}, r_{out}, and c_{in} are known.

As a concrete example, consider a bay with a capacity of 5 million liters, which contains no phosphates at time 0. Then, with $x(t)$ being the amount of phosphate in the bay at any time t, we have

$$V = 5,000,000 \quad \text{and} \quad x(0) = 0.$$

Let the inflow and outflow be at the rate of 50,000 liters per day so that

$$r_{in} = r_{out} = 50,000.$$

Furthermore, let the inflow contain 0.001 grams of phosphate per liter, that is,

$$c_{in} = 0.001.$$

Because the inflow and outflow rates are equal, the volume of the bay remains constant. So, the differential equation (3.103) takes the form

$$\frac{dx}{dt} + 0.01x = 50, \quad x(0) = 0. \tag{3.104}$$

The integrating factor for this equation is $e^{0.01t}$. Multiply both sides of (3.104) by this function. The left-hand side of the resulting equation is an exact derivative of the integrating factor and the unknown:

$$\frac{d}{dt}\left(e^{0.01t}x(t)\right) = 50e^{0.01t}. \tag{3.105}$$

After integrating this equation and using the initial condition, we end up with the solution

$$x(t) = 5000 - 5000e^{-0.01t}. \tag{3.106}$$

The limiting value of $x(t)$ is 5,000, implying that eventually 5,000 grams of phosphate remains present permanently in the entire bay. It is interesting to note that we could have gotten this information directly from the differential equation without solving for the solution (3.106). Assuming that $x(t)$ has a limit as t goes to infinity, its derivative x' must become "flat" with $t \to \infty$. Therefore, the limit of x' is zero as t approaches infinity. Setting x' to zero in (3.104) leads to the equation $0.01x = 50$ for the limiting value of x, from which we deduce the same conclusion we reached from (3.106).

Example 3.5.7 (Bernoulli Equations)

Sometimes it is possible to convert a nonlinear differential equation to a linear form by a change of variables. The Bernoulli equation

$$x' + p(t)x = q(t)x^n \tag{3.107}$$

is an example of such an equation. Because the right-hand side of this equation depends on x^n, this equation is nonlinear if n is different from 0 or 1. The change of variables that renders (3.107) linear is

$$y = x^{1-n}. \tag{3.108}$$

Clearly this transformation is not useful when n is either 0 or 1. We leave it as an exercise to show that if x satisfies (3.108), then y satisfies the linear first-order equation

$$\frac{dy}{dt} + (1-n)p(t)y = (1-n)q(t). \tag{3.109}$$

The main tool in the computation that leads to (3.109) is the chain rule. This type of algebra can be readily performed in either *Mathematica* or *MATLAB*. Here is how one transforms the differential equation (3.107) to (3.109) in *Mathematica* (note that (3.108) is equivalent to $x = y^{\frac{1}{1-n}}$):

```
L[x_] := D[x, t] + p*x - q*x^n;
x = y[t]^(1/(1-n));
a1 = L[x];
a2 = Simplify[a1];
a3 = a2 / y[t]^(n/(1-n));
a4 = PowerExpand[a3];
Simplify[a4]
```

Here, L is defined using := rather than a simple equality. (Why?) The output of a2 suggests dividing it by $y^{\frac{n}{1-n}}$. The use of PowerExpand simplifies the exponent of one of the terms in a3, making it easier for Simplify to cancel the necessary terms.

The following program transforms (3.107) to (3.109) in *MATLAB*:

```
syms x y p q t n a
x = 'y^(1/(1-n))';
x=subs(x,y,'y(t)');
op=diff(x,t)+p*x-q*x^n;
a=op/'y(t)^(n/(1-n))';
simplify(a)
```

Apply this program with a fixed n, say $n = 3$, and compare the output with (3.109).

Problems

1. Use an editor to create a file that contains the program(s) in Example 3.5.2. Verify the findings in that example.

2. Verify the statements in Example 3.5.3.

3. Use DSolve in *Mathematica* or dsolve in *MATLAB* with (3.93). What is the response of these softwares? Now use NDSolve or ode45 with this equation to plot the solution curve on the interval $(0,6)$. How does the graph compare with the graph of the solution to $x' + \exp(-t^2)x = \sin t, x(0) = 1$ (see Figure 3.17)?

4. In each of the following problems find the general solution when the antiderivatives can be written in terms of elementary functions, apply the initial condition to determine the specific solution, and plot this solution in the given interval. When determining antiderivatives in the definition of $P(t)$ (see (3.82)) or evaluating the integral in (3.87) are prohibitive, use the numerical approximation discussed in Examples 3.5.3 and 3.5.4.

   (a) $3x' + 2x = e^{-t}, \quad x(0) = 1, t \in (0,5)$

   (b) $-x' + 2x = \sin t, \quad x(0) = 1, t \in (0,5)$

   (c) $x' + 2x = e^{-t} \sin t, \quad x(0) = 1, t \in (0,5)$

   (d) $x' + tx = \sin t, \quad x(0) = 1, t \in (0,3)$

   (e) $x' + \exp(-t^3)x = -1, \quad x(1) = -1, t \in (0,4)$

5. Apply the method of this section on the following initial value problems. In each case the constants a and b are positive.

   (a) $y' + ay = 1, \quad y(0) = 0$

   (b) $y' + ay = b, \quad y(0) = 0$. What happens to the solution y when x approaches infinity?

   (c) $xy' + y = x^2, \quad y(1) = a$

   (d) $xy' - y = x^2, \quad y(1) = a$. Draw the the graphs of the solutions with $a = 1$ and 2 on the same screen.

   (e) $\sin t \, z' + \cos t \, z = a \sin t, \quad z(\frac{\pi}{2}) = b$

6. Do the change of variables in Example 3.5.7, first by hand, and next using either *Mathematica* or *MATLAB* as described in this example.

7. Apply the change of variables described for the Bernoulli equation (see Example 3.5.7) to find the general solution to the following equations. In each case use *Mathematica*'s DSolve or *MATLAB*'s dsolve with each equation.

   (a) $y' + y = y^2$. What happens to the solutions as the independent variable approaches infinity?

   (b) $y' + ay = y^2$, where $a > 0$ is a constant. What is the asymptotic behavior of the solutions?

   (c) $y' + ty = y^2$, $y(0) = 1$. What happens to this solution for $t \in (0, 2)$?

8. Solve the following differential equations analytically. Check the answers using either DSolve in *Mathematica* or dsolve in *MATLAB*. Draw the graph of each solution for large values of t and determine from each graph what the behavior of the solution is as the independent variable approaches infinity.

   (a) $x' + x = 2e^{-t}$, $x(0) = 0$

   (b) $x' + 2x = 2e^{-t}$, $x(0) = 0$. What happens to the solution with initial condition $x(0) = -\frac{1}{3}$?

   (c) $x' + 2x = \sin t$, $x(0) = 0$

   (d) $x' = 3x + 2 - \cos t$, $x(0) = 0$

9. State carefully the hypotheses on p and q in (3.76) to guarantee the existence and uniqueness of solutions to this initial value problem.

10. A body of mass m falls to the ground from a height h under the influence of gravity. Suppose that the body experiences air resistance proprtional to its velocity. Show that the equation of motion of this body is (distance is measured positively downward)

$$m\frac{dv}{dt} = -kv + mg, \quad v(0) = v_0, \tag{3.110}$$

where v is the velocity of the body and k is constant. Find the solution to this initial value using the method in this section. Check the answer in either *Mathematica* or *MATLAB*. Find the terminal velocity (i.e., $\lim_{t \to \infty} v(t)$) of the body as a function of m and g.

11. A person of mass 70 kilogram is in a free fall from a height of 1500 meters. Suppose that the constant of proportionality of air resistance is 5 kg/s. What is the person's terminal velocity? At what height does the person reach 50% of his terminal velocity?

12. A water tank has a capacity of 200 liters and is filled with fresh water. It has two outlets through which fluid can flow in or out. At time zero a fluid mixture containing 2 grams of salt per liter enters the tank at a rate of 3 liters per minute through one of the outlets. The mixture is then stirred well and exits from the second outlet at the same rate of 3 liters per minute. Find the amount of salt present in the tank at any time. What is the steady-state solution of this problem?

13. A 100-gallon tank is initially half-full with fresh water. At time zero a fluid mixture containing 0.5 lbs of salt per gallon is added to the tank at the rate of 8 gallons per hour. At the same time the well-mixed fluid is removed from the tank at the rate of 6 gallons per hour. Find the amount of salt in the tank at the instant that the tank overflows. (Hint: Note that the volume of fluid inside the tank is a function of time. Observe that the term V in (3.103) should be replaced by $V(t) = 100 - (r_{in} - r_{out})t$.)

3.6 Exact Equations

An **exact** differential equation is a first-order equation of the form

$$M(t, x) + N(t, x)\frac{dx}{dt} = 0, \qquad (3.111)$$

where the functions M and N are related to a single function F through the relations

$$M = \frac{\partial F}{\partial t}, \qquad N = \frac{\partial F}{\partial x}. \qquad (3.112)$$

Exact equations appear naturally in many areas of mathematical physics, an example of which will be treated in Example 3.6.3. Analogous to separable and first-order linear equations, we are able to give an algorithm that constructs the general solution of exact equations. The structure of this algorithm lends itself well to both the symbolic as well as the graphics capabilities of *Mathematica* and *MATLAB*.

First, we show the relations in (3.112) lead to the general solution of the differential equation in (3.111). Let $x(t)$ be a solution of this equation. Evaluate the relations in (3.112) along this solution, and replace the functions M and N in the differential equation by their counterparts in terms of F. We have

$$\frac{\partial F(t, x(t))}{\partial t} + \frac{\partial F(t, x(t))}{\partial x}\frac{dx}{dt} = 0. \qquad (3.113)$$

Equation (3.113) is equivalent to (using the chain rule of differentiation)

$$\frac{d}{dt}(F(t, x(t))) = 0, \qquad (3.114)$$

from which we conclude

$$F(t, x(t)) = C, \qquad (3.115)$$

where C is a constant. The above equation provides us with an **implicit** formula for the solution $x(t)$ of (3.111).

It is worth noting that the pair of equations (3.111), (3.112) is formally equivalent to

$$\frac{\partial F}{\partial t}\, dt + \frac{\partial F}{\partial x}\, dx = 0. \tag{3.116}$$

We recognize the left-hand side of (3.116) as the (total) **differential** of F. So, (3.111) is equivalent to

$$dF = 0, \tag{3.117}$$

which brings us again to the conclusion that F remains constant along a solution of the differential equation (3.111).

Example 3.6.1

Consider the differential equation

$$\frac{dx}{dt} = \frac{2\,t\,x}{2x - t^2}. \tag{3.118}$$

This equation is exact. To see that, rewrite it as

$$2tx + (t^2 - 2x)\frac{dx}{dt} = 0, \tag{3.119}$$

and let $M(t, x) = 2tx$ and $N(t, x) = t^2 - 2x$. Note that the function $F(t, x) = -x^2 + t^2 x$ satisfies the relations in (3.112). It then follows from (3.115) that

$$-x^2 + t^2 x = C \tag{3.120}$$

is a general solution to (3.118).

The relations in (3.112) determine whether an equation is exact or not. To derive a necessary condition for exactness of a differential equation, differentiate these relations with respect to x and t, respectively:

$$\frac{\partial M}{\partial x} = \frac{\partial^2 F}{\partial x \partial t}, \qquad \frac{\partial N}{\partial t} = \frac{\partial^2 F}{\partial t \partial x}. \tag{3.121}$$

Recall that if F is a function whose second derivatives are continuous in a small neighborhood of a point $P = (t, x)$, then its mixed partial derivatives at P are equal. Let us assume the function F is such a function. Accordingly, if (3.111) is exact, then the appropriate partial derivatives of M and N in (3.121) must be equal, that is,

$$\frac{\partial M}{\partial x} = \frac{\partial N}{\partial t}. \tag{3.122}$$

This expression is a **necessary** condition for (3.111) to be exact. This identity gives us an analytic tool to test for this property. For example, in Example 3.111 $M(t, x) = 2tx$ and $N(t, x) = t^2 - 2x$. Applying (3.122) to this pair, we arrive at

$$\frac{\partial M}{\partial x} = 2t = \frac{\partial N}{\partial t}. \tag{3.123}$$

Now that we have a necessary condition to verify whether an equation is exact, how do we actually determine F? The relations in (3.112) are the defining equations for F. So, to determine F, we begin with the first equation in (3.112), namely,

$$\frac{\partial F}{\partial t} = M(t, x), \tag{3.124}$$

and integrate both sides with respect to t:

$$F(t, x) = f(x) + \int M(t, x)dt. \tag{3.125}$$

The function $f(x)$ in (3.125) is the constant of integration. (Why does it depend on x?)

Thus far, F is determined up to a function of a single variable. We use the second relation in (3.112), $N = \frac{\partial F}{\partial x}$, to find f. Differentiate (3.125) with respect to x and equate the result to N:

$$N(t, x) = f'(x) + \frac{\partial}{\partial x}\left(\int M(t, x)dt\right). \tag{3.126}$$

After rearranging (3.126), we find that f must satisfy

$$f'(x) = N(t, x) - \frac{\partial}{\partial x}\left(\int M(t, x)dt\right). \tag{3.127}$$

Finally, we need to integrate the right-hand side of (3.127) with respect to x and determine f. But what guarantee do we have that the right-hand side is independent of t? This is where the necessary condition (3.122) enters the analysis: For the right-hand side of (3.127) to be independent of t, its derivative with respect to that variable must be zero, or

$$\frac{\partial N}{\partial t} - \frac{\partial}{\partial t}\left(\frac{\partial}{\partial x}\left(\int M(t, x)dt\right)\right) \tag{3.128}$$

must vanish. Let us concentrate on the mixed derivatives in (3.128). Assuming the function M is continuously differentiable in the region of interest, we exchange the order of the partial derivatives in (3.128) and arrive at

$$\frac{\partial}{\partial t}\left(\frac{\partial}{\partial x}\left(\int M(t, x)dt\right)\right) = \frac{\partial}{\partial x}\left(\frac{\partial}{\partial t}\left(\int M(t, x)dt\right)\right) = \frac{\partial M}{\partial x}, \tag{3.129}$$

since $\frac{\partial}{\partial t}\left(\int M(t, x)dt\right) = M(t, x)$. Next, combine (3.128) and (3.129) and arrive at

$$\frac{\partial N}{\partial t} - \frac{\partial M}{\partial x} \tag{3.130}$$

for the derivative with respect to t of the right-hand side of (3.127). This expression vanishes because the differential equation is exact. Hence, the function f in (3.127) is truly a function of x alone and can be determined through a single integration of that expression. We have proved the following theorem.

Theorem 3.6.1

Suppose that $M(t, x)$ and $N(t, x)$ are continuously differentiable in a region (say, a rectangle) about a point (t, x). Then the differential equation

$$M(t, x) + N(t, x)\frac{dx}{dt} = 0 \qquad (3.131)$$

is exact if and only if

$$\frac{\partial M}{\partial x} = \frac{\partial N}{\partial t}. \qquad (3.132)$$

The algorithm for determining the function F described in the preceding discussion finds this function up to a constant of integration. In many applications the value of this constant is not significant and is often chosen to be zero.

Mathematica and *MATLAB* are quite helpful with solving exact differential equations. They help in carrying out the necessary symbolic calculations to find out if an equation is exact; if it is, implement the previously described algorithm to determine F. These programs are also quite useful in plotting the graphs of solutions, since the implicit solution to (3.111) is given by (3.115), or in other words, a trajectory (t, x) of (3.111) is a contour or level curve of the function F. Therefore, when a formula for F is available, we can use `ContourPlot` in *Mathematica* or `contour` in *MATLAB* and draw typical trajectories of the differential equation.

Example 3.6.2

Consider the differential equation

$$x' = \frac{2t + x}{3x - t}. \qquad (3.133)$$

Here $M(t, x) = 2t + x$ and $N(t, x) = t - 3x$. Since $\frac{\partial M}{\partial x} = \frac{\partial N}{\partial t} = 1$, (3.133) is exact. We need to determine $F(t, x)$ such that $F_t = M$ and $F_x = N$, or

$$\frac{\partial F}{\partial t} = 2t + x, \qquad \frac{\partial F}{\partial x} = t - 3x. \qquad (3.134)$$

Integrate the first equation in (3.134) with respect to t:

$$F(t, x) = t^2 + tx + f(x). \qquad (3.135)$$

Next, differentiate F with respect to x and equate the result to the right-hand side of the second relation in (3.134), from which we obtain

$$f'(x) = -3x. \qquad (3.136)$$

So $f(x) = -\frac{3}{2}x^2$ (let the integration constant be zero) and

$$F(t, x) = t^2 + tx - \frac{3}{2}x^2. \qquad (3.137)$$

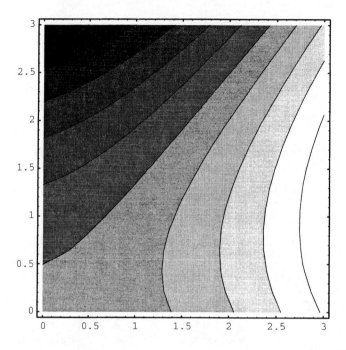

Figure 3.19: Solution curves of the differential equation $x' = \frac{2t+x}{3x-t}$.

The solution curves (trajectories) of (3.133) are the level curves of this function (see Figure 3.19).

All of this analytic work can be carried out in *Mathematica* or *MATLAB*. The program in *Mathematica* that begins with M and N and ends up with Figure 3.19 is as follows:

```
m = 2 t + x;  n = t - 3 x;
a=TrueQ[D[m, x] == D[n, t]];
If[a == True, Print["The equation is exact"];
          F = Integrate[m, t] + f[x];
          derfx = Solve[D[F, x] == n, f'[x]];
          f[x] = Integrate[f'[x] /. derfx, x];
          F = First[F]; Print["F = ", F];
          Print["Trajectories of F ..."];
          output = ContourPlot[F, {t, 0, 3}, {x, 0, 3}],
          Print["The equation is not exact"]
];
```

Remark 3.6.1: The second line of the program tests to see if the differential equation with the given M and N is exact. The third line determines F when the equation is exact or returns "The equation is not exact" otherwise. When the differential equation is exact, the value of a is True . In that case, the next several statements in If are executed. Note also that these lines are all separated by a semicolon and not a comma, so they get executed before the True part of the If statement is complete. The Print statements are added because a semicolon at the end of a line suppresses the output from being printed on the screen.

A small number of alterations to the above program leads to a program that is interactive and more robust when it is run repeatedly in the same session in *Mathematica*. The new feature in the following improved version is the usage of Block:

```
Block[{m,n,F,a,f,output,derfx,b1,b2,c1,c2,t,x},
  m = Input["Input m  "]; n = Input["Input n  "];
  a=TrueQ[D[m, x] == D[n, t]];
  If[a == True, Print["The equation is exact"];
    F = Integrate[m, t] + f[x];
    derfx = Solve[D[F, x] == n, f'[x]];
    f[x] = Integrate[f'[x] /. derfx, x];
    F = First[F]; Print["F = ", F];
    Print["Trajectories of F ...."];
    {b1, b2} = Input["Input the t interval (e. g., {1, 2}) "];
    {c1, c2} = Input["Input the x interval  "];
    output = ContourPlot[F, {t, b1, b2}, {x, c1, c2}],
    Print["The equation is not exact"]
  ];
]
```

The following program in *MATLAB* starts with M and N and determines the function F by carrying out the algorithm we described previously. Once F is known, we can then generate an M-file that defines F as a numeric function to *MATLAB*. To draw the contours of F is then a simple matter.

```
syms m n x t f F
m = '2*t+x'; n = 't - 3*x';
test=diff(m,'x')-diff(n,'t');
F=int(m,'t')
derf = n-diff(F,'x');
f=int(derf,'x');
F=F+f
```

It is not possible to use the contour command directly on F since F is a symbolic function and contour requires a numeric object for its input. In this example F is t^2 + x*t-3/2*x^2. So, we create an M-file, with the name FF.m, and containing the following lines:

```
function y=FF(t,x);
y = t.^2 + x.*t-3/2*x.^2;
```

We are now in a position to use contour in *MATLAB*. First we use meshdom and define the domain:

```
[T,X]=meshdom(0:0.1:3,0:0.1:3);
```

and then use `contour`:

```
contour(FF(T,X), 10)
```

The output shows 10 contoues of `FF`. Alternatively, we could define the domian of `T` and `X` as follows:

```
[T,X]=meshdom(0:0.1:3,3:-0.1:0);
```

with the remaining lines unchanged. The contours of the new output more closely resemble the output of *Mathematica.*

Example 3.6.3 (Exact Equations and Stream Functions)

Fluid flows are characterized by velocity fields $\mathbf{v} = \langle v_1, v_2, v_3 \rangle$, where v_i, the ith component of the velocity vector $\mathbf{v}$, is the derivative of the ith component of position with respect to time. In Section 3.1 we introduced an example of the velocity field of a flow past a cylinder of radius one. The trajectories of this flow, being two-dimensional, are solution curves of the system of differential equations

$$x' = 1 + \frac{y^2 - x^2}{(x^2 + y^2)^2}, \qquad y' = -\frac{2xy}{(x^2 + y^2)^2}, \tag{3.138}$$

where $\langle x(t), y(t) \rangle$ is the position vector of a given particle. Let $y(x)$ be a particle path in this flow. Since

$$\frac{dy}{dx} = \frac{dy}{dt} \bigg/ \frac{dx}{dt} = \frac{y'}{x'}, \tag{3.139}$$

using the expressions in (3.138) and (3.139), y satisfies the first-order differential equation

$$\frac{dy}{dx} = \left(\frac{-2xy}{(x^2 + y^2)^2} \right) \bigg/ \left(1 + \frac{y^2 - x^2}{(x^2 + y^2)^2} \right). \tag{3.140}$$

Compare (3.140) with (3.111). In (3.140) x and y are playing the roles of t and x, respectively. Hence, M and N are functions of x and y and are defined by

$$M(x, y) = \frac{2xy}{(x^2 + y^2)^2}, \qquad N(x, y) = 1 + \frac{y^2 - x^2}{(x^2 + y^2)^2}. \tag{3.141}$$

It is not difficult to see that (3.140) is exact and that $F(x, y)$ is given by

$$y - \frac{y}{x^2 + y^2}. \tag{3.142}$$

(see Problem 11 at the end of this section). This function is called the Stokes stream function of the flow past the cylinder, and following (3.115), its contours give the particle paths of flow.

It is sometimes possible to convert a nonexact differential equation to an exact one by multiplying the equation by an **integrating factor**. For instance, the equation

$$2x + tx' = 0 \tag{3.143}$$

is not exact because $M = 2x$, $N = t$, and $M_x - N_t = 1$ rather than zero. But, if we multiply this equation by the integrating factor $\frac{1}{\sqrt{x}}$, we get a new differential equation

$$\frac{2x}{\sqrt{x}} + \frac{t}{\sqrt{x}}x' = 0, \tag{3.144}$$

which is exact, and the theory developed in this section will go through. So, even though (3.143) is not exact, we are still able to obtain a general solution of this equation by "guessing" an integrating factor for it.

Equation (3.143) had an integrating factor that depended on a single variable (in this case x). As it turns out this fact is crucial in guessing the form of the integrating factor and serves as a basis for constructing an algorithm that is quite useful for a class of nonexact differential equations. The argument proceeds as follows. Consider the differential equation

$$M(t, x) + N(t, x)x' = 0, \tag{3.145}$$

which may not be exact. Multiply this equation by a differentiable function (and as yet undetermined) $\rho(t, x)$:

$$\rho(t, x)M(t, x) + \rho(t, x)N(t, x)x' = 0. \tag{3.146}$$

A necessary condition for (3.146) to be exact is that

$$(\rho M)_x = (\rho N)_t, \tag{3.147}$$

which is equivalent to

$$\rho_x M - \rho_t N = (N_t - M_x)\rho. \tag{3.148}$$

In general, (3.148) is a PDE in ρ, whose solution is often difficult to find. But we can find a solution to this equation in certain special cases. The simplest case is when (3.145) is already exact, in which case the right-hand side of (3.148) is automatically zero and we can choose $\rho = 1$ as a solution to the latter equation. In other words, no integrating factor in necessary when the original differential equation is exact. Another special case occurs when (3.145) has a form that leads to a single-variable integrating factor. For example, under what conditions on M and N is ρ a function of x alone? To answer this question we go back to (3.148). If $\rho = \rho(x)$, then $\rho_t = 0$ and (3.148) reduces to the ODE

$$\rho' = \frac{(N_t - M_x)}{M}\rho, \tag{3.149}$$

where the prime denotes differentiation in x. Since the left-hand side of (3.149) is independent of t, the right-hand side should be as well. So, assuming that this condition on M and N holds, we find that ρ satisfies the separable differential equation (3.149), whose solution is

$$\rho(x) = \exp\left(\int \frac{N_t - M_x}{M}dx\right). \tag{3.150}$$

Equation (3.143) is an example of a nonexact equation for which the right-hand side of (3.149) is indenpendent of x.

Returning to (3.148), if the integrating factor ρ is only a function of t, then this equation reduces to

$$\rho' = \frac{(M_x - N_t)}{N} \rho, \tag{3.151}$$

where now the prime denotes differentiation in t. Again, the constraint on ρ forces the right-hand side of (3.151) to be independent of x. When that is the case we are able to determine ρ by a single integration:

$$\rho(t) = \exp\left(\int \frac{M_x - N_t}{N} dt\right). \tag{3.152}$$

Example 3.6.4
Consider the differential equation

$$x' = \frac{tx}{t^2 - x^2}. \tag{3.153}$$

First, we write this equation as

$$tx + (x^2 - t^2)\frac{dx}{dt} = 0 \tag{3.154}$$

and let $M = tx$ and $N = x^2 - t^2$. Since $M_x - N_t = 3t$, (3.153) is not exact. But, the ratio

$$\frac{M_x - N_t}{M} = \frac{3}{x} \tag{3.155}$$

is independent of t. Hence, following (3.150), the integrating factor for (3.153) is

$$\rho(x) = \exp\left(\int -\frac{3}{x} dx\right) = \frac{1}{x^3}. \tag{3.156}$$

Multiply (3.154) by ρ to get the differential equation

$$\frac{t}{x^2} + \frac{x^2 - t^2}{x^3}\frac{dx}{dt} = 0, \tag{3.157}$$

which is exact. The function F for (3.157) is $\frac{t^2}{2x^2} + \ln(x)$, so an implicit solution of (3.153) is given by

$$\frac{t^2}{2x^2} + \ln(x) = C. \tag{3.158}$$

The idea of an integrating factor for a nonexact differential equation is also very useful for another class of differential equations, those for which $\rho(x, y) = \rho(x^2 + y^2)$, that is, when ρ depends on x and y through the polar radius $r = \sqrt{x^2 + y^2}$. We have chosen x and y as the independent and dependent variables for this class because such equations often represent steady-state motion of a fluid in which time does not explicitly enter in the formulation of the problem. In this setting consider a possibly nonexact differential equation

$$M(x, y) + N(x, y)\frac{dy}{dx} = 0. \tag{3.159}$$

We saw in (3.148) that the integrating factor $\rho(x, y)$ must satisfy

$$\rho_y M - \rho_x N = (N_x - M_y)\rho. \tag{3.160}$$

Now, suppose that

$$\rho(x, y) = \rho(R), \qquad \text{with } R = x^2 + y^2. \tag{3.161}$$

Apply the chain rule to ρ and obtain the relations

$$\rho_x = 2x\rho_R, \qquad \rho_y = 2y\rho_R.$$

Equation (3.160) then becomes

$$\frac{d\rho}{dR} = \frac{N_x - M_y}{2yM - 2xN}\rho. \tag{3.162}$$

This equation in $\rho(R)$ is an ODE if

$$\frac{N_x - M_y}{2yM - 2xN} \tag{3.163}$$

is a function of R alone. In that case (3.162) is separable and its solution is

$$\rho(x, y) = \exp\left(\int \frac{N_x - M_y}{2yM - 2xN}\, dR\right). \tag{3.164}$$

Example 3.6.5
Consider the differential equation

$$\frac{dy}{dx} = \frac{-2xy}{y^2 - x^2 + (x^2 + y^2)^2}. \tag{3.165}$$

Here $M = 2xy$ and $N = y^2 - x^2 + (x^2 + y^2)^2$. Note that $M_y = 2x$ and $N_x = -2x + 4x(x^2 + y^2)$, so (3.165) is not exact. However,

$$\frac{N_x - M_y}{2yM - 2xN} = -\frac{2}{x^2 + y^2} = -\frac{2}{R}. \tag{3.166}$$

Hence, $\rho(R)$ satisfies

$$\frac{d\rho}{dR} = -\frac{2\rho}{R}, \tag{3.167}$$

whose solution is $\rho(R) = \frac{1}{R^2} = \frac{1}{(x^2+y^2)^2}$. Thus, multiplying (3.165) by this integrating factor renders it an exact equation. It is interesting to note that (3.165) is equivalent to (3.140) before the latter was simplified by eliminating the integrating factor.

Problems

1. Create either of the two *Mathematica* or the *MATLAB* programs in this section and execute them with the differential equation in Example 3.6.2 and generate Figure 3.19.

2. Show that the following equations are exact, determine the function F in each case, and draw some of the typical solution curves.

   (a) $\frac{dx}{dt} = -\frac{t}{x}$

   (b) $\frac{dx}{dt} = -\frac{2t+x}{t}$

   (c) $\frac{dy}{dx} = \frac{x^2}{y^2}$

   (d) $\frac{dx}{dt} = \frac{\sin t \sin x}{\cos t \cos x}$

   (e) $y' = \frac{2xy}{y^2 - x^2}$

   (f) $(x + \frac{1}{2}e^{2x} \sin y)\frac{dy}{dx} - e^{2x} \cos y + y = 0$

   (g) $(\sin y + \frac{1}{4}e^{-4x} \cos y)\frac{dy}{dx} - e^{-4x} \sin y + 3x = 0$

   (h) $\cosh x \sin y + \sinh x \cos y \frac{dy}{dx} = 0$

   (i) $\sinh t \sin x + t^3 + (\cosh t \cos x - x)\frac{dx}{dt}$.

3. Use the integrating factor method to find a general solution to each of the following equations.

   (a) $t - x + 2tx' = 0$

   (b) $t + x + 2xx' = 0$

   (c) $t^2 + (x^2 - t)x' = 0$

   (d) $y' = \frac{y^2 - x^2 - (x^2+y^2)^2}{2xy}$. (Hint: Look for an integrating factor ρ as a function of $R = x^2 + y^2$.)

   (e) $y' = \frac{4\,x\,y}{2\,x^2 - 3\,x^4 - 2\,y^2 - 6\,x^2\,y^2 - 3\,y^4}$

4. Show that any separable equation of the form

   $$\frac{dx}{dt} = \frac{f(t)}{g(x)}$$

   is also exact.

5. Show that any first-order linear homogeneous differential equation is exact.

6. Consider the differential equation $M(x, y) + N(x, y)\frac{dy}{dx} = 0$. Suppose that this equation is not exact.

   (a) Find conditions on M and N so that the equation has an integrating factor ρ that is a function of xy.

   (b) Find conditions on M and N so that the equation has an integrating factor ρ that is a function of $\frac{x}{y}$.

   (c) Find conditions on M and N so that the equation has an integrating factor ρ that is a function of $\frac{y}{x}$.

   (d) Find conditions on M and N so that the equation has an integrating factor ρ that is a function of $x - y$.

7. A two-dimensional fluid flow's velocity field is characterized by the equations

$$v_1(x, y) = \frac{\partial \psi}{\partial y}, \quad v_2(x, y) = -\frac{\partial \psi}{\partial x},$$

where the function ψ is called the stream function of such a flow.

   (a) Starting with the relationship $\langle \frac{dx}{dt}, \frac{dy}{dt} \rangle = \langle v_1, v_2 \rangle$ bewteen the velocity and the position of a particle, show that a typical particle path $y(x)$ satisfies the exact differential equation

$$\frac{\partial \psi}{\partial x} + \frac{\partial \psi}{\partial y}\frac{dy}{dx} = 0.$$

   (b) Show that the particle path of a typical fluid particle is a level curve of the stream function ψ.

   (c) Use `ContourPlot` in *Mathematica* or `contour` in *MATLAB* and draw the particle paths of the following stream functions:

      i. $\psi(x, y) = \sin \pi x \sin \pi y$ in the region $\{(x, y)|\, 0 < x < 3, 0 < y < 1\}$. This flow is known as the Rayleigh–Bénard flow. Determine the level curve $\psi(x, y) = 0$ by hand. Did *Mathematica* or *MAT-LAB* draw this level curve?

      ii. $\psi(x, y) = \sinh y \sin x$ in the region $\{(x, y)|\, -10 < x < 0, 0 < y < \pi\}$. These contours model the flow in a rectangular bay. Determine the level curves $\psi(x, y) = 0$ by hand.

      iii. $\psi(x, y) = (-1 + 0.220715\, e^{-100.249\, x} + 0.779285\, e^{0.249378\, x}) \sin \pi y$ in the rectangular region $\{(x, y)|\, 0 < x < 1, 0 < y < 1\}$. This stream function was first derived by H. Stommel in 1948 as a model that exhibits many of the attributes and familiar features of the Gulf stream. The model takes into account the Coriolis force as well as the prevalent wind curl in the Atlantic ocean. Figure 3.20 shows *Mathematica*'s output.

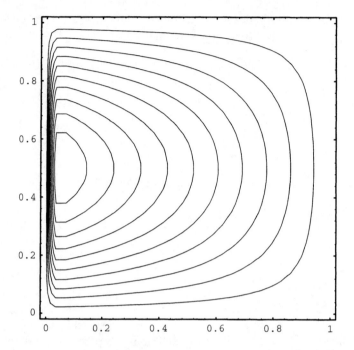

Figure 3.20: The output of *Mathematica* using the "Gulf Stream" stream function. This figure was obtained by setting the `PlotPoints` option to 30 and `ContourShading` to `False`.

8. Find the syntax errors in the following *Mathematica* statements:

    (a) `ContourPlot[ x^2 + t, {t, 0, 1, x, 0, 1}];`

    (b) `m = x + t; n == x- t; TrueQ[D[m, x] = D[n, t]];`

    (c) `Solve[Sin[x] - x = 0, x];`

    (d) `If[t < 0, D[Sin[x], x];`

9. Find the syntax errors in the following *MATLAB* statements:

    (a) `f='x+t'; subs(f, 'x', 2);`

    (b) `m='exp(x) sin(t)'; der = diff(m, x);`

    (c) `(t,x) = meshdom(0:0.1:1;-1:0.1:1);`

10. Write a *Mathematica* or *MATLAB* program that begins with the functions M and N of (3.111) and checks first to see if the equation is exact or not. If it is exact, the program should print a sentence to that effect. If the equation is not exact, it should proceed to check whether there is an integrating factor that depends on a single variable alone and report its findings.

11. Verify the statements in Example 3.6.3.

3.7 Project A—field.m and *xprime.m* in *MATLAB*

In this project we develop the necessary code in *MATLAB* to draw the direction field of a first-order differential equation of the form

$$x' = f(t, x).$$

Use an editor to create and save the following M-files in the main *MATLAB* directory. These M-files are developed for (3.33) but can be readily modified to accommodate any first-order differential equation.

The reader may not be familiar with some of the syntax in `field.m` at this point because they deal with matrix manipulation. The code is written in such a way to take advantage of *MATLAB*'s special computational ability when dealing with vectors.

The M-file `field.m` is as follows:

```
clf;
clear all;
%
%    window size parameters
%
a1=0;  a2=3;  b1=-2;  b2=2;
%
%    increments in t and x directions
%
dt=0.3;  dx=0.2;
%
%    discretized t and x axes
%
t=a1:dt:a2;  x=b1:dx:b2;
%
axis([a1 a2 b1 b2])
hold on
T=ones(size(x))'*t;
X=x'*ones(size(t));
M=xprime(T,X);
Scale=0.1./(2*sqrt(1+M.*M));
Tleft=T-Scale;Tright=T+Scale;
newScale=M.*Scale;
Xleft=X-newScale;Xright=X+newScale;
newTleft=Tleft(:);newTright=Tright(:);
newT=[newTleft newTright];
newXleft=Xleft(:);newXright=Xright(:);
newX=[newXleft newXright];
plot(newT',newX')
```

The four parameters `a1`, `a2`, `b1`, `b2` define the window size. `dt` and `dx` define the increments in the t and x directions where the tangent lines are drawn. The line that begins with `Scale` scales each tangent line to have a uniform length of 0.1. All of these parameters can be altered to get a picture that best suits the differential equation under study. The M-file `field.m` calls on the M-file `xprime.m`, which is written specifially for the differential equation

$$x' = t + x^2.$$

`xprime.m`'s listing is as follows:

```
function prime=xprime(t,x);
prime=t+x.*x;
```

Assuming these files are saved in the main directory of *MATLAB*, we obtain the direction field in Figure 3.10 by executing

```
field
```

in *MATLAB*.

3.8   Project B—Motion in Earth's Gravitational Field

In the examples of motions of falling bodies that we have cited so far we have always assumed that the acceleration due to gravity is constant. This assumption is reasonable for objects that are close to the surface of the Earth, but it is not a valid representation of the gravitational force an object experiences when it is sufficiently far away from the Earth's surface. In fact, Newton's law of gravitation states that two objects with masses M and m exert a force $\mathbf{F}$ on each other whose direction is along the line that passes through the masses and whose magnitude is given by

$$F = \frac{GMm}{r^2}, \tag{3.168}$$

where G is the gravitational constant and r is the distance between the masses. This distance is measured between the centers of mass of the bodies. An object is relatively near the surface of the Earth if its distance to the surface of the Earth is much smaller than the radius of the Earth. In such a case we make the following approximation in (3.168):

$$F = \frac{GMm}{r^2} \approx \frac{GMm}{R^2} \equiv mg, \tag{3.169}$$

where R is the Earth's radius, M its mass, m is the mass of the body, and we have neglected the distance of the body to the surface of the Earth in deference to the radius of our planet. The last statement in (3.169) defines g in terms of the other parameters:

$$g = \frac{GM}{R^2}.\tag{3.170}$$

The parameter g equals 9.8 m/s^2 when we use the standard values for G, M, and R.

1. Let $a = \frac{dv}{dt}$ be the acceleration of an object of mass m moving in a vertical direction with respect to the surface of the Earth. Let $x(t)$ be its distance from the center of the Earth. Use Newton's second law to show that x must satisfy (no air resistance)

$$m\frac{d^2 x}{dt^2} = \pm\frac{GMm}{x^2}.\tag{3.171}$$

Explain the relation between the direction of motion and the appropriate sign in (3.171). What is the order of this differential equation? Is it linear or nonlinear? How many initial conditions are needed to find the solution to this equation? What does the existence and uniqueness theorem (Theorem 3.3.1) say about this equation?

2. Consider the differential equation (3.171) with the minus sign. Show that v satisfies

$$\frac{dv}{dt} = -\frac{GM}{x^2}, \quad v(0) = v_0.\tag{3.172}$$

Assume that the object is located on the surface of the Earth at time zero, so that $x(0) = R = 6370$ kilometers. Show that this equation is equivalent to the first-order equation

$$v\frac{dv}{dx} = -\frac{GM}{x^2}, \quad v(R) = v_0.\tag{3.173}$$

(Hint: Assume that $v = v(x(t))$ and use the chain rule on this expression.) Identify the type of the differential equation in (3.173) (such as separable or first order). Use the appropriate method from this chapter and solve this equation. Check your answer using *Mathematica* or *MATLAB*.

3. The **escape velocity** of a rocket launched from the surface of the Earth is by definition the initial velocity v_0 for which $v(t)$ remains positive for all time. Show that the escape velocity on Earth is given by

$$v_0 = \sqrt{2Rg}.\tag{3.174}$$

Find a numerical value for v_0.

4. Try two different values for v_0: 5 km/s below the escape velocity and 1 km/s above it. Draw the graphs of both $v(t)$s on the same screen. Compare your graph with Figure 3.21.

5. Use `Solve` of *Mathematica* or `solve` of *MATLAB* to determine the value of x at which each velocity vanishes.

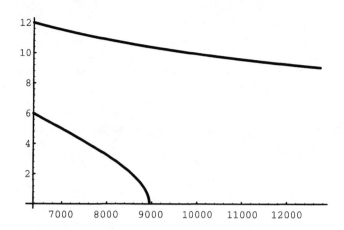

Figure 3.21: The graphs of the velocity of two rocket launches with initial velocities 6 and 12.

3.9 Project C—Picard's Method in *Mathematica* and *MATLAB*

In Section 3.3 we stated a theorem on the existence and uniqueness of solutions of initial value problems for a rather large class of ODEs. We did not attempt to prove this theorem since its details are somewhat beyond the scope of this text, although the general idea behind the proof is quite simple and lends itself to a useful algorithm for obtaining an approximate solution of differential equations. In this project (which is continued of the next chapter with another project in the same spirit) we outline the basic concept of the proof and show how one can take advantage of the symbolic power of *Mathematica* and *MATLAB* and program portions of this proof.

Let us consider the initial value problem

$$x' = f(t, x), \quad x(t_0) = x_0. \tag{3.175}$$

Integrate this equation to get the **integral equation**

$$x(t) = x_0 + \int_{t_0}^{t} f(s, x(s)) \, ds. \tag{3.176}$$

We have not accomplished much so far, other than to convert one equation into another. Let $T[x]$ denote the "operator" on the right-hand side of (3.176)

$$T[x] = x_0 + \int_{t_0}^{t} f(s, x(s)) \, ds. \tag{3.177}$$

The domain of this operator consists of functions x for which the integration on the right-side of (3.177) is feasible. Note that in terms of T, our original differential equation is now equivalent to

$$x(t) = T[x(t)]. \tag{3.178}$$

Typically, x and $T[x]$ will turn out to be different functions, but (3.178) requires that we seek a function x that remains unchanged under the action of T. For this reason we say x is a **fixed point** of the operator T.

The fundamental idea behind Picard's method for obtaining the solution to (3.175) is to generate the fixed point of (3.178) by an **iteration** process. One starts by guessing what the solution may be, say $x(t) = x_1(t)$. The first guess in then substituted in T in (3.177). We perform the integration and get

$$x_2(t) = T[x_1(t)] \tag{3.179}$$

(compare with (3.178)). We hope that x_2 is a better approximation to the fixed point of T than x_1 is. To get an even better approximation, we repeat the process and evaluate T at x_2 to get a third approximate of the solution:

$$x_3(t) = T[x_2(t)]. \tag{3.180}$$

Continuing in this fashion we generate a sequence of functions

$$x_1(t), \quad x_2(t), \quad x_3(t), \quad \dots, \quad x_n(t), \quad \dots \tag{3.181}$$

The hope is that as n gets large, the sequence x_n gets progressively closer to the exact solution. The bulk of the proof of the existence and uniqueness theorem lies in showing that the sequence $\{x_n(t)\}$ converges to a solution of (3.175) in an appropriate sense and under the hypotheses on f that we stated in Theorem 3.3.1.

In this project we implement the above algorithm for the differential equation

$$x' + ax = f(t), \quad x(0) = x_0. \tag{3.182}$$

1. Determine the exact solution of this initial value problem. Check the answer in either *Mathematica* or *MATLAB*.

2. The following program in *Mathematica* starts with the input data of a, f, x_0 and an initial guess of $x_1(t) = 7 + 8t$ and determines the first twelve iterates of T. This operator is defined as op in the program.

The logic of the program directly follows the argument we presented above. We have used a few programming tools to make the output of the program easier to read. First, the expression If[Mod[i,3] < 1, ...] enables us to display every third output of the program produced by Do. The expressions ToString and StringJoin allow us to pass information to the label of the graphs that will be displayed later. The term DisplayFunction -> Identity prohibits the graphs from being displayed until we are ready to see them. With Partition[graph,2] we are able to partition the contents of the list graph into pairs of graphs that are later processed by GraphicsArray. Finally, DisplayFunction -> $ DisplayFunction displays the graphs that have heretofore been hidden from us.

3. Use an editor and create a file called `picard.m` that contains the following lines. Study each line very carefully and make sure that the logic and the syntax of each statement is clear to you.

```
Print["This program implements Picard's"]
Print["Method for the differential equation"]
Print["x' + a x = f(t)"]
Print["with initial condition x(0) = x0."]
Print[""]
Print["In this example"]
Print["x0 = 1; a = 2; f(t) = sin(t)"];
x0 = 1; a = 2; f[t_] = Sin[t];
op[x_] := x0 - a*Integrate[x, {s, 0, t}] + Integrate[f[s],
        {s, 0, t}];
exact = DSolve[{x'[t] + a*x[t] == f[t], x[0] == x0}, x[t], t];
x = 7+8 t;
xx=ToString[x];
aa=ToString[a];
ff=ToString[f[t]];
Print[""]
g=x /. t->s;
Print["Start of Iteration ...."]
Print["Initial guess:  x[t] =", x]
graph={};
Do[S = op[g];
    si = ToString[i];
    label=StringJoin["Exact and Approx solns when i = ", si];
    If[Mod[i,3] < 1,graph=Append[graph,
            Plot[{x[t]/. exact, S}, {t, 0, 1},
                PlotLabel->label,
              PlotRange->All, DisplayFunction->Identity]]
    ];
    g = S /. t-> s,
{i, 12}]
graph1 = Partition[graph,2];
llabel=StringJoin["Picard's Method for x' + ",aa,
    " x = ", ff, "
    and initial guess ",xx]
output=Show[GraphicsArray[graph1],
    DisplayFunction->$DisplayFunction,
    PlotLabel->llabel]
```

Execute this program and get a hardcopy of its output (compare with Figure 3.22).

4. Change the appropriate lines in the program so that 16 iterations are carried out and after every 4 iterations the output is displayed by `GraphicsArray`.

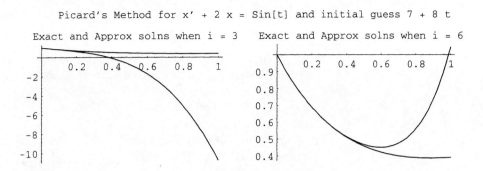

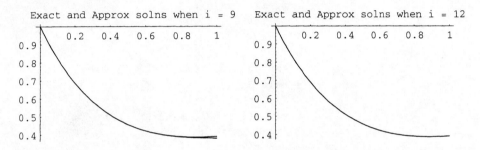

Figure 3.22: Picard's method applied to the differential equation $x' + 2x = \sin t$, $x(0) = 1$.

5. Revise the program so that the initial data can be specified in the form

$$x(t_0) = x_0,$$

where t_0 is not necessarily zero. Add this information to the the figure caption.

6. The program is written to plot the graphs on the interval $(0, 1)$. Revise the program to draw the graphs on the interval $(0, 2)$. Run the program and compare the results with Figure 3.22. What can one conclude about the quality of convergence of Picard's method in this case? Try graphing the outputs on the interval $(0, 3)$? What seems to be happening?

7. Apply this program with the initial guess $x_1(t) = 0$. Get a hardcopy of the graph when $x(t)$ is drawn on the intervals $(0, 1)$, $(0, 2)$, and $(0, 3)$. Compare these results with the case when the initial guess was $x_1(t) = 7 + 8t$. Does it seem that the choice of the initial guess has an effect on how fast the method converges?

8. Revise the program in order to apply it to

$$x' + x^2 = 0, \quad x(0) = 1. \tag{3.183}$$

Experiment with several initial choices of x_1 and report on the rate of convergence of the algorithm.

9. The following *MATLAB* program performs essentially the same tasks as the previously described *Mathematica* program.

```
syms s f diffeqn x f1 x0
x0='1';
f='7+8*t';diffeqn='Dx+2*x=sin(t)';
x=dsolve(diffeqn,'x(0)=c');
x=subs(x,x0,'c');
f1=subs(f,'s','t');
for i=1:7
    f1=-2*f1+sin(s);
    f1=int(f1,'s',0,'t');
    f1=x0+f1;
    subplot(4,2,i);
    ezplot(x,[0,1]);
    hold on
    ezplot(f1,[0,1]);
    title(['i = ', num2str(i)]);
    f1=subs(f1,'s','t');
end
```

Use this program in place of the *Mathematica* program to answer the questions raised in this project.

References

[1] Boyce, W. E., and R. C. DiPrima, *Elementary Differential Equations and Boundary Value Problems*, John Wiley and Sons, New York, 1986.

[2] Nagle, R. K., and E. B. Saff, *Fundamentals of Differential Equations*, Addison–Wesley, Reading MA, 1986.

[3] Edwards, C. H., and D. E. Penney, *Elementary Differential Equations with Boundary Value Problems*, 3rd Edition, Prentice-Hall, Englewood Cliffs, NJ, 1993.

[4] Hale, J. K., *Ordinary Differential Equations*, John Wiley and Sons, New York 1974.

Chapter 4

Linear Differential Equations of Higher Order

4.1 Introduction

In this chapter we study differential equations of the form

$$\frac{d^n x}{dt^n} + a_{n-1}(t)\frac{d^{n-1}x}{dt^{n-1}} + \cdots + a_1(t)\frac{dx}{dt} + a_0 x(t) = f(t), \qquad (4.1)$$

where x is the dependent variable and the **coefficients** a_i and the **forcing term** f are given functions of their arguments. The main part of the chapter is concerned with **constant coefficient** differential equations, where the a_is in (4.1) are real numbers. We will develop a rather complete picture of the behavior of solutions of these equations, at least when n is not too large. Another segment of the chapter will be devoted to the solution of **nonconstant coefficient** differential equations, where power series approximations and numerical methods constitute the main tools of analysis.

Equation (4.1) is often complemented by a set of n initial conditions

$$\frac{d^{n-1}x}{dt^{n-1}}|_{t=t_0} = x_{n-1}, \ldots, \frac{dx}{dt}|_{t=t_0} = x_1, \, x(t_0) = x_0. \qquad (4.2)$$

One of our goals is to determine the solution x to the initial value problem (4.1), ch4intro.1a) and study the behavior of $x(t)$ for various choices of the coefficients and forcing terms.

The bulk of the chapter will be restricted to **second-order** equations of the form

$$x'' + p(t)x' + q(t)x = f(t), \qquad (4.3)$$

although many of the ideas we develop for (4.3) generalize in a straightforward manner to higher order equations.

We saw in the previous chapter (see Theorem 3.3.1) that as long as the coefficients $a_{n-1}, \ldots, a_0$ and the forcing term f in (4.1) and p, q, and f in (4.3)

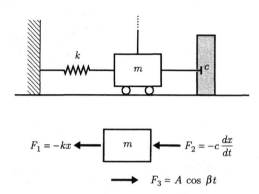

Figure 4.1: A mass-spring-dashpot system.

are continuous functions of their arguments, there is a unique solution to any initial value problem we pose for these equations. The objective of this chapter is to determine this solution analytically when possible and approximately or numerically in cases where it is too tedious or impossible to find the analytic solution.

We call (4.1) **homogeneous** if $f(t) \equiv 0$. Otherwise, the equation is called **nonhomogeneous**. An important property of linear homogeneous equations is that a linear combination of its solutions constitutes another solution of this equation; that is, if $x_1(t)$, $x_2(t), \ldots, x_m(t)$ are solutions of (4.1) or (4.3) with $f \equiv 0$, then so is the function $x(t)$ given by

$$x(t) \equiv c_1 x_1(t) + c_2 x_2(t) + \cdots + c_m x_m(t), \tag{4.4}$$

for any arbitrary set of constants $c_1, c_2, \ldots, c_m$. This statement is false in general for **nonlinear** as well as nonhomogeneous differential equations.

Our strategy in determining the solution to the homogeneous case of (4.1) or (4.3) is to find certain special solutions of these equations and to generate the **general solution** of the equation as a linear combination of these solutions. If there is a sufficient number of such solutions, we are then able to choose the constants in (4.4) appropriately so that all of the initial conditions in (4.2) are satisfied. We will refer to this method as the **principle of superposition**.

We have already encountered examples of higher order differential equations in the previous chapter. These equations appear naturally in mechanical settings where Newton's second law of motion is relevant or in the description of an electrical circuit where a capacitor, a resistor, and an inductor are present. The following two examples review the typical way second-order equations appear in the forementioned settings.

Example 4.1.1 (Mass-Spring-Dashpot System)

Consider a body of mass m attached to a spring with the spring constant k and a dashpot that exerts a resistive force on the system (see Figure 4.1). We denote by $x(t)$ the distance the body has deviated from its equilibrium position.

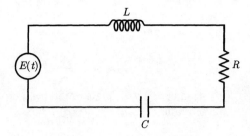

Figure 4.2: An electrical circuit.

According to Hooke's law the force that a spring exerts on a body is proportional to how far the body has moved from its position of equilibrium, that is,

$$F_1 = -kx(t). \tag{4.5}$$

Similarly, the force the dashpot exerts on the body is proportional to how fast the body moves, that is,

$$F_2 = -c\frac{dx}{dt}, \tag{4.6}$$

where c is a nonnegative constant.

The two forces in (4.5) and (4.6) must be balanced by the rate of change of the linear momentum of the body, which is

$$m\frac{d^2x}{dt^2}. \tag{4.7}$$

Putting the latter three statements together, we end up with the following second-order linear homogeneous differential equation for $x(t)$:

$$mx'' + cx' + kx = 0. \tag{4.8}$$

In certain cases an external force acts on the body. Typically, a mass-spring-dashpot system may be attached to a motor that induces an independent oscillatory force of amplitude A and frequency β on the system. Equation (4.8) must then be altered to take into account this additional force:

$$mx'' + cx' + kx = A\cos\beta t. \tag{4.9}$$

Thus, the addition of an external force to our problem causes the the differential equation to become nonhomogeneous.

Example 4.1.2 (Electrical Circuits)

We now consider an electrical circuit with a resistor of R ohms, an inductor of L henrys, and a capacitor of C farads. The circuit is driven by an electromotive force $E(t)$ (see Figure 4.2). Kirchhoff's law states that the sum of the voltage drops across the accessories must balance the input voltage. Since LI', RI, and

$\frac{1}{C}q$, where q is the charge accumulated in the capacitor, are the voltage drops across the inductor, the resistor, and the capacitor, respectively, we have

$$LI' + RI + \frac{1}{C}q = E(t). \tag{4.10}$$

Equation (4.10) is a single differential equation in the two variables I and q. The charge q and the current I are related through Faraday's law:

$$\frac{dq}{dt} = I. \tag{4.11}$$

This equation suggests differentiating (4.10) and using the relation in (4.11) to eliminate q. We get

$$LI'' + RI' + \frac{1}{C}I = E'(t). \tag{4.12}$$

As in the previous example, this physical problem's mathematical description has led to a second-order linear (nonhomogeneous) differential equation.

In the previous example we originally ended up with two simultaneous first-order differential equations, (4.10) and (4.11), with I and q as unknowns. We proceeded to eliminate one of the variables (in this case q) to get a single second-order equation in the other variable. This procedure is particularly convenient when the system of linear differential equations has constant coefficients, since we can always reduce such a system to an nth-order linear differential equation with constant coefficients. We will develop quite a complete mathematical analysis of constant coefficient equations in this chapter and then extend the analysis in a straightforward manner to systems as well.

We point out that the converse of the above procedure (writing a higher order equation as a first-order system) is also possible and, in certain circumstances, quite beneficial.

Example 4.1.3 (Higher Order Equations and Systems)
Consider the first-order system

$$x' = 2x - y, \quad y' = -y + 3x. \tag{4.13}$$

Here x and y are functions of t. The first equation is equivalent to $y = 2x - x'$, whose derivative with respect to t is $y' = 2x' - x''$. The last two expressions define y and y' in terms of x. Use them to eliminate y from the second-equation in (4.13). This leads the following second-order equation in x:

$$x'' - x' + x = 0. \tag{4.14}$$

We could have just as easily obtained an equation in y by eliminating x from (4.13).

Now consider the second-order (nonlinear) equation

$$y'' + 2y' + \sin y = 0.1 \cos 3t. \tag{4.15}$$

One way to convert (4.15) into a system is to introduce a new variable z that is related to y through

$$z = \frac{dy}{dt}.$$ (4.16)

Equation (4.15) is now equivalent to the system

$$y' = z, \quad z' = -2z - \sin y + 0.1\cos 3t,$$ (4.17)

since $z' = y''$.

Certain second-order differential equations can be converted to first-order equations, which in turn may be solved by the methods we discussed in Chapter 3.

Example 4.1.4 (Reduction to First-Order Equations)
Consider the second-order equation

$$y'' + y = 0.$$ (4.18)

We leave it to the reader to show by inspection that $y_1(t) = \sin t$ and $y_2(t) = \cos t$ are solutions of this equation. Since (4.18) is linear and homogeneous, any linear combination of y_1 and y_2, given by

$$y(t) = c_1 \sin t + c_2 \cos t,$$ (4.19)

is also a solution of this equation.

It is possible to arrive at (4.19) by converting (4.18) to a first-order equation. To see this, let z be a new dependent variable that is related to y through

$$z = y'.$$ (4.20)

Then $\frac{d^2y}{dt^2} = \frac{dz}{dt} = \frac{dz}{dy}\frac{dy}{dt} = \frac{dz}{dy}z$, where we have tacitly assumed that z can be thought of as a function of y and then applied the chain rule to obtain y''. Since $y'' = -y$ (from (4.18)), we have

$$z\frac{dz}{dy} = -y.$$ (4.21)

The preceding differential equation in z is first-order and separable. After writing it as $zdz = -ydy$ and integrating the result once, we have

$$z^2 = -y^2 + c^2,$$ (4.22)

where c^2 is a constant of integration. Since $z = y'$, (4.22) is equivalent to the first-order separable equation

$$y'^2 = -y^2 + c^2$$ (4.23)

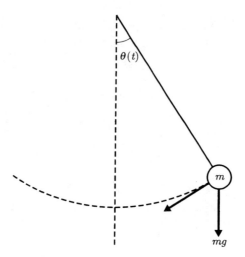

Figure 4.3: A schematic of an oscillating pendulum.

or

$$\frac{dy}{\sqrt{c^2 - y^2}} = \pm dt. \tag{4.24}$$

Integrate both sides to get

$$\frac{1}{c} \sin^{-1}\left(\frac{y}{c}\right) = \pm t + d. \tag{4.25}$$

Solve (4.25) for y and obtain

$$y(t) = c \sin(\pm ct + cd). \tag{4.26}$$

It is not difficult to show that (4.26) is equivalent to (4.19) with appropriate relations between c_1, c_2 and c, d.

In Example 4.1.4 we were able to reduce a second-order equation to two first-order equations, each of which we were able to solve using the methods of Chapter 3. As it turns out this is actually a roundabout way of solving (4.18). In the remainder of this chapter we will develop methods that will lead to the solution (4.19) in a simpler and more straightforward fashion.

In the next example we derive the equation of motion of a pendulum. The resulting differential equation is second order and nonlinear.

Example 4.1.5 (Nonlinear Pendulum)

We consider a pendulum whose bob has mass m and is attached to an inextensible string of length L (see Figure 4.3).

At any time t the pendulum makes an angle $\theta(t)$ with its equilibrium position. We assume that only three forces act on the pendulum: a) the force of gravity, b) air resistance, and c) an external periodic force of amplitude A and frequency ω.

The force of gravity is related to the angle θ according to the relation

$$F_1 = -mg \sin\theta, \tag{4.27}$$

since the string is inextensible and only the component of the gravitational force perpendicular to the string contributes to the motion. We assume that the force of air resistance is linearly proportional to the velocity of the bob so that

$$F_2 = -cL\frac{d\theta}{dt}, \tag{4.28}$$

where c is nonnegative. Finally the external forcing is modeled by

$$F_3 = A\cos\omega t. \tag{4.29}$$

The resultant of the three forces must be balanced by the rate of change of linear momentum $mL\frac{d^2\theta}{dt^2}$. We thus have the following differential equation for the variation in θ:

$$mL\theta'' + cL\theta' + mg\sin\theta = A\cos\omega t. \tag{4.30}$$

Equation (4.30) is in general extremely difficult to solve. Because this equation is nonlinear, most of the methodology we develop in this chapter will not apply to it and so we will eventually resort to approximate methods to gain insight into how its solutions behave.

We saw in Chapter 3 how useful the internal functions DSolve and NDSolve in *Mathematica*, and their counterparts dsolve and ode45 in MATLAB, were in determining exact and approximate solutions of first-order differential equations. These functions are equally powerful tools for higher order equations, and as the complexity of the structure of the solutions of these equations increases, the utility of symbolic calculations and numerical solvers will become more apparent. We end this section by reminding the reader about the syntax of the differential equation solvers in *Mathematica* and *MATLAB*.

Example 4.1.5 (DSolve and NDSolve)
Consider the differential equation

$$y'' + 3y' + 2y = \sin t. \tag{4.31}$$

We apply DSolve to (4.31) as follows:

```
DSolve[y''[t] + 3 y'[t] + 2 y[t] == Sin[t], y[t], t]
```

Mathematica responds with

```
          C[1]    C[2]    Cos[t] - 2 Sin[t]    -Cos[t] + Sin[t]
{{y[t] -> ---- + ---- + ----------------- + ----------------}}
           2 t     t             5                   2
            E       E
```

We are given the exact solution of the differential equation with two arbitrary constants of integration in the form of `C[1]` and `C[2]`. In a later section we will see how to get the same answer analytically using the methods described in the next few sections.

The differential equation

$$y'' + ay' + 2y = \sin t, \tag{4.32}$$

where a is a parameter, is a bit more challenging equation for this software. The syntax to handle this equation is no different from the previous case:

```
b=DSolve[y''[t] + a y'[t] + 2 y[t] == Sin[t], y[t], t];
Simplify[b]
```

which results in

```
{{y[t] ->

                                                          2
                           2              Sqrt[-8 + a ] t
>      (-a Cos[t]) + ((1 + a ) (C[1] + E                C[2]) +
                           2
         1/2 (a + Sqrt[-8 + a ]) t
>       E                           Sin[t]) / 2
                           2
         -(-(1/2) (a + Sqrt[-8 + a ]) t)           2
>       E                             ) / (1 + a )}}
```

It should now be clear why it is necessary to let *Mathematica* know that t is the independent variable and `a` just a parameter. `Mathematica` struggles with this example somewhat but eventually gives a reasonable candidate for the solution. Now, let us try the differential equation for the pendulum with the following parameter values

$$y'' + \sin y = 0. \tag{4.33}$$

When we try

```
DSolve[y''[t] + Sin[y[t]] == 0, y[t], t]
```

Mathematica responds with

```
Solve::ifun: Inverse functions are being used by Solve, so some
     solutions may not be found.

Solve::verif: Potential solution {y[t] -> ComplexInfinity}
     cannot be verified automatically. Verification may require
     use of limits.
```

The final output is a restatement of the input. NDSolve, which is designed to give accurate *approximate* solutions to differential equations, handles the pendulum equation with relative ease. Let us consider the same equation with initial conditions

$$y(0) = 0, \quad y'(0) = 1. \tag{4.34}$$

These constraints define the initial configuration of the pendulum—the pendulum is passing through its equilibrium position at time 0 with velocity 1. We use NDsolve to get the position of the pendulum during $t \in (0, 2\pi)$:

```
NDSolve[{y''[t] + Sin[y[t]] == 0, y[0] == 0, y'[0] == 1}, y[t],
          {t, 0, 2*Pi}]
```

Mathematica's response is

```
{{y -> InterpolatingFunction[{0., 6.28319}, <>]}}
```

The approximate solution y[t] is now available for plotting and other manipulations.

Example 4.1.6 (dsolve and ode45)

We now go through the same examples from Example 4.1.5 and demonstrate the syntax of dsolve and ode45. We start with the differential equation (4.31) and pass along to dsolve as follows:

```
dsolve('D2y + 3*Dy + 2*y = sin(t)')
```

MATLAB responds with

```
-1/10*(3*cos(t)*exp(t)-sin(t)*exp(t)-10*C1-10*C2*exp(-2*t)*
     exp(t))/exp(t)
```

MATLAB has given us the exact solution of the differential equation with two arbitrary constants of integration in the form of C1 and C2.

The second differential equation, (4.32), has a as a parameter in it. Nevertheless, the syntax for inputting this equation to dsolve remains the same:

```
b=dsolve('D2y + a*Dy + 2*y = sin(t)');
simplify(b)
```

MATLAB responds with

```
(-cos(t)*a+sin(t)+C1*exp(-1/2*(a-(a^2-8)^(1/2))*t)*a^2+C1*
exp(-1/2*(a-(a^2-8)^(1/2))*t)+C2*exp(-1/2*(a+(a^2-8)^(1/2))*t)*
a^2+C2*exp(-1/2*(a+(a^2-8)^(1/2))*t))/(a^2+1)
```

Now, let us try the differential equation for the pendulum (4.33). When we try

```
dsolve('D2y + sin(y) = 0')
```

MATLAB responds with

```
Warning: Explicit solution could not be found.
```

The function **ode45**, however, can handle the pendulum equation. To illustrate, we consider the same initial conditions (4.34). The syntax of **ode45** requires a call to an M-file that contains the definition of the differential equation. Before creating such an M-file, we convert (4.33) to a first-order system (see Example 4.1.4). Let $z = y'$. Equation (4.33) is equivalent to the system

$$y' = z, \quad z' = -\sin(y), \qquad y(0) = 0, \quad z(0) = 1. \tag{4.35}$$

Next, create an M-file called **pend.m** consisting of the following lines:

```
function xprime=pend(t,x);
xprime=[x(2); -sin(x(1))];
```

Here, **x** is a vector whose first component is y and second component is z; that is, **x(1)** stands for y and **x(2)** for z. The following call to **ode45** now computes an approximate solution to (4.35) in the interval $(0, 2\pi)$:

```
[t, x] = ode45('pend',[0 2*pi],[0 1]);
```

The two columns of **x** contain the numerical values for y and z (to actually see these columns of **x**, run the above line in *MATLAB* without the semicolon). To draw the graph of y versus t, for example, input

```
plot(t,x(:,1))
```

So far we have looked at examples of higher order differential equations in the context of physical problems that lead to *initial* value problems. In these cases all of the side conditions in (4.2) are given at a single "time" t_0. There are, however, numerous important problems in mathematical physics where the side conditions are evaluated at different values of the independent parameter. A typical example arises in the description of the transverse deflection of a uniform elastic beam. With $y(x)$ denoting the deflection at x, it can be shown that y satisfies the fourth-order differential equation

$$EIy^{(iv)} = w, \tag{4.36}$$

where E is the Young's modulus, I indicates the moment of inertia of the cross section of the beam, w is the force per unit length acting on the beam (for instance, w may be the beam's weight), and $y^{(iv)}$ denotes the fourth derivative of y with respect to x. Because (4.36) is fourth-order, we need to specify four side conditions to determine its solution uniquely. These side conditions are determined from the way the beam is supported at its two ends. If the two ends of the beam are clamped to a wall, then

$$y(0) = y'(0) = y(L) = y'(L) = 0, \tag{4.37}$$

where L is the beam's length. This way the two ends of the beam are not allowed to move and are constrained to remain perpendicular to the wall. If the beam is simply supported on its ends, then

$$y(0) = y(L) = 0, \quad y''(0) = y''(L) = 0. \tag{4.38}$$

The second derivatives vanishing at $x = 0$ and $x = L$ guarantee that the beam's curvature vanishes at the two ends. If the beam is in the cantilever position, then

$$y(0) = y'(0) = 0, \quad y''(L) = y'''(L) = 0. \tag{4.39}$$

The last boundary condition on y states that the beam is free of stress at $x = L$.

Equation (4.36) is unusual in its simplicity in that when w is a known function of x, four quadratures lead to its solution. As we will see in the exercises, the four boundary conditions in each case of (4.37)–(4.39) lead to a unique solution for the four constants of integrations.

Problems

1. Verify the information in Example 4.1.5.

2. Determine in each of the following problems if the differential equation is linear. If the equation is linear, is it homogeneous? What is the order of each equation? Use *Mathematica* or *MATLAB* to verify that the designated functions are solutions.

   (a) $y'' + 4y' + 3y = 0$, $\quad y(t) = e^{-3t}$

   (b) $x'' + x'^2 = 0$, $\quad x(t) = \ln|1 + t|$

   (c) $x''' + 3x'' + 3x' + x = 0$, $\quad x_1(t) = e^{-t}, x_2(t) = te^{-t}, x_3(t) = t^2 e^{-t}$. Is $x(t) = 2e^{-t} + 3te^{-t}$ a solution? Why?

   (d) $y^{(iv)} - 16y = 0$, $\quad y(x) = \sinh 2x$ ($y^{(iv)}$ denotes the fourth derivative of y.) Determine ω so that $y(x) = \sin \omega x$ is also a solution.

   (e) $x^2 y'' + xy' + y = 0$, $\quad y(x) = \cos(\ln x)$

   (f) $x^2 y'' + xy' - y = 0$, $\quad y(x) = \frac{1}{x}$

3. Determine the constant in each problem so that the given function is a solution of the differential equation.

(a) $x'' + ax = 0$, $x(t) = \sin 3t$

(b) $y'' + 3y' + 2y = 0$, $y(x) = e^{mx}$

(c) $t^2 x'' + 3tx' - x = 0$, $x(t) = t^r$

(d) $x''' - x'' - 2x' + 6x = 0$, $x(t) = e^{\lambda t}$

(e) $x''' + 2x'' - 3x' + 17x = 0$, $x(t) = e^{\lambda t}$ (Use solve in *MATLAB* or Solve in *Mathematica* to determine the roots of the polynomial in λ.)

4. Apply the method of reduction of order to determine a nontrivial solution to the following second-order differential equations.

(a) $y'' + y = 1$

(b) $y'' + \sin y = 0$

(c) $y'' + \cos y = 0$

(d) $y'' - \sin y = 1$

(e) $y'' + \frac{1}{y} = 0$

(f) $y'' + \frac{1}{y^2 + 1} = 0$

(g) $y'' + y^2 = 0$

(h) $y'' - y^2 = 1$

(i) $y'' + y'^2 = 0$

(j) $y'' + y'^2 = 1$

(k) $ay'' + by = c$, where a, b, and c are real constants.

(l) $y'' - 3y^2 = 2$

(m) $x^{(iv)} - 2x'' - 14x = 0$, $x(t) = e^{\lambda t}$

5. The deflection of a beam of length L, Young's modulus E, and cross-sectional moment of inertia I is given by (4.36). Assume that the beam is made of a homogeneous material and that w is the weight per unit length of the beam. Show by direct integration that the defelction y is given by

$$y(x) = \frac{w}{24EI} x^4 + c_1 x^3 + c_2 x^2 + c_3 x + c_4, \tag{4.40}$$

where the cs are arbitrary constants.

(a) Use the boundary conditions in (4.37) to determine the constants c_is in (4.40). (Ans: $y(x) = \frac{w}{24EI} x^2 (x - L)^2$)

(b) Use *Mathematica* or *MATLAB* to determine y in the previous setting. (Warning: E and I have special connotations in *Mathematica*. You may wish to use e and i instead.)

(c) Find the value of maximum deflection and the point on the beam at which it occurs. (Ans: $x = \frac{L}{2}$ and $\frac{wL^4}{384EI}$) Use *Mathematica* or *MATLAB* to verify the answers.

(d) Do the previous parts for the boundary conditions (4.38) and (4.39).

6. Consider the pendulum equation

$$y'' + ay' + \sin y = A \cos \omega t.$$

Following Exmaple 4.1.6, let $z = y'$ and consider the system of differential equations

$$y' = z, \quad z' = -az - \sin y + A \cos \omega t.$$

(a) Let $a = A = 0$.

    i. Plot the graphs of the solution pair $(y(t), z(t))$ for the various initial data $(y(0), z(0))$ given by

$$(-\pi - 0.5, 0), \quad (-\pi + 0.50), \quad (-0.5, 0), \quad (0.5, 0),$$
$$(\pi - 0.5, 0), \quad (\pi + 0.5, 0).$$

    Plot each solution for $t \in (0, T)$ with T sufficiently large to capture the general behavior of each solution (you may need to choose different Ts for different solutions.)

    ii. Next, plot the graphs of the solutions through the points $(0, 0)$, $(-\pi, 0)$, and $(\pi, 0)$. What is the physical interpretation of each of the latter solutions in the context of an oscillating pendulum?

(b) Let $A = 0$ and $a = 0.1$. Regenerate the plots of the solutions through the initial data given in part a). What is the principal difference you observe in the behavior of the solutions that one can attribute to the addition of ay' term?

(c) Let $A = 0$ and $a = -0.1$. Repeat the analysis of the previous part.

(d) Let $a = 0$, $A = 1$, and $\omega = 2$. Plot all solutions with initial data given in part a). What is the influence of the forcing term on the behavior of the solutions?

(e) Repeat the analysis of part d) when $\omega = 4$. Report on the changes you observe in the graphs of the resulting solutions when compared with the case $\omega = 2$.

(f) Let $a = 0.1$, $A = 1$, and $\omega = 2$. Plot all solutions of part a). Again, report on the change of behavior of solutions.

4.2 Linear Independence and Wronskian

In Section 3.3 we stated conditions under which the initial value problem of a typical nth-order differential equation has a unique solution. We now look at the consequences of this result for the special case of linear equations. In particular, we are interested in obtaining conditions under which we can find all solutions of linear ODEs.

The analysis we describe now is outlined in the context of the second-order linear differential equation

$$x'' + p(t)x' + q(t)x = 0, \quad x(t_0) = x_0, \quad \frac{dx}{dt}\Big|_{t=t_0} = x_1; \qquad (4.41)$$

although the general ideas are rather easily extended to higher order differential equations. Theorem 3.3.1 contains the basic assumptions on p and q that guarantee the existence and uniqueness of a solution to (4.41). With $f(t, x, x') = -q(t)x - p(t)x'$, the hypotheses of this theorem are satisfied if p and q are continuous. We state this fact as a corollary to Theorem 3.3.1.

Corollary 4.2.1

Suppose that the functions p and q are continuous in the interval $(0, T)$. The second-order differential equation (4.41) has a unique solution in that interval.

This corollary assures us that the solutions to the initial value problem (4.41) exist and are unique. It does not, however, point us in the direction of how to construct these solutions. As it will turn out, we will be able to know quite a bit more about the structure of the solutions to (4.41). When p and q are constant, for example, we will be able to write down the general solution explicitly in terms of elementary functions of calculus, although the situation is rather challenging when we deal with the higher order analogue of (4.41). When p and q are nonconstant, still quite a bit of information can be deduced from (4.41) and, on occasion, we are able to write down the solution explicitly. We will employ approximate techniques later in this chapter and get comfortably close to the exact solution to (4.41) guaranteed by Theorem 3.3.1.

Before addressing the relevant theorems for determining the general solution to (4.41), we state the following definition.

Definition 4.2.1 (Wronskian)

Let x_1 and x_2 be two differentiable functions. The wronskian of x_1 and x_2, denoted by $w(x_1, x_2)$, is defined as the determinant of the 2×2 matrix

$$\begin{bmatrix} x_1(t) & x_2(t) \\ x_1'(t) & x_2'(t) \end{bmatrix}, \tag{4.42}$$

that is,

$$w(x_1, x_2) = x_1(t)x_2'(t) - x_2(t)x_1'(t). \tag{4.43}$$

Example 4.2.1

Direct differentiation shows that the wronskian of $x_1(t) = e^{-2t}$ and $x_2(t) = \sinh t - 2\cosh 2t$ is

$$w(x_1, x_2) = -4 - \frac{1}{2}e^{-3t} + \frac{3}{2}e^t. \tag{4.44}$$

We obtain the same result by computing the above wronskian in *Mathematica*:

```
x1 = Exp[-2 t]; x2 = Sinh[t] - 2 Cosh[2 t];
w = Det[{{x1, x2}, {D[x1, t], D[x2, t]}}];
Simplify[%]
```

Mathematica returns

```
         1       3
-4 - ------- + ----
        3 t       t
       2 E       2 E
```

A somewhat more compact way of entering the definition of wronskian in *Mathematica* is

```
a = {Exp[-2 t], Sinh[t] - 2 Cosh[2 t]};
output = Det[{a, D[a, t]}];
Simplify[output]
```

MATLAB handles the wronskian in much the same way:

```
a = '[exp(-2*t) sinh(t)-2*cosh(2*t)]';
b = diff(a, 't');
c = [a;b];
d = det(c);
simplify(d)
```

MATLAB's output is

```
ans =
```

```
exp(-2*t)*cosh(t)-4*exp(-2*t)*sinh(2*t)+2*exp(-2*t)*sinh(t)-
        4*exp(-2*t)*cosh(2*t)
```

which, although equivalent to *Mathematica*'s output, looks different from it.

The next theroem shows the connection between the wronskian and the general solution of differential equations.

Theorem 4.2.1 (General Solution)

Let x_1 and x_2 be two solutions of (4.41). Suppose that there exists a point t_0 for which the wronskian of x_1 and x_2 is nonzero. Let $x(t)$ be any solution of (4.41). Then there exist constants c_1 and c_2 such that

$$x(t) = c_1 x_1(t) + c_2 x_2(t), \qquad (4.45)$$

for all t in the domain of x.

Proof: Let x be any solution of (4.41). We need to find c_1 and c_2 so that (4.45) holds. To that end, evaluate (4.45) at t_0. Next, differentiate (4.45) and evaluate the result at t_0. We obtain the following two algebraic equations in the two unknowns c_1 and c_2:

$$x(t_0) = c_1 x_1(t_0) + c_2 x_2(t_0), \quad x'(t_0) = c_1 x_1'(t_0) + c_2 x_2'(t_0). \qquad (4.46)$$

The above equations have a unique solution (c_1, c_2). We leave it to the reader to show that this solution is given by

$$c_1 = \frac{w(x(t_0), x_2(t_0))}{w(x_1(t_0), x_2(t_0))}, \quad c_2 = \frac{w(x_1(t_0), x(t_0))}{w(x_1(t_0), x_2(t_0))}. \tag{4.47}$$

The above pair (c_1, c_2) exists because, by hypothesis, the denominator of (4.47) is not zero. Hence, $x(t)$ and $c_1 x_1(t) + c_2 x_2(t)$ and their derivatives agree at t_0. Since solutions of (4.41) are unique, these two functions must agree for all values of t, which completes the proof of the theorem.

Definition 4.2.2 (Linear Independence)

Two functions x_1 and x_2 are said to be linearly independent on the interval (a, b) if

$$c_1 x_1(t) + c_2 x_2(t) \equiv 0 \quad \text{implies} \quad c_1 = c_2 = 0. \tag{4.48}$$

According to this definition two functions x_1 and x_2 are linearly independent if the only way one can write the zero function as a linear combination of x_1 and x_2 is by choosing c_1 and c_2 to be zero. Putting this in a different way, two functions are linearly independent if one can be written in terms of the other. Two functions that are not linearly independent are called **linearly dependent**.

Example 4.2.2

Let $x_1(t) = t$ and $x_2(t) = t^2$. To see that these functions are linearly independent on the real line, we form the identity

$$c_1 t + c_2 t^2 = 0. \tag{4.49}$$

This expression must hold for all t. If we divide by t and set $t = 0$, we get $c_1 = 0$. This, when combined with (4.49), implies that c_2 must be zero.

The concept of linear independence is intimately related to the wronskian, although the two are not equivalent.

Theorem 4.2.2 (Wronskian and Linear Independence)

Let x_1 and x_2 be two continuously differentiable functions on the interval (a, b). Suppose that at some point $t_0 \in (a, b)$ the wronskian of this set is nonzero. Then x_1 and x_2 are linearly independent.

We leave the proof of this result to the reader, since it is very similar to the reasoning that followed (4.47).

It is important to note that it is possible to have a set of functions that are linearly independent on an interval (a, b) but whose wronskian vanishes at some point in (a, b). For example, the functions t and t^2 are linearly independent on the interval $(-1, 1)$, and yet their wronskian vanishes at $t = 0$. The next theorem shows that this cannot happen for a set of functions that are solutions of the differential equation (4.41).

Theorem 4.2.3

Let $\{x_1, x_2\}$ be two solutions of

$$x'' + p(t)x' + q(t)x = 0, \tag{4.50}$$

with p and q continuous on the interval (a, b).

1. *If x_1 and x_2 are linearly independent, then the wronskian is never zero on (a, b).*

2. *If x_1 and x_2 are linearly dependent, then the wronskian is identically zero on (a, b),*

Proof: If the two functions x_1 and x_2 are linearly dependent, then there is a constant k such that $x_1(t) = kx_2(t)$, for all $t \in (a, b)$. It then follows from the definition of wronskian in (4.43) that the wronskian of x_1 and x_2 is identically zero.

Now suppose that x_1 and x_2 are two linearly independent solutions of (4.41). Suppose, by way of contradiction, that there is $t_0 \in (a, b)$ at which the wronskian is zero. Then the system of linear equations

$$c_1 x_1(t_0) + c_2 x_2(t_0) = 0, \quad c_1 x_1'(t_0) + c_2 x_2'(t_0) = 0 \tag{4.51}$$

has a nontrivial solution (c_1, c_2). Now, the function $x(t) = c_1 x_1(t) + c_2 x_2(t)$ satisfies (4.50) and has zero initial data, by (4.51). It follows from the uniqueness theorem that $x(t) \equiv 0$, which contradicts the linear independence of x_1 and x_2. This completes the proof of the theorem.

The results of this section show that the general solution of (4.50) is a linear combination of any two of its linearly independent solutions. Most of our effort in the remainder of this chapter will be spent on methods for determining linearly independent solutions of (4.50).

Example 4.2.3

Consider the differential equation

$$y'' + 4y' + 3y = 0. \tag{4.52}$$

The functions $y_1(t) = e^{-3t}$ and $y_2(t) = e^{-t}$ are linearly independent solutions of (4.52), as can be easily checked. So, any solution y of (4.52) can be written as a linear combination of y_1 and y_2:

$$y(t) = c_1 e^{-3t} + c_2 e^{-t}. \tag{4.53}$$

For example, we can now determine the solution to the initial value problem (4.52) and

$$y(0) = -1, \quad y'(0) = 1 \tag{4.54}$$

by finding the appropriate c_1 and c_2. Relations (4.54) lead to the algebraic equations

$$c_1 + c_2 = -1, \quad -3c_1 - c_2 = 1, \tag{4.55}$$

from which we deduce that $c_1 = 0$ and $c_2 = -1$. Thus, the solution to (4.52)–(4.54) is $y(t) = -e^{-t}$.

Similar calculations determine the solution to the boundary value problem (4.52) and

$$y(0) = -1, \quad y(1) = 1. \tag{4.56}$$

Now c_1 and c_2 satisfy

$$c_1 + c_2 = -1, \quad c_1 e^{-3} + c_2 e^{-1} = 1, \tag{4.57}$$

from which we get that $c_1 = \frac{e^2}{1-e}$ and $c_2 = \frac{1+e^3}{e^2-1}$. So, the solution to (4.52)–(4.56) is

$$y(t) = \frac{e^2}{1-e}e^{-3t} + \frac{1+e^3}{e^2-1}e^{-t}. \tag{4.58}$$

The reader is strongly encouraged to check the results of these calculations using DSolve of *Mathematica* or dsolve of *MATLAB*.

A convenient way of generating linear independent solutions of (4.41) is to look for two solutions of this equation subject to the initial conditions

$$x_1(t_0) = 1, \quad x_1'(t_0) = 0 \quad \text{and} \quad x_2(t_0) = 0, \quad x_2'(t_0) = 1. \tag{4.59}$$

Such solutions will be linearly independent since their wronskian is nonzero at t_0 (see Theorem 4.2.1).

So far we have stated the definitions of wronskian and linear independence in the context of second-order linear homogeneous equations. These definitions generalize to higher order equations in a straightforward manner.

Definition 4.2.3
 Let $\{x_1, x_2, \ldots, x_n\}$ be a set of functions defined in the interval (a, b) and having at least $n-1$ derivatives in their their domain. The wronskian of this set of functions is denoted by $w(x_1, x_2, \ldots, x_n)$ and is defined by

$$w(x_1, x_2, \ldots, x_n) = \det \begin{bmatrix} x_1 & x_2 & \cdots & \cdots & x_n \\ x_1' & x_2' & \cdots & \cdots & x_n' \\ \cdot & \cdot & & \cdot & \\ \cdot & \cdot & & \cdot & \\ x_1^{(n-1)} & x_2^{(n-1)} & \cdots & \cdots & x_n^{(n-1)} \end{bmatrix}. \tag{4.60}$$

Example 4.2.4

Consider the functions $x_1(t) = e^{-t}$, $x_2(t) = \sin 2t$, and $x_3(t) = t \cos t$. The wronskian of these functions is the determinant of the 3×3 matrix

$$\begin{bmatrix} e^{-t} & \sin 2t & t \cos t \\ -e^{-t} & 2 \cos 2t & \cos t - t \sin t \\ e^{-t} & -4 \cos 2t & -2 \sin t - t \cos t \end{bmatrix}, \tag{4.61}$$

which is

$$\frac{1}{2} e^{-t}(-2 \cos t - 9t \cos t + 2 \cos 3t + t \cos 3t + 9 \sin t + 3t \sin t + \sin 3t$$
$$+ 3t \sin 3t).$$

This wronskian can be readily determined in *Mathematica* by executing the lines

```
a={Exp[-t], Sin[2 t], t Cos[t]};
wronskian= Det[{a, D[a, t], D[a, {t, 2}]}];
Simplify[wronskian]
```

and, equivalently, in *MATLAB*

```
a='[exp(-t) sin(2*t) t*cos(t)]';
b=diff(a);
c=diff(b);
d=[a;b;c];
wronskian=det(d);
simplify(wronskian)
```

The concept of linear independence also generalizes naturally to a set of n functions.

Definition 4.2.4

Let $\{x_1, x_2, \cdots, x_n\}$ be a set of functions defined on the interval (a, b). We say that this set is linearly independent in (a, b) if

$$c_1 x_1(t) + c_2 x_2(t) + \cdots + c_n x_n(t) \equiv 0, \quad \text{for all } t \in (a, b), \quad \text{implies that}$$
$$c_1 = c_2 = \cdots = c_n = 0. \tag{4.62}$$

Otherwise the functions are said to be linearly dependent.

According to this definition, a set of functions is linearly dependent if the only way one can generate a linear combination of these functions and obtain the identically zero function is by taking all of the coefficients—the c_is—to be zero.

Example 4.2.5

Let us check that the set of functions defined in Example 4.2.4 are linearly independent in the interval $(-\pi, 2\pi)$. We first look at a typical linear combination of x_1, x_2, and x_3 and set it equal to zero:

$$c_1 e^{-t} + c_2 \sin 2t + c_3 t \cos t = 0, \quad \text{for all } t \in (-\pi, 2\pi). \qquad (4.63)$$

We take advantage of the fact that (4.63) holds for all t and evaluate it at three different values of t: $t = 0$, $t = \frac{\pi}{4}$, and $t = \pi$. We thus obtain three equations in the three unknowns c_1, c_2, and c_3:

$$c_1 = 0, \quad c_2 + \frac{\pi\sqrt{2}}{8} c_3 = 0, \quad -c_3 = 0, \qquad (4.64)$$

from which we deduce that $c_1 = c_2 = c_3 = 0$. Therefore, the set of functions $\{x_1, x_2, x_3\}$ is linearly independent.

At the beginning of this section we presented the theorems that related the concepts of wronskian, linear independence of solutions, and general solution of a second-order equation. We now generalize these concepts to higher order equations.

Theorem 4.2.4
Consider the nth-order equation

$$x^{(n)} + a^{(n-1)}(t)x^{(n-1)} + \cdots + a_1(t)x' + a_0(t)x = 0, \qquad (4.65)$$

where all of the coefficients are continuous in the interval (a, b). Let x_1, x_2, ..., x_n be n solutions of this equation. If x_1, x_2, ..., x_n are linearly independent, then $w(x_1, x_2, \ldots, x_n)$ is not zero for any t in (a, b) and

$$x(t) = c_1 x_1(t) + c_2 x_2(t) + \cdots + c_n x_n(t), \qquad (4.66)$$

is the general solution of (4.65). On the other hand, if these solutions are linearly dependent, then their wronskian is identically zero.

We use the above theorem in the same way we used it in the context of second-order equations: To generate the general solution of the homogeneous equation (4.65), we construct n linearly independent solutions of this equation by considering the initial conditions

$$x_1(t_0) = 1, \quad x_1'(t_0) = 0, \quad \ldots, \quad x_1^{(n-1)}(t_0) = 0, \qquad (4.67)$$

$$x_2(t_0) = 0, \quad x_2'(t_0) = 1, \quad \ldots, \quad x_2^{(n-1)}(t_0) = 0, \qquad (4.68)$$

and in general

$$x_i(t_0) = 0, \quad x_i'(t_0) = 0, \quad \ldots, \quad x_i^{(i-1)}(t_0) = 1, \quad \ldots, \quad x_i^{(n-1)}(t_0) = 0. \quad (4.69)$$

This way the wronskian of these solutions is 1 at $t = t_0$; and therefore, by the above theorem, they form a linearly independent set of solutions of (4.65). The general solution is then a linear combination of the x_is.

The previous discussion was focused on linear homogeneous differential equations. It turns out that the information we have obtained so far is quite useful in determining the general solutions of nonhomogeneous differential equations as well. First we note that if y_1 and y_2 are two solutions of the nonhomogeneous differential equations

$$x^{(n)} + a^{(n-1)}(t)x^{(n-1)} + \cdots + a_1(t)x' + a_0(t)x = f(t), \qquad (4.70)$$

then their difference, $x(t) = y_2(t) - y_1(t)$, satisfies the homogeneous differential equation (4.65). Given that x_1, x_2, $\cdots$, x_n form a set of linearly independent solutions of the homogeneous problem, we see that

$$y_2 - y_1 = c_1 x_1(t) + c_2 x_2(t) + \cdots + c_n x_n(t). \qquad (4.71)$$

The strategy for finding the general solution of the nonhomogeneous problem is now clear. First we need to determine a **particular solution** $x_p(t)$ of the nonhomogeneous equation (4.70), by whatever method possible. The general solution of this equation is then given by

$$x(t) = x_p(t) + c_1 x_1(t) + c_2 x_2(t) + \cdots + c_n x_n(t), \qquad (4.72)$$

where x_1, x_2 , $\cdots$, x_n form a linearly independent set of solutions of the homogeneous equation. We summarize these findings in the following theorem.

Theorem 4.2.5 (General Solution of Nonhomogeneous Equations)
The general solution of the nonhomogeneous equation

$$x^{(n)} + a^{(n-1)}(t)x^{(n-1)} + \cdots + a_1(t)x' + a_0(t)x = f(t)$$

is given by

$$x(t) = x_p(t) + x_c(t),$$

where

$$x_c(t) = c_1 x_1(t) + c_2 x_2(t) + \cdots + c_n x_n(t).$$

It is the sum of a **particular solution** x_p *of (4.70) and* x_c, *the general solution of the homogeneous equation (4.65), known as the* **complementary solution** *and given by (4.66).*

Proof: Let $x(t)$ be the general solution to (4.70). Let x_p be a particular solution to this equation, that is, a solution to (4.70) that we have obtained by some procedure (guessing, perhaps). Define the function z_c by

$$z_c(t) \equiv x(t) - x_p. \qquad (4.73)$$

Because both x and x_p satisfy (4.70), the function z_c satisfies the homogeneous equation (4.65) (since, for any linear operator L we have that $L[z_c] = L[x - x_p] = L[x] - L[x_p] = f - f = 0$). Hence, z_c can be expressed in terms of the general solution to (4.65), which completes the proof.

Example 4.2.6

Consider the second-order linear nonhomogeneous equation

$$x'' + 9x = 3 - 2t. \tag{4.74}$$

It is not difficult to show that $x_p(t) = \frac{1}{3} - \frac{2}{9}t$ is a particular solution of this equation and that $x_1(t) = \sin 3t$ and $x_2(t) = \cos 3t$ form a linearly independent set of solutions of the homogeneous equation. It then follows that

$$x(t) = c_1 \sin 3t + c_2 \cos 3t + \frac{1}{3} - \frac{2}{9}t \tag{4.75}$$

is the general solution of (4.74). The constants c_1 and c_2 are determined from the initial conditions. For instance, let

$$x(0) = 1 \quad \text{and} \quad x'(0) = -1. \tag{4.76}$$

The constants c_1 and c_2 now satisfy the two simultaneous equations

$$c_2 + \frac{1}{3} = 1, \quad 3c_1 - \frac{2}{9} = -1, \tag{4.77}$$

so $c_1 = -\frac{7}{27}$ and $c_2 = \frac{2}{3}$. Hence, the solution to the initial value problem (4.74)–(4.76) is

$$x(t) = -\frac{7}{27} \sin 3t + \frac{2}{3} \cos 3t + \frac{1}{3} - \frac{2}{9}t. \tag{4.78}$$

Mathematica and *MATLAB* give us the same answer. To obtain the solution in *Mathematica*, we enter

```
DSolve[{x''[t] + 9 x[t] == 3 - 2 t, x[0] == 1, x'[0] == -1},
    x[t], t]
```

which leads to (4.78), while in *MATLAB* the command

```
dsolve('D2x+9*x=3-2*t','x(0)=1','Dx(0)=-1')
```

produces the same result.

Problems

1. Compute the wronskian of the following functions. Check the computations in *Mathematica* or *MATLAB*:

   (a) $x_1(t) = 1 - t - t^2$, $\quad x_2(t) = t^3 + 1$

   (b) $x_1(t) = \log t$, $\quad x_2(t) = t^2 - t$

   (c) $x_1(t) = \frac{1}{1-t^2}$, $\quad x_2(t) = 1 - t^2$

   (d) $y_1(x) = \sin x$, $\quad y_2(x) = \sin 2x$, $\quad y_3(x) = \sin 3x$

   (e) $y_1(x) = \cos x$, $\quad y_2(x) = \cos 2x$, $\quad y_3(x) = \cos 3x$

   (f) $x_1(t) = \sin 2t$, $\quad x_2(t) = \cos 2t$, $\quad x_3(t) = 1$

(g) $w_1(x) = \exp(x)$, $\quad w_2(x) = \exp(2x)$, $\quad w_3(x) = \exp(3x)$

(h) $x_1(t) = \exp(at)\sin bt$, $\quad x_2(t) = \exp(at)\cos bt$, where a and b are constants

(i) $x_1(t) = \exp(at)$, $\quad x_2(t) = t\exp(at)$, where a is a constant

(j) $y_1(x) = \cos(\omega \ln x)$, $\quad y_2(x) = \sin(\omega \ln x)$, where ω is a constant

2. Determine whether the following set of functions are linearly independent.

   (a) $x_1(t) = \sin t$, $\quad x_2(t) = \cos t$, and $(a, b) = (0, \pi)$

   (b) $x_1(t) = e^t$, $\quad x_2(t) = e^{2t}$, and $(a, b) = (0, 3)$

   (c) $f(t) = e^t \sin 2t$, $\quad g(t) = e^t \sin t \cos t$, and $(a, b) = (0, \pi)$

   (d) $f(t) = 1$, $\quad g(t) = -t + 1$, $\quad h(t) = 3t^2 + 2t - 4$, and $(a, b) = (-1, 2)$

   (e) $f(t) = 1 - t$, $\quad g(t) = t^2 + t + 1$, $\quad h(t) = t^3 - 3t^2 + 2t - 4$, and $(a, b) = (-2, 2)$

   (f) $y_1(x) = \sin(2 \ln x)$, $\quad y_2(x) = \cos(2 \ln x)$, $\quad y_3(x) = x$, and $(a, b) = (1, \pi)$

   (g) $y_1(x) = e^x$, $\quad y_2(x) = xe^x$, $\quad y_3(x) = x^2 e^x$, and $(a, b) = (0, \infty)$

3. In each of the following problems show that the given functions are solutions of the differential equation; determine if they form a set of linearly independent functions; and if they do, construct the general solution. Use *Mathematica* or *MATLAB* to check the answers.

   (a) $x'' + 4x = 0$, $\quad x_1(t) = \sin 2t$, $\quad x_2(t) = \cos 2t$

   (b) $y'' - 3y' - 4y = 0$, $\quad y_1(t) = e^{4t}$, $\quad y_2(t) = e^{-t}$

   (c) $x''' + 3x'' + 3x' + x =$, $\quad x_1(t) = e^{-t}$, $\quad x_2(t) = te^{-t}$, $\quad x_3(t) = t^2 e^{-t}$

   (d) $t^2 x'' - 2tx' + 3x = 0$, $\quad x_1(t) = t^{\frac{3}{2}} \sin(\frac{\sqrt{3}}{2} \ln t)$, $\quad x_2(t) = t^{\frac{3}{2}} \cos(\frac{\sqrt{3}}{2} \ln t)$

4. Show that the following functions are linearly independent solutions of the given differential equations. Find the solution to each initial value problem. Check each answer in *Mathematica* or *MATLAB*.

   (a) $x'' + 6x' + 9x = 0$, $\quad x_1(t) = e^{-3t}$, $\quad x_2(t) = te^{-3t}$; $x(0) = -2$, $x'(0) = 3$

   (b) $x'' + 2x = 0$, $\quad x_1(t) = \sin \sqrt{2}t$, $\quad x_2(t) = \cos \sqrt{2}t$; $x(0) = 1$, $x'(0) = -1$

   (c) $x'' + 2x' + 2x = 0$, $\quad x_1(t) = e^{-t} \sin t$, $\quad x_2(t) = e^{-t} \cos t$; $x(0) = -2$, $x'(0) = 2$

   (d) $x''' + 6x'' + 11x' + 6x = 0$, $\quad x_1(t) = e^{-t}$, $\quad x_2(t) = e^{-2t}$, $\quad x_3(t) = e^{-3t}$; $x(0) = 0$, $x'(0) = -2$, $x''(0) = 1$

(e) $x''' - 3x'' + x' - 3x = 0,$ $x_1(t) = e^{3t},$ $x_2(t) = \sin t,$ $x_3(t) = \cos t;$
   $x(\pi) = 0,$ $x'(\pi) = -2,$ $x''(\pi) = 1$

(f) $x^{(iv)} - x = 0,$ $x_1(t) = e^{-t},$ $x_2(t) = e^t,$ $x_3(t) = \sin t,$ $x_4(t) = \cos t;$ $x(0) = 1,$ $x'(0) = 0,$ $x''(0) = -1,$ $x'''(0) = 0$

(g) $x^{(iv)} + x = 0,$ $x_1(t) = e^{\frac{t}{\sqrt{2}}} \cos \frac{t}{\sqrt{2}},$ $x_2(t) = e^{-\frac{t}{\sqrt{2}}} \cos \frac{t}{\sqrt{2}},$ $x_3(t) = e^{\frac{t}{\sqrt{2}}} \sin \frac{t}{\sqrt{2}},$ $x_4(t) = e^{-\frac{t}{\sqrt{2}}} \sin \frac{t}{\sqrt{2}};$ $x(0) = 0,$ $x'(0) = x''(0) = x'''(0) = 0.$

5. Determine all values of a so that the functions $x_1(t) = 1,$ $x_2(t) = a - a^2 t,$ $x_3(t) = at^2$ are linearly independent.

6. Determine all values of a, b, and c so that $x_1(t) = a + bt + ct^2,$ $x_2(t) = c + bt + at^2,$ and $x_3(t) = b + at + ct^2$ are linearly independent.

7. Let f be a continuously differentiable function.

   (a) Show that $w(f(t), tf(t)) = f^2(t).$

   (b) Use *Mathematica* or *MATLAB* to prove the preceding result.

8. Let f be a twice continuously differentiable function.

   (a) Show that $w(f(t), tf(t), t^2 f(t)) = 2f^3(t).$

   (b) Use *Mathematica* or *MATLAB* to prove the preceding result.

   (c) Generalize the preceding formula to that case of an $(n-1)$-times continuously differentiable function f and $w(f(t), tf(t), \cdots, t^{n-1}f(t)).$

9. Let $f,$ $g,$ and h be continuously differentiable functions.

   (a) Show that $w(f + h, g + h) = w(f, g) + w(f, h) + w(h, g).$

   (b) Use *Mathematica* or *MATLAB* to prove the preceding result.

10. Let x_1, x_2, x_3, and h be twice continuously differentiable functions.

   (a) Show that $w(x_1 + h, x_2 + h, x_3 + h) = w(x_1, x_2, x_3) + w(h, x_2, x_3) + w(x_1, h, x_3) + w(x_1, x_2, h).$

   (b) Use *Mathematica* or *MATLAB* to prove the preceding result.

11. Let $\{x_1, x_2, \cdots, x_n, h\}$ be a set of functions that are $(n-1)$-times continuously differentiable. Show that

$$w(x_1 + h, x_2 + h, \cdots, x_n + h) = w(x_1, x_2, \cdots, x_n) + \sum_{i=1}^{n} w_i(x_1, x_2, \cdots, x_n),$$

where $w_i = w(x_1, x_2, \cdots, h(t), \cdots, x_n),$ that is, w_i is obtained from w by evaluating $h(t)$ at the ith position.

12. Show that the set of functions $x_1(t), x_2(t), \cdots, x_n(t)$ is linearly dependent if one of the functions x_is is identically zero.

13. Show that there is no p and q, both continuous in the interval $(-1, 1)$, such that (4.41) would have $x_1(t) = t$ and $x_2(t) = t^2$ as solutions (see Example 4.2.2).

14. Show that if y_1 and y_2 are solutions of the nonhomogeneous equation (4.70), then their difference is a solution of the homogeneous equation (4.65).

15. Consider the differential equation $y'' + 4y = 0$.

    (a) Show that $y_1(x) = \sin 2x$ and $y_2(x) = \cos 2x$ are two linearly independent solutions of this equation. According to Theorem 4.2.1, any solution $y(x)$ can be written as

    $$y(x) = c_1 y_1(x) + c_2 y_2(x),$$

    with appropriate constants c_1 and c_2.

    (b) Show that $y_3(x) = 2\sin 2x - 3\cos 2x$ and $y_4(x) = 3\sin 2x + 2\cos 2x$ are also linearly independent solutions of the differential equation. Hence, any solution y can be written as

    $$y(x) = d_1 y_3(x) + d_2 y_4(x),$$

    for appropriate constants d_1 and d_2.

    (c) Find a relationship between the pairs $\{c_1, c_2\}$ and $\{d_1, d_2\}$.

4.3 Second-Order Equations with Constant Coefficients—the Homogeneous Case

Linear second-order differential equations with constant coefficients are equations of the form

$$x'' + ax' + bx = f(t), \tag{4.79}$$

where a and b are constants in R, the set of real numbers, and f is a given function of t. In this section we will be concerned with an algorithm that leads to the general solution to (4.79) when $f \equiv 0$, that is, the homogeneous case.

We saw in the previous section that the general solution of a second-order homogeneous differential equation is a linear combination of any two linearly independent solutions of that equation. As it turns out, it is relatively easy to construct such solutions, since the bulk of the analysis reduces to finding roots of second-order polynomials.

It seems intuitively clear that since a and b are constants, the solution to the differential equation

$$x'' + ax' + bx = 0 \tag{4.80}$$

and its first and second derivatives must be comparable functionally, that is, they must belong to a class of functions that allows for the type of algebraic cancellation one sees in (4.80). Exponential and trigonometric functions ($\sin ct$

and $\cos ct$) are examples of such functions. In fact, since $\sin ct$ and $\cos ct$ are related to the exponential function through Euler's formula, $e^{ict} = \cos ct + i \sin ct$, exponential functions are the only candidates for solutions of (4.80). With this in mind, we seek solutions to (4.80) of the form

$$x(t) = e^{mt}, \tag{4.81}$$

and our goal is to choose m, in terms of a and b, so that (4.81) satisfies (4.80). We begin by differentiating (4.81) and substituting the result into (4.80), which results in

$$e^{mt}(m^2 + ma + b) = 0. \tag{4.82}$$

Since (4.82) must be zero for all t, the quadratic expression in (4.82) must be zero:

$$m^2 + ma + b = 0. \tag{4.83}$$

This polynomial is called the **characteristic equation** of the differential equation (4.80). The roots of this quadratic polynomial are

$$m_1 = \frac{-a + \sqrt{a^2 - 4b}}{2}, \quad m_2 = \frac{-a - \sqrt{a^2 - 4b}}{2}. \tag{4.84}$$

The general solution of (4.80) depends on whether m_1 and m_2 are real or complex, distinct or repeated. This classification depends on whether $a^2 - 4b$ is positive, zero, or negative. We describe each possibility in a separate case.

Case I. $a^2 - b > 0$—Distinct Real Roots

We get two distinct roots from (4.83) when $a^2 - 4b$ is positive. It is not difficult to show that these solutions are linearly independent. The general solution to (4.80) is any linear combination of $e^{m_1 t}$ and $e^{m_2 t}$. We have the following theorem.

Theorem 4.3.1 (General Solution—Distinct Roots)

When $a^2 - 4b > 0$, the general solution of (4.80) is given by

$$x(t) = c_1 e^{m_1 t} + c_2 e^{m_2 t}, \tag{4.85}$$

where c_1 and c_2 are constants in R and m_1 and m_2 are given by (4.84). Moreover, the unique solution to (4.80) with

$$x(t_0) = x_0, \quad \frac{dx}{dt}\Big|_{t=t_0} = x_1 \tag{4.86}$$

is given by (4.85) with

$$c_1 = \frac{m_2 x_0 - x_1}{m_2 - m_1} e^{-m_1 t_0} \quad \text{and} \quad c_2 = \frac{-m_1 x_0 + x_1}{m_2 - m_1} e^{-m_2 t_0}. \tag{4.87}$$

The proof of this result is straightforward and is left to the reader. We point out that one does not memorize these formulas since it is relatively an easy matter to determine c_1 and c_2 by direct application of the initial data, as the following example shows.

Example 4.3.1
Consider the differential equation

$$x'' + 3x' + 2x = 0, \quad x(0) = 1, x'(0) = -1. \tag{4.88}$$

The characteristic equation of (4.88) is the quadratic equation

$$m^2 + 3m + 2 = 0, \tag{4.89}$$

whose roots are $m_1 = -1$ and $m_2 = -2$. The general solution to (4.88) is

$$x(t) = c_1 e^{-t} + c_2 e^{-2t}. \tag{4.90}$$

Now substitute the initial data $x(0) = 1$ and $x'(0) = -1$ into (4.90), which results in the following system of two algebraic equations in c_1 and c_2:

$$c_1 + c_2 = 1, \quad -c_1 - 2c_2 = -1. \tag{4.91}$$

Its solution is given by $(c_1, c_2) = (1, 0)$. Thus, the function $x(t) = e^{-t}$ is the solution to (4.88). The same exact information is available through *Mathematica* and *MATLAB*. In *Mathematica*

```
DSolve[{x''[t] + 3 x'[t] + 2 x[t] ==0, x[0] == 1, x'[0] == -1},
    x[t], t]
```

shows that $x(t) = e^{-t}$ is the solution, while in *MATLAB*

```
dsolve('D2x + 3*Dx +2*x=0','x(0)=1','Dx(0)=-1')
```

leads to the desired result.

Example 4.3.2
The computation that led to the formulas for c_1 and c_2 in the preceding example can be carried out rather easily in *Mathematica* or *MATLAB*. Given that we had already determined the general solution, we proceed as follows in *Mathematica* to get c_1 and c_2:

```
x  = c1 Exp[-t] + c2 Exp[-2 t];
eqns = {x == 1/. t -> 0, D[x, t] == -1/. t-> 0};
a=Solve[eqns, {c1, c2}];
Simplify[a]
```

This program takes the following form in *MATLAB*:

```
func='c1*exp(-t)+c2*exp(-2*t)';
eqn1=subs(func,0,'t');
eqn2=subs(diff(func),0,'t');
[c1 c2]=solve(eqn1-1, eqn2+1)
```

Case II. $a^2 - 4b = 0$—Repeated Roots

When the roots of the characteristic equation (4.83) are repeated, this algorithm provides us with only one exponential solution. We obtain a second solution to this problem by applying a limiting process to the general solution of Theorem 4.3.1, as the proof of the next theorem demonstrates.

Theorem 4.3.2 (Repeated Roots)

Suppose that the roots of (4.83) are $m_1 = m_2 = m$. Then the general solution to (4.83) is given by

$$x(t) = c_1 e^{mt} + c_2 t e^{mt} \tag{4.92}$$

and the solution to (4.80) and (4.86) is given by

$$x(t) = (x_0 - (t - t_0)(mx_0 - x_1))e^{m(t-t_0)}. \tag{4.93}$$

Proof: It follows from (4.85) and (4.87) that the solution to (4.80) and (4.87) is given by

$$x(t, m_1, m_2) \equiv \frac{(m_2 x_0 - x_1)e^{m_1(t-t_0)} - (m_1 x_0 - x_1)e^{m_2(t-t_0)}}{m_2 - m_1}, \tag{4.94}$$

when the roots of the characteristic equation are distinct.

We obtain the solution to our problem by taking the limit of (4.94) when m_1 approaches m_2. Since this limit is indeterminate, we apply L'Hôpital's rule and find that

$$\lim_{m_1 \to m_2} x(t, m_1, m_2) = \lim_{m_1 \to m_2} x_0 e^{m_1(t-t_0)} - (t - t_0)(m_2 x_0 - x_1)e^{m_2(t-t_0)} \tag{4.95}$$

or

$$x(t) = x_0 e^{m(t-t_0)} - (t - t_0)(m_2 x_0 - x_1)e^{m(t-t_0)}, \tag{4.96}$$

which is equivalent to (4.93). The reader can verify, by direct differentiation, that (4.96) satisfies (4.80) and (4.86) and that e^{mt} and te^{mt} are linearly independent. This completes the proof.

Example 4.3.3

Consider the differential equation

$$x'' + 4x' + 4 = 0. \tag{4.97}$$

The characteristic equation of this equation is $m^2 + 4m + 4 = 0$, which has the repeated roots -2 and -2. Therefore, the general solution of this equation is

$$x(t) = c_1 e^{-2t} + c_2 t e^{-2t}. \tag{4.98}$$

The constants c_1 and c_2 are either directly determined from the initial conditions or from (4.93).

The analytic work that was done in the proof of Theorem 4.3.2 can be reproduced in *Mathematica*, as the next example shows.

Example 4.3.4
Here we carry out the limiting process in the proof of Theorem 4.3.2 in *Mathematica*:

```
f = ((m2*x0 - x1)*Exp[m1*(t - t0)] -
            (m1*x0 - x1)*Exp[m2*(t - t0)])/(m2 - m1);
Limit[f, m1 -> m2]
```

Remark 4.3.1: It is interesting to note that the second line in this program can be replaced by

```
Limit[Limit[f, m1 -> m], m2 -> m]
```

Case III. $a^2 - 4b < 0$—Complex Roots
The resolution of this case hinges on the statement of the formula due to Euler, which provides the real and imaginary parts of the complex exponential function

$$f(z) = e^z, \tag{4.99}$$

where $z = a + bi$. Here, a and b are real numbers and i is the complex number defined by $i^2 = -1$. We will have more to say about (4.99) and other complex-valued functions in the chapter on complex functions. We now present a formal derivation of Euler's formula that demonstrates the origins of all the parts in this formula. We begin by recalling the Taylor series of the real-valued functions e^x, $\sin x$, and $\cos x$ expanded about $x = 0$:

$$e^x = 1 + x + \frac{x^2}{2!} + \frac{x^3}{3!} + \cdots = \sum_{n=0}^{\infty} \frac{x^n}{n!}, \tag{4.100}$$

$$\sin x = x - \frac{x^3}{3!} + \frac{x^5}{5!} - \cdots = \sum_{n=0}^{\infty} \frac{(-1)^n x^{2n+1}}{(2n+1)!}, \tag{4.101}$$

and

$$\cos x = 1 - \frac{x^2}{2!} + \frac{x^4}{4!} + \cdots = \sum_{n=0}^{\infty} \frac{(-1)^n x^{2n}}{(2n)!}, \tag{4.102}$$

where, by definition, $0! = 1$. Replace x with bi in (4.100), employ the definition of i (that $i^2 = -1$, $i^3 = -i$, and $i^4 = 1$), to arrive at

$$e^{bi} = \left(1 - \frac{b^2}{2!} + \frac{b^4}{4!} + \cdots\right) + i\left(b - \frac{b^3}{3!} + \frac{b^5}{5!} - \cdots\right) = \cos b + i \sin b, \quad (4.103)$$

where we have used (4.101) and (4.102). It then follows that

$$e^{(a+bi)x} = e^{ax}e^{(bx)i} = e^{ax}(\cos bx + i \sin bx). \quad (4.104)$$

We summarize these calculations in the following theorem.

Theorem 4.3.3 (Euler's Formula)
The complex-valued function e^{x+iy} is equivalent to

$$e^{x+iy} = e^x(\cos y + i \sin y). \quad (4.105)$$

One of the consequences of Euler's formula (4.105) is that it enables us to write a complex number $a + bi$ in its **polar form**

$$a + bi = re^{i\theta}. \quad (4.106)$$

The quantity r is called the **modulus** or the **absolute value** of the complex number and θ its **argument**. After comparing (4.106) with (4.105), we see that

$$a = r\cos\theta, \quad b = r\sin\theta \quad (4.107)$$

and

$$r = \sqrt{a^2 + b^2}, \quad \tan\theta = \frac{b}{a}. \quad (4.108)$$

Euler's formula is the key ingredient in determining the general solution of (4.80) when the roots of the characteristic equation are complex numbers. We emphasize that the coefficients a and b in (4.80) are real numbers. This has the consequence that if $m = m_1 + m_2 i$ is a root of the characteristic equation, then so is $\bar{m} = m_1 - m_2 i$, the complex conjugate of m. We leave the proof of this fact to the reader.

Theorem 4.3.4 (Complex Roots)
Let $m = m_1 + im_2$ with

$$m_1 = \frac{a}{2}, \quad m_2 = \frac{\sqrt{4b - a^2}}{2} \quad (4.109)$$

be a root of (4.83). Then the general solution to (4.80) is

$$x(t) = e^{m_1 t}\left[c_1 \cos(m_2 t) + c_2 \sin(m_2 t)\right]. \quad (4.110)$$

As before, the (real) constants c_1 and c_2 are calculated from the initial conditions.

Proof: We begin by using Euler's formula and obtaining the real and imaginary parts of $x(t) = e^{(m_1 + m_2 i)t}$:

$$x_1(t) \equiv e^{m_1 t} \cos(m_2 t), \quad x_2(t) \equiv e^{m_1 t} \sin(m_2 t). \tag{4.111}$$

A straightforward calculation shows that $x_1(t)$ and $x_2(t)$ are solutions of (4.80). We substitute x_1 into the differential equation (4.80) and obtain

$$\begin{aligned} x_1'' + ax_1' + bx = (m_1^2 - m_2^2 + am_1 + b)e^{m_1 t} \cos(m_2 t) \\ + (-2m_1 m_2 - a)e^{m_2 t} \sin(m_2 t) \end{aligned} \tag{4.112}$$

and, using the definitions of m_1 and m_2 from (4.109), reach the conclusion that the coefficients of the two exponential functions in (4.112) are zero. Similarly, one shows that x_2 satisfies (4.80).

Let $x(t) = c_1 x_1(t) + c_2 x_2(t)$ and let (4.86) be typical initial data. To show that x is the general solution of (4.80), we need to show that c_1 and c_2 can be computed from the initial data (alternatively, we may demonstrate that x_1 and x_2 are linearly independent). Differentiate $x(t)$, evaluate it at t_0 and use (4.86) to get

$$\begin{cases} a_1 c_1 + a_2 c_2 = b_1, \\ a_3 c_1 + a_4 c_2 = b_2, \end{cases} \tag{4.113}$$

where

$$\begin{aligned} a_1 &= \cos m_2 t_0, & a_3 &= m_1 \cos m_2 t - m_2 \sin m_2 t, & b_1 &= x_0 e^{-m_1 t}, \\ a_2 &= \sin m_2 t_0, & a_4 &= m_1 \sin m_2 t + m_2 \cos m_2 t, & b_2 &= x_1 e^{-m_1 t}. \end{aligned} \tag{4.114}$$

We leave it to the reader to show that the solution to (4.113) is

$$c_1 = \frac{b_1 a_4 - a_2 b_2}{m}, \quad c_2 = \frac{a_1 b_2 - b_1 a_3}{m}. \tag{4.115}$$

This completes the proof of the theorem.

Example 4.3.5
Consider the differential equation

$$x'' + x' + x = 0. \tag{4.116}$$

The characteristic equation of this equation is $m^2 + m + 1 = 0$, which has complex roots with real and imaginary parts

$$m_1 = -\frac{1}{2}, \quad m_2 = -\frac{\sqrt{3}}{2}. \tag{4.117}$$

Hence, the general solution of the equation is (cf. (4.110))

$$x(t) = e^{-\frac{1}{2}t}\left[c_1 \cos\left(\frac{\sqrt{3}}{2}t\right) + c_2 \sin\left(\frac{\sqrt{3}}{2}t\right) \right]. \tag{4.118}$$

Suppose now that $x(0) = 1$ and $x'(0) = 2$. Substituting this information into (4.118), we have

$$c_1 = 1 \qquad \text{and} \qquad -\frac{c_1}{2} + \frac{\sqrt{3}c_2}{2} = 2, \tag{4.119}$$

which gives us $c_2 = \frac{5\sqrt{3}}{3}$. Thus,

$$x(t) = e^{-\frac{1}{2}t}\left[\cos\left(\frac{\sqrt{3}}{2}t\right) + \frac{5\sqrt{3}}{3}\sin\left(\frac{\sqrt{3}}{2}t\right)\right] \tag{4.120}$$

is the solution to the initial value problem.

Example 4.3.6

In this example we carry out the computations of Example 4.3.5 in *Mathematica*. We begin by defining the differential operator $x'' + x' + x$

```
eqn[x_] := D[x, {t, 2}] + D[x, t] + x;
```

Note the use of :=). Next, we derive the characteristic polynomial as follows

```
soln = Exp[m t];
char = Simplify[ eqn[soln] / soln]
```

We have divided `eqn[soln]` by `soln`, anticipating that `Exp[m t]` will be a common factor. The output is

```
          2
1 + m + m
```

Next, we determine the roots of `char` via

```
roots = Roots[ char == 0, m]
```

which results in

```
          1/3                2/3
m == -(-1)      || m == (-1)
```

Let `firstRoot` denote the first root from the above expression, that is,

```
firstRoot = roots[[1,2]]
```

We need to extract the real and imaginary parts of `firstRoot` in order to construct two real and linearly independent solutions:

```
a = Re[firstRoot]; b = Im[firstRoot];
```

The general solution of the differential equation can now be determined.

```
genSolution = Exp[a t]*(c1 Cos[b t] + c2 Sin[b t]);
```

The constants c1 and c2 are determined from the initial data $x(0) = 1$ and $x'(0) = 2$. First, we define eqn1 and eqn2 as follows:

```
eqn1 = (genSolution /. t -> 0) == 1;
eqn2 = (D[genSolution, t] /. t -> 0) == 2;
```

Next, we use Solve with eqn1 and eqn2 and determine c1 and c2 via

```
coeffs = Solve[{eqn1, eqn2}, {c1, c2}];
```

Mathematica returns

```
            -5
{{c2 -> -------, c1 -> 1}}
         Sqrt[3]
```

The solution to the initial value problem is now at hand:

```
solution = genSolution /. coeffs
```

returns

```
      Sqrt[3] t              Sqrt[3] t
Cos[---------]      5 Sin[---------]
        2                      2
{-------------- + ----------------}
      t/2                  t/2
     E             Sqrt[3] E
```

Compare this expression with (4.120).

The procedure described in this section clearly has an algorithmic flavor to it. One can imagine putting together the various cases of solving second-order homogeneous equations with constant coefficients in a program that determines the general solution or the solution to an initial value problem automatically. This is the beginning of what has been incorporated into *Mathematica*'s DSolve and *MATLAB*'s dsolve. Although we have already encountered the syntax of these commands, in the next two examples we review these commands and use their outputs for typical operations, such as plotting and integration.

Example 4.3.7 (*Mathematica* and DSolve)
The syntax for obtaining the solution to

$$x'' + 2x' + x = 0$$

is

```
DSolve[x''[t] + 2 x'[t] + x[t] == 0, x[t], t]
```

Mathematica returns

```
            C[1]    t C[2]
{{x[t] ->  ----  + ------}}
             t       t
            E       E
```

One gets the solution to the initial value problem

$$x'' + 2x' + x = 0, \quad x(0) = 1.1, \quad x'(0) = -1.2$$

via

```
output = DSolve[{x''[t] + 2 x'[t] + x[t] == 0,
               x[0] == 1.1, x'[0] == -1.2}, x[t], t]
```

The output of this computation is stored in output. We can plot the solution $x(t)$ by entering

```
Plot[x[t] /. output, {t, 0, 3}]
```

or integrate $x^2(t)$ by

```
NIntegrate[(x[t])^2 /. output, {t, 0, 4}]
```

Example 4.3.8 (dsolve and *MATLAB*)

The syntax for obtaining the solution to

$$x'' + 2x' + x = 0$$

in *MATLAB* is

```
dsolve('D2x + 2*Dx + x = 0')
```

MATLAB's output is

```
(C1+C2*t)/exp(t)
```

One determines the solution to the initial value problem

$$x'' + 2x' + x = 0, \quad x(0) = 1.1, \quad x'(0) = -1.2$$

via

```
output = dsolve('D2x + 2*Dx + x = 0', 'x(0)=1.1', 'Dx(0)=-1.2')
```

The output of this computation is stored in output. We can now get a plot of the solution $x(t)$ by entering

```
ezplot(output)
```

or integrate $x^2(t)$ by

```
int(output^2,'t',0,3)
```

Problems

1. Write each expression in terms of its real and imaginary parts.

   (a) e^{2i}

   (b) $e^{\pi i}$

   (c) e^{-2+3i}

   (d) $e^{(1+2i)t}$

   (e) e^{bit}

   (f) $e^{\sqrt{2}it}$

   (g) $\exp(\frac{(1+\sqrt{3}i)t}{2})$

   (h) $\exp(\frac{(a+\sqrt{b}i)t}{a})$

2. Find the general solution to each differential equation. Verify the answer in *Mathematica* or *MATLAB*.

   (a) $x'' - 3x' + 2x = 0$

   (b) $x'' + 4x' - x = 0$

   (c) $x'' + 9x' - 7x = 0$

   (d) $x'' + 4x' + 4x = 0$

   (e) $y'' - 4y' + 4y = 0$

   (f) $x'' - x = 0$

   (g) $x'' + x' + x = 0$

   (h) $z'' + z = 0.$

3. Find the unique solution to the following second-order homogeneous equations. Verify each answer in *Mathematica* or *MATLAB*.

   (a) $x'' + 4x = 0, \quad x(0) = 1, \quad x'(0) = -2$

   (b) $x'' + 4x = 0, \quad x(1) = 1, \quad x'(1) = 2$

   (c) $y'' + 3y' + 4y = 0, \quad y(0) = 1, \quad y'(0) = -1$

   (d) $z'' + 6z' + 9z = 0, \quad z(2) = 0, \quad z'(2) = 1$

   (e) $x'' + 4x = 0, \quad x(0) = a, \quad x'(0) = 0$

   (f) $x'' + 16x = 0, \quad x(0) = 0, \quad x'(0) = a$

   (g) $x'' + 4x' + 4x = 0, \quad x(0) = a, \quad x'(0) = 0$

   (h) $x'' + 4x' + 8x = 0, \quad x(0) = a, \quad x'(0) = 0$

4. Use the technique described in Example 4.3.6 to solve the following initial value problems and plot each solution in an interval containing the initial data. For each solution compute

$$\int_a^{a+1} x^2(t) + (x'(t))^2 \, dt$$

where a is the initial time.

(a) $x'' + x' + x = 0;$ $x(0) = 2,$ $x'(0) = -1$

(b) $x'' - x' + x = 0;$ $x(1) = 0,$ $x'(1) = 2$

(c) $x'' + 2x' + 4x = 0;$ $x(0) = -1,$ $x'(0) = 1$

(d) $x'' + 4x = 0;$ $x(0) = x_0,$ $x'(0) = x_1$

5. In each problem below use DSolve or dsolve to obtain the solutions to the various

$$x(0) = \pm i, \quad i = 0, 1, 2, \qquad x'(0) = 0.$$

Plot these solutions on the same screen. Determine a general trend as to how the solutions of each differential equation behave from these graphs.

(a) $2x'' + 3x' + x = 0$

(b) $3x'' + 0.1x' + x = 0$

(c) $3x'' + x' - 2x = 0$

(d) $x'' - 2x' + 7x$

6. Modify the procedure in Example 4.3.6 and determine the solution to the boundary value problem

$$y'' + y' + 4y = 0, \qquad y(0) = 1, \quad y(2) = 3.$$

7. In the following problems a is a real number.

(a) Consider the differential equation $x'' + ax' + ax = 0$, where $a > 0$. Find all as for which this equation has real exponential solutions. How does the general solution behave as t approaches infinity for a typical a in this range? Find the range of a for which the general solution has an oscillatory behavior. Again, how does a typical general solution behave as t approaches infinity? Use *Mathematica* and draw the graph of a typical solution for the two classes of a.

(b) Repeat (a) for $x'' + (a - 1)x = 0$. Use *Mathematica* or *MATLAB* to plot the graphs of a typical solution for the three cases $a < 1$, $a = 1$, and $a > 1$.

8. Find the solution to $y'' + a^2 y = 0$, $y(0) = 1, y'(0) = 1$, with a a positive constant.

9. Determine the solution to $y'' + (1 - a)y' + ay = 0$, $y(0) = 1, y'(0) = 1$, with a a positive constant. For which values of a does a typical solution of this equation have oscillatory behavior?

10. Solve $x'' + 7x' + 2x = 0$, $x(3) = 0, x'(3) = 0$. (Hint: What does the existence and uniqueness theorem say about the solution to this problem?)

11. In each problem first find the general solution and then use it to solve the **boundary** value problem. Plot the solution and observe that the boundary conditions are satisfied.

(a) $y'' + 3y = 0$, $\quad y(0) = 1$, $\quad y(\pi) = 2$

(b) $y'' + 3y' + 2y = 0$, $\quad y(1) = 0$, $\quad y(2) = 2$

(c) $x'' + 4x = 0$, $\quad x(0) = 1$, $\quad x(2) = 1$

(d) $x'' + a^2 x = 0$, $\quad x(0) = 1$, $\quad x(2) = 1$

(e) $x'' + ax' + ax = 0$, $\quad x(0) = 0$, $\quad x(2) = 1$

(f) $x'' + x' = 0$, $\quad x(0) = 1$, $\quad x(4) = 1$

12. Consider the third-order differential equation

$$y''' + y = 0. \tag{4.121}$$

The technique for solving this equation is the same as that which we employed for the second-order equations we considered in this section. Begin by seeking solutions of the form $y(t) = e^{mt}$. Show that the characteristic equation is

$$m^3 + 1 = 0. \tag{4.122}$$

Use *Mathematica* or *MATLAB* to find the three roots of this polynomial. Let m_1, m_2, and m_3 be the three roots of this polynomial. Write down the general solution using the m_is. Compare the answer with the output of DSolve or dsolve.

13. Let $f(x) = ax^2 + bx + c$, where a, b and c are real numbers. Show that if z is a complex root of f, then so is $\bar{z}$.

14. Derive the formulas in (4.87).

15. Write a *Mathematica* or *MATLAB* program that begins with a and b in (4.80), x_0 and x_1 in (4.86), and ends up with the coefficients (4.87).

16. (**Heave motion**) A solid object floating in a fluid typically undergoes rather complicated motions. These motions are often linear superposition of simple motions about individual axes. In this problem we derive and solve the differential equation that governs one such simple motion.

Consider a cylindrical object of denisty ρ_s, height H, and cross-sectional radius a, floating in a fluid of density ρ_f, with $\rho_f > \rho_s$ (see Figure 4.4).

(a) Let h be the height of the submerged segment of the cylinder at equilibrium. Show that h and H are related by the formula

$$H\rho_s = h\rho_f. \tag{4.123}$$

(Hint: Balance the buoyancy force (due to the weight of the displaced fluid) with the weight of the cylinder.)

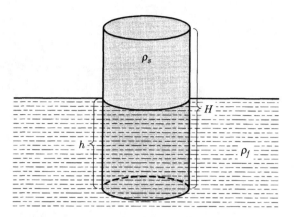

Figure 4.4: A submerged cylinder.

(b) The heave motion of the cylinder is defined as the vertical deviation away from its equilibrium. Let $x(t)$ denote this deviation from the depth h computed from (4.123). Show that x satisfies the second-order equation

$$\rho_s H x'' + \rho_f g x = 0, \qquad (4.124)$$

when frictional and external force are ignored. (Hint: Use Newton's second law.) Show that the two terms in (4.124) have the same dimensions.

(c) With parameter values $H = 10$ centimeters, $\rho_s = 0.5$ gm/cm^3, $\rho_f = 1$ gm/cm^3, $g = 0.098$ cm/s^2, and initial conditions

$$x(0) = 1, \qquad x'(0) = 0,$$

determine the frequency of oscillation of the cylinder.

(d) The heave motion of the cylinder is generally slowed down by the frictional force generated between the cylinder and the fluid. We model this force by a linear term proportional to the velocity. Equation (4.124) now becomes

$$\rho_s H x'' + \alpha x' + \rho_f g x = 0. \qquad (4.125)$$

Determine the solution $x(t, \alpha)$ to (4.125) in terms of α with the initial data and parameter values given in the previous part.

(e) Suppose that after 7 oscillations the maximum of the solution $x(t, \alpha)$ in a half-cycle is 1% of the initial displacement. Determine α.

4.4 Higher Order Homogeneous Equations with Constant Coefficients

In many respects the treatment of higher order homogeneous differential equations with constant coefficients is remarkably similar to second-order equations. We construct the general solutions of higher order equations as linear combination of solutions of exponential type. As was the case with second-order equations, we seek exponential solutions of the form e^{mt} and find that m must be the root of the polynomial with real coefficients. The main source of mathematical complication arises in determining roots of characteristic equations with degrees larger than four, where we no longer possess analytic formulas for the roots in terms of the coefficients and often have to resort to numerical approximation. Even when the charactersitics polynomial is third or fourth ordered, the analytic representation of the roots is sometimes too complicated to be useful, and in practice we apply numerical approximate methods to determine the roots.

We approach finding the general solution of a typical nth-order equation

$$x^{(n)} + a_{n-1}x^{(n-1)} + \cdots + a_1 x' + a_0 x = 0 \tag{4.126}$$

in the same way we started with (4.80); that is, we seek solutions of (4.126) in the form

$$x(t) = e^{mt}. \tag{4.127}$$

Substituting this form into (4.126) and canceling out e^{mt} from the resulting expression yield the **characteristic equation**

$$m^n + a_{n-1}m^{n-1} + \cdots + a_1 m + a_0 = 0. \tag{4.128}$$

The character of the solutions of (4.126) will depend on whether the roots of the characteristic equation are real and distinct, repeated, or complex. When all of the roots are distinct, say $m_1, m_2, \cdots, m_n$, the set of functions $\{e^{m_1 t}, e^{m_2 t}, \cdots, e^{m_n t}\}$ is linearly independent and the general solution of (4.126) takes the form

$$x(t) = c_1 e^{m_1 t} + c_2 e^{m_2 t} + \cdots + c_n e^{m_n t}. \tag{4.129}$$

Example 4.4.1
Consider the differential equation

$$x''' - 3x'' + 2x' + 6x = 0. \tag{4.130}$$

Its characteristic equation is

$$m^3 - 3m^2 + 2m + 6 = 0, \tag{4.131}$$

whose roots are (check this fact in *Mathematica* or *MATLAB*)

$$m_1 = -1, \quad m_2 = 2 - \sqrt{2}i, \quad m_3 = 2 + \sqrt{2}i. \tag{4.132}$$

(Why are the complex roots complex conjugates?)   We therefore have three linearly independent solutions

$$x_1(t) = e^{-t}, \quad x_2(t) = e^{2t} \sin \sqrt{2}t, \quad x_3(t) = e^{2t} \cos \sqrt{2}t, \tag{4.133}$$

with the general solution being the linear combination of these functions. Let us use `DSolve` in *Mathematica* and verify this solution:

```
DSolve[x'''[t] - 3 x''[t] + 2 x'[t] + 6 x[t] == 0, x[t], t]
```

yields

```
         C[1]     (2 - I Sqrt[2]) t
{{x[t] -> ---- + E                      C[2] +
           t
          E


     (2 + I Sqrt[2]) t
   E                      C[3]}}
```

which is equivalent to (4.133). Similarly, in *MATLAB*, we enter

```
dsolve('D3x-3*D2x+2*Dx+6*x=0')
```

MATLAB responds with

```
(C1+C2*exp(2*t)*cos(2^(1/2)*t)*exp(t)+C3*exp(2*t)*sin(2^(1/2)*t)*
    exp(t))/exp(t)
```

which is exactly the same as (4.133).

When a root m is repeated, the situation for higher order equations is similar to the second-order equations in that e^{mt} and te^{mt} are linearly independent solutions of (4.126). It can happen, however, that a higher order equation's characteristic equation has a root with multiplicity k larger than 2. To understand how one generates k linearly independent solutions in such a case, we could follow the reasoning of the previous section and apply L'Hôpital's rule $k - 1$ times to the general solution. Instead of following this procedure, we present a somewhat different argument based on the concept of the differential operator $D = \frac{d}{dt}$ that not only leads to the solution of our present problem but also introduces a mathematical tool that has its own inherent benefits.

In the differential operator notation, x' is replaced by Dx and x'' by D^2x and, in general,

$$x^{(i)} = D^i x. \tag{4.134}$$

So now the differential equation (4.126) can be written as

$$L[x] \equiv D^n x + a_{n-1} D^{n-1} x + \cdots + a_1 Dx + a_0 x = 0. \tag{4.135}$$

$L[x]$ is a **linear operator**, in the language that was introduced in Section 3.2. We will refer to it as the differential operator of (4.126).

The differential operation D has one very useful property, namely, that it allows us to think of a differential equation in algebraic terms and in particular, to bring the algebraic concept of factoring into the realm of differential equations. For example, let a and b be arbitrary constants; then

$$L[x] \equiv x'' + (a+b)x' + ab\,x = 0 \tag{4.136}$$

can also be written as

$$(D+a)(D+b)x = 0 \tag{4.137}$$

or as

$$(D+b)(D+a)x = 0. \tag{4.138}$$

In writing (4.138), we used the fact that the differential operators $(D+a)$ and $(D+b)$ commute, the proof of which we leave to the reader. The characteristic equation of $L[x] = 0$ is $(m+a)(m+b) = 0$, whose relation to (4.137) and (4.138) should be clear. Therefore, once a differential equation is *factored* in the sense of (4.137) or (4.138), its characteristic equation is obtained by merely replacing D with m, and vice versa, when the characteristic equation of a differential equation is known, the differential equation itself can be reconstructed by replacing m with D.

Returning to the nth-order equation and the issue of repeated roots, suppose that the characteristic equation (4.128) has a root $\bar{m}$ with multiplicity k. This equation can be factored as

$$(m - m_1) \cdots (m - m_i)(m - \bar{m})^k, \tag{4.139}$$

where, for simplicity, we assume that the other roots—m_1 through m_i—are not repeated. Then (4.126) is equivalent to

$$(D - m_1)(D - m_2) \cdots (D - m_i)(D - \bar{m})^k\, x = 0. \tag{4.140}$$

We know of course that $x_1(t) = e^{\bar{m}t}$ is a solution of (4.126). We guess, based on our experience with second-order equations, that

$$x_2(t) = t e^{\bar{m}t} \tag{4.141}$$

is a second linearly independent solution of (4.126). This fact can be verified directly by substituting (4.141) into (4.140). If k, the multiplicity of $\bar{m}$, is 2, then we produced two linearly independent solutions and need not go further. But if $k > 2$ we need to produce as many as k linearly independent solutions from the root $\bar{m}$. The success of (4.141) suggests looking for solutions of the form

$$x(t) = y(t)e^{\bar{m}t}. \tag{4.142}$$

Substitute (4.142) into (4.140). We note that if we are able to determine y such that

$$(D - \bar{m})^k (y(t)e^{\bar{m}t}) = 0, \tag{4.143}$$

then (4.140) is automatically satisfied. First, note that

$$(D - \bar{m})(y(t)e^{\bar{m}t}) = D(y(t)e^{\bar{m}t}) - \bar{m}y(t)e^{\bar{m}t}$$
$$= e^{\bar{m}t}Dy + \bar{m}y(t)e^{\bar{m}t} - \bar{m}y(t)e^{\bar{m}t} = e^{\bar{m}t}Dy. \tag{4.144}$$

Similarly,

$$(D - \bar{m})^2(y(t)e^{\bar{m}t}) = e^{\bar{m}t}D^2y, \tag{4.145}$$

and, in general,

$$(D - \bar{m})^k(y(t)e^{\bar{m}t}) = e^{\bar{m}t}D^ky. \tag{4.146}$$

So (4.143) holds if y satisfies the kth-order differential equation

$$D^k y = 0. \tag{4.147}$$

Fortunately, this equation is easy to solve. We find its solution by integrating it k times, resulting in

$$y(t) = c_1 + c_2t + c_3t^3 + \cdots + c_kt^{k-1}. \tag{4.148}$$

Therefore, when $\bar{m}$ is a repeated root with multiplicity k, we generate k linearly independent solutions of this equation in the form

$$x_1(t) = e^{\bar{m}t}, \quad x_2(t) = te^{\bar{m}t}, \quad x_3(t) = t^2e^{\bar{m}t}, \quad \cdots, \quad x_k(t) = t^{k-1}e^{\bar{m}t}. \tag{4.149}$$

The proof of the fact that these solutions are linearly independent is left for the reader. We summarize these findings in the following theorem:

Theorem 4.4.1 (Repeated Roots)

Suppose that the characteristic equation of (4.126) has a root $\bar{m}$ with multiplicity k. Then the functions defined by (4.149) are k linearly independent solutions associated with $\bar{m}$.

Example 4.4.2

Consider the differential equation

$$x''' + 3x'' + 3x' + x = 0. \tag{4.150}$$

The characteristic polynomial of this equation is

$$m^3 + 3m^2 + 3m + 1, \tag{4.151}$$

which has $m = -1$ for its root with multiplicity 3. Therefore, the general solution of (4.150) is

$$x(t) = c_1e^{-t} + c_2te^{-t} + c_3t^2e^{-t}. \tag{4.152}$$

Another possibility that could occur for higher order equations is that a complex root may be repeated (such a possibility does not exist for second-order equations with real constant coefficients. Why?) Suppose that $m = a + bi$ is a root of the characteristic equation with multiplicity 2. Assuming that the coefficients of (4.126) are real, the complex conjugate of m, $a - bi$, is also a root of the characteristic equation with multiplicity 2. We therefore need to construct four linearly independent solutions out of these roots:

$$x_1(t) = e^{at}\cos bt, \quad x_2(t) = e^{at}\sin bt, \quad x_3(t) = te^{at}\cos bt, \quad x_4(t) = te^{at}\sin bt.$$
$$(4.153)$$

We can now imagine how to generate the necessary number of linearly independent solutions when the multiplicity of the complex root is larger than 2.

Example 4.4.3
Consider the fifth-order differential equation

$$x^{(v)} - 4x^{(iv)} + 18x''' - 34x'' + 45x' - 25x = 0, \qquad (4.154)$$

with its characteristic equation

$$m^5 - 4m^4 + 18m^3 - 34m^2 + 45m - 25 = 0. \qquad (4.155)$$

The polynomial in Equation (4.155) is fifth-order. In general *Mathematica* and *MATLAB* are unable to determine roots of such polynomials analytically. This polynomial, however, yields to these programs. For example, `Solve` of *Mathematica* shows that (4.155) has the following roots

$$m = 1, \quad m = 1 \pm 2i \quad \text{(each with multiplicity 2).} \qquad (4.156)$$

The general solution of (4.154) is therefore

$$x(t) = c_1 e^t + c_2 e^t \cos 2t + c_3 t e^t \cos 2t + c_4 e^t \sin 2t + c_5 t e^t \sin 2t. \qquad (4.157)$$

Problems

1. Show that $x_1(t) = te^{at}\sin bt$ and $x_2(t) = te^{at}\cos bt$ are linearly independent for all nonzero constants a and b. Verify this statment in *Mathematica* or *MATLAB*.

2. Find the general solution of the following equations. In each problem use *Mathematica* or *MATLAB* to check for the roots as well as to check for the general solution.

   (a) $x''' + 3x'' - x = 0$

   (b) $3y''' - 4y'' + 4y' - y = 0$

   (c) $y^{(iv)} - 4y'' + y = 0$

   (d) $x^{(iv)} + x = 0$

   (e) $x^{(iv)} - x = 0$

(f) $y^{(iv)} - 8y''' + 42y'' - 104y' + 169y = 0$

(g) $x^{(iv)} - 4x''' + 16x' - 16x = 0$.

3. Find the solution to the following initial value problems. Use DSolve or dsolve to check your answers.

(a) $y''' - 3y'' - 3y' + y = 0$, $y(0) = y'(0) = 0, y''(0) = -2$

(b) $y^{(iv)} - 4y' + y = 0$, $y(0) = 1, y'(0) = 0, y''(0) = 1, y'''(0) = 0$

(c) $x''' + x = 0$, $x(0) = x'(0) = 0, x''(0) = 1$

(d) $y^{(iv)} + 64y = 0$, $y(0) = y'(0) = 0, y''(0) = 1, y'''(0) = 1$

(e) $y^{(iv)} - 64y = 0$, $y(0) = 1, y'(0) = 0, y''(0) = 1, y'''(0) = 0$

(f) $x^{(iv)} - 4x''' + 16x' - 16x = 0$, $x(0) = 1, x'(0) = 1, x''(0) = x'''(0) = 0$.

4. Solve the following boundary value problems. In each problem first find the general solution and then solve for the particular coefficients. Use *Mathematica* or *MATLAB* in the evaluation of the coefficients as well as to check your answers.

(a) $x''' - x = 0$, $x(0) = 0, x'(0) = 1, x(1) = 3$

(b) $x^{(iv)} + 16x = 0$, $x(0) = x''(0) = 0, x(\pi) = 1, x''(\pi) = 0$

(c) $x^{(iv)} - 16x = 0$, $x(0) = x''(0) = 0, x(\pi) = 0, x''(\pi) = 1$.

(d) $y''' + 3y'' + 3y' + y = 0$, $y(0) = 0,   y(1) = 1,   y'(1) = 0$

5. Consider the fourth-order differential equation

$$y^{(iv)} + y'' + y = 0.$$

Discuss the behavior of a typical solution of this equation: Is it oscillatory? Does it approach zero as the independent variable approaches infinity?

6. Consider the fourth-order differential equation

$$y^{(iv)} + y'' + a^2 y = 0.$$

For what values of a does a solution of this equation oscillate? Is there an interval I in the set of real numbers such that if $a \in I$ then a typical solution of this equation decays to zero as the independent variable approaches infinity?

7. Show that $e^{at}, te^{at}, \cdots, t^k e^{at}$ are linearly independent for all as on the real line and $k \geq 1$.

8. Consider the third-order polynomial $p(m) = m^3 + am^2 + bm + c$, where a, B, and c are real numbers. Show that p cannot have a repeated complex root.

9. Let D be the differentiation operator and a a constant. Prove the following identities.

(a) $(D+a)^2 = D^2 + 2aD + a^2$

(b) $(D+a)^3 = D^3 + 3aD^2 + 3a^2D + a^3$

(c) Let f be a differentiable function. Show that $D(f^2) = 2fDf$. Determine formulas for $D^2(f^2)$ and $D^3(f^3)$.

(d) Let f and g be differentiable functions of a single variable. Show that $(D+a)(fg) = g(D+a)f + fDg$.

10. Show that the differential operators $(D+a)$ and $(D+b)$ commute if a and b are constants.

11. Let a be a differentiable function. Is it true that $(D+a)^2 = D^2 + 2aD + a^2$?

12. Let a and b be differentiable functions. Show that the differential operators $D+a$ and $D+b$ commute if and only if $Da - aD = Db - bD$.

4.5 Nonhomogeneous Differential Equations—Undetermined Coefficients

In the previous section we saw how to find the general solution to the second-order linear differential equation

$$x'' + ax' + bx = 0, \tag{4.158}$$

when a and b are constants in R. We now use the algorithm described there and find the general solution to the nonhomogeneous problem

$$x'' + ax' + bx = f(t) \tag{4.159}$$

when f belongs to a special class of functions. We present the main ideas in the context of second-order equations, but as the examples at the end of this section demonstrate, these ideas generalize in a straightforward manner to higher order equations.

We saw in Section 3.3 that the initial value problem for the nonhomogeneous equation

$$L[x] = f(t) \tag{4.160}$$

has a unique solution when L is a linear operator with constant coefficients and f is a reasonably smooth function. We also showed in Theorem 4.2.6 that the general solution to (4.160) is of the form

$$x_c(t) + x_p(t), \tag{4.161}$$

where x_c, the complementary solution, is the general solution of the homogeneous problem and x_p is a particular solution of (4.160). Our goal in this section is to introduce an algorithm that allows us to construct a particular solution of (4.159) when f belongs to the class of polynomials, trigonometric expressions consisting of sines and cosines functions, and exponential functions. This method also works well with certain combinations of functions in the previous class, such as $e^{at} \sin bt$ or $p(t) \cos t$, where p is a polynomial. Several examples will make clear the extent of applicability of this method.

In Section 4.3 we classified the different cases that one encounters when generating linearly independent solutions of (4.158). We noted that all the possible solutions consisted of product of polynomials in t and exponential solutions, real- or complex-valued. It turns out that when the forcing term in (4.159) has a similar character, we are able to determine a particular solution to this equation by fitting a linear combination of f and its derivatives to (4.159). We demonstrate this idea by an example.

Example 4.5.1
Consider the differential equation

$$x'' + 3x' + 2x = \sin t. \tag{4.162}$$

Following the ideas of the previous section, we find that

$$x_c(t) = c_1 e^{-2t} + c_2 e^{-t} \tag{4.163}$$

is the general solution of the homogeneous part $x'' + 3x' + 2x = 0$. Since the nonhomogeneity in (4.162) is $\sin t$, for x_p we seek a function of the form

$$x_p(t) = A \sin t + B \cos t, \tag{4.164}$$

where A and B are as yet undetermined. Note that the right-hand side of (4.164) consists of $\sin t$ and all of its derivatives with respect to t—$\cos t$ and $\sin t$ in this case—in an attempt to create an expression whose derivatives up to order two are all comparable to each other.

Substitute x_p in (4.164) into (4.162). We obtain the identity

$$(A - 3B - 1) \sin t + (B + 3A) \cos t = 0. \tag{4.165}$$

This equation must hold for every value of t. Therefore, and the reader should justify this step in detail, the coefficients of $\sin t$ and $\cos t$ must vanish, that is,

$$A - 3B = 1, \quad B + 3A = 0. \tag{4.166}$$

The pair $(A, B) = (\frac{1}{10}, -\frac{3}{10})$ is the solution to (4.166), so the particular solution x_p is

$$x_p(t) = \frac{1}{10} \sin t - \frac{3}{10} \cos t. \tag{4.167}$$

Expressions (4.163) and (4.167) combine to define the general solution to (4.162):

$$x(t) = c_1 e^{-2t} + c_2 e^{-t} + \frac{1}{10} \sin t - \frac{3}{10} \cos t. \tag{4.168}$$

One can readily verify this answer using DSolve in *Mathematica* or dsolve in *MATLAB*, which we leave as an exercise for the reader.

This example lays out the blueprint for the **method of undetermined coefficients**. Based on the form of the forcing term f, we make an educated guess as to the form of the particular solution x_p. The function x_p generally contains several coefficients that must be determined by directly substituting x_p into the differential equation. As a result of the subsitution, we obtain a system of simultaneous algebraic equations for the coefficients. Obtaining a solution to the latter system is usually straighforward.

This method is particularly successful when the underlying homogeneous differential equation has constant coefficients and the forcing term f is a linear combination of sines and cosines, exponentials and polynomials, and product of polynomials with the exponentials or sines and cosines functions. The educated guess for x_p always consists of starting with a general expression based on f and augmenting the trial solution with all of its possible derivatives. For instance, if $f(t) = e^{3t} \cos 2t$, then

$$x_\mathrm{p}(t) = Ae^{3t} \cos 2t + Be^{3t} \sin 2t \tag{4.169}$$

would be an appropriate choice for the particular solution. This method is specially effective when x_p consists of a *finite* number of terms *and* all of the derivatives of x_p have the same form as x_p, as in (4.169). Hence, this technique is the method of choice when f belongs to the class of functions described previously (such as polynomials, sines, and cosines) but not particularly useful for functions such as $\tan t$, $\frac{1}{t+1}$, or $\ln t$.

In general, *assuming that the forcing term is **not** a solution of the homogeneous equation*, we will use the following recipe to choose candidates for the particular solution x_p. We will present shortly a general method in terms of annihilators that applies to other cases of interest, including the case where the forcing term is a solution of the associated homogeneous equation.

4.5.1 Trial Solutions when $L[f] \neq 0$

1. If $f(t)$ is a simple trigonometric function in terms of sines and cosines, say

$$f(t) = c \sin \omega t + d \cos \omega t,$$

with c, d, and ω known, we try an expression of the form

$$x_\mathrm{p}(t) = A \sin \omega t + B \cos \omega t,$$

for the particular solution x_p, and determine A and B by substituting x_p in (4.159).

2. If $f(t)$ is a polynomial in t such as

$$f(t) = a_n t^n + a_{n-1} t^{n-1} + \cdots + a_1 t + a_0, \tag{4.170}$$

then x_p has the form

$$x_\mathrm{p}(t) = A_n t^n + A_{n-1} t^{n-1} + \cdots + A_1 t + A_0.$$

3. If $f(t)$ is an exponential function

$$f(t) = ae^{mt},$$

then our guess for x_p is

$$x_p(t) = Ae^{mt}.$$

4. Finally, if $f(t)$ is a product of the polynomial (4.170) and one of the trigonometric functions $\sin \omega t$ or $\cos \omega t$, we try

$$x_p(t) = \left(A_n t^n + A_{n-1} t^{n-1} + \cdots + A_1 t + A_0\right) \sin \omega t$$
$$+ \left(B_n t^n + B_{n-1} t^{n-1} + \cdots + B_1 t + B_0\right) \cos \omega t.$$

Similarly, if f is a product of the polynomial (4.170) and the exponential function ae^{mt}, we try

$$x_p(t) = \left(A_n t^n + A_{n-1} t^{n-1} + \cdots + A_1 t + A_0\right) e^{mt}.$$

Example 4.5.2
To find a particular solution of

$$2x'' + 3x' - x = -3e^{2t}, \tag{4.171}$$

we try $x_p(t) = Ae^{2t}$ (note that e^{2t} is not a solution of $2x'' + 3x' - x = 0$). After substituting x_p into (4.171) and carrying out the necessary simplifications, we get that A satisfies

$$13A = 3.$$

So $x_p(t) = \frac{3}{13} e^{2t}$ is a particular solution of (4.171).

Example 4.5.3
To determine a particular solution of

$$4y'' + y = x^3, \tag{4.172}$$

we try

$$y(x) = Ax^3 + Bx^2 + Cx + D. \tag{4.173}$$

The choice of this polynomial ensures that x^3 and all of its derivatives are represented in (4.173). Then substituting x_p into the differential equation yields the identity

$$x^3 = Ax^3 + Bx^2 + (6A + C)x + (D + 2B).$$

The set of functions $\{x^3, x^2, x, 1\}$ is linearly independent. Therefore, we can equate the coefficients of respective terms on either side of the above expression and end up with

$$A = 1, \quad B = 0, \quad 6A + C = 0, \quad D + 2B = 0,$$

from which we determine the values for C and D. We obtain that $x_p(t) = x^3 - \frac{1}{6}x$.

An important aspect of the method of undetermined coefficients is that it is a linear algorithm; in the sense that when the forcing term f is a linear combination of functions in the class of polynomials, exponentials, or sines and cosines, we can construct a trial solution x_p as a linear combination of trial solutions associated with individual terms in f. For example, we are able to construct a particular solution of

$$L[x] = f_1(t) + f_2(t) \tag{4.174}$$

by first constructing a particular solution $x_{p_1}(t)$ of

$$L[x] = f_1(t) \tag{4.175}$$

and a particular solution $x_{p_2}(t)$ of

$$L[x] = f_2(t). \tag{4.176}$$

Then the sum

$$x_p(t) = x_{p_1}(t) + x_{p_2}(t) \tag{4.177}$$

is a particular solution of (4.174). We leave the proof of this result to the reader. The next example shows how we carry out this idea in practice.

Example 4.5.4
Consider the equation

$$x'' - 3x' + 2x = e^{-t} + 7e^{3t}. \tag{4.178}$$

Here $f_1(t) = e^{-t}$ and $f_2(t) = 7e^{3t}$. Each exponential function in the forcing function gives rise to a term in our trial solution:

$$x_p(t) = Ae^{-t} + Be^{3t}. \tag{4.179}$$

We substitute (4.179) into (4.178) and end up with the relation

$$6Ae^{-t} + 2Be^{3t} = e^{-t} + 7e^{3t}. \tag{4.180}$$

Since the functions e^{-t} and e^{3t} are linearly independent, the coefficients of the respective terms in (4.180) must be equal (why?). Hence,

$$6A = 1, \quad 2B = 7.$$

It then follows that a particular solution to (4.178) is

$$x_p(t) = \frac{1}{6}e^{-t} + \frac{7}{2}e^{3t}.$$

Next, consider the differential equation

$$x'' + x' + x = \sin 3t + 4e^{-2t}. \tag{4.181}$$

The functions f_1 and f_2 in (4.174) are $\sin 3t$ and $4e^{-2t}$. We then choose $x_{p_1}(t) = A \sin 3t + B \cos 3t$ and $x_{p_2}(t) = Ce^{-2t}$ as the individual trial particular solutions. After combining these particular solutions, our guess for x_p is

$$x_p(t) = A \sin 3t + B \cos 3t + Ce^{-2t}.$$

Substituting x_p into (4.181) leads to the three simultaneous equations

$$-8A - 3B = 1, \quad -8B + 3A = 0, \quad 3C = 4,$$

from which we deduce that $A = -\frac{8}{73}$, $B = \frac{3}{73}$, and $C = \frac{4}{3}$. Therefore, $x_p(t) = -\frac{8}{73} \sin 3t + \frac{3}{73} \cos 3t + \frac{4}{3}e^{-2t}$.

4.5.2   The Annihilator Method

An alternative, and more general, way of viewing the method of undetermined coefficients is to use the concept of the **annihilator**. To introduce this idea, we start by casting the nonhomogeneous differential equation in its differential operator form

$$L[x] = f(t) \qquad\qquad (4.182)$$

and ask if there is a differential operator N that annihilates f, that is, yields

$$N[f] = 0.$$

Having such an operator in hand, we apply N to both sides of (4.182). We obtain the higher order *homogeneous* differential equation

$$N[L[x]] = 0,$$

whose general solution we will be able to determine by the techniques of the previous section. The appropriate form of the particular solution of the lower order equation $L[x] = f$ then appears as part of the general solution of the higher order equation $N[L[x]] = 0$.

It is not difficult to guess what form N should take if f belongs to the class of functions we alluded to before. For instance, if $f(t) = e^{mt}$, then

$$N = (D - m),$$

since $(D - m)(e^{mt}) = 0$. Similarly, if $f(t) = e^{at} \sin bt$, then

$$N = (D - (a + bi))(D - (a - bi)) = (D^2 - 2aD + (a^2 + b^2)). \qquad (4.183)$$

This choice of N should be clear from our experience with second-order differential equations whose characteristic polynomial has complex roots. Returning now to the differenetial equation in Example 4.5.1, we note that (4.162) is equivalent to $L[x] = f$, where

$$L = D^2 + 3D + 2,$$

and $f(t) = \sin t$. According to (4.183), $a = 0$ and $b = 1$. Thus, the operator

$$N = (D^2 + 1)$$

annihilates $\sin t$. Next, we apply N to $L[x] = f$, obtaining

$$N[L[x]] = (D^2 + 1)(D^2 + 3D + 2) = 0. \tag{4.184}$$

Thus, the general solution of (4.162) is also a solution of the fourth-order equation. Its characteristic polynomial is

$$(m^2 + 1)(m^2 + 3m + 2)$$

with roots $\pm i$ and -2 and -1. Thus,

$$x(t) = c_1 e^{-2t} + c_2 e^{-t} + c_3 \sin t + c_4 \cos t. \tag{4.185}$$

The first two terms of (4.185) form the complementary solution of (4.162). Therefore, the remaining part, $c_3 \sin t + c_4 \cos t$, is the appropriate trial solution for x_p of the latter equation, as we guessed in Example 4.5.1.

4.5.3 The Case of Resonance

In all of the preceding examples the forcing term f and the complementary solution x_c of the associated homogeneous equation did not have any terms in common. Putting it slightly differently, the operator L is not an annihilator of f. There are important examples in physics where mathematical models naturally lead to differential equations of the type $L[x] = f$, where f is a linear combination of terms, one of which is annihilated by L. This phenomenon, which we call **resonance**, occurs in the case of the mass-spring system

$$mx'' + kx = \sin \omega t,$$

when the natural frequency of the system, $\sqrt{\frac{k}{m}}$, is the same as the forcing frequency ω. In such a case, the recipe we stated prior to the definition of annihilator fails, as the next example demonstrates.

Example 4.5.5
Consider the differential equation

$$x'' + x = \sin t. \tag{4.186}$$

Following the recipe outlined after Example 4.5.1, we try

$$x_p(t) = A \sin t + B \cos t. \tag{4.187}$$

The function x_p is not a suitable candidate for a particular solution to the differential equation since $L[x] = x'' + x$ already annihilates x_p. Hence, substituting (4.187) in (4.186) leads to the expression

$$0 = \sin t,$$

from which we cannot determine A and B.

The concept of annihilator is particularly useful with the equation in the previous example. To illustrate, let us return to the differential equation in (4.186). We note that this equation can be written as

$$L[x] = (D^2 + 1)x = \sin t. \tag{4.188}$$

Here the forcing term $\sin t$ is a solution of the homogeneous equation

$$(D^2 + 1)x = 0;$$

so, in this special case, $N = L = D^2 + 1$. Applying N to (4.188), we get

$$N[L[x]] = (D^2 + 1)^2 x = (D^2 + 1)(\sin t) = 0. \tag{4.189}$$

The general solution of the fourth-order equation $(D^2 + 1)^2 x = 0$ is determined from its characteristic polynomial

$$(m^2 + 1)^2 = 0,$$

whose roots are $\pm i$ with multiplicity 2. Thus,

$$x(t) = c_1 \sin t + c_2 \cos t + c_3 t \sin t + c_4 t \cos t$$

is the general solution of (4.189). Since the first two terms on the right-hand side form the complementary solution of (4.186), the remaining two terms (where we have replaced c_3 and c_4 by A and B)

$$x_{\mathrm{p}}(t) = At \sin t + Bt \cos t \tag{4.190}$$

forms the appropriate guess for the particular solution of (4.188). After substituting (4.190) into (4.188) and solving for A and B, we arrive at the particular solution

$$x_{\mathrm{p}}(t) = -\frac{1}{2} t \cos t. \tag{4.191}$$

We labeled a system (4.159) in resonance if a part of the forcing term f is already part of the complementary solution—the solution of the associated homogeneous equation. As (4.188) and its particular solution demonstrate, when the input frequency of the forcing term and the natural frequency of the vibrating system match, it is then possible to generate solutions that have unbounded amplitude, which in many practical problems lead to disastrous consequences. In the case of (4.188) the amplitude of the input function is 1, while the output function (4.191) achieves as large a value as we wish, positive or negative, as time evolves. We will have more to say about resonance in mechanical systems in the projects at the end of the chapter.

Example 4.5.6
We seek a particular solution to the differential equation

$$x'' + 3x' + 2x = 4e^{-t} \tag{4.192}$$

whose complementary solution is given by

$$x_c(t) = c_1 e^{-t} + c_2 e^{-2t}. \tag{4.193}$$

Because e^{-t} is a part of the complementary solution, it would be useless to try Ae^{-t} as a trial particular solution of (4.192). Instead we write this equation in its differential operator form

$$(D^2 + 3D + 2)x = 4e^{-t} \tag{4.194}$$

and note that we can write $L = D^2 + 3D + 2$ as $(D+2)(D+1)$. The part $D+1$ of L is responsible for the function e^{-t} in the complementary solution (4.193). Hence, the annihilator operator N in this case is

$$N = D + 1. \tag{4.195}$$

Apply N to both sides of (4.192) and arrive at the third-order homogeneous equation

$$(D + 3)(D + 1)^2 x = 0. \tag{4.196}$$

The general solution of this equation is $c_1 e^{-t} + c_2 e^{-2t} + c_3 t e^{-t}$, since -1 is a repeated root of the characteristic polynomial. Recalling (4.193), we try

$$x_p(t) = Ate^{-t} \tag{4.197}$$

as a candidate for a particular solution of (4.192). After substituting this function into (4.192), we find that $A = 4$. So, the general solution of (4.192) is

$$x(t) = c_1 e^{-t} + c_2 e^{-2t} + 4te^{-t}. \tag{4.198}$$

Note that unlike the case of the general solution of (4.188), the solution in (4.198) still approaches zero as t approaches infinity. Although the effect of resonance is not nearly as dramatic as in the case of (4.191), its effect is felt nevertheless since the function te^{-t} increases before it is damped out by the effect of e^{-t}. In some practical situations this initial increase in x_p is large enough to cause concern for the overall safety performance of the mechanical system this equation models.

It is possible that the forcing term f in (4.159) is not part of the complementary solution, but a derivative of it is. In such cases the annihilation method we described still leads to the correct choice for the ansatz for the particular solution, even though "resonance," at least as we have defined it, does not occur. The next example points to such a case.

Example 4.5.7
Consider the differential equation

$$x'' + 4x = t \sin 2t. \tag{4.199}$$

The complementary solution is $x_c(t) = c_1 \sin 2t + c_2 \cos 2t$. Even though the forcing term $t \sin 2t$ is linearly independent from the complementary solution, its first derivative, $\sin 2t + 2t \cos 2t$, is not. This is cause for concern because the choice $x_p(t) = At \sin 2t + Bt \cos 2t$ will not lead to a successful determination of A and B (we leave it to the reader to verify this fact). Note that L in this case is $D^2 + 4$ and that the annihilator of the forcing term is $N = (D^2 + 4)^2$. Applying N to (4.199) leads to

$$(D^2 + 4)^3 x = 0. \tag{4.200}$$

The general solution of this equation is ($\pm 2i$ are roots of the characteristic polynomial with multiplicity 3)

$$c_1 \sin 2t + c_2 \cos 2t + c_3 t \sin 2t + c_4 t \cos 2t + c_5 t^2 \sin 2t + c_6 t^2 \cos 2t. \tag{4.201}$$

Compare this solution with the complementary solution of (4.197). It is clear that a good choice for the particular solution of (4.197) is

$$x_p(t) = At \sin 2t + Bt \cos 2t + Ct^2 \sin 2t + Dt^2 \cos 2t. \tag{4.202}$$

Example 4.5.8 (The algorithm in *Mathematica* and *MATLAB*)

It should be clear by now that the method of undetermined coefficients requires a judicious initial guess at the particular solution combined with routine and tedious algebra that eventually leads to the unknown coefficients. In this example we point out how we can use *Mathematica* or *MATLAB* effectively in the latter aspect of the computation. Consider the differential equation

$$x'' + 3x' - 4x = 3 \sin t + 7e^{-4t} - e^{2t}. \tag{4.203}$$

Following the arguments presented in the past few examples, we choose

$$x_p(t) = A \sin t + B \cos t + Cte^{-4t} + De^{2t} \tag{4.204}$$

as the candidate for the particular solution. After substituting x_p in (4.203), we end up with an expression that is an identity in t and, therefore, valid for all $t \in R$. One way to obtain four equations in the four unknowns A, B, C, and D is to take advantage of the linear independence of the functions $\sin t$, $\cos t$, te^{-4t}, and e^{2t}. This is the approach we took throughout the previous examples. A second way, which is somewhat more suitable for the computer programs we will present shortly, is to evaluate this identity at four distinct value of t, say $t = 0$, 1, -1, and 2. This way we obtain four algebraic equations in the unknowns. A third way, used in the programs that follow, is to obtain four equations by evaluating the identity at $t = 0$, then differentiate the identity and evaluate the result at $t = 0$, and continue to differentiate the identity (in this case three times) until a sufficient number of equations are collected. When we carry out the latter idea on $L[x_p] = 3 \sin t + 7e^{-4t} - e^{2t}$, we get the following four simultaneous equations for the unknowns A, B, C, and D:

$$\begin{aligned}
3A - 5B - 5C + 6D &= 6, \\
-5A - 3B + 20C + 12D &= -27, \\
-3A + 5B - 80C + 24D &= 108, \\
5A + 3B + 320C + 48D &= 459.
\end{aligned} \qquad (4.205)$$

This system has the unique solution $A = -\frac{15}{16}$, $B = -\frac{9}{34}$, $C = -\frac{7}{5}$, and $D = -\frac{1}{6}$.

The following two programs in *Mathematica* and *MATLAB* begin with the differential equation (4.203) and end with the exact solution of the algebraic system (4.205). We give two slightly different versions of these programs in *Mathematica*.

```
L[x_] := D[x, {t, 2}] + 3 D[x, t] -4 x - 3 Sin[t] - 7 Exp[-4 t] +
     Exp[2 t];
xp[t_] = a Sin[t] + b Cos[t] + c t Exp[-4 t] + d Exp[2 t];
operator = L[xp[t]];
eqns = {operator/. t->0, D[operator, t] /. t->0,
     D[operator, {t,2}]/. t-> 0, D[operator, {t,3}]/. t-> 0};
coeffs = Solve[eqns == 0, {a, b, c, d}];
Print[coeffs]
DSolve[x''[t] + 3 x'[t] - 4 x[t] == 3 Sin[t] +
     7 Exp[-4 t] - Exp[2 t], x[t], t]
```

The last line is put in the program to check the answer we get from `coeffs`.

A second way of achieving the same result in *Mathematica* is to use `CoefficientList`:

```
L[x_] := D[x, {t, 2}] + 3 D[x, t] -4 x - 3 Sin[t] - 7 Exp[-4 t] +
     Exp[2 t];
xp[t_] = a Sin[t] + b Cos[t] + c t Exp[-4 t] + d Exp[2 t];
operator = L[xp[t]]   /. {Sin[t] -> 1, Cos[t] -> s,
        Exp[-4 t] -> s^2, Exp[2 t] -> s^3};
eqns = CoefficientList[operator, s];
coeffs = Solve[eqns == 0, {a, b, c, d}];
Print[coeffs]
DSolve[x''[t] + 3 x'[t] - 4 x[t] == 3 Sin[t] +
     7 Exp[-4 t] - Exp[2 t], x[t], t]
```

In the third line in this program we have taken advantage of the linear independence of the functions $\sin t$ and $\cos t$, for example, and converted `operator` to a polynomial in a singe variable s. We can then apply `CoefficientList` to this polynomial.

Here is how this algorithm is implemented in *MATLAB*:

```
syms t xp A B C D
xp = A*sin(t)+B*cos(t)+C*t*exp(-4*t)+D*exp(2*t);
```

```
l = diff(xp,2,t)+3*diff(xp,t)-4*xp;
eqn=l-3*sin(t)-7*exp(-4*t)+exp(2*t);
eqn1=subs(eqn,0,t);
eqn2=subs(diff(eqn,t),0,t);
eqn3=subs(diff(eqn,t,2),0,t);
eqn4=subs(diff(eqn,t,3),0,t);
[A B C D]= solve(eqn1,eqn2,eqn3,eqn4)
dsolve('D2x+3*Dx-4*x=3*sin(t)+7*exp(-4*t)-exp(2*t)')
```

Again, the last line is used to check the answer one gets from `coeffs`.

The generalization of these ideas to higher order equations is straightforward and no special complications present themselves. The main source of consternation is that the level of tediousness of the algebra seems to grow rather fast with the degree of the differential equation. The next example shows that the method of undetermined coefficients is equally efficient with third- and higher order equations.

Example 4.5.9
Consider the third-order equation

$$x''' + x'' = 3 + t. \tag{4.206}$$

In operator notation this equation is $L[x] = 3 + t$, where $L = D^3 + D^2$, and its complementary solution is $x_c(t) = c_1 + c_2 t + c_3 e^{-t}$. The operator $N = D^2$ is the annihilator of the forcing term $3 + t$; so if x satisfies (4.206), it also satisfies the fifth-order equation

$$D^2(D^3 + D^2)x = 0. \tag{4.207}$$

The latter equation's characteristic polyomial is $m^5 + m^4 = 0$, whose roots are $m = 0$ with multiplicity 4 and $m = -1$. Hence, its general solution is

$$x(t) = c_1 + c_2 t + c_3 t^2 + c_4 t^3 + c_5 e^{-t}, \tag{4.208}$$

which, upon comparison with the complementary solution of (4.206), suggests that the candidate for a particular solution of this equation is

$$x_p(t) = A + Bt + Ct^2 + Dt^3. \tag{4.209}$$

Problems

1. Verify the contents of Examples 4.5.1 through 4.5.4 in *Mathematica* or *MATLAB*.

2. Find the annihilator for each of the following functions. Check each answer in *Mathematica* or *MATLAB*.

   (a) $\sin 2t$

   (b) $t^2 \sin 2t$

   (c) t^4

   (d) $e^{-3t} + 2e^t$

   (e) $t^2 e^{-3t}$

   (f) $t^2 + \sin 5t$

   (g) $e^t \sin 2t$

   (h) $te^{-t} \cos 3t$

   (i) $t^3 e^t + \sin 2t$

3. Find the general solution to the following differential equations. Verify each answer in *Mathematica* or *MATLAB*.

   (a) $x'' + x = e^t$

   (b) $y'' + y' + y = \sin 2t$

   (c) $x'' + x = \cos t$

   (d) $x'' - 4x = 2\sin t + \cosh 2t$

   (e) $y'' + 6y' + 9y = \sin t + 2e^{-3t}$

   (f) $x'' - x = e^t \cos t$

   (g) $x''' - 3x'' + 3x' + x = 1 + t$

   (h) $x''' + x'' = 1 - t$

   (i) $y^{(iv)} + y = \sin x$

4. Find the solution of each initial value problem and plot the graph of the solution over a suitable interval in t.

   (a) $x'' + 4x = \sin t$, $x(0) = 0, x'(0) = 0$

   (b) $x'' + 4x = \sin 2t$, $x(0) = 0, x'(0) = 0$. Discuss the difference between the behavior of the solution of this problem and that of (a).

   (c) $y'' + 2y' + y = 1$, $y(0) = y'(0) = 0$

   (d) $y'' + 3y' + 2y = \sin t$, $y(0) = y'(0) = 0$

   (e) $z'' + z' = 1$, $z(0) = -1$, $z'(0) = 0$

   (f) $y'' + 4y' + 4y = e^{-2t}$, $y(0) = y'(0) = 0$

   (g) $y'' - 2y' - y = \sin 3t + 2\cos t$, $y(0) = y'(0) = 0$

   (h) $y'' + 5y' + 6y = te^{-2t}$, $y(0) = y'(0) = 0$

   (i) $2w'' + w' + 3w = t^2 - 3t + 1$, $w(0) = w'(0) = 1$

5. Find the solution of each initial value problem and plot the graph of the solution over a suitable interval in t.

   (a) $y''' - 8y = 1$, $y(0) = y'(0) = y''(0) = 0$. What is the asymptotic behavior of the solution?

   (b) $y''' + y'' = 1 - 3t$, $y(0) = y'(0) = y''(0) = 0$

(c) $w''' + 4w'' = \sin t, \quad w(0) = w'(0) = w''(0) = 0$

(d) $z^{(iv)} + z = \sin t, \quad z(0) = z'(0) = z''(0) = z'''(0) = 0$

(e) $y''' + 4y'' + 4y' = 2t - 3, \quad y(0) = y'(0) = y''(0) = 0$

(f) $y''' + 3y'' + 3y' + y = t\sin t, \quad y(0) = y'(0) = y''(0) = 0$

(g) $u^{(iv)} - u = t\cos t, \quad u(0) = 0, \quad u'(0) = 1, u''(0) = u'''(0) = 0$

(h) $y''' + y' = 1 + e^{-t}\sin 3t, \quad y(0) = y'(0) = y''(0) = 0$

6. Find the solution of each boundary value problem and plot the graph of the solution over the given interval in t.

(a) $y'' + y = t, \quad y(0) = 0, \quad y(1) = 1$

(b) $y'' + 2y = t, \quad y(0) = 0, \quad y(1) = 1$

(c) $y'' + y = t + \cos t, \quad y(0) = 0, \quad y(1) = 1$

(d) $z'' + 2z' + z = e^{-t}, \quad z(0) = 1, \quad z(1) = 0$

(e) $y''' - y = 1, \quad y(0) = 0, \quad y'(0) = 0, \quad y(1) = 1$

(f) $w''' + w' = a, \quad w(0) = 0, \quad w'(0) = 1, \quad w(1) = 1$

7. Complete the computations in Example 4.5.9.

8. Consider the forced mass-spring system modeled by the differential equation $3x'' + 2x = 4\sin\omega t$.

(a) Find all values of ω that lead to resonance in this system.

(b) Let ω_r be a resonant frequency. Let $x(0) = x'(0) = 0$. Find the solution to the initial value problem. Draw the graph of the solution for $t \in (0, 10)$. How does this graph compare with the graph of the solution to the initial value problem with $\omega = \omega_r - 0.1$? How does it compare with the solution to the initial value problem when $\omega = \omega_r + 0.1$? Plot all three solutions on the same screen.

9. Consider the damped mass-spring system $x'' + ax' + bx = e^{-0.1t}\cos 2t$, where a and b are nonnegative constants. Find all values of a and b for which the system is in resonance.

4.6 Variation of Parameters Formula

The technique we describe now is another method for obtaining a particular solution to an ODE with a forcing term. The main difference between this method and the method of undetermined coefficients is that the present method will also apply to differential equations with nonconstant coefficients as well as to forcing terms that do not belong to the special class of sine, cosine and exponential functions.

We outline the method of variation of parameters in the context of second-order differential equations. We emphasize, however, that this method is quite general and can be applied to higher-order differential equations.

Consider the differential equation

$$x'' + p(t)x' + q(t)x = f(t), \tag{4.210}$$

where p, q, and f are smooth functions. We assume that the homogeneous equation has the general (complementary) solution

$$x_c(t) = c_1 x_1(t) + c_2 x_2(t). \tag{4.211}$$

The **method of variation of parameters** seeks a particular solution to (4.210) in the form (4.211) where the coefficients c_1 and c_2 are allowed to depend on t. Specifically, we look for a particular solution x_p of (4.210) in the form

$$x_p(t) = C_1(t)x_1(t) + C_2(t)x_2(t). \tag{4.212}$$

We begin by substituting (4.212) into (4.210). After carrying out the necessary differentiations and using the fact that x_1 and x_2 satisfy the homogeneous equation, we end up with the following equation in C_1 and C_2:

$$(C_1'' x_1 + C_2'' x_2) + 2(C_1' x_1' + C_2' x_2') + p(t)(C_1' x_1 + C_2' x_2) = f(t). \tag{4.213}$$

Equation (4.213) represents one equation in the two functions C_1 and C_2. We need to supplement this equation with another one involving these unknowns. Traditionally, the second equation is chosen in such a way so as to simplify the computation of C_1 and C_2. Keeping this point in mind, we choose the second relation so that the third term on the left side of (4.213) is zero, that is,

$$C_1' x_1 + C_2' x_2 = 0. \tag{4.214}$$

We differentiate (4.214) once and find a second relation between C_1 and C_2:

$$(C_1'' x_1 + C_2'' x_2) + (C_1' x_1' + C_2' x_2') = 0 \tag{4.215}$$

The above equation further simplies (4.213): Substituting (4.215) into (4.213) leads to

$$C_1' x_1' + C_2' x_2' = f(t). \tag{4.216}$$

Equations (4.214) and (4.216) define two simultaneous equations in the unknowns C_1' and C_2' in terms of the known functions x_1, x_2, and f. The solution to this system is

$$C_1'(t) = -\frac{f(t)x_2(t)}{w(t)}, \qquad C_2'(t) = \frac{f(t)x_1(t)}{w(t)}, \tag{4.217}$$

where $w(t)$ is the wronskian of (4.210), that is,

$$w(t) = \det \begin{bmatrix} x_1(t) & x_2(t) \\ x_1'(t) & x_2'(t) \end{bmatrix} = x_1(t)x_2'(t) - x_2(t)x_1'(t). \tag{4.218}$$

We note that w is not identically zero if the two complementary solutions x_1 and x_2 are linearly independent. Equations (4.217) yield C_1 and C_2 after an integration. Putting this information together with (4.212), we obtain the following formula for a particular solution to (4.210):

$$x_{\mathrm{p}}(t) = -x_1(t) \int \frac{f(t)x_2(t)}{w(t)} dt + x_2(t) \int \frac{f(t)x_1(t)}{w(t)} dt. \qquad (4.219)$$

We summarize the results of these computations in the following theorem.

Theorem 4.6.1 (Variation of Parameters Formula)
Formula (4.219) provides a particular solution to (4.210) once two linearly independent solutions x_1 and x_2 of the homogeneous part of (4.210) are known.

Example 4.6.1
Consider the differential equation

$$x'' + x = \sin t. \qquad (4.220)$$

We could find a particular solution of this equation by the method of undetermined coefficients, as described in the previous section. However, to illustrate the various steps in the method of undetermined coefficients, we now apply this method to (4.220) and obtain its general solution. First we need to find two linearly independent solutions to the homogeneous equation

$$x'' + x = 0. \qquad (4.221)$$

They are

$$x_1(t) = \cos t, \quad x_2(t) = \sin t. \qquad (4.222)$$

The wronskian of (4.222) is

$$w(t) = \det \begin{bmatrix} \cos t & \sin t \\ -\sin t & \cos t \end{bmatrix} = 1. \qquad (4.223)$$

Equation (4.219) now yields

$$x_{\mathrm{p}}(t) = -\cos t \int \sin^2 t\, dt + \sin t \int \sin t \cos t\, dt. \qquad (4.224)$$

The integrations in (4.224) are standard. We end up with

$$x_{\mathrm{p}}(t) = -\frac{t}{2} \cos t + \frac{1}{2} \sin t. \qquad (4.225)$$

Remark 4.6.1: The particular solution we obtained in (4.225) is not the same as that we would have obtained had we used the method of undetermined coefficients. Nevertheless, what we have computed is a particular solution to

(4.220), and as the results of previous sections showed, when this x_p, or the one we obtain from the method of undetermined coefficients, is added to the complementary solution of (4.220), we arrive at the general solution to (4.220).

Example 4.6.2
Consider the differential equations

$$t^2 x'' + t x' - x = \sin t. \tag{4.226}$$

This equation is a second-order differential equation with nonconstant coefficients. We have not yet developed an algorithm to find the general solution of the homogeneous part $t^2 x'' + t x' - x = 0$, nevertheless the reader can verify by direct differentiation that

$$x_1(t) = t, \quad x_2(t) = \frac{1}{t} \tag{4.227}$$

satisfy the homogeneous part of (4.226). The wronskian of (4.227) is

$$w(t) = -\frac{2}{t}. \tag{4.228}$$

The nonhomogeneous part of (4.226) is

$$f(t) = \frac{\sin t}{t^2} \tag{4.229}$$

(compare (4.226) with (4.210)). Then, a particular solution to (4.226) is given by

$$x_p(t) = 2t \int \frac{\sin t}{t^4} dt - \frac{2}{t} \int \frac{\sin t}{t} dt. \tag{4.230}$$

Equation (4.230) gives a formula for x_p, up to the integration of the preceding terms. Unfortunately, these integrations, although defined in terms of standard mathematical functions, are nontrivial. The definitions of these antiderivatives are best described in the language of complex analysis, which we will take up in a later chapter. It suffices to say at this point that the functions that constitute x_p are available in both *Mathematica* and *MATLAB*, of which we now take advantage in order to learn about the long-term behavior of x_p. Figure 4.5 depicts the graph of x_p. This graph was obtained by the following set of *Mathematica* commands:

```
C1[t_] := NIntegrate[Sin[u]/u^4, {u, 1, t}];
C2[t_] := NIntegrate[Sin[u]/u, {u, 1, t}];
xp[t_] := 2 t C1[t] - 2/t C2[t];
Plot[xp[t], {t, 1, 10}]
```

Equivalently, we could obtain Figure 4.5 by executing the following code in *MATLAB*:

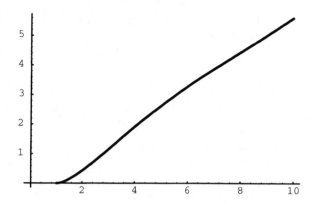

Figure 4.5: The graph of the particular solution to $t^2x'' + tx' - x = \sin t$.

```
syms t u c1 c2
c1 = int(sin(u)/(u^4),u,1,t);
c2 = int(sin(u)/u,u,1,t);
xp = 2*t*c1-2/t*c2;
ezplot(xp, [1, 10])
```

Example 4.6.3

Mathematica and *MATLAB* are ideal tools for carrying out the details of the variation of parameters formula (4.219). Once the functions x_1 and x_2 are entered into *Mathematica*, the following program shows how one proceeds with the algorithm. The program is written for

$$x'' + 3x' + 2x = \tan t \qquad (4.231)$$

but the reader may easily modify it for any other equation.

```
x1[t_] = Exp[-2 t]; x2[t_] = Exp[- t]; f[t_] = Tan[t];
w[t_] = Det[{{x1[t], x2[t]}, {D[x1[t],t], D[x2[t],t]}}];
xp[t_] := - x1[t] NIntegrate[f[u]*x2[u]/w[u], {u, 0, t}] +
        x2[t] NIntegrate[f[u]*x1[u]/w[u], {u, 0, t}];
Plot[xp[t], {t, 0, 1}]
```

The output of this program should be compared with Figure 4.6. The *MATLAB* program that leads to the same conclusion is

```
syms  t x1 x2 f a b w integrand1 u term1 integrand2 term2 xp
x1=exp(-2*t); x2=exp(-t); f=tan(t);
a=[x1; x2];
b=diff(a,t);
w=det([a b]);
integrand1=x2*f/w; integrand1=subs(integrand1,u,t);
term1=int(integrand1,u,0,t);
```

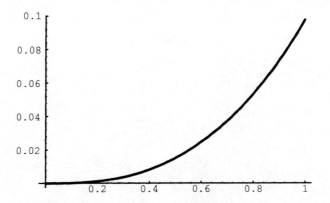

Figure 4.6: The graph of the particular solution to $x'' + 3x' + 2x = \tan t$.

```
integrand2=x2*f/w; integrand2=subs(integrand2,u,t);
term2=int(integrand2,u,0,t);
xp=x2*term2-x1*term1;
graph=[];
for i=1:100
    graph=[graph numeric(subs(xp,0.01*i,t))];
end
plot(graph)
```

Problems

1. Use an editor and create a file called **variat.m** for either of the programs in Example 4.6.2.

   (a) Run this program and compare the output with Figure 4.6.

   (b) What is the response of **DSolve** of *Mathematica* or **dsolve** of *MAT-LAB* to the differential equation in (4.231)?

   (c) Consider this differential equation with the initial data

   $$x(0) = 0, \quad x(0) = 1.$$

   Does the presence of initial data affect the response of **DSolve** or **dsolve**?

   (d) Try either **NDSolve** or **ode45** with this initial value problem. Plot the solution in the interval $(0, 1)$.

2. Use the variation of parameters formula and find a particular solution for each of the following differential equations. Check each answer in either *Mathematica* or *MATLAB*.

(a) $x'' + x = \sin 2t$

(b) $x'' + 4x = \sin 2t$

(c) $x'' + 3x' + 2x = 3e^{-t}$

(d) $y'' + y = \tan t$

(e) $y'' + 4y = \sin^2 t$

(f) $x^2 y'' + xy' - y = 1$, where the complementary solution is $y(x) = c_1 x + \frac{c_2}{x}$

(g) $x^2 y'' + xy' - y = x^2$, where the complementary solution is stated in (f)

(h) $x'' - x = \sinh t$

(i) $x'' + 4x' + 4x = e^{-2t}$.

3. Find the solution to the initial value problem

$$x'' + x = \frac{1}{t}, \quad x(1) = x'(1) = 1. \tag{4.232}$$

Use the program in Example 4.6.2 and plot the graph of the solution. Compare your solution to the output of DSolve.

4. The method of variational parameters applies equally well to higher-order equations. Consider the third-order equation

$$x''' + a_1 x'' + a_2 x' + a_3 x = f(t). \tag{4.233}$$

We seek a particular solution of (4.233) in the form

$$x_p(t) = c_1(t)x_1(t) + c_2(t)x_2(t) + c_3(t)x_3(t), \tag{4.234}$$

where the functions x_1, x_2, and x_3 are linearly independent solutions of the homogeneous equation in (4.234).

(a) Show that if c_is satisfy the equations

$$\begin{aligned}
c_1' x_1 + c_2' x_2 + c_3' x_3 &= 0, \\
c_1' x_1' + c_2' x_2' + c_3' x_3' &= 0, \\
c_1' x_1'' + c_2' x_2'' + c_3' x_3'' &= f(t),
\end{aligned} \tag{4.235}$$

then the function x_p given by (4.234) is a particular solution of (4.233).

(b) Show that the solution $\{c_1', c_2', c_3'\}$ of (4.235) is given by

$$c_i' = \frac{w_i}{w}, \tag{4.236}$$

where w is the wronskian of the solution set $\{x_1, x_2, x_3\}$ and w_i is the determinant of the same matrix we construct to determine the wronskian except that its ith column is replaced by the right-hand side of (4.235)

$$\begin{bmatrix} 0 \\ 0 \\ f(t) \end{bmatrix}.$$ (4.237)

So, for example, w_2 is defined as

$$w_2 = \det \begin{bmatrix} x_1 & 0 & x_3 \\ x_1' & 0 & x_3' \\ x_1'' & f(t) & x_3'' \end{bmatrix}.$$ (4.238)

Note that the particular solution x_p is then known once the three integrations in (4.236) are carried out.

5. In each of the following problems find the particular solution of the third-order equation. Check each result in *Mathematica* or *MATLAB*. Draw the graph of each particular solution and comment on its behavior as the independent variable approaches infinity.

   (a) $x''' - 3x'' + 3x' - x = e^{-4t}$

   (b) $x''' - x = e^{-t}$

   (c) $y''' = e^{-2t} + 3t$

   (d) $y''' + 27y = \sin 3t$

   (e) $x^3 y''' + 5x^2 y'' + 2xy' - 2y = \frac{1}{x}$. First show that $y_1(x) = x$, $y_2(x) = \frac{1}{x}$, and $y_3(x) = \frac{1}{x^2}$ are three linearly independent complementary solutions of this equations.

6. Find the solution to the each initial value problem and determine its behavior as the independent variable approaches infinity.

   (a) $y''' + 3y'' + 9y' + 27y = \sin 2t$, $y(0) = y'(0) = y''(0) = 0$

   (b) $y''' - y = \sin t$, $y(0) = y'(0) = y''(0) = 0$

   (c) $3y''' + 5y'' = t + \sin 2t$, $y(0) = y'(0) = y''(0) = 0$

   (d) $y''' + y' = 1$, $y(0) = y'(0) = y''(0) = 0$.

7. Write a *Mathematica* or *MATLAB* program that begins with three linearly independent complementary solutions of the differential equation

$$x''' + ax'' + bx' + cx = f(t)$$

and f and proceeds to compute c_1, c_2, and c_3 as described in Problem 4. Apply your program to

$$x''' + x'' = \sin 2t.$$

Check your result against the answer from either DSolve or dsolve.

8. Generalize the analysis of Problem 4) to an nth-order differential equation

$$x^{(n)} + a_{n-1}x^{(n-1)} + \cdots + a_1 x' + a_0 x = f(t).$$

4.7 Reduction of Order

The method of reduction of order applied to a second-order linear differential equation provides a useful and important technique for determining a second solution of the differential equation once one solution of this equation is known. The two solutions obtained by this method will be linearly independent. The technique we present is particularly significant in the context of symbolic algebra systems, since the algorithm depends solely on integration of functions, an operation that can be carried out in closed form for a large class of elementary functions in *Mathematica* and *MATLAB*. For those functions whose antiderivatives are difficult to determine, the numerical integration operations of *Mathematica* and *MATLAB* will be quite useful, because they are capable of determining accurate approximations of many such antiderivatives. Moreover, as we will see in Section 4.10, the reduction of order method is an effective tool when we find one solution of a second-order differential equation using a series method. We are then able to determine a second solution by incorporating the series solution in the general formula we obtain in this section.

Let $x_1(t)$ be a (nonzero) solution of a general second-order differential equation of the form

$$x'' + p(t)x' + q(t)x = 0. \tag{4.239}$$

We seek a second solution $x_2(t)$ of (4.239) of the form

$$x_2(t) = x_1(t)f(t). \tag{4.240}$$

As it turns out the function f will satisfy a special second-order linear differential equation, a solution of which we can always determine.

We begin by differentiating (4.240) twice and substituting the result into (4.239). After rearranging the terms, we get

$$(x_1'' + px_1' + qx_1)f + (2x_1' + px_1)f' + x_1 f'' = 0. \tag{4.241}$$

Since x_1 is solution of (4.239), the first term on the left-hand side of (4.241) vanishes. Since the function x_1 does not vanish identically, we can divide (4.241) by x_1 and obtain a first-order linear differential equation for $g \equiv f'$:

$$g' + \left(p + 2\frac{x_1'}{x_1}\right)g = 0. \tag{4.242}$$

Note that the coefficient of g in (4.242) is a known function of t. Equation (4.242) is of the type (3.76), which we encountered in Section 3.5. Its general solution is

$$g(t) = k \exp\left(-\int^t p(s) + 2\frac{x_1'(s)}{x(s)}ds\right). \tag{4.243}$$

Without loss of generality we set $k = 1$. Introduce the change of variables $z = x_1(t)$. Then the second integral in (4.243) reduces to

$$\exp\left(-\int 2\frac{x_1'(s)}{x_1(s)}ds\right) = \exp\left(-2\int \frac{dz}{z}\right) = \frac{1}{z^2} = \frac{1}{[x_1(t)]^2}. \tag{4.244}$$

We use (4.244) in (4.243) and recall the relation between f and g:

$$f'(t) = \frac{1}{[x_1(t)]^2} \exp\left[-\int^t p(s)ds\right].$$

(4.245)

Integrating (4.245) and multiplying by $x_1(t)$ lead to the desired second solution of (4.239).

$$x_2(t) = x_1(t) \int^t \left[\frac{1}{[x_1(u)]^2} \exp\left[-\int^u p(s)ds\right]\right] du.$$

(4.246)

Note that the two functions x_1 and x_2 are linearly independent, because, if not, the function multiplying x_1 in (4.246) must be constant, requiring the integrand to be zero, which is not possible since x_1 is nonzero. We have thus proved the following theorem.

Theorem 4.7.1
Let x_1 be a (nonzero) solution of the second-order equation (4.239). Then x_2 given by (4.246) is a second linearly independent solution of this equation.

Example 4.7.1
Consider the differential equation

$$tx'' + 2x' + tx = 0.$$

(4.247)

It can be shown by direct differentiation that

$$x_1(t) = \frac{\sin t}{t}$$

(4.248)

is a solution of (4.247). We use the method of reduction of order to get a second solution. We look for a solution x_2 in the form

$$x_2(t) = f(t)x_1(t) = f(t)\frac{\sin t}{t}.$$

(4.249)

Differentiate (4.249) and substitute into (4.247):

$$f''\frac{\sin t}{t} + 2f'\frac{\cos t}{t} = 0.$$

(4.250)

Introduce the change of variable $f' = g$, which converts (4.250) to the first-order equation

$$g' + (2\cot t)g = 0$$

(4.251)

for g. Compare (4.251) with (4.242). The integrating factor for (4.251) is $\sin^2 t$. After applying the standard integrating factor algorithm to this equation, we end up with

$$g(t) = \frac{1}{\sin^2 t}.$$

(4.252)

Since $f' = g$, an integration of (4.252) leads to

$$f(t) = -\cot t.$$

Hence, the second solution to (4.247) is

$$x_2(t) = x_1(t)f(t) = -\frac{\cos t}{t}. \tag{4.253}$$

Putting (4.248) and (4.253) together, we see that the general solution to (4.247) is

$$x(t) = c_1 \frac{\sin t}{t} + c_2 \frac{\cos t}{t}. \tag{4.254}$$

The main step that could possibly be a stumbling block in implementing this method is in the integration of f'. This is the point in the algorithm that we often resort to a symbolic manipulator.

Carrying out the steps of the reduction of order method in either *Mathematica* or *MATLAB* is straightforward. We show the necessary steps for the example at hand and leave its variations to other equations to the reader. The *Mathematica* program is

```
p = 2/t;
x1 = Sin[t]/t;
x2 = x1 * Integrate[1/(x1)^2 *
            Exp[-Integrate[p, t]], t]
```

whose output is identical with (4.253). The equivalent *MATLAB* program reads

```
syms t x1 p x2
p=2/t;
x1 = sin(t)/t;
x2=x1*int((1/x1)^2*exp(-int(p,t)),t)
```

As we pointed out earlier, one of the main applications of the method of reduction of order concerns obtaining a second solution to a second-order differential equation when a first solution is constructed by a standard method. For example, the differential equation

$$x'' + 2x' + x = 0$$

has the characteristic equation $m^2 + 2m + 1 = 0$, which has a multiple root at $m = -1$, from which we deduce that $x_1(t) = e^{-t}$ is a solution. Of course, we know from previous analysis that a second linearly independent solution is $x_2(t) = te^{-t}$. The method of reduction of order is an alternative way of reaching the same conclusion, as some of the exercises will demonstrate.

Among other second-order differential equations to which this method applies is the class of equations known as the **Euler–Cauchy** equations. These equations are characterized by the fact that each derivative is multiplied by the independent variable raised to the same power as the order of the derivative; that is, a Euler–Cauchy equation has the form

$$at^2 x'' + btx' + cx = 0, \tag{4.255}$$

where a, b, and c are real constants. Because of the special way that x and t enter into (4.255), we look for solutions of this equation among functions of the form

$$x(t) = t^r. \tag{4.256}$$

After substituting this function into (4.255), we arrive at the **characteristic equation**

$$ar(r-1) + b(r-1) + cr = 0. \tag{4.257}$$

This quadratic equation generally has two roots, r_1 and r_2. When these roots are real, they lead to the two linearly independent solutions

$$x_1(t) = t^{r_1}, \quad x_2(t) = t^{r_2}. \tag{4.258}$$

When r_1 and r_2 are complex, they must be complex conjugates. Let $r = \lambda + \mu i$ be one of the roots. Then $t^r = t^{(\lambda + \mu i)} = t^\lambda t^{\mu i}$. But note that we can rewrite $t^{\mu i}$ as

$$t^{\mu i} = \exp(\ln(t^{\mu i})) = \exp(\mu i \ln t). \tag{4.259}$$

The last expression in (4.259) is now written in a form to which we can apply Euler's formula ($e^{iy} = \cos y + i \sin y$). We have

$$\exp(\mu \ln t\, i) = \cos(\mu \ln t) + i \sin(\mu \ln t). \tag{4.260}$$

It is not difficult to show that the real and imaginary parts of $\exp(\mu \ln t\, i)$ are linearly independent. We have thus constructed two linearly indepenent solutions of (4.255) when the roots of its characteristic equation are complex:

$$x_1(t) = t^\lambda \cos(\mu \ln t), \quad x_2(t) = t^\lambda \sin(\mu \ln t). \tag{4.261}$$

Finally, when the roots of the characteristic equation are equal to r, we end up with one solution $x_1(t) = t^r$ and resort to the method of reduction of order to construct a second solution $x_2(t)$. We leave it to the reader to show that this method leads to

$$x_2(t) = t^r \ln t. \tag{4.262}$$

We summarize these findings in the following theorem.

Theorem 4.7.2 (The Euler–Cauchy Equation)

The solution of the Euler–Cauchy equation (4.255) depends on the following classification of the roots of its characteristic equation

$$ar(r-1) + br + c = 0. \tag{4.263}$$

1. *If the roots r_1 and r_2 of (4.263) are real and distinct, then the general solution of (4.255) is given by*

$$x(t) = c_1 t^{r_1} + c_2 t^{r_2}. \tag{4.264}$$

2. *If (4.263) has multiple roots r, then*

$$x(t) = c_1 t^r + c_2 t^r \ln t \tag{4.265}$$

is the general solution.

3. *When the roots of the characteristic equation are complex with $r = \lambda \pm \mu i$, then*

$$x(t) = c_1 t^\lambda \cos(\mu \ln t) + c_2 t^\lambda \sin(\mu \ln t) \tag{4.266}$$

is the general solution of (4.255).

Problems

1. Verify the statements in Example 4.7.1.

2. Revise the programs in Example 4.7.1 so that each program begins with the functions p and x_1 and ends up graphing the two linearly independent solutions x_1 and x_2 on the same screen.

3. Determine the relationship between a, b, and c in (4.255) that leads to multiple roots in its characteristic equation. Let r be this repeated root and $x_1(t) = t^r$ the solution of (4.255) associated with this root. Use the method of reduction of order and prove that $x_2(t) = t^r \ln t$ is a second solution.

4. Find a second solution to the following equations using the method of reduction of order and the given solution. In each problem check your answer and the calculations that lead to it in either *Mathematica* or *MATLAB*:

   (a) $x'' + x = 0, \quad x_1(t) = \sin t$

   (b) $x'' + 4x' + 4x = 0, \quad x_1(t) = e^{2t}$

   (c) $(1 - t^2)x'' - 2tx' + 2x = 0, \quad x_1(t) = t$

   (d) $3x'' + 2x' + x = 0, \quad x_1(t) = \exp(-\frac{t}{3}) \cos \frac{\sqrt{2}t}{3}$

   (e) $(1 - x^2)y'' + 2xy' - 2y = 0, \quad y_1(x) = x$

   (f) $(1 - x^2)y'' - 2xy' + 6y = 0, \quad y_1(x) = \frac{1}{2}(x^2 - 1)$

5. Determine the parameters a, b, and c so that the function $x_1(t) = at^2 + bt + c$ is a solution to

$$(1 - t^2)x'' - 2tx' + 6x = 0. \tag{4.267}$$

Use this solution and determine a second solution of this equation.

6. Use `DSolve` in *Mathematica* and find the general solution of the differential equation in Example 4.7.1. How does the solution given by *Mathematica* relate to that obtained in the example? (Recall Euler's formula: $e^{it} = \cos t + i \sin t$.)

7. Generalize the method we described for the Euler–Cauchy equation to the third-order equation

$$at^3 x''' + bt^2 x'' + ctx' + dx = 0.$$

Apply this method to the following differential equations. Use *Mathematica* or *MATLAB* to aid with the analysis.

(a) $t^3 x''' - 3t^2 x'' + x = 0$

(b) $x^3 y''' + xy' + 4y = 0$

(c) $3x^3 y''' - 2xy' + y = 0$

(d) $x^3 y''' + x^2 y'' - 3xy' + y = 0$

(e) $x^3 y''' + 3x^2 y'' - 3xy' + y = 0$

(f) $t^3 x''' + x = 0$

8. Consider the Euler–Cauchy equation

$$x^2 y'' + xy' + ay = 0.$$

Find all values of a for which the solution to the equation approaches zero as x goes to infinity.

4.8 Series Solutions of Differential Equations

In the past several sections we developed analytic methods for obtaining the general solution of linear second-order differential equations. These methods are especially effective when the differential equation has constant coefficients but become cumbersome, and at times intractable, when the coefficients of the differential equation depend on the independent variable. In this section we will develop an alternate method based on the series expansion of functions of one variable which will prove effective for linear differential equations with or without constant coefficients. An equally important aspect of the series method is that this method is at times an effective way of obtaining approximate solutions for nonlinear equations.

First let us recall some of the basic properties of **Taylor series** expansions of elementary functions. Let x be a smooth function of a single variable t. Assuming this function has at least $n + 1$ continuous derivatives at the point t_0 in its domain, x can be approximated by the polynomial P_n of degree n in the following manner:

$$P_n(t) = a_0 + a_1(t - t_0) + a_2(t - t_0)^2 + \cdots + a_n(t - t_0)^n, \qquad (4.268)$$

where the coefficients a_i are calculated from the **Taylor formula**

$$a_i = \frac{1}{i!} \frac{d^i x(t)}{dt^i}\bigg|_{t=t_0}, \tag{4.269}$$

and $0! = 1$, by convention. How closely P_n approximates x depends on t and n. The error one makes by replacing x with P_n is quantified by the **remainder formula**

$$|x(t) - P_n(t)| \leq \frac{d^{n+1} x}{dt^{n+1}}\bigg|_{t=\xi} \frac{|t - t_0|^{n+1}}{(n+1)!}, \tag{4.270}$$

where ξ is a point between t_0 and t. Hence, because of the $(n+1)!$ factor in the denominator of (4.270) the size of the error one makes in using P_n in place of x decreases as n increases. This error also decreases the closer t is chosen to t_0 due to the presence of the $(t - t_0)$ term in the numerator of (4.270), while it is sensitive to the level of oscillations in x, because of the dependence of the error term on the derivatives of x. By a Taylor **series expansion** of x we mean

$$\lim_{n \to \infty} P_n(t),$$

where the limit is taken in the pointwise sense.

It turns out that many functions of interest in physical applications have Taylor polynomial approximations of any order, and these approximations get better with larger n. In fact, many of the special applications we encounter later in this text involve functions with Taylor series expansions that converge to the original function x on rather large domains. In most cases the interval of convergence of the series is a symmetric interval of the real line about the point t_0, while on occasion the interval of convergence is the half-line $t > 0$ or even the entire real line. In many applications it is possible, and often quite desirable, to replace the function x by its series representation.

We think of P_n as an approximation of x and expect to be able to use it wherever x is needed, as long as t is reasonably close to t_0 or n is large enough. For many functions the Taylor polynomial is such a good approximation of x that one can comfortably replace x by P_n in such computations as integration and differentiation. Because P_n is a polynomial, these operations become routine, albeit tedious, calculations that are often well-suited for symbolic and numerical packages.

When the function x is known, the computation of its Taylor polynomial (4.268) reduces to calculating the necessary derivatives in (4.269) and deducing the values of the coefficients a_i's. When x is a solution of a differential equation, however, one is not able to compute these derivatives explicitly and must resort directly to the differential equation for that purpose. The next example shows the procedure for a concrete differential equation.

Example 4.8.1
Consider the differential equation

$$x' + 2x = 0, \quad x(0) = 1. \tag{4.271}$$

Equation (4.271) is a first-order equation whose solution is

$$x(t) = e^{-2t}. \tag{4.272}$$

Applying the Taylor polynomial formula (4.268)–(4.269) to x in (4.272) shows that it has the following series representation about $t_0 = 0$:

$$x(t) = 1 - 2t + 2t^2 - \frac{4}{3}t^3 + \cdots + (-1)^n \frac{2^n}{n!}t^n + \cdots = \sum_{i=0}^{\infty}(-1)^i \frac{2^i}{i!}t^i. \tag{4.273}$$

It is not difficult to use the ratio test and confirm from (4.273) that the radius of convergence of (4.273) is infinite.

We can also obtain the series solution (4.273) by directly applying the differential equation (4.271) to (4.268). Note that the differential equation (4.271) provides a relationship between x' and x, namely

$$x'(t) = -2x(t). \tag{4.274}$$

We know from the initial datum that $x(0) = 1$, which implies $a_0 = 1$ from (4.269). Moreover, it follows from (4.274) that

$$a_1 = x'(0) = -2x(0) = -2. \tag{4.275}$$

To get a_2, we need information about $x''(0)$. The relation (4.274) provides that piece of information; differentiate this identity with respect to t and evaluate it at $t = 0$:

$$x''(t) = -2x'(t) \tag{4.276}$$

which implies

$$x''(0) = -2x'(0) = 4.$$

Hence, $a_2 = \frac{x''(0)}{2!} = 2$. Other coefficients such as a_3 and a_4 are obtained similarly, by differentiating (4.274) an appropriate number of times and evaluating the result at $t = 0$. We find that

$$x'''(t) = -2x''(t), \quad \text{so } a_3 = \frac{y'''(0)}{3!} = -\frac{4}{3}. \tag{4.277}$$

This iterative procedure can be readily programmed in *Mathematica*. What follows contains the syntax of a program that begins with the differential equation (4.271) and computes the Taylor polynomial approximation of $x(t)$ for prescribed $n = 6$:

```
x'[t_] := - 2 x[t];
x[0] := 1;
nn = 6
nthder[t_, n_] := D[x'[t], {t, n-1}];
a = Table[nthder[t, i] /. t -> 0, {i, 1, nn}];
sol[t_] = x[0] + Sum[a[[i]]/i! t^i, {i, 1, nn}]
```

Executing this program results in

```
                3     4     5     6
          2   4 t   2 t   4 t   4 t
1 - 2 t + 2 t  - ---- + ---- - ---- + ----
                3     3    15    45
```

Compare this series with the output of *Mathematica*'s internal function `Series`

`a = Series[Exp[-2 t], {t, 0, 6}]`

The function `Normal` removes the remainder in the above output.

Example 4.8.2

Consider the second-order differential equation

$$x'' + 2x' + x = 0, \quad x(0) = 1, \ x'(0) = -1. \tag{4.278}$$

The unique solution to this problem is $x(t) = e^{-t}$, with a series expansion given by (4.273) but with t replaced by $\frac{t}{2}$.

The technique described in the previous example yields the series solution to (4.278). Note first that the initial data in (4.278) imply

$$a_0 = 1 \quad \text{and} \quad a_1 = -1.$$

Also, the differential equation is equivalent to

$$x''(t) = -2x'(t) - x(t), \tag{4.279}$$

so that $x''(0) = 1$ and $a_2 = \frac{x''(0)}{2} = \frac{1}{2}$. Moreover, $x''' = \frac{dx''}{dt}$, when combined with (4.279) yields

$$x'''(t) = -2x''(t) - x'(t). \tag{4.280}$$

After using (4.279) and the initial data we arrive at $x'''(0) = -1$, which leads to $a_3 = -\frac{1}{3}$. It should be clear that this process can be repeated ad infinitum and that all coefficients a_i are computable in this manner. In practice one stops after a finite number of a_i have been obtained.

The following *Mathematica* program is based on this algorithm and is a slight modification of that given in the previous example. Its output is the Taylor polynomial of x of degree nn specified in the program (start a new session of *Mathematica*)

```
x''[t_] := - 2x'[t] - x[t];
x[0] := 1; x'[0] = -1;
nn = 6;
nthder[t_, n_] := D[x''[t], {t, n - 2}];
a = Table[nthder[t, i] /. t-> 0, {i, 2, nn}];
sol[t_] = x[0] + x'[0]*t + Sum[a[[i - 1]]/i! t^i, {i, 2, nn}]
```

It results in the output

```
      2    3    4    5     6
      t    t    t    t     t
1 - t + -- - -- + -- - --- + ---
      2    6    24   120   720
```

Example 4.8.3

Consider the differential equation

$$x'' + tx = 0, \quad x(0) = 1, x'(0) = 0. \tag{4.281}$$

This equation, which is called **Airy**'s equation, is different from those we considered in the previous examples since (4.281) has nonconstant coefficients. Hence, the methods we studied in the previous sections do not apply and we are unable to find an explicit solution. However, Theorem 3.3.1 guarantees that (4.281) has a unique solution. We now construct a series approximation to this solution. Note that from (4.281)

$$x''(t) = -tx(t), \tag{4.282}$$

with the initial data that accompanies (4.281). The initial data state that a_0 and a_1 of (4.268) are 1 and 0, respectively. To get a_2, we evaluate (4.282) at $t = 0$ to get

$$x''(0) = 0 \quad \text{or} \quad a_2 = 0.$$

To find a_3, we first differentiate (4.282) with respect to t

$$x'''(t) = -x(t) - tx'(t) \tag{4.283}$$

and then evaluate the result at $t = 0$, obtaining $x'''(0) = -1$. Hence, $a_3 = -\frac{1}{6}$. Continuing in this fashion, we find that the following Taylor polynomial (which is P_6 in the notation of (4.268)) approximates the unique solution to (4.281):

$$1 - \frac{1}{6}t^3 + \frac{1}{180}t^6. \tag{4.284}$$

In this approximation we stopped with $n = 6$. The process, however, can continue indefinitely until one obtains a series solution to (4.281). With the denominators in the coefficients of (4.284) being as large as they are, one can feel comfortable that the series solution to (4.281) has a large radius of convergence. In fact, it can be shown that its radius of convergence is infinite.

The following is the *Mathematica* program that computes the coefficients of (4.268) for this differential equation to any order n ($n = 12$ is chosen for the sake of concreteness).

```
x''[t_] := - t* x[t];
x[0] := 1; x'[0] = 0;
nn = 12;
```

```
nthder[t_, n_] := D[x''[t], {t, n-2}];
a = Table[nthder[t, i] /. t -> 0, {i, 2, nn}];
sol[t_] = x[0] + x'[0]*t + Sum[a[[i - 1]]/i! t^i, {i, 2, nn}]
```

Mathematica's output is

$$
1 - \frac{t^3}{6} + \frac{t^6}{180} - \frac{t^9}{12960} + \frac{t^{12}}{1710720}
$$

Example 4.8.4

The strength of the algorithm we described in the preceding examples becomes more apparent when we apply it to a *nonlinear* differential equation such as

$$
x'' + x' + \sin x = \cos t, \quad x(0) = 0, \, x'(0) = 0. \tag{4.285}
$$

A unique solution to this initial value problem is guaranteed by Theorem 3.3.1. However, because the equation in (4.285) is nonlinear, the methodology we developed in the previous sections does not apply and a numerical approximation of the unique solution to (4.285) is the only option open to us at this point. Assuming that the unique solution to (4.285) does have a Taylor polynomial expansion (4.268) of arbitrary order, we are able to find the coefficients a_i in much the same spirit as we did in the prior example. Clearly, $a_0 = a_1 = 0$ in this case. Since from (4.285) we have

$$
x''(t) = -x'(t) - \sin x(t) + \cos t, \tag{4.286}
$$

using the initial data from (4.285) we obtain

$$
a_2 = \frac{x''(0)}{2} = \frac{1}{2}. \tag{4.287}
$$

Similarly, $a_3 = \frac{x'''(0)}{3!}$. We compute $x'''(t)$ by differentiating (4.286) with respect to t:

$$
x'''(t) = -x''(t) - x'(t) \cos x(t) - \sin t, \tag{4.288}
$$

which when evaluated at $t = 0$ leads to $x'''(0) = -1$ and $a_3 = -\frac{1}{6}$. Substituting the information we gathered so far into (4.268), we find

$$
P_3(t) = \frac{1}{2}t^2 - \frac{1}{6}t^3 \tag{4.289}
$$

which provides an approximate solution to the unique solution to (4.285).

As we mentioned earlier, the higher the order of the Taylor polynomial the closer this function approximates the true solution to the differential equation. With only four terms in the approximate solution (4.289), we do not expect that (4.289) remains close to the solution to (4.285) when t is far away from 0. We can construct better and better approximate solutions to this equation by carrying out the algorithm that led to (4.289) a few more steps and obtaining a higher

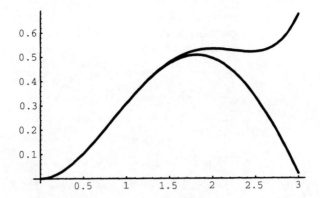

Figure 4.7: A comparison of two approximate solutions to $x'' + x' + \sin x = \cos t, x(0) = x'(0) = 0$.

order Taylor polynomial. Because this computation is tedious, we compute the higher order approximations in *Mathematica* using the following program:

```
x''[t_] := - x'[t] - Sin[x[t]] + Cos[t];
x[0] := 0; x'[0] = 0;
nn = 6;
nthder[t_, n_] := D[x''[t], {t, n - 2}];
a = Table[nthder[t, i] /. t-> 0, {i, 2, nn}];
sol[t_] = x[0] + x'[0]*t + Sum[a[[i - 1]]/i! t^i, {i, 2, nn}];
solution = NDSolve[{xx'[t] == z[t], z'[t] == - z[t] - Sin[xx[t]] +
        Cos[t], xx[0] == 0, z[0] == 0}, {xx, z}, {t, 0, 3}];
plot1 = Plot[Evaluate[xx[t] /. solution], {t, 0, 3}];
plot2 = Plot[sol[t], {t, 0, 3}]
Show[plot1, plot2]
```

This program computes the sixth-order Taylor polynomial approximation to (4.285) and compares it to the approximate solution to this equation that we obtain using NDSolve in *Mathematica*. The latter function uses an accurate numerical technique based on the method of Runge–Kutta to solve ODEs and, in general, is considerably more accurate than the technique we are using now. Still, as Figure 4.7 shows, the Taylor polynomial agrees with the solution obtained through NDSolve, as long as t remains relatively close to 0, but these two approximations disagree for large values of t.

Problems

1. In each of the following problems find the solution to the initial value problem analytically. Also find a Taylor polynomial approximation of the same solution with the specified n. Use the appropriately altered *Mathematica* program in Example 4.8.1 and compute the Taylor polynomial

solution. Draw the analytic and the approximate solutions on the same graph with t in a neighborhood of the initial time. When appropriate, use `Normal[Series[...]]` of *Mathematica* to check your results.

(a) $x' - 3x = 0, \quad x(0) = 2, \quad n = 6$

(b) $x' - 3x = 0, \quad x(1) = 2, \quad n = 6$

(c) $x' - tx = 0, \quad x(0) = 2, \quad n = 6$

(d) $x' - tx = 0, \quad x(1) = 2, \quad n = 6$

(e) $y' - 3y = \sin t, \quad y(0) = 2, \quad n = 12$

(f) $2y' - 3y = \sin t, \quad y(1) = 2, \quad n = 12$

(g) $y' - y = e^t, \quad y(0) = 2, \quad n = 6$

(h) $3y' + y = t, \quad y(1) = 1, \quad n = 6$

(i) $y' + y = t^2, \quad y(0) = 2, \quad n = 12$

(j) $ty' + y = \sin t, \quad y(1) = 2, \quad n = 12$

2. Find the solution to each of the following second-order linear differential equations. As in the previous problem, compare the analytic solution to the Taylor polynomial approximation (see the program in Example 4.8.2) with the specified order. Draw the exact and the approximate solutions in a neighborhood of the initial data on the same graph.

(a) $y'' + y = 0, \quad y(0) = 1, y'(0) = 0, \quad n = 6$

(b) $y'' + y' + 4y = 0, \quad y(0) = 1, y'(0) = 0, \quad n = 6$

(c) $y'' + 3y' + y = 0, \quad y(0) = 1, y'(0) = -1, \quad n = 6$

(d) $y'' + y = \sin 2t, \quad y(0) = 0, y'(0) = 0, \quad n = 12$

(e) $y'' + y = \sin t, \quad y(0) = 0, y'(0) = 0, \quad n = 12$

(f) $y'' + y = \sin t, \quad y(0) = 0, y'(0) = 0, \quad n = 24$

(g) $y'' + 0.1y' + y = \sin t, \quad y(0) = 0, y'(0) = 0, \quad n = 24$

(h) $y'' + 0.01y' + y = \sin t, \quad y(0) = 0, y'(0) = 0, \quad n = 24$

(i) $y'' - y' + y = \sin t, \quad y(0) = 0, y'(0) = 0, \quad n = 12$

(j) $y'' + y = t, \quad y(0) = 0, y'(0) = 0, \quad n = 12$

(k) $y'' + 0.1y' + y = t, \quad y(0) = 0, y'(0) = 0, \quad n = 12$

(l) $y'' - 0.1y' + y = t, \quad y(0) = 0, y'(0) = 0, \quad n = 12$

3. Use the program in Example 4.8.3 and find the approximate solution to the initial value problems. Then use `NDSolve` and get a second approximate solution to these problems. For each problem draw the two solutions on the same graph.

(a) $y'' + ty = 0, \quad y(0) = 0, y'(0) = 1, \quad n = 6$

(b) $y'' + ty = 0, \quad y(0) = 1, y'(0) = 1, \quad n = 6$

(c) $y'' + 0.1y' + ty = 0, \quad y(0) = 1, y'(0) = 1, \quad n = 6$

(d) $y'' + \sin t y' + ty = 0, \quad y(\frac{\pi}{4}) = 0, y'(\frac{\pi}{4}) = 1, \quad n = 6$

(e) $t y'' + \sin t y' + ty = 0, \quad y(\frac{\pi}{4}) = 0, y'(\frac{\pi}{4}) = 1, \quad n = 6$

(f) $y'' + 0.1ty' + ty = 0, \quad y(0) = 0, y'(0) = 1, \quad n = 6$

(g) $y'' + e^{-t}y = 0, \quad y(0) = 1, y'(0) = 1, \quad n = 12.$

4. The function e^{-t} has a second-order Taylor polynomial $1 - t + \frac{t^2}{2}$.

   (a) Compare the graphs of the solution in (g) with the Taylor polynomial solution of $y'' + p(t)y = 0$ with initial data $y(0) = 1$ and $y'(0) = 1$ and $n = 12$, where $p(t) \equiv 1 - t$, that is, the first two terms of the Taylor polynomial approximation of e^{-t}.

   (b) Compare the approximate solutions of (g) and (g)i with the approximate solution to $y'' + (1 - t + \frac{t^2}{2})y = 0, \quad y(0) = 1, y'(0) = 1$.

   (c) $y'' + e^{t}y = 0, \quad y(0) = 1, y'(0) = 1, \quad n = 12$

5. For the following problems use the program in Example 4.8.3 and find a Taylor polynomial approximation of the solution. Note that this program needs to be altered appropriately to allow for the initial data, the order of the polynomial, and the constants in the equation that enter as parameters.

   (a) **The Legendre equation** $(1 - t^2)y'' - 2ty' + \alpha(\alpha + 1)y = 0, \quad y(0) = y_0, y'(0) = y_1$. Let α take on values 1, 2, and 3.

   (b) **The Hermite equation** $y'' - 2ty' + 2\alpha y = 0, \quad y(0) = y_0, y'(0) = y_1$. Let α be an integer between 1 and 3.

   (c) **The Bessel equation** $t^2 y'' + ty' + (t^2 - m^2)y = 0, \quad y(1) = y_0, y'(1) = y_1$. Let m be an integer between 1 and 5.

6. Modify the program in Example 4.8.4 and find an approximate solution to the following nonlinear differential equations.

   (a) $y'' + y^2 = 0, \quad y(0) = 1, y'(0) = 0, \quad n = 6$

   (b) The **pendulum equation** $y'' + \sin y = 0, \quad y(0) = 1, y'(0) = 0, \quad n = 12$

   (c) The **damped pendulum equation** $y'' + 0.1y' + \sin y = 0, \quad y(0) = 1, y'(0) = 0, \quad n = 12$

   (d) $y' + y^2 = 0, \quad y(0) = 1, \quad n = 10$

7. Modify the program in Example 4.8.4 and find an approximate solution to the following higher order differential equations.

   (a) $y''' + y'' + y' + y = 0, \quad y(0) = y'(0) = 0, y''(0) = 1, \quad n = 10$

   (b) $y''' + y^2 = 0, \quad y(0) = y'(0) = 0, y''(0) = 1, \quad n = 10$

   (c) $y^{iv} + y' + y^2 = 0, \quad y(0) = y'(0) = y''(0) = 0, y'''(0) = 1, \quad n = 10$

8. Explain how one would generalize the Taylor polynomial method to obtain an approximate solution to the third-order equation

$$x''' + ax'' + bx' + cx = f(t), \quad x(0) = x_0, x'(0) = x_1, x''(0) = x_2.$$

Assume that a, b, and c are constants and that f and its derivatives are defined at $t = 0$. Write down the fifth-order Taylor polynomial of this equation.

9. Apply the Taylor polynomial method to

$$x''' + x'' = 1, \quad x(0) = x'(0) = x''(0) = 0,$$

and determine $P_7(t)$. Draw the graph of the exact and approximate solutions for $t \in (0, 10)$ on the same screen.

4.9   The Power Series Method

It is probably clear from the previous section that one can continue the process of obtaining a polynomial solution of a second-order differential equation

$$x'' + p(t)x' + q(t)x = 0 \tag{4.290}$$

for any degree that one wishes, the only obstacles being the degree of smoothness of the functions p and q and the computing resources available to us. In many cases of physical significance we can in fact seek an infinite series solution of (4.290) and obtain a representation of the solution that is often every bit as useful as a closed-form and analytic representation of the solution. The **power series method**, as this method is called, has played a very significant role in the development of science and engineering. In this section we will review the standard definitions and properties of power series and discuss an algorithm in *Mathematica* that effectively leads to a power series solutions of (4.290) in many cases of interest.

A **power series** is an infinite series of the form

$$a_0 + a_1(t - t_0) + a_2(t - t_0)^2 + \cdots + a_n(t - t_0)^n + \cdots = \sum_{n=0}^{\infty} a_n(t - t_0)^n. \tag{4.291}$$

We say a power series **converges** at t if the limit of the **partial sums** p_N defined by

$$p_N(t) = \sum_{n=0}^{N} a_n(t - t_0)^n \tag{4.292}$$

converges and write

$$\sum_{n=0}^{\infty} a_n(t - t_0)^n = \lim_{n \to \infty} p_N(t). \tag{4.293}$$

We recall from calculus that many elementary functions such as e^t, $\sin t$, $\cos t$, $\ln t$, and $\sqrt{t}$ have power series expansions about a point $t = t_0$. Given a function $f(t)$, we construct a power series expansion of that function in the form (4.291), where

$$a_n = \left.\frac{f^{(n)}(t)}{n!}\right|_{t=t_0} \tag{4.294}$$

is the familiar **Taylor series** formula. Needless to say, a function f must be infinitely many times differentiable at $t = t_0$ for the formula (4.294) to make sense. When the power series (4.291)–(4.294) converges in an open interval (a, b) that contains t_0, then f is **analytic** in that interval.

We also recall from elementary calculus the role the **ratio test** plays in determining the **radius of convergence** of a power series. A series $\sum_{n=0}^{\infty} b_n$ converges if $r \equiv \lim_{n \to \infty} \frac{|b_{n+1}|}{|b_n|}$ is less than unity. The series diverges if r is greater than one, and the ratio test is inconclusive if $r = 1$. If we allow $b_n = a_n(t - t_0)^n$, we are able to extract information about the convergence of power series, which we summarize in the next theorem.

Theorem 4.9.1 (Radius of Convergence)
Consider the power series $\sum_{n=0}^{\infty} a_n(t - t_0)^n$. Let

$$\rho = \lim_{n \to \infty} \frac{|a_n|}{|a_{n+1}|}. \tag{4.295}$$

Suppose that ρ is finite. Then the following propositions hold

1. *If $\rho \in (0, \infty)$, the power series converges for $|t - t_0| < \rho$ and diverges for $|t - t_0| > \rho$*

2. *If $\rho = \infty$, the power series converges for all t.*

3. *If $\rho = 0$, the power series converges only at $t = t_0$.*

Example 4.9.1
The function $x(t) = \frac{t}{t+1}$ has the power series expansion

$$x(t) = t - t^2 + t^3 - t^4 + t^5 - t^6 + \cdots = \sum_{n=1}^{\infty} (-1)^{n-1} t^n, \tag{4.296}$$

with $t_0 = 0$. Here, $a_n = (-1)^{n-1}$ and $\rho = 1$. Hence, according to Theorem 4.9.1, the series converges for $|t| < 1$ and diverges for $|t| > 1$. It diverges for $t \pm 1$. (Why?)

As we saw earlier, in *Mathematica* the statement

```
Series[t/(1+t), {t, 0, 10}]
```

produces the first 10 terms of the power series of x, plus an additional term that gives information about the power of the next nonzero term in the series. The function `Normal`, when applied to the previous output, removes the remainder. We get the same information from $MATLAB$ by entering

```
taylor('t/(1+t)',10)
```

Note that `taylor` assumes the power series is being expanded about the point $t_0 = 0$.

We can carry out the standard algebraic operations on power series. For example, if f and g have power series representations $\sum_{n=0}^{\infty} a_n(t - t_0)^n$ and $\sum_{n=0}^{\infty} b_n(t - t_0)^n$, respectively, that converge for t in an open interval that contains t_0, then $\sum_{n=0}^{\infty}(a_n + b_n)(t - t_0)^n$ is the power series expansion of $f + g$ in that interval. Similar considerations apply to $f - g$, while the power series expansions of fg and $\frac{f}{g}$ are determined as one suspects, at least formally. The series one obtains by multiplying the series for f and g converges in the common interval of convergence of these series, while the power series for $\frac{f}{g}$ generally converges in an interval that is a subset of the intervals of convergence of f and g. The culprit in reducing the size of the interval of convergence in the latter case is usually the set of zeros of g.

Analytic operations of differentiation and integration can also be carried out term by term on power series, much like the way we would differentiate and integrate a polynomial. These operations, together with the algebraic operations, allow us to substitute the power series representation of a solution directly into a differential equation. Having reviewed some of the rudimentary properties of power series, we are now ready to investigate how the power series method works in constructing the solution of a differential equation.

Consider the initial value problem

$$x'' + 4x' + 3x = 0, \quad x(0) = 1, x'(0) = -1. \tag{4.297}$$

We start with the power series representation of x

$$x(t) = a_0 + a_1 t + a_2 t^2 + \cdots = \sum_{n=0}^{\infty} a_n t^n, \tag{4.298}$$

where the coefficients a_n are as yet unknown. Since $x(0) = 1$ and $x'(0) = -1$, we deduce easily that $a_0 = 1$ and $a_1 = -1$. We differentiate and substitute (4.298) into (4.297), write the result as a single power series, and obtain the identity

$$\begin{aligned} 0 = (-1 + 2a_2) + (-3 + 8a_2 + 6a_3)t + \cdots \\ + (3a_{n-2} + 4na_{n-1} + n(n-1)a_n)\, t^{n-2} + \cdots . \end{aligned} \tag{4.299}$$

This identity must hold for all t. Since the left-hand side vanishes for all t, we conclude that the coefficient of each t^i on the right-hand side must vanish as well. So

$$-1 + 2a_2 = 0, \tag{4.300}$$

or $a_2 = \frac{1}{2}$. Similarly,

$$-3 + 8a_2 + 6a_3 = 0, \qquad\qquad (4.301)$$

so $a_3 = -\frac{1}{6}$ and so on. The coefficient of t^{n-2} gives the **recurrence relation**

$$a_n = -\frac{3a_{n-2} + 4a_{n-1}}{n(n-1)}, \quad n > 1. \qquad\qquad (4.302)$$

It is easy to check that $x(t) = e^{-t}$ is the exact solution to (4.297) and that a_ns are the coefficients of its power series.

Example 4.9.2

One of the features of the above discussion that makes the power series method appealing is its algorithmic nature. Block, Sum, Expand, Coefficient, Append, and Solve are among *Mathematica*'s internal functions that are quite useful in such an algorithm. One programs this method in *Mathematica* as follows:

```
sersol[n_, a0_, a1_]:=Block[{L,x,term,b,coeff},
    L[x_] := D[x, {t, 2}] + 3 D[x, t] + 4 x;
    x = Sum[a[i] t^i, {i, 0, n}];
    coeflist = Table[a[i], {i, 2, n}];
    term = Expand[L[x]];
    b0 = term /. t-> 0; b0 = {b0 == 0};
    b=Table[Coefficient[term,t^i]==0, {i, 1, n-2}];
    b=Append[b0,b]; b=Flatten[b,1];
    b = b/. {a[0] -> a0, a[1] -> a1};
    coeff=Solve[b,coeflist];
    coeff=Flatten[{a[0] -> a0, a[1] -> a1, coeff}]; Print[coeff];
    f[t_] = Sum[a[i] t^i /. coeff, {i, 0, n}];
    Plot[f[t], {t, 0, 3}, PlotLabel ->"The Approximate Solution"];
    Print[f[t]]
]
```

This program's output with initial conditions $x(0) = 1$, $x'(0) = 4$ is shown in Figure 4.8. The program was executed with $n = 3, 7$, and 10 and the results were compared with the solution one obtains from DSolve (shown with the thicker line).

The method we described previously works equally well for nonhomogeneous initial value problems as well as those with nonconstant coefficients. What is required of these additional terms is that they all have power series that converge about the initial point t_0. The algebra involved often becomes tedious and cumbersome when these problems are attacked by hand, but it does not take much effort to revise the program in Example 4.9.1 to handle the necessary calculations in *Mathematica*. Several of the problems at the end of this section address nonhomogeneous and nonconstant coefficient differential equations.

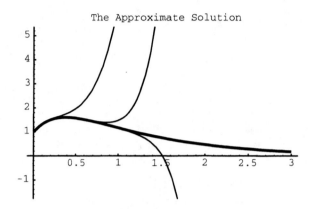

The Approximate Solution

Figure 4.8: Exact and approximate series solutions of $x'' + 3x' + 4x = 0, x(0) = 1, x'(0) = 4$. The exact solution is shown with the thicker line.

Problems

1. Create a file called `sersol.m` containing the program in Example 4.9.2. Use this program to verify the contents of Figure 4.8.

2. Find a series solution to the following initial value problems. For each equation determine its exact and series solutions using either *Mathematica* or *MATLAB*. In each case use $n = 3$, 7, and 10 and draw the graphs of the partial sums on an appropriate interval of the independent variable.

   (a) $x'' + 4x = 0$, $x(0) = 0, x'(0) = 1$

   (b) $x'' + x' + 4x = 0$, $x(0) = 1, x'(0) = 0$

   (c) $2y'' - 3y' + y = 0$, $y(0) = 0, y'(0) = 1$

   (d) $y'' + 4y' + 4y = 0$, $y(0) = -1, y'(0) = 1$

3. Explain the new features that are added to the power series method in order to handle a nonhomogeneous differential equation of the form

$$x'' + ax' + bx = f(t).$$

4. Find an approximate series solution to the following nonhomogeneous equations. Use *Mathematica* or *MATLAB* to determine the series solution of f in each case. Modify the program in Example 4.9.2 to handle the nonhomogeneous term.

   (a) $x'' + 4x = \sin t$, $x(0) = 0, x'(0) = 0$

   (b) $x'' + x' + 4x = e^{-t}$, $x(0) = 0, x'(0) = 0$

   (c) $x'' + x = \ln t$, $x(1) = 0, x'(1) = 0$

5. Find a series solution of the following initial value problems.

   (a) $x'' + tx = 0$, $x(0) = 1, x'(0) = 0$
   (b) $y'' + xy' + y = 0$, $y(0) = -1, y'(0) = -2$
   (c) $y'' + (1 + x)y' - y = 0$, $y(0) = 0, y'(0) = 1$

6. Generalize the power series method to higher order equations. Apply this method to the following initial value problems. Experiment with various values of n to obtain a good approximation to the exact solution.

   (a) $x''' + 2x' + x = 0$, $x(0) = 1, x'(0) = -1, x''(0) = 0$
   (b) $y''' - y = 0$, $y(0) = 0, y'(0) = 1, y''(0) = 0$
   (c) $x^{(iv)} - x = 0$, $x(0) = 1, x'(0) = 0, x''(0) = 0, x'''(0) = 0$
   (d) $x''' + 4x'' = \sin t$, $x(0) = 0, x'(0) = 0, x''(0) = 0$
   (e) $w''' + 3w'' + 3w' + w = t + \cos t$, $w(0) = 0, w'(0) = 0, w''(0) = 0$

4.10 Special Functions

The technique we described in Section 4.8 works well for many differential equations, linear or nonlinear, as along as the equation can be used to obtain the coefficients of the Taylor polynomial by evaluating the differential equation at the initial data. Similarly, the power series method is a convenient way of obtaining an approximate solution to a linear differential equation as long as the coefficients in the differential equation have reasonably well-behaved power series expansions about the point of interest. Unfortunately, the Taylor polynomial and the power series methods fail to work for some differential equations of special interest in mathematical physics, as the following example shows.

Example 4.10.1
Consider the differential equation

$$t^2 y'' + \frac{1}{4}y = 0, \quad y(0) = 1, y'(0) = 1. \tag{4.303}$$

Following the procedure of Section 4.8, we need to compute a_2 of (4.268) by evaluating y'' at zero. From (4.303) we have

$$y''(t) = -\frac{y(t)}{4t^2}, \tag{4.304}$$

so $y''(0) = \frac{1}{0}$, which is undefined. We conclude that (4.303) does not have a second- (or higher) order Taylor polynomial approximation.

The point $t = 0$ is called a singular point of (4.303). We now give a formal definition of such a point, confining our treatment to second-order differential equations.

Definition 4.10.1 (Ordinary and Singular Points) *Consider the second-order differential equation*

$$y'' + p(t)y' + q(t)y = 0, \quad y(a) = y_0, y'(a) = y_1. \tag{4.305}$$

*We say that $t = a$ is an **ordinary** point of (4.305) if both p and q have convergent power series expansions about $t = a$. Otherwise, $t = a$ is called a **singular** point of (4.305).*

With the notation of (4.305) the functions p and q of Example 4.10.1 are 0 and $\frac{1}{4t^2}$. The function p, of course, has a power series expansion about any $t = a$. The function q, on the other hand, fails to have a power series expansion about $t = 0$, the point at which the initial data of (4.303) are specified. The method we described in Section 4.8 will work for any t_0 other than $t_0 = 0$.

In this section we will modify the method of Section 4.8 and obtain a series (but, in general, not a power series) approximate solution to equations of the form (4.305), even when there is a singularity in the problem.

Definition 4.10.2 (Regular Singular Points)

*A point $t = a$ is called a **regular singular point** of (4.305) if it is a singular point and if*

$$(t - a)p(t) \quad and \quad (t - a)^2 q(t) \tag{4.306}$$

have power series expansions about $t = a$.

According to Definition 4.10.2, the differential (4.303) in Example 4.10.1 has a regular singular point at $t = 0$ since $t^2 q(t) = \frac{1}{4}$, which clearly has a power series expansion about $t = 0$.

Our first theorem in this section provides a method for obtaining approximate solutions of equations of the form (4.305) when $t = a$ is a regular singular point.

Theorem 4.10.1

Let $t = a$ be a regular singular point of (4.305). Let p_0 and q_0 be defined by

$$p_0 = \lim_{t \to a}(t - a)p(t) \quad and \quad q_0 = \lim_{t \to a}(t - a)^2 q(t). \tag{4.307}$$

*Suppose that the quadratic polynomial, called the **indicial polynomial** equation,*

$$r(r - 1) + p_0 r + q_0 = 0, \tag{4.308}$$

has only real roots r_1 and r_2, with $r_1 > r_2$. Furthermore, suppose that the difference between these roots is not a positive integer. Then there are two linearly independent approximate solutions to (4.305) given by

$$y_1(t) = (t - a)^{r_1} \sum_{i=0}^{n} b_i(t - a)^i \quad and \quad y_2(t) = (t - a)^{r_2} \sum_{i=0}^{n} c_i(t - a)^i. \tag{4.309}$$

As always, the larger n is, the better the functions in (4.309) approximate the solutions to (4.305). When n is infinite we obtain a series solution to (4.305).

The proof of Theorem 4.10.1 can be found in many books on ODEs and will not be repeated here (see, e.g., the text by Boyce and DiPrima, *Elementary Differential Equations and Boundary Value Problems*, 2nd Edition, John Wiley and Sons, New York, 1966, pp. 171–195). The proof relies on a method known as the **Frobenius method**, which we will elaborate on shortly in the context of an example. The main reason we have chosen not to present the Frobenius method in its generality in this text is that the great majority of the special differential equations of science and engineering are accessible by the modern software packages of *Mathematica* and *MATLAB*. Our approach will be to use the solutions of these equations (often called **Special Functions**) in application settings and to concentrate on what role these functions play in the physical settings they model. Still, to gain an idea of how the proof of a theorem such as Theorem 4.10.1 proceeds and where an equation such as (4.308) comes from, in the next example we will present the main ideas of the proof of this theorem for the **Bessel equation of order one-half**. As we will see in this example, the Frobenius method is an iterative method that is rather easy to implement on a numerical software package. This method also lends itself well to a symbolic manipulator software package since its basis depends on carrying out L'Hôpital's rule and being able to solve algebraic equations. It is the latter implementation, carried out in *Mathematica*, that is described in detail in the following example.

Example 4.10.2 Bessel's equation of order one-half

The Bessel equation of order $\frac{1}{2}$ is

$$t^2 y'' + ty' + \left(t^2 - \frac{1}{4}\right)y = 0. \tag{4.310}$$

Since $p(t) = \frac{1}{t}$ and $q(t) = \frac{t^2 - 1/4}{t^2}$, the only singular point of (4.310) is $t = 0$. After checking the conditions of Definition 4.10.2, we see that this point is a regular singular point with

$$p_0 = 1 \quad \text{and} \quad q_0 = -\frac{1}{4}. \tag{4.311}$$

We seek solutions to (4.310) of the form

$$y(t) = t^r f(t). \tag{4.312}$$

Our strategy is to choose r in such a way that the differential equation satisfied by f has ordinary points only. The function f can then be determined by the method outlined in Section 4.8. We begin by differentiating (4.312) twice and substituting the results into (4.310). We find that f satisfies the second-order differential equation

$$t^2 f'' + (2r + 1)tf' + \left[r(r-1) + r + \left(t^2 - \frac{1}{4}\right)\right]f = 0. \tag{4.313}$$

It is not obvious that we have accomplished anything at this point because the equation for f seems as complicated at the equation for y. The parameter r, however, is as yet undetermined: We have the option of choosing r to simplify the equation for f. Note that the coefficient of f in (4.313), when evaluated at $t = 0$, is precisely the indicial polynomial in (4.308) (see (4.311)), namely,

$$r(r - 1) + r - \frac{1}{4}. \tag{4.314}$$

This quadratic polynomial has the two roots

$$r_1 = -\frac{1}{2} \quad \text{and} \quad r_2 = \frac{1}{2}. \tag{4.315}$$

When we choose $r_1 = -\frac{1}{2}$, (4.313) reduces to

$$f'' + f = 0, \tag{4.316}$$

which is a second-order differential equation with constant coefficients, whose general solution is easily found to be $c_1 \sin t + c_2 \cos t$. We thus obtain two linearly independent solutions of (4.307):

$$y(t) = c_1 \frac{\sin t}{\sqrt{t}} + c_2 \frac{\cos t}{\sqrt{t}}. \tag{4.317}$$

The first function in (4.317) is traditionally denoted by $J_{\frac{1}{2}}$, while the second by $J_{-\frac{1}{2}}$.

Remark 4.10.1: The differential equation (4.316) is simple enough that we could essentially guess its general solution. In other problems, the analogue of (4.316) may be more complicated. One can, however, find an approximate solution to this equation using the methods of Sections 4.8 and 4.9.

In the preceding example we found two linearly independent solutions of (4.307) using only the smaller of the roots of the indicial polynomial (4.314). Let us confirm that we do not gain any new information from the second root $r_1 = \frac{1}{2}$. With this value for r, (4.313) becomes

$$tf'' + 2f' + tf = 0. \tag{4.318}$$

Although this equation looks as if it has a singularity at $t = 0$, it has one solution that does not exhibit this singularity. Going back to the method of Section 4.8, let us find a Taylor polynomial approximate solution to (4.318), satisfying the initial conditions

$$f(0) = 1 \quad \text{and} \quad f'(0) = 0. \tag{4.319}$$

As we will see, the choice of $f'(0) = 0$ is crucial. From (4.318) we have that

$$f''(t) = -\frac{2f'(t) + tf(t)}{t}. \tag{4.320}$$

We cannot directly find $f''(0)$ from (4.320) because the evaluation of the right-hand side of (4.320) leads to an indeterminate case of the form $\frac{0}{0}$. We chose the initial condition $f'(0) = 0$ precisely to cause this indeterminacy; with $f'(0) \neq 0$ the quantity $f''(t)$ does not have a limit at 0. Applying L'Hôpital's rule to (4.320), we obtain

$$f''(0) = -2f''(0) - f(0), \tag{4.321}$$

or, after using $f(0) = 1$,

$$f''(0) = -\frac{1}{3}. \tag{4.322}$$

Hence, the second-order Taylor polynomial approximation of the solution to (4.318), (4.319) is

$$p_2(t) = 1 - \frac{1}{6}t^2. \tag{4.323}$$

We can get higher order approximations by repeating this procedure. We find that the first five terms of the Taylor polynomial approximation are

$$a_0 = 0, \quad a_1 = 0, \quad a_2 = -\frac{1}{6} = -\frac{1}{3!}, \quad a_3 = 0, \quad a_4 = \frac{1}{120} = \frac{1}{5!}. \tag{4.324}$$

Eventually, one arrives at the formula

$$a_n = \frac{(-1)^n}{(2n+1)!},$$

so

$$p_m(t) = \sum_{n=0}^{m} \frac{(-1)^n}{(2n+1)!} t^{2n}. \tag{4.325}$$

It is not difficult to show that p_m is in fact the mth partial sum of the series expansion of $\frac{\sin t}{t}$ about $t = 0$. Putting this together with the $t^{\frac{1}{2}}$ factor in the solution y, we find that the Taylor polynomial solution to y corresponding to the larger root r_1 of the indicial polynomial gives us nothing but the approximation to $J_{\frac{1}{2}}$ referred to previously.

The evaluation of the series solution to (4.318) offers quite a challenge to a symbolic manipulator because not only did we use L'Hôpital's rule in the evaluation of $f''(0)$ but this evaluation needed an additional algebraic operation (cf. (4.321)). One can obtain (4.325) in *Mathematica* via the program (start a new session in *Mathematica*)

```
y[0] = 1; y'[0] = 0;
f[t_] = Series[y[t], {t, 0, 12}];
g[t_] = -(t f[t] + 2 f'[t])/t;
nthdery[t_, n_] = D[y[t], {t, n}];
Do[
    a = Solve[Limit[g[t], t -> 0] == nthdery[0, n],
            nthdery[0, n]][[1,1,2]];
    Derivative[n][y][0] = a;
    f[t_] = Series[y[t], {t, 0, 12}];
```

```
    Print[f[t]];
    Print[a];
    g[t_] = D[g[t], t], {n, 2, 11}
]
```

The final output of this program is the Taylor polynomial whose coefficients are listed in (4.324). Its intermediate output consist of the output of Do as it performs its iterations.

The method we outlined in the prior example, and the *Mathematica* program that goes with it, finds one solution of a second-order linear differential equation that has a regular singular point at a point t_0, the point at which the initial data is specified. A second approximate solution can now be obtained using the method of reduction of order described in Section 4.7. The *Mathematica* program in the latter section can easily be combined with the preceding one to yield the two linearly independent solutions of the differential equation.

The previous example gives a good indication as to the magnitude of the difficulties one encounters when dealing with singular differential equations. The theorem we quoted in the beginning of this section gives a constructive answer to the question of existence of linearly independent solutions of a singular equation, although it should be noted that the conclusions of the theorem are obtained under strong hypotheses on the roots of the indicial polynomial. This theorem can be strengthened sufficiently to encompass the other cases of interest (repeated and complex roots), especially for those differential equations that commonly appear in applications. The mathematics of the nineteenth and early twentieth centuries gained a rich history from the efforts of numerous mathematicians whose work eventually resolved these and other fundamental issues in the theory of special functions.

The analytic and series representations of most of the special functions we encounter in mathematical physics have been programmed and are available in several of the popular software packages in the market. We finish this section with a list of such functions and their internal labels in *Mathematica*. These functions have very similar names in *MATLAB*. In general, a simple query through the help mode of either of these programs will give the reader quite a bit more information.

1. **Bessel's Equation:**

$$t^2 y'' + ty' + (t^2 - \nu^2)y = 0. \tag{4.326}$$

ν is called the order of the Bessel function. The two linearly independent solutions of this equation are denoted by $J_\nu(t)$ and $Y_\nu(t)$. In *Mathematica* these functions are labeled

BesselJ[n, t] and BesselY[n, t]

2. Legendre's Equation of order α:

$$(1 - t^2)y'' - 2ty' + \alpha(\alpha + 1)y = 0. \tag{4.327}$$

The solutions to this equation are denoted by

LegendreP[n, t] and LegendreQ[n, t].

3. **Chebychev's Equation:**

$$(1 - t^2)y'' - ty' + \lambda y = 0, \qquad (4.328)$$

with solutions

ChebyshevT[n, t] and ChebyshevU[n, t].

4. **Laguerre's Equation:**

$$ty'' + (1 - t)y' + \lambda y = 0, \qquad (4.329)$$

whose polynomial solutions are denoted by

LaguerreL[t, n]

in *Mathematica*.

5. **Hermite's Equation:**

$$y'' - 2ty' + \lambda y = 0. \qquad (4.330)$$

A Hermite polynomial is obtained when $\lambda = 0, 2, 4, 6, \cdots$, and is denoted by $H_n(t)$. This solution is obtained by entering HermiteH[n, t].

Problems

1. Verify the statements in Example 4.10.2. Create a file that contains the *Mathematica* lines in this example. Execute this program and compare its output with the assertions in that example.

2. Find the singular points of the Bessel, Hermite, Laguerre, Legendre, and Chebychev equations. Identify which of these points are regular singular points.

3. Find a series solution to the Hermite equation. Show that if λ is an even integer this solution terminates. Compare your answers with HermiteH[t, n] of *Mathematica*.

4. Find a series solution to Bessel's equation of order one when $t_0 = 0$. Find the indicial polynomial in this case and the series solution y for which $y(0) = 0$. Compare your solution to BesselJ[1,t] of *Mathematica* by drawing the graphs of the two functions on the same screen. Find a second solution by the method of reduction. How does this solution compare with BesselY[1,t]?

5. Draw the graph of $J_1(t)$, the Bessel function of order 1. Use FindRoot of *Mathematica* or roots of *MATLAB* and find its first five zeros.

6. Draw the graphs of J_0' and J_1 on the same screen. Does the graph suggest an algebraic relationship between them?

7. Use *Mathematica* or *MATLAB* and determine the value of the following integrals.

   (a) $\int_0^1 J_0(t)\,dt$

   (b) $\int_0^{10} J_0(t)\,dt$

   (c) $\int_0^\infty J_0(t)\,dt$

   (d) $\int_0^1 t J_0(t)\,dt$

   (e) $\int_0^\infty J_0^2(t)\,dt$

   (f) $\int_0^\infty J_1^2(t)\,dt$

   (g) $\int_0^1 J_0(t)Y_0(t)\,dt$

8. Find a series solution about $t_0 = 0$ to Legendre's equation when $\alpha = 4$. Show that t_0 is an ordinary point and use the method of the previous section to get the Taylor polynomial approximate solution. Show that the fourth-order Taylor polynomial is an exact solution to Legendre's equation. Compare your solution to `LegendreP[2,t]` or `LegendreD[2,t]`.

4.11 Project A—Harmonic Oscillations

One of the fundamental applications of linear differential equations occurs in the description of the vibration of the simple mechanical system consisting of a mass m attached to an elastic spring and a dashpot as shown in Figure 4.9. In this project we will use the basic techniques we developed for solving linear second-order differential equations and obtain the solution of this mechanical system in terms of all of the parameters involved.

Let $x(t)$ be the displacement of the mass from its equilibrium.

The forces acting on the mass are due to the restoring force of the spring, which, after appealing to Hooke's law, is given by

$$F_1 = -kx, \tag{4.331}$$

where k is the **spring constant**, and the frictional force due to the dashpot

$$F_2 = -c\frac{dx}{dt}, \tag{4.332}$$

where c is the **damping constant**. In certian circumstances an **external force** $f(t)$ is added to this system as well.

As a consequence of Newton's second law of motion, the resultant of these forces must be balanced by the time rate of change of linear momentum, $\frac{d}{dt}\left(m\frac{dx}{dt}\right)$. Since the mass of the spring does not change with time, the time rate of change of

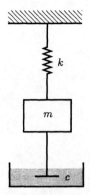

Figure 4.9: A mass-spring-dashpot system.

linear momentum reduces to mx'', mass times acceleration. We apply Newton's law and end up with the following linear nonhomogeneous second-order equation

$$mx'' + cx' + kx = f(t). \tag{4.333}$$

The spring is said to undergo **free** vibrations when the external force is not present in (4.333). When the dashpot is not present in this equation (that is, when $c = 0$), the motion is said to be **undamped**.

1. Consider the undamped and free vibration of a body of mass m and a spring with constant k. Show that the general solution of the governing equation

$$mx'' + kx = 0 \tag{4.334}$$

   is

$$x(t) = c_1 \sin \omega t + c_2 \cos \omega t, \quad \text{with} \quad \omega = \sqrt{\frac{k}{m}} \tag{4.335}$$

   (see Example 4.3.10). The parameter ω is called the **natural frequency** or **circular frequency** of the system. Show that (4.334) can also be written as

$$x(t) = A \sin(\omega t + \alpha), \tag{4.336}$$

   where

$$A = \sqrt{c_1^2 + c_2^2} \quad \text{and} \quad \tan \alpha = \frac{c_2}{c_1}. \tag{4.337}$$

The parameter A is called the **amplitude** of the wave and α its **phase angle**. The **period** of the solution (4.335), denoted by T, is given by $T = \frac{2\pi}{\omega}$ with units generally measured in seconds, minutes, or hours. The system's **frequency**, defined as $\frac{1}{T}$, is $\frac{\omega}{2\pi}$. The natural frequency ω and frequency have units of radians per seconds and hertz (Hz), respectively. One hertz is one cycle per second.

2. Let $m = 20$ kilograms and $k = 6$ newtons per meter (N/m). Determine the natural frequency of vibration, the period, and the amplitude of the motion.

3. Show that the solution to the initial value problem

$$mx'' + kx = 0, \quad x(0) = x_0, \quad x'(0) = x_1, \qquad (4.338)$$

is given by

$$x(t) = \frac{x_1}{\omega} \sin \omega t + x_0 \cos \omega t. \qquad (4.339)$$

Write (4.339) in the form given by (4.336).

4. A body weighing 4 pounds is attached to a spring that would stretch 3 feet when a weight of 200 pounds hangs from it. Suppose that at time zero the body-spring system is stretched 6 inches from its equilibrium and given an initial velocity of 3 ft/s toward its equilibrium. Find the amplitude, phase angle, period, and the frequency of this vibration. (Note that the mass of a body that weighs w pounds is $m = \frac{w}{g}$, where $g = 32$ ft/s^2.) Plot the graph of the wave over an interval of length three times its period.

5. The equation for a forced undamped vibration is given by

$$mx'' + kx = f(t). \qquad (4.340)$$

We consider the special case of

$$f(t) = A \sin \beta t. \qquad (4.341)$$

(a) Show that when $\beta^2 \neq \frac{k}{m}$, the general solution to (4.340) is given by

$$x(t) = x_c(t) + x_p(t), \qquad (4.342)$$

where x_c is the same as $x(t)$ in (4.335) and

$$x_p(t) = \frac{A}{k - m\beta^2} \sin \beta t. \qquad (4.343)$$

Verify this computation in *Mathematica* or *MATLAB*.

(b) Show that when $\beta^2 = \frac{k}{m}$, the particular solution (4.343) should be replaced by

$$x_p(t) = -\frac{A}{2} \sqrt{\frac{m}{k}} \, t \cos \sqrt{\frac{k}{m}} t. \qquad (4.344)$$

The significance of this result lies in the fact that the forcing function, $A \sin \beta t$, is periodic and has finite amplitude, while the output function, at least the part represented by x_p in (4.344), is not periodic and its "amplitude" grows without bound as time evolves. This important phenomenon is called **resonance**, which typically occurs when the frequency of the input, in this case determined by β, matches the natural frequency of the vibrating body, given by $\sqrt{\frac{k}{m}}$. In rare cases when these frequencies match, the crests and troughs of the input and output functions are synchronous, leading to (often disastrous) outputs with unbounded amplitude.

6. Find the solution of the following initial value problems with given m, k, and $f(t)$. Determine in each case whether the system is in resonance or not. Check each result in *Mathematica* or *MATLAB*. Draw the graph of the solution over an interval of time that is representative of its behavior as time approaches infinity.

   (a) $m = 4$, $k = 16$, $f(t) = \sin t$; $x(0) = 1$, $x'(0) = -1$

   (b) $m = 1$, $k = 16$, $f(t) = \sin 2t + \cos t$; $x(0) = 0$, $x'(0) = 0$

   (c) $m = 4$, $k = 16$, $f(t) = \sin 2t$; $x(0) = 1$, $x'(0) = -1$ (Draw the graph of this solution together with the graph of (a) on the same screen for comparison.)

   (d) $m = 3$, $k = 27$, $f(t) = t \cos 3t$; $x(0) = 0$, $x'(0) = 0$

7. Take a careful look at the following program in *Mathematica*. What is its output going to be? What is the value of f[t] in the first line labeled by a? How about in b? Are the equations in resonance?

   Use an editor and make a file called **projmech.m** that contains these lines. Execute the program in *Mathematica* and report on its output(s).

```
a := DSolve[{x''[t] + x[t] == f[t], x[0] == 0, x'[0] == 0},
       x[t], t];
b = a /. f[t] -> Sin[2 t];
Plot[x[t]/.b, {t, 0, 4 Pi}]
c =  a /. f[t] -> Sin[t];
Plot[x[t]/. c, {t, 0, 4 Pi}]
```

8. Make appropriate changes in the preceding program and use **Block** to create a *Mathematica* program that takes m, k, f, x_0, x_1, and t_{final} as its input values and plots the solution of the initial value problem (4.340) with initial conditions $x(0) = x_0$, $x'(0) = x_1$ on the interval $(0, t_{\text{final}})$. Apply this program to Problem 6.

9. The analogue of the preceding program in *MATLAB* is as follows:

```
a=dsolve('D2x+x=f(t)','x(0)=0','Dx(0)=0')
b=subs(a,'sin(2*u)','f(u)')
c=symop(b)
d=ezplot(c)
```

Apply this program to Problem 6.

4.12 Project B—Damped Harmonic Oscillations

In this project we continue with the study of harmonic oscillations in the presence of a damping device such as a dashpot. The free damped oscillation of a mass m is governed by (4.333) with $f \equiv 0$.

1. Show that the characteristic equation of

$$mx'' + cx' + kx = 0 \tag{4.345}$$

   is given by

$$mr^2 + cr + k = 0, \tag{4.346}$$

   with discriminant $c^2 - 4mk$. The value of the damping c at which the discriminant vanishes is called the **critical damping** . Show that the critical damping c^* is related to m and k according to the relation

$$c^* = 2\sqrt{mk}. \tag{4.347}$$

2. Show that when $c = c^*$, the general solution of (4.346) is

$$x(t) = c_1 e^{-\frac{c^*}{2}t} + c_2 t e^{-\frac{c^*}{2}t}. \tag{4.348}$$

   Check this result in *Mathematica* or *MATLAB*.

3. Show that when the system is **overdamped**, that is, when $c > c^*$, then the general solution of (4.346) is

$$x(t) = c_1 e^{-\alpha_1 t} + c_2 e^{-\alpha_2 t}. \tag{4.349}$$

   How are α_1 and α_2 related to m, c, and k? What sign do they have? What happens to $x(t)$ as t approaches infinity?

4. Show that when $c < c^*$, so that the system is **underdamped**, then the general solution of (4.346) takes the form

$$x(t) = c_1 e^{-\alpha_1 t} \cos(\alpha_2 t) + c_2 e^{-\alpha_1 t} \sin(\alpha_2 t). \tag{4.350}$$

   How do α_1 and α_2 depend on m, c, and k? What is the sign of α_1 for physically reasonable parameter values? What happens to x as t approaches infinity?

5. In each of the following problems determine whether the following systems are critically damped, underdamped, or overdamped. Use the initial data to determine the unique solution. Check your result in *Mathematica* or *MATLAB*.

   (a) $m = 2$, $c = 0.1$, $k = 3$; $x(0) = 1$, $x'(0) = -1$

   (b) $m = 1$, $c = 0.1$, $k = 1$; $x(0) = 0$, $x'(0) = -1$

(c) $m = 4$, $c = 4$, $k = 16$; $x(0) = 1$, $x'(0) = 1$

(d) $m = 1$, $c = 4$, $k = 16$; $x(0) = 1$, $x'(0) = 1$

(e) $m = 4$, $c = 1$, $k = 1$; $x(0) = 1$, $x'(0) = 1$.

6. The damped forced oscillation of a vibrating mass-spring-dashpot system is governed by

$$mx'' + cx' + kx = f(t). \tag{4.351}$$

We consider the case where

$$f(t) = A \cos \beta t. \tag{4.352}$$

Show that the general solution of (4.351) is

$$x(t) = x_c(t) + x_p(t), \tag{4.353}$$

where x_c is given by (4.348), (4.349), or (4.350) and

$$x_p(t) = \frac{A}{(k - m\beta^2)^2 + c^2\beta^2} \left(c\beta \sin \beta t + (k - m\beta^2) \cos \beta t \right). \tag{4.354}$$

Use this information to show that no matter what positive values m, k, c, and β take, the solution to the initial value problem (4.351), (4.352) will eventually be dominated by (4.354). For this reason x_p is called the **steady-state** solution of (4.351).

7. Find the solution of the initial value problem (4.351), (4.352) with the specified parameter values. Use *Mathematica* or *MATLAB* to check your answers and to draw the graphs of each solution in a suitable interval of time.

(a) $m = 4$, $c = 0.1$, $k = 16$, $A = 1$, $\beta = 2$; $x(0) = 0$, $x'(0) = 0$

(b) $m = 1$, $c = 1$, $k = 1$, $A = 1$, $\beta = 1$; $x(0) = 0$, $x'(0) = 0$

(c) $m = 1$, $c = 1$, $k = 4$, $A = 1$, $\beta = 1$; $x(0) = 0$, $x'(0) = 0$

(d) $m = 1$, $c = 1$, $k = 16$, $A = 1$, $\beta = 1$; $x(0) = 0$, $x'(0) = 0$

8. The preceding graphs and discussions seem to indicate that solutions of damped harmonic oscillators with a trigonometric forcing such as (4.352) eventually die out. This fact does not mean that the mechanical oscillations cannot do the kind of damage we saw in the case of undamped harmonic oscillators where the presence of resonance is a major design issue with which one must reckon. Resonance can still occur in the case of damped oscillations but not quite with the severity that one experiences in the undamped case. This point is particularly important because it is difficult in nature to find a mechanical system that does not enjoy some form of damping, however small its effects may be. The damping term, with its coefficient c, tends to reduce the inertial effect of the initial data, as can

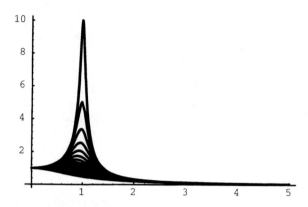

Figure 4.10: Practical resonance for a forced damped harmonic oscillator.

be seen from the complementary solutions x_c in (4.348)–(4.350). But its effect might take time to express itself in the particular solution x_p. It is possible that the maximum "amplitude" of this portion of the solution is mechanically unacceptable at a certain input frequency β even when damping is present. Keep all other parameters fixed but allow β to vary. It is clear that the maximum value x_p can achieve will depend on the specific value of β. Let $\gamma(\beta)$ denote this function. We call the **practical resonance frequency** of the mechanical system that value of β at which $\gamma(\beta)$ reaches a maximum. Let $m = 1$ and $k = \beta_0^2$ in (4.351), (4.352). Show that x_p can be written as

$$x_p(t) = \gamma \cos(\beta t - \delta), \tag{4.355}$$

where

$$\gamma(\beta) = \frac{A}{\sqrt{(\beta_0^2 - \beta^2)^2 + c^2 \beta^2}} \quad \text{and} \quad \tan \delta(\beta) = \frac{2c\beta}{\beta_0^2 - \beta^2}. \tag{4.356}$$

9. Show that the function $\gamma(\beta)$ reaches its maximum when

$$\beta = \sqrt{\beta_0^2 - \frac{c^2}{2}}. \tag{4.357}$$

Verify this result in *Mathematica* or *MATLAB*.

10. Let $A = 1$ and $\beta_0 = 1$. Draw the graph of $\gamma(\beta)$ for $\beta \in (0,5)$ as c varies from 0.1 to 2 at increments of 0.1 (see Figure 4.10). In *Mathematica* use `Plot[Evaluate[Table[ ...]]]` and `PlotRange -> All` to get the

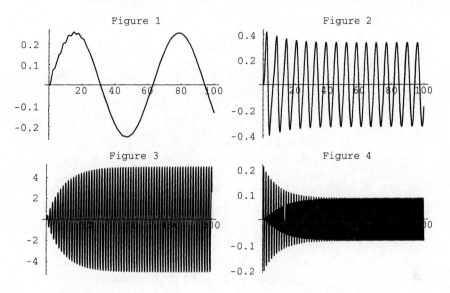

Figure 4.11: Practical resonance for $y'' + 0.1y' + y = \sin \omega t$.

analogue of Figure 4.10. In *MATLAB* use `for ... end` to create a loop that generates the plots and use `hold on` to keep the graphs on the same screen.

11. Draw the graph of $\delta(\beta)$ for the same values of the parameters described.

12. Consider the differential equation

$$y'' + 0.1y' + 4y = \sin \omega t, \quad y(0) = 0, y'(0) = 0. \qquad (4.358)$$

(a) Show that when the frictional term is absent from (4.358) and when $\omega = 2$,

$$y(t) = \frac{1}{8} \sin 2t - \frac{1}{4} t \cos 2t \qquad (4.359)$$

is the unique solution. Check your answer in *Mathematica* or *MATLAB*. Draw the graph of this solution for $t \in (0, 4\pi)$.

(b) Let ω range over the values in the set $\{0.1, 1, 2, 4\}$. Use `DSolve` or `dsolve` and find the solution to the initial value problem (4.358) for each frequency ω in the preceding set. Use `GraphicsArray` in *Mathematica* or `subplot` in *MATLAB* and draw the four solutions on the same screen (see Figure 4.11). First, what is the main difference between these graphs and the graph of the solution obtained when friction was not present? Second, what is the main difference among the latter four graphs when the frictional term is present? What is the approximate value of the resonance frequency?

4.13 Project C—Kinetic Energy and the Forced Damped Harmonic Oscillator

We saw in the previous two projects that the solution of the forced damped harmonic oscillator

$$mx'' + cx' + kx = A\cos\beta t, \quad x(0) = x_0, x'(0) = x_1, \qquad (4.360)$$

is the sum of two functions x_c and x_p. The function x_c is the complementary part, which contains information from the initial data and decays to zero exponentially. The particular solution x_p, given by (4.355) and (4.356), is a periodic trigonometric function. Because x_c converges to zero as t approaches infinity, this part of the solution is called the transient solution of (4.360). The function x_p, on the other hand, is called the steady-state solution.

We saw in the previous two projects that:

a) When the forcing term and friction are both absent from (4.360), the motion of the harmonic oscillator is governed by the natural frequency $\omega = \sqrt{\frac{k}{m}}$.

b) When the forcing term is absent, the critical damping given by $c^* = 2\sqrt{\frac{k}{m}}$ determines the cutoff point as to the behavior of solutions as t varies in $(0, \infty)$. When $c < c^*$, the underdamped case, the solution exponentially decays but oscillates with frequency

$$\omega = \sqrt{\frac{k}{m} - 4c^2}. \qquad (4.361)$$

c) When forcing and friction are present, the resonant frequency that leads to the maximum amplitude of the steady-state solution is given by

$$\omega = \sqrt{\frac{k}{m} - \frac{c^2}{2}}. \qquad (4.362)$$

In this project we will determine the analogue of part c) for the kinetic energy of the steady-state solution. First, we define the kinetic energy of a T-periodic solution $f(t)$ of (4.360) by

$$KE = \frac{m}{2T} \int_0^T f'^2(t)\, dt, \qquad (4.363)$$

where m is the mass of the body.

1. Show that the kinetic energy of the steady-state solution of (4.360) is given by

$$KE(\beta) = \frac{mA^2\beta^2}{4(k/m - \beta^2)^2 + c^2\beta^2}. \qquad (4.364)$$

(Hint: The period of (4.355) is $\frac{2\pi}{\beta}$.)

2. Draw the graph of $KE(\beta)$ when $m = k = 1$, $A = 1$ for $\beta \in (0,5)$ as c varies from 0.1 to 2 at increments of 0.1. By examining the graph, determine at which value of β is KE reaching its maximum? How do these values compare with Figure 4.10?

3. Show that the maximum of $KE(\beta)$ occurs at

$$\beta = \sqrt{\frac{k}{m}}, \tag{4.365}$$

independent of c. Note that (4.365) is the same as the natural frequency of the **undamped** and **unforced** harmonic oscillator.

4.14 Project D—Motion in Earth's Gravitational Field

We saw in Project B of Chapter 3 that the motion of a body in a gravitational field is governed by the second-order differential equation

$$\frac{d^2x}{dt^2} = -\frac{GM}{x^2}. \tag{4.366}$$

with $GM = gR^2$, where $g = 9.8$ m/s^2 is the acceleration of gravity and $R = 6370$ kilometers is the Earth's radius. The function $x(t)$ measures the distance of the body to the center of the planet.

1. Supply initial conditions to this equation to simulate launching a rocket in vertical motion. Use `NDSolve` of *Mathematica* or `ode45` of *MATLAB* to determine an approximate solution to this equation on an interval of t of your choice. Try two different values for v_0: one 5 km/s below the escape velocity and one 1 km/s below it. Draw the graphs of both $x(t)$s on the same screen.

2. Air resistance is generally not a significant factor for launching a rocket when it has reached high elevation. At lower elevations, however, this force could be significant. As we have seen in other models of air friction, this force depends on the velocity of the moving object, except that in the present case the frictional force's magnitude decreases the farther the rocket gets from the Earth. We model the magnitude of this force by

$$ke^{R-x}\frac{dx}{dt}, \quad k > 0.$$

Use `NDSolve` or `ode45` and solve the initial value problem

$$m\frac{d^2x}{dt^2} + ke^{R-x}\frac{dx}{dt} + m\frac{gR^2}{x^2} = 0, \quad x(0) = R, \quad \frac{dx}{dt}(0) = v_0, \tag{4.367}$$

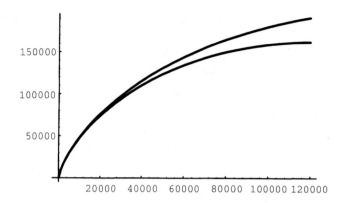

Figure 4.12: The output of `gravity.m` without and with friction ($k = 100$). The values of `tfinal` and `initvel` are 120,000 and 11, respectively.

where the rocket's mass is 2,500 kilograms, its initial speed is 11 km/s, and R is the Earth's radius. Solve (4.367) first with $k = 0$ and next with $k = 100$. Plot the graphs of each $x(t)$ for $t \in (0, 120,000)$ on the same screen.

3. The following program in *Mathematica* was used to obtain Figure 4.12. Study this program carefully for its syntax and then create a file called `gravity.m` that contains its content. Execute this program for the parameter values (see the figure caption) that led to Figure 4.12. Combine the graphical outputs on the same screen.

```
gravity[k_,tfinal_,initvel_]:= Block[{sol, m},
R = 6370 (* Earth radius in kilometers*);
m = 2500; (* Rocket's mass *)
g = 0.0098 (* kilometer/sec*);
sol=NDSolve[{m x''[t] +k*Exp[R-x[t]]*x'[t] +
        m*g*R^2/(x[t]^2) == 0,
           x[0] == R, x'[0] == initvel}, x[t], {t, 0, tfinal}];
Plot[x[t]-R/. sol, {t, 0, tfinal}, PlotRange->All]]
```

4.15 Project E—Couette Flow

The Couette flow is a specialized fluid flow between two concentric cylinders. We assume that two infinitely long cylinders of radii R_1 and R_2 are concentric and contain a viscous fluid of density ρ and viscosity ν between them. A fundamental problem in fluid dynamics is to understand the behavior of the fluid particles as the cylinders rotate. The mathematical model of the fluid motion is described by a set of PDEs known as the **Navier–Stokes** equations that balance the change in the linear momentum against the viscous and pressure forces (we will study these equations in detail in a later chapter). In the special case of the Couette flow the velocity field assumes the form

$$\mathbf{v} = v(r, t)\mathbf{e}_\theta, \tag{4.368}$$

where v is expressed in cylindrical coordinates in which a typical particle position has coordinates (r, θ, z). In (4.368) $\mathbf{e}_\theta$ is a unit vector in the direction of change in θ, which in rectangular coordinates becomes

$$\mathbf{e}_\theta = -\sin\theta\,\mathbf{i} + \cos\theta\,\mathbf{j}, \tag{4.369}$$

where $\mathbf{i}$ and $\mathbf{j}$ are the standard unit vectors in the x and y directions. Equation (4.368) shows how special and idealized the Couette flow is. In assuming (4.368) we are stating that the fluid particles do not move in the directions parallel and perpendicular to the rotational axis of the cylinders, the z and r directions, respectively. With these assumptions it can be shown that the Navier–Stokes equations reduce to a single PDE

$$\rho v_t = \nu\left(v_{rr} + \frac{1}{r}v_r - \frac{v}{r^2}\right). \tag{4.370}$$

In this project we are interested in determining the set of all possible **steady-state** solutions of (4.370) when the velocity of the fluid is specified at the walls $r = R_1$ and $r = R_2$.

1. Let $v(r)$ be the steady motion of a fluid in motion between two cylinders of radii R_1 and $R_2 > R_1$, rotating with angular velocities Ω_1 and Ω_2. Show that v satisfies the boundary value problem

$$v'' + \frac{1}{r}v' - \frac{v}{r^2} = 0, \quad v\big|_{r=R_1} = \Omega_1 R_1, \quad v\big|_{r=R_2} = \Omega_2 R_2. \tag{4.371}$$

2. Note that the differential equation in (4.371) is of the Euler–Cauchy type (see Section 4.7). Determine the general solution of this equation. Apply the boundary conditions in (4.371) to show that

$$v(r) = Ar + \frac{B}{r}, \tag{4.372}$$

where

$$A = \frac{\Omega_2 R_2^2 - \Omega_1 R_1^2}{R_2^2 - R_1^2}, \quad B = \frac{\Omega_1 - \Omega_2}{R_2^2 - R_1^2}R_1^2 R_2^2. \tag{4.373}$$

3. Verify in *Mathematica* or *MATLAB* that the preceding v is the solution to the boundary value problem in (4.371).

4. Suppose that $R_1 = 1$, $R_2 = 2$, $\Omega_1 = 1$, and $\Omega_2 = \omega$. Determine v in this case and draw its graph for ω ranging from 0.5 to 1.5 at increments of 0.1.

5. Let $R_1 = 1$, $R_2 = 2$, $\Omega_1 = -1$, and $\Omega_2 = \omega$. Determine v in this case and draw its graph for ω ranging from 0.5 to 1.5 at increments of 0.1.

6. Show that the velocity profile is a monotone function if $R_2 > R_1$ for any Ω_1 and Ω_2.

7. Show that the velocity profile does not have an inflection point when $R_2 > R_1$. How would you prove this statement in *Mathematica* or *MATLAB*?

4.16 Project F—Shooting Method in *Mathematica*

In this project we develop an algorithm, known as the **shooting method**, for solving **boundary value problems** for linear or nonlinear second-order differential equations. Such problems are of the form

$$y'' = f(x, y, y'), \quad y(a) = y_0, \quad y(b) = y_1. \tag{4.374}$$

Boundary value problems appear naturally and frequently in many areas of mathematical physics. For example, the study of buckling of beams made of elastic material, as well as the structure of the so-called boundary layers in the motion of viscous fluids near a solid wall, lead to boundary value problems of the form (4.374). In Chapter 6 we will address the same problem for first-order systems of differential equations.

Unlike an initial value problem, where the values of y and y' are known at the same value of x, in a boundary value problem we are given two pieces of information at two *different* values of x. If the differential equation in (4.374) is linear, we would first find its general solution and then use (4.374b) and (4.374c) and determine the two constants in the general solution. This idea does not work for nonlinear differential equations because, in general, it is quite difficult to find any or all solutions of such an equation in closed form. We therefore resort to approximate methods to construct solutions to boundary value problems. One such method is the shooting method.

The idea behind the shooting method is quite simple. We start with the point $(a, y(a))$ and solve the differential equation (4.374a), for $x \in (a, b)$, with **initial values**

$$y(a) = y_0, \quad y'(a) = z, \tag{4.375}$$

for some z. If we are extremely lucky we choose the correct z and hit $y = y_1$ at $x = b$. As you can imagine this scenario rarely happens and we either **undershoot** or **overshoot** the target. What is important is that we have the capability of solving the differential equation (4.374a), (4.375) and evaluating $y(b)$. Evidently this value of y depends on the slope z that we chose in (4.375). Let F be this function; that is, $F(z)$ is the end value of y we obtain by using the shooting value of z (see Figure 4.13).

We would solve our problem if we could choose z such that

$$F(z) - y_1 = 0, \tag{4.376}$$

that is, if we could determine a root of the function $G(z) = F(z) - y_1$. The following program shows how to use NDSolve with (4.374a) - (4.375) and FindRoot

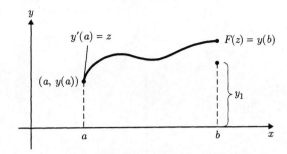

Figure 4.13: The shooting method.

with the function in (4.376) to get the desired result. This program is written for the boundary value problem

$$y'' + \exp(y) = 0, \quad y(0) = 1, \quad y(1) = 1. \tag{4.377}$$

```
f[x_, y_, p_] = Exp[y]; b = 1; a = 0; y0 = 1; y1 = 1;
F[shoot_] := Block[{sol}, sol = NDSolve[{y''[x] ==
    f[x, y[x], y'[x]], y[a] == y0, y'[a] == shoot}, {y},
      {x, a, b}];
    out = Evaluate[y[x] /. sol/. x -> b]; out[[1]]];
output = FindRoot[F[shoot] - y1, {shoot, 1, 0.9}];
solution = NDSolve[{y''[x] ==  f[x, y[x],  y'[x]],
    y[a] == y0, y'[a] == shoot /. output}, {y}, {x, a, b}];
Plot[Evaluate[y[x] /. solution], {x, a, b}]
```

The `Block` function is used in this program to keep `sol` internal to that portion of the program. This is not necessary, but it is convenient. The `FindRoot` function is based on a routine called the bisection method that requires two starting values to proceed, hence the need for 1 and 0.9 in the `output` statement. This would be one of the trouble spots with which to experiment when the shooting method does not seem to converge. The last two commands of the program solve the intial value problem (4.374a), (4.376) with the value of z that has just been determined. The graph of the solution to the boundary value problem is then plotted. One use of graphing the solution is to check visually that the solution satisfies the correct boundary conditions.

1. Create a file (say `shooting.m`) that contains the preceding program. Input this file to *Mathematica* and obtain the same graph as in Figure 4.14.

2. Modify this program for the differential equation

$$y'' + y = 0, \quad y(0) = 1, \quad y(1) = -1. \tag{4.378}$$

Find the exact solution to (4.378) and plot it against the solution you find by the shooting method.

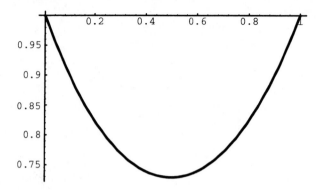

Figure 4.14: The shooting method solution to $y'' + \exp(y) = 0, y(0) = y(1) = 1$.

3. Modify the previous program and plot the solution to

$$y'' + y = x, \quad y(0) = 1, \quad y(1) = -1. \tag{4.379}$$

Again, compare your solution to the exact solution.

4. Apply this method to

$$y'' + \sin y = 0, \quad y(0) = 1, \quad y(3) = -1. \tag{4.380}$$

5. Modify the previous program to the boundary value problem (4.374a,b) but with

$$y'(b) = y_1. \tag{4.381}$$

Apply the new program to

$$y'' + y = 0, \quad y(0) = 1, \quad y'(2\pi) = 0. \tag{4.382}$$

Plot your result against the exact solution of the this problem.

4.17 Project G—Eigenvalue Problems

In the context of differential equations, an **eigenvalue** problem equations is typically a **boundary value problem** for which more than one solution exists. For example, consider the problem

$$x'' + \pi^2 x = 0, \quad x(0) = x(1) = 0. \tag{4.383}$$

Evidently this problem has at least two solutions: one is the trivial solution $x_1(t) \equiv 0$, and the other $x_2(t) = A \sin \pi t$. When this problem is cast in the context of the one-parameter family of problems

$$x'' + \lambda x = 0, \quad x(0) = x(1) = 0, \tag{4.384}$$

where λ is a parameter whose control is in our hands, then the behavior of the solution(s) to (4.384) depend crucially on λ. When $0 \leq \lambda < \pi^2$, the boundary value problem (4.384) has the unique solution $x(t) \equiv 0$ (why?). But when λ hits the critical value π^2, this problem has the two solutions x_1 and x_2 stated previously. Such a critical value of the parameter λ is called the **eigenvalue** of (4.384) and the corresponding function x its associated **eigenfunction**. As we will see, there are infinitely many such eigenvalues for (4.384), whose knowledge is critical in many areas of physics and mechanics.

1. Show that the general solution of the differential equation in (4.384) is given by
$$x(t) = c_1 \sin \sqrt{\lambda} t + c_2 \cos \sqrt{\lambda} t. \tag{4.385}$$

2. Apply the boundary conditions in (4.384) to the general solution to show that when $(n-1)^2 \pi^2 < \lambda < n^2 \pi^2$, $x(t) \equiv 0$ is the only solution of (4.384).

3. Use (4.385) to determine all solutions of (4.384) when λ reaches an eigenvalue $n^2 \pi^2$.

4. Apply `DSolve` of *Mathematica* or `dsolve` of *MATLAB* to (4.383). Does the response from these packages indicate that $\lambda = \pi^2$ is an eigenvalue in (4.383)?

5. Apply `DSolve` or `dsolve` to (4.384) with λ as a parameter. Does the response identify which values of λ are critical?

We next consider a **nonlinear** eigenvalue problem that has many similarities to (4.384) and some important differences. Consider the pendulum problem
$$x'' + \lambda \sin x = 0, \quad x(0) = 0, \quad x(1) = 0. \tag{4.386}$$

Note that when x is small, such that $\sin x$ can be approximated by x, (4.386) reduces to (4.384).

1. Show that $x(t) \equiv 0$ is a solution of (4.383) no matter what real value λ takes.

2. Show that every solution of (4.386) satisfies the identity
$$\frac{1}{2} x'^2 - \lambda \cos x = \text{const.} \tag{4.387}$$

(Hint: Multiply (4.386) by x' and integrate the result with respect to t. Note that $x'' x' = \frac{1}{2} \frac{d}{dt}(x'^2)$.)

3. The function
$$E(x, x') = \frac{1}{2} x'^2 - \lambda \cos x \tag{4.388}$$

is called a **first integral** of (4.386). The relation (4.387) shows that every solution of (4.386) is a contour level of E. Use *Mathematica* or *MATLAB* to

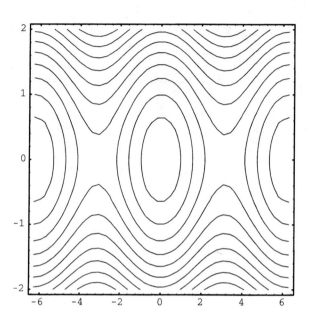

Figure 4.15: The contours of the first integral of equation for the oscillation of a pendulum when $\lambda = 1$.

draw the contours of E when $\lambda = 1$ and then again when $\lambda = 6$ (compare the graphs with Figure 4.15). These graphs, which show how x and x' are related as the pendulum swings back and forth, are called the phase diagrams of the differential equation. Note that each solution of (4.386) defines a closed curve in the (x, x')-plane. In solving the boundary value problem in (4.386) we are interested in those nontrivial trajectories that begin on the vertical axis at time zero (because $x(0) = 0$) and swing to the other side of the vertical axis in **exactly** one unit of time (because $x(1) = 0$). It is not clear that such a solution should exist for any value of λ unless we choose the trivial trajectories $x(t) \equiv 0$, or $x(t) \equiv \pi$, or in general, $x(t) = n\pi$ and let λ be arbitrary.

4. Let x_m be the x coordinate of the point of intersection of each trajectory of (4.386) in the (x, x')-plane. We note that if a trajectory satisfies the boundary conditions in (4.386), the following relation between x and x_m must hold:

$$x\left(\frac{1}{2}\right) = x_m. \tag{4.389}$$

Use this fact with (4.387) to show that the trajectory of the solution to the boundary value problem satisfies the identity

$$\frac{1}{2}x'^2 - \lambda \cos x = -\lambda \cos x_m \tag{4.390}$$

and, moreover,

$$\frac{\sqrt{\lambda}}{2} = \int_0^{x_m} \frac{1}{\sqrt{\cos x - \cos x_m}} \, dx. \tag{4.391}$$

(Hint: To get (4.391) solve (4.390) for x' and note that $dt = \int \frac{1}{x'} dx$.) What has been shown so far is that if there is a nontrivial solution to (4.386), then λ and x_m must satisfy (4.391).

5. The integral in (4.391) is an example of an **improper** integral, that is, one for which the integrand becomes infinite at a point in the interval of integration. Recall from calculus that such integrals are well defined when the integrand has an appropriate rate of growth or decay near the point of singularity. For example, $\int_0^1 \frac{1}{t} \, dt$ does not exist but $\int_0^1 \frac{1}{\sqrt{t}} \, dt$ does exist, as can easily be shown. The integral in (4.391) is of the latter type, although this fact is rather difficult to show. The determination of this integral has a rich history in mathematics. We will now use a well-known change of variables and convert (4.391) to an integral that is available in *Mathematica* and *MATLAB*. We introduce the new dependent variable u which is related to x by

$$\sin \frac{x_m}{2} \sin u = \sin \frac{x}{2}. \tag{4.392}$$

Show that

$$dx = \frac{\sqrt{2} du}{\sqrt{1 - \sin^2 x_m / 2 \sin^2 u}} \tag{4.393}$$

and that (4.391) becomes

$$\sqrt{\lambda} = 2f(x_m), \tag{4.394}$$

where

$$f(x_m) = \int_0^{\frac{\pi}{2}} \frac{du}{\sqrt{1 - k^2 \sin^2 u}}, \quad \text{with} \quad k = \sin \frac{x_m}{2} \tag{4.395}$$

The integral in (4.395) is called the **complete Elliptic Integral**. In *Mathematica* this function is denoted by EllipticK and in *MATLAB* by ellipke.

6. Use *Mathematica* or *MATLAB* and draw the graph of the eigenvalue λ versus x_m. Compare the graph with Figure 4.16.

Remark 4.17.1: The graph in Figure 4.16 shows that as x_m approaches zero the value of $\sqrt{\lambda}$ approaches π, which is the first eigenvalue of (4.384). So, when x is small, the boundary value problem (4.384) closely approximates (4.386). This statement can be made mathematically rigorous in the context of **linearization**: Equation (4.384) is the linearization of (4.386)

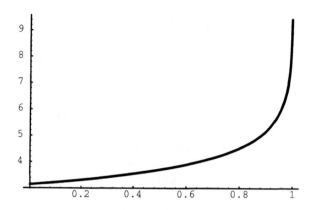

Figure 4.16: The bifurcation curve for $x'' + \lambda \sin x = 0$ with x_m and the square root of λ on the horizontal and vertical axes, respectively.

about the solution $x(t) \equiv 0$. If this concept and in general the ideas that have been presented in this project are of interest, the reader may consult E. L. Reiss, "Column Buckling—An Elementary Example of Bifurcation," in [1], pp. 1–16, for more information.

Remark 4.17.2: Note an important difference between (4.384) and (4.386) that is typical between linear and nonlinear differential equations. What we found for (4.384) was that any function of the form $A \sin \pi t$ is an eigenfunction of (4.384) for the eigenvalue $\lambda = \pi^2$. When (4.384) models compression loading of a column, this statement's interpretation is that we can generate the configuration $A \sin \pi t$ *of any amplitude* A with a single load proportional to π^2, which is an unrealistic scenario. The nonlinear problem (4.386) gives a different interpretation as Figure 4.16 shows. The graph of λ versus x_m is a one-to-one function.

Remark 4.17.3: The graph in Figure 4.16 is one **branch** of the **bifurcation diagram** of the boundary value problem (4.384). This particular branch emanates from the point $(0, \pi)$ in the $(x_m, \sqrt{\lambda})$-plane. Other similar branches also bifurcate from the other eigenvalues of the linearized problem (4.384), namely, from $(0, n\pi)$, $n = 2, \dots$. For more information see the aforementioned article of Reiss.

7. Use **Solve** of *Mathematica* or **solve** of *MATLAB* to determine λ when $x_m = 0.1$ and 0.9. In each case find the value $x_1 \equiv x'(0)$.

8. Use **NDSolve** of *Mathematica* or *ode45* of *MATLAB* with the preceding values of λ and x_1 and solve the **initial value problem**

$$x'' + \lambda \sin x, \quad x(0) = 0, \quad x'(0) = x_1. \tag{4.396}$$

Determine the value of x at $t = 1$ and verify that it satisfies the boundary conditions in (4.386). Draw the graph of the solution in each case. Finally, plot the two graphs on the same screen, using different graphical styles to identify them.

4.18 Project H—Eigenvalues and Column Buckling

The buckling of a vertical column under its own weight, or when it is subjected to external forces, is an old problem in applied mathematics. The study of this problem has led to the development of many analytical and numerical tools in this discipline. In this project we analyze an eigenvalue problem that is intimately related to the phenomenon of buckling, whose eigenvalue relates to the critical loading and whose eigenfunction describes the shape of the buckled column. In a series of papers C. Y. Wang (see, e.g., J. Appl. Mech., vol. 50, 1983, pp. 311–314) studied variations of the buckling problem when the column is emersed in a dense fluid. This particular setting is important to offshore drilling and to our understanding of the stability of man-made structures in the oceans.

Consider the differential equation

$$x'' + \lambda s x = 0, \quad x'(0) = 0, \quad x(1) = 0. \tag{4.397}$$

Here $x(s)$ is the angle the column makes with the vertical, s is the arclength parameter along the column, and λ is a parameter proportional to the weight of the column (or the force due to the buoyancy effect, in the case of an immersed column). See Figure 4.17. The differential equation is obtained by balancing the momentum of the column at each point s (see Wang's article for a careful derivation of this equation). The first boundary condition allows the end $s = 0$ of the column to be moment-free (and, hence, allowed to move), while the second boundary condition keeps the other end of the column stationary.

1. Identify the type of this differential equation (for example, Bessel or Airy) by a comparison with the list of the special differential equation discussed in Section 4.10.

2. Show that no matter what value λ takes the function $x(s) \equiv 0$ is a solution of (4.397).

3. Apply `DSolve` of *Mathematica* or `dsolve` of *MATLAB* to find the general solution of this equation.

4. Apply the first boundary condition in (4.397) to eliminate one of the constants c_1 or c_2 from the general solution. Next, apply the second boundary condition. Let $f(\lambda) = 0$ be the resulting equation in λ. Draw the graph of f and compare with Figure 4.18.

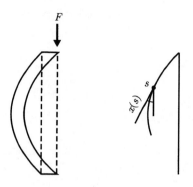

Figure 4.17: The buckling column schematic.

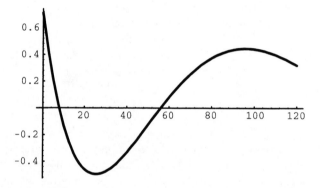

Figure 4.18: The bifurcation equation for $x'' + \lambda s x = 0$.

5. The zeros of f are the eigenvalues of this boundary value problem. Apply **FindRoot** of *Mathematica* or **roots** of *MATLAB* to determine the first five eigenvalues λ_1, λ_2, λ_3, λ_4, and λ_5 of this problem.

6. Apply **DSolve** or **dsolve** with each λ_i to determine the solution to (4.397). Combine the graphs of all five buckled columns in one graph.

7. Study carefully the syntax of the following program in *Mathematica*. This program begins with the boundary value problem (4.397) and ends with the first eigenvalue of the problem. It is written with an option for the length of the column (denoted by **length**, in the program).

```
g[length_]:=Block[{sol, lambda, s, x, eqn, newsol, d},
    sol=DSolve[x''[s] + lambda s x[s] == 0, x[s], s];
    eqn[1] = D[x[t]/.sol, t]/. t->0;
    c2 = Solve[eqn[1][[1]] == 0, C[2]][[1,1,2]];
    newsol = x[s] /. sol /. C[2] -> c2;
```

```
    newsol = Simplify[newsol/C[1]];
    f[lambda_] = newsol /. s->length;
    Plot[f[lambda], {lambda, 0, 120},
    PlotLabel->"Buckling Loads"];
    answer = FindRoot[First[d[lambda]]==0, {lambda, 0.1}]
]
```

Create a file called `column.m` containing the above program. Execute `column.m` and compare the output with the value of λ_1.

8. Revise `column.m` by adding appropriate lines to it so that the final output is the value of the eigenvalue and the graph of the solution (eigenfunction) associated with λ.

9. What changes to `column.m` are necessary to get all five eigenvalues and eigenfunctions of (4.397)?

4.19 Project I—Perturbation Method in *Mathematica*

In many problems in mathematical physics we encounter differential equations with small parameters in them. Examples of such problems are

$$u'' + 2\epsilon u' + u = 0, \tag{4.398}$$

which models the motion of a mass-spring system with weak damping, and

$$u'' + u = a + \epsilon u^2, \tag{4.399}$$

which is known as the **Einstein's two-body equation** for the precession of the planet Mercury. In (4.399) the ϵ term takes into account the relativistic effects in the motion of that planet (see Donald Smith, *Singular-Perturbation Theory*, Springer-Verlag, New York, 1985, for motivation and references). In this project we wish to study the solutions of equations such as (4.398) and (4.399) and their dependence on the parameter ϵ.

We will concentrate on (4.398) and develop the perturbation method for this example. The techniques and programs developed here will apply to (4.399) and others with minor changes.

The perturbation method seeks a series solution of the form

$$u(t, \epsilon) = \sum_{i=0}^{\infty} u_i(t)\epsilon^i \tag{4.400}$$

for the solution $u(t, \epsilon)$ of an equation such as (4.398). The basic idea behind this series expansion is that, when it converges, the process will give us a hierarchy of terms that represent the influence of ϵ, whose importance enters the picture at different rates as ϵ approaches zero. We expect that the zeroth-order term u_0

captures the main features of the solution u when ϵ is near 0, that u_1 captures the order of ϵ features, and so on. As we will see shortly, the u_0 term of (4.398) satisfies this equation with $\epsilon = 0$, so it represents the purely oscillatory part of the solution, while u_1 brings the frictional features of the problem into view, however small they may be.

1. Consider (4.398) with initial conditions

$$u(0) = a, \quad u'(0) = b. \tag{4.401}$$

   Show that

$$u(t, \epsilon) = e^{-\epsilon t}\left(a \cos \beta t + \frac{b + \epsilon a}{\beta} \sin \beta t \right), \tag{4.402}$$

   with $\beta = \sqrt{1 - \epsilon^2}$, is the exact solution to (4.398)–(4.401).

2. Begin by substituting the series (4.400) into (4.398), collect coefficients of equal powers of ϵ, and by setting them equal to zero derive the following equations for $u_0, u_1, \dots$:

$$
\begin{aligned}
u_0'' + u_0 &= 0, & u_0(0) = a,\, u_0'(0) = b, \\
u_1'' + u_1 &= -2 & u_0, u_1(0) = 0,\, u_1'(0) = 0, \\
u_2'' + u_2 &= -2 & u_1, u_2(0) = 0,\, u_2'(0) = 0,
\end{aligned}
\tag{4.403}
$$

   and, in general,

$$u_i'' + u_i = -2u_{i-1}, \quad u_i(0) = 0,\, u_i'(0) = 0, \tag{4.404}$$

   when $i \geq 1$.

3. Show that

$$u_0(t) = a \cos t + b \sin t \tag{4.405}$$

   is the solution to (4.403a). Use this solution in (4.403b) and show that

$$u_1(t) = a \sin t - t(a \cos t + b \sin t) \tag{4.406}$$

   is the solution to (4.403b).

4. Let $P_i(t, \epsilon)$ be the ith partial sum of the series (4.400). Let $\epsilon = 0.1$. Let $a = b = 1$. Use *Mathematica* to draw the graphs of the exact solution $u(t, \epsilon)$, $P_0(t, \epsilon)$, and $P_1(t, \epsilon)$ on the interval $t \in (0, 3)$. Repeat these graphs for $\epsilon = 0.01$.

5. The procedure we described previously has an iterative flavor to it and can be implemented in a symbolic manipulator. The following program in *Mathematica* does precisely that. This program is written to compute P_5 but is capable of obtaining any order perturbation u. The program computes each u_i and displays it as `U[i][t]` = :

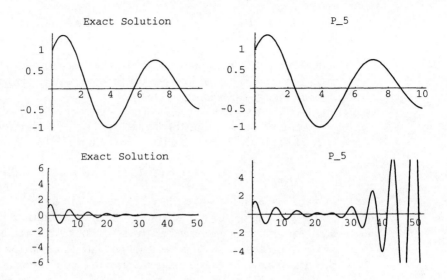

Figure 4.19: Exact and perturbation solutions of $u'' + 2\epsilon u' + u = 0$ with $\epsilon = 0.01$ and 0.1.

```
n = 5;
L[u_] := D[u, {t, 2}] + 2 eps D[u, t] + u;
P[t_, eps_] = Sum[U[i][t] eps^i, {i, 0, n}];
term = L[P[t, eps]];
sol = DSolve[{term == 0 /. eps -> 0, U[0][0] == a,
        U[0]'[0] == b}, U[0][t], t];
U[0][t_] = sol[[1, 1, 2]];
Print["U[0][t] = ", U[0][t]];
Do[
    sol = DSolve[{D[term, {eps, i}] == 0 /. eps -> 0,
              U[i][0] == 0, U[i]'[0] == 0}, U[i][t], t];
    U[i][t_] = Simplify[sol[[1,1,2]]];
              Print["U[",i,"][t] = ", U[i][t]], {i, 1, n}
];
```

Apply this program and compare the results for u_0 and u_1 with those derived analytically.

6. Execute the preceding program and draw the graphs of P_5 and the exact solution on the same screen for $t \in (0, 1)$ and $\epsilon = 0.1$. Enlarge the domain of u until a substantial change in the behavior of the exact solution versus the perturbation solution is noticed (see Figure 4.19). Repeat this part for $\epsilon = 0.01$. What changes come about when the time interval is increased to $(0, 10)$? What about $(0, 100)$?

7. Write up the findings of this project in a paper. In particular, describe carefully the behavior of P_5 in relation to the exact solution when $t \in$

$(0, 100)$ (Hint: Look at the terms of the form $\epsilon^j t^i \sin t$ and $\epsilon^j t^i \cos t$ in P_5 and study what happens to these terms for large t and small ϵ.)

8. Repeat this entire set of exercises for the Einstein equation (4.399). Describe the modifications one needs to make in the program to accommodate the nonlinear equation (4.399).

Remark 4.19.1: The differential equations we discussed are examples of a large class of initial value problems of the form

$$x''(t) = f(x(t), x'(t), \epsilon), \quad x(0) = a, \quad x'(0) = b, \qquad (4.407)$$

to which we can apply the preceding procedure and obtain a series solution in ϵ. It is possible to prove that under reasonable conditions on the function f the perturbation series converges for each fixed t to the unique solution of (4.407) as ϵ approaches zero. The behavior we have seen in the preceding examples, that the quality of convergence deteriorates as t becomes large, is typical for this method. This difficulty is ostensibly due to the presence of what has often been referred to as the "spurious secular terms" in the series. In the case of (4.398), these secular terms appear in the form of the resonance terms, $t \sin t$ for instance, that contribute to the slow convergence of the infinite series for large t. A different way of putting this is to say that the prior series converge **pointwise** and not **uniformly**. If we were to apply the $\delta - N$ definition of convergence we would be obliged to choose N as a function t. Attempts to improve on this deficiency of the previous algorithm began with the great mathematician Henri Poincaré at the end of the nineteenth century who suggested a "multiscale" series expansion of $u(t, \epsilon)$, which often leads to a uniform approximation of u in t. An interested reader will find a wealth of worked examples and general theory on the multiscale method in the text referenced above.

Chapter 5

Laplace Transform and Differential Equations

5.1 Introduction

The Laplace transform offers a powerful technique for solving ordinary and partial differential equations. One of the basic features of this operation is its ability to transform a differentiation operation into an algebraic one. A second feature of the Laplace transform is its linearity which renders this tool very effective in solving linear differential equations. Additionally, the Laplace transform is invertible, which will enable us recover the solution to a linear differential equation in a somewhat roundabout way after having gone through a series of algebraic operations. This transform, precisely because of these properties, is very adaptable for use with a symbolic manipulator. The main goal of this chapter is to develop the Laplace transform, itemize its operational properties, make use of it in solving linear ODEs and systems, and finally develop computer programs that carry out the necessary (and sometimes tedious) algebraic work.

The Laplace transform method is often preferable to the analytic methods we developed in the previous chapters when dealing with linear differential equations with *discontinuous* or *periodic* forcing terms. These equations arise naturally in many physical settings, including in forced vibrations where the forcing term undergoes an abrupt change in a short time span or distance and in electrical circuits where the applied electromotive force is turned on and off periodically. Although these equations can be treated by the methods of the previous chapters, their resolution is often cumbersome and tedious. The Laplace transform method offers a natural alternative in these cases.

As we will see shortly, the Laplace transform of a function f involves the integral of this function against a kernel, e^{-st}, where the integration is performed over the semi-infinite interval $(0, \infty)$. The calculation of such an **improper** integral requires two limiting processes: one associated with estimating the semi-infinite interval with a set of finite intervals of the form $(0, n)$ and letting n approach infinity and the other the usual discretization of the integral defined on the finite interval. This aspect of the definition of Laplace transform must be kept in mind when determining the analytic properties of this operation. Our treatment in this text, however, will emphasize the operational aspects of this transform, especially in the context of symbolic manipulators such as *Mathematica* and *MATLAB*. We will, therefore, not develop the analytic tools needed to prove some of the main theorems that will be stated in the sequel.

5.2   Definition of Laplace Transform

In the entire chapter we will assume that f is an integrable function on the interval $(0, \infty)$. Typically, f will be a *piecewise continuous* function defined on the interval $(0, \infty)$; that is, f will be continuous on $(0, \infty)$ except perhaps at a finite number of points at which it is discontinuous but with the proviso that the limits from the left and right exist at each such point.

Example 5.2.1

1. The function

$$f(x) = \begin{cases} 1 & \text{for} \quad 0 \le x \le 1, \\ \cos x & \text{for} \quad x > 1 \end{cases} \tag{5.1}$$

   is piecewise continuous because the functions 1 and $\cos x$ are continuous on their respective domains and the limits from the left and right at $x = 1$ exist.

2. Consider the function

$$g(x) = \begin{cases} 0 & \text{for} \quad x = 1, \\ \sin \frac{1}{x-1} & \text{for} \quad x \ne 1. \end{cases} \tag{5.2}$$

   This function is not piecewise continuous because the limits of g from the left and right do not exist at $x = 1$.

Definition 5.2.1
Let f be a piecewise continuous function defined on the semi-infinite interval $[0, \infty)$. The Laplace transform of f, denoted by $L[f]$, is a function with domain (s^*, ∞) defined by

$$L[f](s) \equiv \int_0^\infty e^{-st} f(t)dt \equiv \lim_{M \to \infty} \int_0^M e^{-st} f(t)dt, \tag{5.3}$$

when this integral exists.

Note that we have used a limiting process in the definition of the Laplace transform, the reason being that the integral in (5.3) is over an infinite interval. The limiting process allows us to obtain this **improper** integral by converting it to an integration over a finite interval. We then have available to us all of the theorems and tools for the manipulation of such integrals.

Example 5.2.2

1. Let $f(t) \equiv 1$. Then

$$L[1](s) = \lim_{M \to \infty} \int_0^M e^{-st} dt = \lim_{M \to \infty} \left[-\frac{1}{M} e^{-Mt} + \frac{1}{s} \right] = \frac{1}{s}, \tag{5.4}$$

since $\lim_{M \to \infty} Me^{-Mt} = 0$ by L'Hôpital's rule. The domain of $L[1]$ is $s > 0$.

2. Let $f(t) = t$. After applying integration by parts once, we have

$$L[t](s) = \lim_{M \to \infty} \int_0^M te^{-st} \, dt$$

$$= \lim_{M \to \infty} \left[-\frac{te^{-st}}{s} |_{t=0}^{t=M} \right] + \lim_{M \to \infty} \left[\frac{1}{s} \int_0^M e^{-st} dt \right]. \tag{5.5}$$

The first limit in (5.5) is zero because the exponential function e^{-Mt} dominates any polynomial in t as t approaches infinity (again, apply L'Hôpital's rule). The second limit yields $\frac{1}{s^2}$. Hence

$$L[t](s) = \frac{1}{s^2}. \tag{5.6}$$

The domain of $L[t]$ is the semi-infinite interval $s > 0$.

In *Mathematica* the Laplace transform is denoted by `LaplaceTransform`. To apply this command one must first input the package that makes this transform and available to *Mathematica*:

`<<Calculus'LaplaceTransform'`

One determines the Laplace transform of the two functions in Example 5.2.2 as follows.

Example 5.2.3

The following sequence of commands given to *Mathematica* determines the Laplace transform of $f(t) = 1$:

`LaplaceTransform[1, t, s]`

which results in

```
1
-
s
```

while

```
LaplaceTransform[t, t, s]
```

returns the Laplace transform of $g(t) = t$

```
    -2
   s
```

Similarly

```
LaplaceTransform[Exp[a t], t, s]
```

gives us the Laplace transform of $h(t) = e^{at}$. What is the domain of $L[e^{at}](s)$?

   The syntax of Laplace transform in $MATLAB$ is very similar. `laplace(f, t, s)` returns the Laplace transform of f. For example,

```
syms t s;
laplace(1, t, s)
```

returns 1/s, and

```
laplace(exp(-t),t,s)
```

returns 1/(s+1). An alternative way of computing the Laplace transform of a function in $MATLAB$ is to appeal directly to the $Maple$ kernel that comes with the $Symbolic\ Math\ TOOLBOX$. One determines the Laplace transform of $f(t) = e^{at} \cos bt$, say, via

```
maple('laplace(exp(a*t)*cos(b*t),t,s)')
```

or, alternatively,

```
f = 'exp(a t)*cos(b*t)';
maple('laplace',f,'t','s')
```

Before proceeding with the properties of the Laplace transform, we state a theorem that guarantees the existence of $L[f]$ for a large class of functions f.

   Theorem 5.2.1
   Let f be a piecewise continuous function on the interval $(0, \infty)$. Suppose further that this function satisfies the bound

$$|f(t)| \leq Ce^{at}, \quad \textit{as long as } t \geq b > 0. \tag{5.7}$$

Then the $L[f]$ exists for $s > a$.

   Proof: From the definition of the Laplace transform we have

$$L[f](s) = \int_0^\infty e^{-st} f(t)\, dt = \int_0^b e^{-st} f(t)\, dt + \int_b^\infty e^{-st} f(t)\, dt. \tag{5.8}$$

The first integral on the right-hand side is finite because both e^{-st} and f are well-behaved and continuous on $(0, b)$. The second integral, however, satisfies (note that $|e^{-st}f(t)| \leq Ce^{-(s-a)t}$ by (5.7))

$$\int_b^\infty e^{-st} f(t)\, dt \leq C \int_b^\infty e^{-(s-a)}\, dt, \tag{5.9}$$

which is finite if $s > a$. This completes the proof.

Remark 5.2.2: It is worth emphasizing that the exponential condition in Theorem 5.2.1 is a sufficient but not necessary condition. There are functions that do not satisfy this estimate and yet have a well-defined Laplace transform.

The first important property of the Laplace operator we investigate is its linearity. Using this property we are able to compute the transform of a large class of complicated functions by breaking them into sums of functions whose Laplace transforms we have previously determined.

Theorem 5.2.2 (Linearity of the Laplace Operator)
The Laplace operator is linear; that is, given two functions f and g and scalars a and b we have

$$L[af + bg] = aL[f] + bL[g] \tag{5.10}$$

for every s in the common domain of $L[f]$ and $L[g]$.

Proof: From the definition of the Laplace operator, the left-hand side of (5.10) is

$$\begin{aligned}
L[af + bg] &= \int_0^\infty e^{-st}[af + bg]\, dt \\
&= a \int_0^\infty e^{-st} f(t)\, dt + b \int_0^\infty e^{-st} g(t)\, dt \\
&= aL[f(t)] + bL[g(t)],
\end{aligned} \tag{5.11}$$

which is the right-hand side of (5.10).

Example 5.2.4
Using this theorem the Laplace transform of $1 + 2t$ is (see Example 5.2.2)

$$L[1 + 2t](s) = L[1](s) + 2L[t](s) = \frac{1}{s} + \frac{2}{s^2}. \tag{5.12}$$

Similarly,

$$L[e^{2t} + 3e^{-t}](s) = L[e^{2t}](s) + 3L[e^{-t}](s) = \frac{1}{s-2} + \frac{3}{s+1}. \tag{5.13}$$

A second important property of the Laplace operator is its invertibility; that is, given a function $F(s)$ (from a rather large class of functions), there is a unique function $f(t)$ for which $L[f(t)](s) = F(s)$. We will not attempt to quantify the class of functions F for which the inversion process is permissible since the mathematical tools needed to characterize this class are beyond the scope of this text. Suffice it to say that for the type of functions $F(s)$ that we will encounter in the applications in this text, we will be able to determine $f(t)$ explicitly. We will refer the reader to [1] for details on the theoretical and analytic aspects of the inversion of the Laplace transform.

Definition 5.2.3

We denote by L^{-1} the inverse to the Laplace operator L, that is, if $L[f] = F$, then

$$L^{-1}[F] = f. \tag{5.14}$$

Theorem 5.2.3 (Linearity of the Inverse Operator)

The inverse Laplace operator L^{-1} is a linear operator; that is, given two functions F and G, and two scalars a and b, we have

$$L^{-1}[aF + bG] = aL^{-1}[F] + bL^{-1}[G]. \tag{5.15}$$

Example 5.2.5

Suppose that $F(s) = \frac{s+3}{s^2}$. We want to find a function f such that $L[f] = F$. Since

$$F(s) = \frac{s+3}{s^2} = \frac{1}{s} + \frac{3}{s^2} \tag{5.16}$$

and $\frac{1}{s} = L[1](s)$ and $\frac{1}{s^2} = L[t](s)$, we have

$$F = L[1] + 3L[t] = L[1 + 3t].$$

Hence

$$f(t) = 1 + 3t. \tag{5.17}$$

Example 5.2.6

The Laplace inverse operator is a standard function in *Mathematica* and *MATLAB*. The following *Mathematica* command results in the same answer as that obtained in the previous example (we are assuming that the package `Calculus'LaplaceTransform'` has already been entered into *Mathematica*):

`InverseLaplaceTransform[(s+3)/s^2, s, t]`

Mathematica's response is

`1 + 3 t`

while, in *MATLAB* we need to input

```
syms s
F=(s+3)/s^2
ilaplace(F)
```

or

```
maple('invlaplace((s+3)/s^2,s,t)')
```

In general the process of computing the inverse of a Laplace transform is quite difficult. There are certain algebraic techniques, however, that are helpful in reducing the effort in computing L^{-1} to a manageable task. The method of partial fractions, with which the reader may be familiar from elementary algebra, is one such technique. This technique applies to Laplace transforms that are rational functions, that is, functions F that have the form

$$F(s) = \frac{a(s)}{b(s)}, \tag{5.18}$$

where $a(s)$ and $b(s)$ are polynomials in s. The method of partial fractions is then used to write such an F in terms of irreducible first- and second-order rational functions, that is, rational functions of the form

$$\frac{a_1}{a_2 s + a_3} \quad \text{or} \quad \frac{a_1 s + a_2}{a_3 s^2 + a_4 s + a_5}. \tag{5.19}$$

Since L^{-1} is a linear operator, the task of computing the Laplace inverse of F reduces to finding the inverses of (5.19) and summing the results. We will give two examples of how one carries out the partial fraction technique. Because, however, both *Mathematica* and *MATLAB* have this technique as a basic tool in computing the inverse of Laplace transforms, we will not dwell on this technique any further. Full details concerning this technique (and its nuances and exceptions) can be found in any standard text on ODEs.

Example 5.2.7

Let $F(s) = \frac{1}{s^2+3s+2}$. To determine the Laplace inverse of F, we reduce F to a sum of first-order fractions. The denominator of F factors into $(s + 1)(s + 2)$ suggests seeking constants A and B, so we can write

$$\frac{1}{s^2 + 3s + 2} = \frac{A}{s + 1} + \frac{B}{s + 2}. \tag{5.20}$$

After taking the common denominator and simplifying the preceding expression, we find that A and B must satisfy

$$A(s + 2) + B(s + 1) = 1 \tag{5.21}$$

for all s. Evaluating this expression first at $s = -2$ and then at $s = -1$ yield

$$A = 1, \quad B = -1, \tag{5.22}$$

so

$$F(s) = \frac{1}{s+1} - \frac{1}{s+2}. \tag{5.23}$$

Finally, using the linearity of L^{-1}, we have

$$f(t) = L^{-1}[F] = e^{-t} - e^{-2t}. \tag{5.24}$$

We note that this is the same output we would get from
`InverseLaplaceTransform[1/(s^2 + 3s + 2), s, t]`.

Example 5.2.8
Let $F(s) = \frac{1}{s^2+s+1}$. Unlike the previous example, the roots of the denominator in F are complex; hence, this polynomial is irreducible. We begin by completing the square in the denominator:

$$F(s) = 1/\left[\left(s + \frac{1}{2}\right)^2 + \frac{3}{4}\right]. \tag{5.25}$$

Note that $L[e^{at} \sin bt] = \frac{b}{(s-a)^2+b^2}$ (see Problem 1(g) or check this fact in *Mathematica* or *MATLAB*). Returning to (5.25), we have

$$a = -\frac{1}{2}, \quad b^2 = \frac{3}{4}, \tag{5.26}$$

so

$$f(t) = L^{-1}[F] = \frac{2}{\sqrt{3}}e^{-\frac{t}{2}} \sin \frac{\sqrt{3}t}{2}. \tag{5.27}$$

Example 5.2.9 (Partial fractions in *Mathematica*)
We now carry out in *Mathematica* the calculations that led to the values of A and B in Example 5.2.7. We start by entering the original expression $\frac{1}{s^2+3s+2}$

```
original = 1/(s^2 + 3s + 2);
```

Next we enter the trial expression

$$\frac{A}{s+1} + \frac{B}{s+2}$$

into *Mathematica* by

```
trial = A/(s+1) + B/(s+2);
```

The difference between `original` and `trial` must be zero; hence,

```
diff = Simplify[original - trial]
```

which results in the output

```
-1 + 2 A + B + A s + B s
-------------------------
              2
    2 + 3 s + s
```

We next isolate the numerator by

```
num = Numerator[diff]
```

and obtain

```
-1 + 2 A + B + A s + B s
```

The simultaneous equations that A and B must satisfy are derived by setting the coefficients of this first-order polynomial to zero. One way of accomplishing this task is

```
eqns = CoefficientList[num, s]
```

which results in

```
{-1 + 2 A + B, A + B}
```

The list of the parts in **eqns** is now ready for use with **Solve**:

```
coeffs=Solve[eqns == 0, {A, B}]
```

Mathematica returns

```
{{A -> 1, B -> -1}}
```

We now pass this information to **trial** via

```
F = trial /. coeffs
```

whose output

```
    1       1
{----- - -----}
 1 + s   2 + s
```

is ready for **InverseLaplaceTransform**:

```
InverseLaplaceTransform[F, s, t]
```

which gives us the final result

```
   -2 t    -t
{-E     + E  }
```

The *Mathematica* internal function **Apart** is designed to carry out the above steps and resolve a rational function in terms of its partial fractions. The command

```
Apart[1/(s^2+3 s + 2)]
```

leads to

```
  1         1
-----  -  -----
1 + s     2 + s
```

Example 5.2.10 (Partial fractions in *Mathematica* (continued))
Consider the function

$$F(s) = \frac{s-3}{(s-2)(s^2 + 2s + 2)}. \tag{5.28}$$

The trial partial fraction expression for this function is

$$\frac{A}{s-2} + \frac{Bs+C}{s^2 + 2s + 2}. \tag{5.29}$$

Proceeding as in the previous example, we define **origianl** and **trial** to *Mathematica* via

```
original = (s-3)/((s-2)*(s^2 + 2 s + 2));
trial = A/(s-2) + (B s + C)/(s^2 + 2 s + 2);
```

Next, we construct the difference between the above expressions and use **together** to force *Mathematica* to combine the fractions by

```
diff = Together[original - trial]
```

This results in

```
                                          2
(3 - 2 A + 2 C + s - 2 A s + 2 B s - C s - A s  -

     2               2
  B s ) / ((-2 + s) (2 + 2 s + s ))
```

We generate **num** and **eqns** as in the previous example.

```
num = Numerator[diff];
eqns = CoefficientList[num, s]
```

which result in

```
{-3 - 2 A + 2 C, 1 - 2 A + 2 B - C, -A - B}
```

we are now ready to solve for A, B and C via

```
coeffs = Solve[eqns == 0, {A, B, C}]
```

The output

```
       1         7         1
{{B -> --, C -> -, A -> -(--)}}
      10         5        10
```

must be substituted in (5.29) which we accomplish by

`F = trial /. coeffs`

The outcome is

```
                         7    s
                       - + --
        -1             5   10
{----------- + ------------}
 10 (-2 + s)              2
               2 + 2 s + s
```

The inverse Laplace transform of (5.28) is now obtained from

`InverseLaplaceTransform[F, s, t]`

It is instructive to compare the latter output with the one from

`InverseLaplaceTransform[original, s, t]`

Problems

1. Use the definition of the Laplace transform and determine the transform of the following functions. In each case check your answer in *Mathematica* or *MATLAB*.

   (a) $f(t) = e^{-at}$ with $a > 0$. What is the domain of the $L[f]$?

   (b) $f(t) = t^2 e^t$

   (c) $f(t) = te^{-at}$

   (d) $f(t) = t^2 e^{-bt}$, with $b > 0$ (Hint: Do two integration by parts.)

   (e) $f(t) = t^n$ with n a positive integer

   (f) $f(t) = a + bt + ct^2$, where a, b, and c are constants (Hint: Use the linearity of the Laplace operator.)

   (g) $f(t) = e^{at} \sin bt$ (Hint: Do two integration by parts. Alternatively, apply the formula $\sin(x) = (e^{ix} - e^{-ix})/2i$.)

   (h) $f(t) = e^{at} \cos bt$

   (i) $f(t) = e^{3t} \sin 2t$

   (j) $f(t) = e^{-3t} \cos 2t$

2. Apply Definition 5.2.1 to find directly the Laplace transform of the following functions.

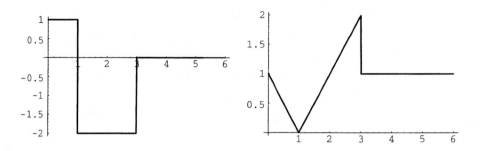

Figure 5.1: Find the Laplace transforms of the above functions from their graphs.

(a) $f(t) = \begin{cases} 1 & \text{if } 0 \le t < 2 \\ 0 & \text{otherwise.} \end{cases}$

(b) $f(t) = \begin{cases} -2 & \text{if } 0 \le t < 3 \\ 0 & \text{if } 3 \le t < 5 \\ 4 & \text{if } 5 \le t < 8 \\ 0 & \text{otherwise.} \end{cases}$

(c) $f(t) = \begin{cases} \sin t & \text{if } 0 \le t < 2\pi \\ 0 & \text{otherwise.} \end{cases}$

(d) $f(t) = \begin{cases} 0 & \text{if } 0 \le t < 2\pi \\ 1 - \cos 2t & \text{if } 2\pi \le t < 4\pi \\ 0 & \text{otherwise.} \end{cases}$

3. Determine the Laplace transform of each function with the given graph in Figure 5.1.

4. Determine the Laplace transform of the following functions (recall that, by definition, $\cosh x = (e^x + e^{-x})/2$ and and $\sinh x = (e^x - e^{-x})/2$.

   (a) $\cosh at$

   (b) $\sinh bt$

   (c) $t \sin t$

   (d) $t \cosh t$

   (e) $\sin^2 t$

   (f) te^{3t}

   (g) $\sin 3t - 2 \cos 2t$

5. (**Gamma function**) The integral

$$\Gamma(x) = \int_0^\infty t^{x-1} e^{-t}\, dt, \tag{5.30}$$

defines the Gamma function.

(a) Draw the graph of this function (using *Mathematica* or *MATLAB*) for $x \in (0.01, 4)$.

(b) Show that $\Gamma(1) = 1$ and that

$$\Gamma(x + 1) = x\Gamma(x). \tag{5.31}$$

(c) Apply the definition in (5.30) to show that

$$L[t^a] = \frac{\Gamma(a + 1)}{s^{a+1}}. \tag{5.32}$$

6. Determine the Laplace transform of each of the following functions.

   (a) $\sqrt{t}$
   (b) $\frac{1}{\sqrt{t}}$
   (c) $t^{3/2}$

7. Apply the method of partial fractions to determine the inverse Laplace transform of the following functions. Apply `Apart` to each F and compare *Mathematica*'s response to yours. Verify each inverse Laplace transform in *Mathematica* or *MATLAB*.

   (a) $F(s) = 2s/(s^2 + 3s - 4)$
   (b) $F(s) = (s - 2)/(s^2 + 3s + 2)$
   (c) $F(s) = (s^2 - 2)/(s^3 - 1)$
   (d) $F(s) = s/(s^2 - 5)$
   (e) $F(s) = (3s - 2)/(10 + s^2 - 7s)$
   (f) $F(s) = (as + b)/(s^2 + 3s + 2)$, where a and b are constants

8. Use *Mathematica* or *MATLAB* to find the inverse Laplace transform of each of the following

   (a) $F(s) = 1/(s^3 + 1)$
   (b) $F(s) = 1/(s^4 + 1)$
   (c) $F(s) = 1/(s^5 + 1)$

9. Define f on *Mathematica* by

   ```
   f[t_] = InverseLaplaceTransform[F[s], s, t]
   ```

   Plot f for the following Fs:

   (a) $F(s) = 1/(s^3 + 1)$
   (b) $F(s) = 1/(s^3 - 1)$
   (c) $F(s) = s/(s^3 + 1)$

10. Determine the Laplace transform of the periodic function

$$f(t) = \begin{cases} 1 & \text{if } 0 \leq t < 1, \\ -1 & \text{if } 1 \leq t < 2, \end{cases}$$

and $f(t+2) = f(t)$. Draw the graph of f and its Laplace transform on the same screen using **GraphicsArray**.

11. (a) Let f be a piecewise continuous function which satisfies the bound $|f(t)| < Me^{at}$, for $t > t_0$. Let g be defined by

$$g(t) = \int_0^t f(u)\, du.$$

Show that g also satisfies the exponential bound $|g(t)| < Ne^{at}$ for some N. (Hint: Use the fact that $|\int f(u)\, du| \leq \int |f(u)|\, du$.)

(b) Let f and g be related as above. Show that

$$L[\int_0^t f(u)\, du] = \frac{F(s)}{s},$$

where $F = L[f]$. (Hint: Use integration by parts.)

(c) Apply the above result to determine the inverse Laplace transforms of the following functions.

   i. $1/s(s+1)$
   ii. $1/s(s^2+1)$
   iii. $(s-1)/s(s^2-a^2)$
   iv. $1/s(s-1)(s-2)$

12. Use integration by parts to show

(a) $L[tf(t)] = -\frac{dF}{ds}$, where $F = L[f]$. Verify this result in *Mathematica*.

(b) $L[t^n f(t)] = (-1)^n F^{(n)}$.

(c) Apply the above results to determine the Laplace transform of the following functions.

   i. te^{at}
   ii. $t^2 \sin t$
   iii. $t \cos 2t$
   iv. $t \sin 2t + \cos 3t$
   v. $t^2 \sinh t$

5.3 Solution of Initial Value Problems

The main property of the Laplace transform that renders this operation so useful when applied to linear differential equations is its ability to relate the transform of the derivative of a function to the transform of the function itself. The following theorem contains this result.

Theorem 5.3.1 (Laplace transform of derivatives)
Let f be a continuous function with piecewise continuous first derivative f'. Let f satisfy the exponential property (5.7) of Theorem 5.2.1. Then $L[f']$ exists for $s > a$ and

$$L[f'] = sL[f] - f(0). \tag{5.33}$$

Proof: Without loss of generality we assume that f' is actually continuous rather than piecewise continuous; otherwise in what follows replace the integrals involving f' by a sum of integrals over the segments in which f' is continuous. It follows from the definition of the Laplace transform that

$$L[f'](s) = \int_0^\infty e^{-st} f'(t)\, dt. \tag{5.34}$$

Integrate (5.34) by parts:

$$L[f'](s) = e^{-st} f(t)|_0^\infty + s \int_0^\infty e^{-st} f(t)\, dt. \tag{5.35}$$

The first term on the right-hand side of this expression is $-f(0)$ since $\lim_{t \to \infty} e^{-st} f(t) = 0$ if $s > a$, where we have exploited the exponential property of f. The second term on the right-hand side is nothing but $L[f]$. This completes the proof.

Theorem 5.3.1 provides us with an alternative method for computing the Laplace transform of the derivative of a function. The formula in (5.33) is particularly useful when we already know the Laplace transform of f, and even more useful when we combine it with information about a function that satisfies a differential equation, as the next example shows.

Example 5.3.1
Consider the initial value problem

$$x' + 3x = 0, \quad x(0) = 1. \tag{5.36}$$

By inspection we see that $x(t) = e^{-3t}$ is the solution to the problem. Let us determine this solution by taking the Laplace transform of both sides of (5.36):

$$L[x'] + 3L[x] = 0. \tag{5.37}$$

From (5.33) we have that $L[x'] = sL[x] - x(0) = sL[x] - 1$. Combine this fact with (5.37). $L[x]$ satisfies the algebraic equation

$$(s + 3)L[x] = 1. \tag{5.38}$$

Hence $L[x] = \frac{1}{s+3}$, from which we deduce that $x(t) = e^{-3t}$.

The preceding example shows clearly how the operation of the Laplace transform is capable of transforming a differential equation into an algebraic one. This procedure can easily be programmed.

Example 5.3.2

The following is the list of commands in *Mathematica* that accomplish what we did in the previous example (these lines are executed after entering `Calculus'LaplaceTransform'`):

```
x[0] = 1;
X[s_] = LaplaceTransform[x[t], t, s];
a = LaplaceTransform[x'[t] + 3 x[t], t, s];
output = Solve[a == 0, X[s]];
x[t_] = InverseLaplaceTransform[X[s] /. output, s, t]
```

The following *MATLAB* program finds the solution to the initial value problem:

```
syms t s lapx
diffeqn='diff(x(t),t)+3*x(t)';
a=laplace(diffeqn,t,s);
a=subs(a,lapx,'laplace(x(t),t,s)');
c=solve(alapx);
gensol=ilaplace(c,s,t);
sol=subs(gensol,'x(0)',1)
```

Remark 5.3.1: Note that, unlike the methods that were presented in the previous chapter, the Laplace transform method finds the solution to a differential equation in one lump sum, not by first computing the general solution and then applying the initial data.

Remark 5.2.2: It may appear from the way we presented the Laplace transform method that it applies only to problems whose initial data is specified at $t = 0$. We leave the necessary calculations to the reader to show that if the initial data are given at $t = a$, then the change of variables $w = t - a$ converts the old differential equation into a new one with $x(w)$ as the unknown and the initial data for the new equation specified at $w = 0$. Therefore, without loss of generality, we may assume that the initial data is always given at $t = 0$.

The method of Example 5.3.1 is effective for higher order differential equations as well. Note that, since $f'' = (f')'$, we can apply (5.33) twice and get an algebraic expression for the Laplace transform of f'' in terms of the Laplace transform of f:

$$L[f''] = L[(f')'] = sL[f'] - f'(0) = s^2 L[f] - sf(0) - f'(0). \qquad (5.39)$$

This relation opens the door for applications of the Laplace transform to second-order equations. Before presenting an example, we state the following corollary, whose proof follows from a finite number of applications of the formula in (5.33).

Corollary 5.2.1 (Laplace transform of higher derivatives)
Let f, f', f'', ..., $f^{(n-1)}$ be piecewise continuous and satisfy the exponential property (5.7). Then the laplace transform of $f^{(n)}$, the n-th derivative of f, exists and is given by

$$L[f^{(n)}] = s^n L[f] - s^{n-1} f(0) - s^{n-2} f'(0) - \ldots - sf^{(n-2)}(0) - f^{(n-1)}(0). \qquad (5.40)$$

Example 5.3.3
Consider the differential equation

$$x'' + 3x' + 2x = \sin t, \quad x(0) = 0, x'(0) = 0. \qquad (5.41)$$

Take the Laplace transform of both sides of (5.41) and use (5.33) and (5.39). We find that $L[x]$ satisfies the algebraic relation

$$s^2 L[x] + 3L[x] + 2L[x] = L[\sin t]. \qquad (5.42)$$

Since $L[\sin t] = \frac{1}{1+s^2}$, equation (5.42) leads to

$$L[x] = \frac{1}{(s^2 + 1)(s^2 + 3s + 2)}. \qquad (5.43)$$

Equation (5.43) can either be inverted using the method of partial fractions or *Mathematica*. In either case the answer is

$$x(t) = -\frac{e^{-2t}}{5} + \frac{e^{-t}}{2} - \frac{3}{10} \cos t + \frac{1}{10} \sin t. \qquad (5.44)$$

Example 5.3.4
The second-order equation (5.41) is solved in *Mathematica* in much the same way that (5.36) was solved. The program is

```
<<Calculus'LaplaceTransform'
x[0] = 0; x'[0] = 0;
X[s_] = LaplaceTransform[x[t], t, s];
f = x''[t] + 3 x'[t] + 2 x[t] - Sin[t];
a = LaplaceTransform[f, t, s];
output = Solve[a == 0, X[s]];
x[t_] = InverseLaplaceTransform[X[s] /. output, s, t]
```

Example 5.3.3 (Laplace transform and resonance)
Consider the initial value problem

$$x'' + 4x' + 3x = e^{-t}, \qquad x(0) = x'(0) = 0. \tag{5.45}$$

We note in passing that the above equation's characteristic polynomial is $m^2 + 4m + 3$ with roots $m = -3$ and $m = -1$. In particular, the forcing term e^{-t} will cause the phenomenon of resonance to occur in this system. We recall that we needed to take special care in such situations when applying the method of undetermined coefficients. As it turns out, when applicable, the Laplace transform method is indifferent to the presence of resonance.
We begin by taking the transform of both sides of (5.45):

$$(s^2 + 4s + 3)L[x] = \frac{1}{s+1},$$

or

$$L[x] = \frac{1}{(s+1)(s^2 + 4s + 3)}.$$

The inverse Laplace transform of the above expression (using *Mathematica*) is

$$\frac{1}{4}e^{-3t} - \frac{1}{4}e^{-t} + \frac{1}{2}te^{-t}.$$

Problems

1. Create a file called `odelap.m` that contains the *Mathematica* or *MATLAB* program in this section and verify that their outputs match the claims made in the examples.

2. How should the *MATLAB* program in Example 5.3.2 be altered to allow for second-order differential equations?

3. Use the method described in this section and find the solution to the following equations. In each case use *Mathematica* or *MATLAB* to verify your answer.

   (a) $x' + 2x = \sin t$, $x(0) = 1$

   (b) $x' - 3x = e^{-2t}$, $x(0) = -2$

   (c) $3x' + 7x = 5\sin t + \cos t$, $x(0) = 1$

   (d) $y' + ay = t$, $y(0) = -1$, and a is a positive constant. What happens to the solution when t approaches infinity?

   (e) $y' + ay = \cos bt$, $y(0) = y_0$

   (f) $y' + ay = e^{-t}\cos bt$, $y(0) = y_0$

   (g) $y' + y = t - \frac{t^3}{3!}$, $y(0) = 0$

   (h) $y' + y = e^{-t}$

4. Use the Laplace transform method and find the solution to the following problems. Use *Mathematica* or *MATLAB* to check your result.

   (a) $y'' + 4y = 0$

   (b) $y'' + 3y' + 2y = 1$, $y(0) = y'(0) = 0$

   (c) $x'' + 6x' + 9x = \sin t$, $x(0) = x'(0) = 0$

   (d) $x'' + 6x' + 9x = e^{-3t}$, $x(0) = x'(0) = 0$

   (e) $x'' + \omega^2 x = \sin \omega t$, $x(0) = x'(0) = 0$

   (f) $y'' + a^2 y = e^{-bt}$, $y(0) = y'(0) = 0$

5. Extend the ideas presented in (5.39) to higher order derivatives and solve the following equations. Use *Mathematica* or *MATLAB* to invert the necessary Laplace transforms.

   (a) $y''' + y = 0$

   (b) $y''' + y = 1$, $y(0) = y'(0) = y''(0) = 0$

   (c) $y''' + y = t^2$, $y(0) = y'(0) = y''(0) = 0$

   (d) $y''' + y = t^4$, $y(0) = y'(0) = y''(0) = 0$

   (e) $y''' + y' + y = 0$

   (f) $y''' + y' + y = t^2$, $y(0) = y'(0) = y''(0) = 0$

6. In a mass-spring-dashpot system a body of mass $m = 3$ kilograms is attached to a spring with spring constant $k = 7$ N/m and a dashpot with the frictional constant $c = 0.4$ kg/s. Suppose that at time zero the system is in equilibrium and subsequently an external force $f(t) = 4 \sin 3t$ N is applied to it. Determine the transient and steady-state solutions of the vibration.

7. In the following problems, the values of mass m, frictional constant c, spring constant k, and forcing term f are given. Solve the initial value problem
$$my'' + cy' + ky = f, \quad y(0) = y'(0) = 0.$$

   (a) $m = 2$, $k = 3$, $c = 0$, $f(t) = 1$

   (b) $m = 1$, $k = 3$, $c = 1$, $f(t) = 1$

   (c) $m = 24$, $k = 6$, $c = 0.1$, $f(t) = \sin t$

   (d) $m = 1$, $k = 9$, $c = 0.1$, $f(t) = \cos 2t$

8. An RLC circuit is put in a series with a 9-volt battery and the circuit is switched on at time zero. Suppose that the resistance is $R = 0.8$ ohms, the capacitance is $C = 0.01$ farad, and the inductance is $L = 2$ henry. Find the transient and the steady-state currents in the circuit. What is the steady-state charge in the capacitor?

9. In each of the following problems the values of the parameters in an RLC circuit are given. Solve the initial value problem with

$$I(0) = q(0) = 0.$$

(Note that $LI' + RI + 1/Cq = e(t)$.)

(a) $R = 100$, $L = 1$, $C = 0.001$, $e(t) = 100$

(b) $R = 50$, $L = 5$, $C = 0.001$, $e(t) = 1$

(c) $R = 1$, $L = 10$, $C = 0.01$, $e(t) = 20\sin 5t$

(d) $R = 300$, $L = 9$, $C = 0.0004$, $e(t) = 100\sin 20t$

5.4 Linear Systems and Laplace Transform

Laplace transform is used frequently in science and engineering to solve systems of linear differential equations with constant coefficients. As in the case of single equations, the transform method reduces a system of differential equations to a set of linear simultaneous equations, with the Laplace transform of the individual components of the solution as the unknowns. The ability of *Mathematica* or *MATLAB* to solve such algebraic systems now plays an important role in making the Laplace transform the method of choice. The following examples elaborate on this process.

Example 5.4.1
Consider the system of linear differential equations

$$x' = -x - y, \qquad y' = -y + x \tag{5.46}$$

subject to the initial conditions

$$x(0) = 1, \qquad y(0) = -2. \tag{5.47}$$

We begin by applying the Laplace transform operator to both sides of each equation in (5.46) and use the initial data (5.47) to get the following two simultaneous equations in $L[x]$ and $L[y]$:

$$\begin{cases} (s+1)L[x] + L[y] & = 1, \\ -L[x] + (s+1)L[y] & = -2. \end{cases} \tag{5.48}$$

This linear system has the unique solution

$$L[x] = \frac{s+3}{(s+1)^2 + 1}, \qquad L[y] = -\frac{2s+1}{(s+1)^2 + 1}. \tag{5.49}$$

Recall that

$$L[e^{-t}\sin t] = \frac{1}{(s+1)^2 + 1}, \quad \text{and} \quad L[e^{-t}\cos t] = \frac{s+1}{(s+1)^2 + 1}.$$

Applying the method of partial fractions, we could now reduce each term in (5.49) to a linear combination of Laplace transforms of $e^{-t}\sin t$ and $e^{-t}\cos t$. Alternatively, one could resort to *Mathematica* or *MATLAB* to determine the inverse Laplace transform of the expressions in (5.49). In either case we end with

$$x(t) = e^{-t}(2\sin t + \cos t), \quad y(t) = e^{-t}(\sin t - 2\cos t).$$

Mathematica and *MATLAB* could be programmed to reach the above conclusion once the differential equations (5.46) are introduced to them. One proceeds in *Mathematica* as follows; After entering the package `Calculus'LaplaceTransform'`, we begin by renaming the Laplace transforms of x and y as X and Y, respectively, via

```
{X[s_], Y[s_]} = LaplaceTransform[x[t], t, s],
    LaplaceTransform[y[t], t, s]};
```

Next, we rearrange the equations in (5.46) to read

$$x' + x + y = 0, \qquad y' + y - x = 0. \tag{5.50}$$

We define the differential operators on the left-hand side of each equations in (5.50) to *Mathematica* by

```
eqns = {x'[t] + x[t] + y[t], y'[t] + y[t] - x[t]}
```

Next, we determine the Laplace transform of `eqns`:

```
eqnsTransformed = LaplaceTransform[eqns, t, s]/. {x0] -> 1,
    y[0] -> -2};
```

Mathematica returns

```
{-1 + LaplaceTransform[x[t], t, s] +
  s LaplaceTransform[x[t], t, s] +
  LaplaceTransform[y[t], t, s],
 2 - LaplaceTransform[x[t], t, s] +
  LaplaceTransform[y[t], t, s] +
  s LaplaceTransform[y[t], t, s]}
```

At this stage we have essentially reproduced the information in (5.48). We now apply `Solve` to `eqnsTransformed` via

```
solutionTransforms = Solve[eqnsTransformed == 0, {X[s], Y[s]}]
```

which results in

```
{{LaplaceTransform[x[t], t, s] ->

        (-1 - s) (1 - 2 (1 + s))
  2 + ------------------------,
                  2
        -1 - (1 + s)


                                    1 - 2 (1 + s)
    LaplaceTransform[y[t], t, s] -> -(-------------)}}
                                          2
                                    -1 - (1 + s)
```

Finally, we obtain x and y by

```
solutions = InverseLaplaceTransform[{X[s], Y[s]}
      /.solutionTransforms, s, t]
```

After simplification, the output is

```
  Cos[t] + 2 Sin[t]     2 Cos[t] - Sin[t]
{{-----------------, -(-----------------)}}
          t                   t
         E                   E
```

In the above example we considered an initial value problem consisting of only two linear equations. It should be clear from the way we constructed the algorithm in *Mathematica* that this approach applies equally well to higher order systems. Its main limitation is in the final step where the inverse Laplace transform of the individual components of the solution must be computed. This limitation is dictated by the complexity of the structure of each transform and by the hardware capability on which *Mathematica* is used. The next example shows a slight alteration of the scheme developed in Example 5.4.1.

Example 5.4.2

Consider the system of two second-order equations

$$\begin{cases} m_1 x_1'' = -(k_1 + k_2)x_1 + k_2 x_2, \\ m_2 x_2'' = k_2 x_1 - (k_2 + k_3)x_2 \end{cases} \tag{5.51}$$

subject to the initial data

$$x_1(0) = 1,\ x_2(0) = x_1'(0) = x_2'(0) = 0. \tag{5.52}$$

The above system describes the governing equations for a typical motion in the coupled mass-spring system shown in Figure 5.2. The equations in (5.51) are derived directly from Newton's second law of motion. Each function x_i denotes the displacement of the ith mass from its equilibrium. As an example, let

$$m_1 = 2,\ m_2 = 1,\quad k_1 = 4,\ k_2 = 2,\ k_3 = 2.$$

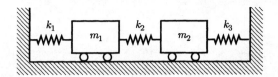

Figure 5.2: A mass-spring system consisting of two masses and three springs.

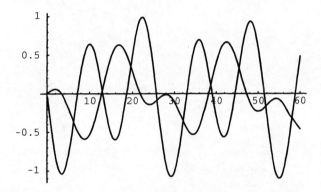

Figure 5.3: The graphs of the displacements x_1 and x_2.

We would like to find the solution to (5.51)–(5.52) with the preceding parameter values. The following program applies the Laplace Transform method to this initial value problem and plots the graphs of x_is on a single coordinate system:

```
k[1] = 4; k[2] = 2; k[3] = 2; m[1] = 2; m[2] = 1;
initialdata = {x[1][0] -> 1, (x[1])'[0] -> 0,
               x[2][0] -> 0, (x[2])'[0] -> 0};
transforms = Table[X[i]=LaplaceTransform[x[i][t], t, s],
     {i, 1, 2}];
eqns = {m[1] x[1]''[t] + (k[1]+k[2]) x[1][t] - k[2] x[2][t],
        m[2] x[2]''[t] - k[2] x[1][t] + (k[2] + k[3]) x[2][t]};
eqnsTransformed = LaplaceTransform[eqns, t, s] /. initialdata;
solTransforms = Solve[eqnsTransformed == 0, transforms];
Xs = transforms /. solTransforms;
Xs = Flatten[Xs, 1];
solution=InverseLaplaceTransform[Xs,s,t];
graph=Plot[Evaluate[solution], {t, 0, 2Pi}]
```

Figure 5.3 shows the output of **graph**. Next, let us consider the forced linear system

$$\begin{cases} m_1 x_1'' & = -(k_1 + k_2)x_1 + k_2 x_2, \\ m_2 x_2'' & = k_2 x_1 - k_2 x_2 + A\sin\omega t \end{cases} \tag{5.53}$$

subject to zero initial data. Here the forcing term $A\sin\omega t$ has replaced the third spring in Figure 5.2. Figure 5.4 shows the graph of x_1 and x_2 where we have

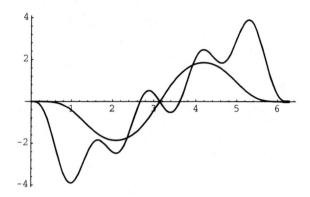

Figure 5.4: The graphs of the displacements x_1 and x_2 for the forces mass-spring system.

assumed the same parameter values for m_1, m_2, k_1, k_2, and let $A = 20$ and $\omega = 5$. Figure 5.3) shows that output of **graph**.

Problems

1. Solve each of the following initial value problem by applying the process described in Example 5.4.1. In each case plot the graphs of all solution components on a single screen.

   (a) $x' = x - y$, $\quad y' = 2x - y$; $x(0) = 0$, $y(0) = 1$

   (b) $x' = 3x - 2y$, $\quad y' = 2x - y$; $x(0) = -1$, $y(0) = 1$

   (c) $x' = 2x + y - \sin t$, $\quad y' = x + y$; $x(0) = 0$, $y(0) = 0$

   (d) $x' = y$, $\quad y' = x - 3y - \cos t$; $x(0) = 0$, $y(0) = 0$

   (e) $x' = y$, $\quad y' = x + y + e^t$; $x(0) = 0$, $y(0) = 0$

   (f) $x' = ay$, $\quad y' = -x + \sin 2t$; $x(0) = 0$, $y(0) = 0$, and a is a positive number. Solve this initial value problem with a as a parameter, then plot the graphs of x and y for a as it varies between 1 and 5. What is the effect of a on the amplitude of x and y?

2. Refer to Example 5.4.2. Each system below is an instance of a coupled mass-spring system. Solve each initial value problem and plot the graphs of the solution components on a single screen.

   (a) $2x_1'' = -3x_1 + 2x_2$, $x_2'' = 2x_1 - 2x_2$, $x_1(0) = 1$, $x_2(0) = x_1'(0) = x_2'(0) = 0$

   (b) $x_1'' = 5x_1 - 2x_2$, $2x_2'' = 2x_1 + x_2$, $x_1(0) = 1$, $x_2(0) = x_1'(0) = x_2'(0) = 0$

   (c) $x_1'' = 3x_1 - 4x_2$, $x_2'' = -2x_1 + x_2$, $x_1(0) = 0$, $x_2(0) = -1$, $x_1'(0) = x_2'(0) = 0$

(d) $x_1'' = -x_1 + x_2 + \sin t$, $x_2'' = x_1 + x_2 + \sin t$, $x_1(0) = x_2(0) = x_3(0) = x_4(0) = 0$

(e) $x_1'' = 2x_1 - 3x_2 + \sin 2t$, $x_2'' = -x_1 + x_2$, $x_1(0) = x_2(0) = x_3(0) = x_4(0) = 0$

3. Consider the mass-spring system consisting of three masses m_1, m_1, and m_3 attached to three springs with spring constants k_1, k_2, and k_3. Show that the functions x_is, $i = 1, 2, 3$, the displacements of the individual masses from equilibrium, satisfy the following system of differential equations

$$\begin{cases} m_1 x_1'' = -(k_1 + k_2)x_1 + k_2 x_2, \\ m_2 x_2'' = k_2 x_1 - (k_2 + k_3)x_2 + k_3 x_3, \\ m_3 x_3'' = k_3 x_2 - k_3 x_3. \end{cases} \tag{5.54}$$

Supply the appropriate set of data to pose an initial value problem.

4. Referring to the above problem, determine the solution to the following coupled mass-spring systems. Plot the graphs of x_is on the same screen.

(a) $m_1 = m_2 = m_3 = 1$, $k_1 = k_2 = k_3 = 1$, $x_1(0) = 1$, $x_2(0) = x_3(0) = x_1'(0) = x_2'(0) = x_3'(0) = 0$

(b) $m_1 = 1$, $m_2 = 2$, $m_3 = 4$, $k_1 = k_2 = k_3 = 1$, $x_1(0) = 1$, $x_2(0) = x_3(0) = x_1'(0) = x_2'(0) = x_3'(0) = 0$

(c) $m_1 = m_2 = m_3 = 1$, $k_1 = 1$, $k_2 = 2$, $k_3 = 4$, $x_1(0) = 1$, $x_2(0) = x_3(0) = x_1'(0) = x_2'(0) = x_3'(0) = 0$

(d) $m_1 = m_2 = m_3 = 1$, $k_1 = k_2 = k_3 = 1$, $x_1(0) = -1$, $x_2(0) = 1$, $x_3(0) = x_1'(0) = x_2'(0) = x_3'(0) = 0$

(e) $m_1 = m_2 = m_3 = 1$, $k_1 = k_2 = k_3 = 1$, $x_1(0) = 1$, $x_2(0) = 1$, $x_3(0) = 1$, $x_1'(0) = x_2'(0) = x_3'(0) = 0$

(f) $m_1 = m_2 = m_3 = 1$, $k_1 = k_2 = k_3 = 1$, $x_1(0) = x_2(0) = x_3(0) = 0$, $x_1'(0) = 1$, $x_2'(0) = x_3'(0) = 0$

5.5 Taylor Series and Laplace Transforms

One of the main impediments to using the Laplace transform method for determining solutions to differential equations lies in our inability to evaluate the inverse Laplace transform of rational functions whose denominators consist of polynomials of large degrees. The ability of symbolic packages in evaluating inverse transforms is certainly a great help in the right direction, although these tools have limitations that become apparent rather quickly.

In this section we consider evaluating the inverse Laplace transform of some common transforms with the aid of Taylor series. The main idea is as follows; given a smooth function F, if we could expand this function as a series in the form

$$F(s) = \frac{a_1}{s} + \frac{a_2}{s^2} + \ldots + \frac{a_n}{s^n} + \ldots = \sum_{n=1}^{\infty} \frac{a_n}{s^n}, \tag{5.55}$$

then because

$$L^{-1}[\frac{1}{s^n}] = (n+1)!\, t^n,$$

we would be able to write, at least formally, the inverse Laplace transform of F as

$$f(t) = \sum_{n=1}^{\infty} a_n (n+1)!\, t^n. \qquad (5.56)$$

The only computationally intensive aspect of the above procedure is in evaluating (5.55) where F must be expanded in a series in $\frac{1}{s}$. Equivalently, (5.55) could be viewed as the Taylor series of the function G, where G and F are related by

$$G(u) = F(1/u).$$

We can therefore determine the Taylor (Maclaurin) series of G about $u = 0$, and subsequently replace u by $\frac{1}{s}$ in the series. *Mathematica* or *MATLAB* turn out to be quite effective here as the examples show.

Example 5.5.1
Consider the transform

$$F(s) = \frac{s}{s^2 + 4}.$$

In this case we know that $f(t) = \cos 2t$. Let us apply the Taylor series method to understand how closely we might be able to approximate f by the latter method. First, we construct G as

$$G(u) = F\left(\frac{1}{u}\right) = \frac{u}{1 + 4u^2}.$$

Using *Mathematica*'s **Series** function, we determine the 19th degree Taylor polynomial approximation of G by

```
G = u/(1+4 u^2);
seriesG = Normal[Series[G, {u, 0, 19}]]
```

We get

```
        3       5       7        9           11
u - 4 u  + 16 u  - 64 u  + 256 u  - 1024 u    +

          13           15           17            19
    4096 u   - 16384 u   + 65536 u    - 262144 u
```

Next, we replace u by $\frac{1}{s}$ via

```
seriesG = seriesG /. u -> 1/s
```

which results in

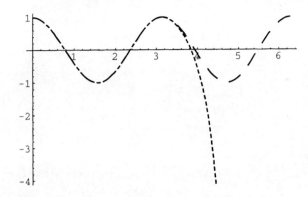

Figure 5.5: The inverse Laplace transform of $\frac{s}{s^2+4}$ computed in two different ways.

```
-262144    65536   16384    4096   1024   256   64
-------- + ----- - ----- + ---- - ---- + --- - -- +
   19        17      15     13     11     9     7
  s         s        s      s      s      s     s

 16   4   1
 -- - -- + -
  5   3   s
 s    s
```

This series is a truncated version of (5.55) for $F(s) = \frac{s}{s^2+4}$. Now, we apply InverseLaplaceTransform to seriesG:

```
f[t_] = InverseLaplaceTransform[seriesG, s, t]
```

We get

```
            4      6      8      10       12
     2    2 t    4 t    2 t    4 t      4 t
1 - 2 t + ---- - ---- + ---- - ----- + ------ -
           3      45     315   14175   467775

        14          16           18
      8 t         2 t          4 t
   -------- + ---------- - -----------
   42567525   638512875   97692469875
```

Figure 5.5 shows the comparison between the $\cos 2t$ and f computed above. This figure is the output of

```
Plot[{Cos[2 t], f[t]}, {t, 0, 2Pi},
    PlotStyle->{Dashing[{0.05}], Dashing[{0.01}]}]
```

Clearly, f is an excellent approximation to $\cos 2t$ as long as t is small. Evidently, one must consider Taylor polynomial approximations of degree higher than 19 to improve on the quality of the approximation.

Example 5.5.2
Consider the transform

$$F(s) = \frac{3 - 2s^3}{3 + 2s^2 + s^4}.\tag{5.57}$$

Let us begin by applying InverseLaplaceTransform to this function:

```
f = InverseLaplaceTransform[(3-2s^3)/(3+2 s^2 + s^4)
```

The output consists of a rather long list of expressions involving the InverseLaplaceTransform and its derivatives, which may not seem useful at first glance. However, when we try

```
N[f /. t->1}
```

Mathematica returns

```
0.356723
```

as a numerical approximation to the value of f at $t = 1$. Let us now apply the Taylor series method as outlined in Example 5.5.1 to the function in (5.57). The output of the program

```
F = (3 - 2 s^3)/(3 +2 s^2+s^4);
G = Together[F/. s->1/u];
seriesG = Normal[Series[G, {u, 0, 19}]];
seriesG = seriesG /. u->1/s;
f[t_] = InverseLaplaceTransform[seriesG, s, t];
N[f[1]]
```

is

```
0.356723
```

The Taylor series method is quite effective when we seek to understand the behavior of functions for small values of t. In some of the problems the reader will be asked to combine this method with the programs developed in the previous section in order to gain insight into the behavior of solutions of mass-spring systems with a moderately large number of components.

Problems

1. Apply the Taylor series method to find an approximation to the inverse transform of the following functions. In each case experiment with the number of terms in the Taylor series and compare graphically the approximation with the exact inverse transform.

(a) $\frac{1}{s-1}$

(b) $\frac{s+2}{s-1}$

(c) $\frac{s-2}{s^2-9}$

(d) $\frac{3s-2}{2s^2+3s+7}$

(e) $\frac{s^2-1}{s(s^2+4)}$

(f) $\frac{s}{(s^2+1)^2}$

2. Apply the Taylor series method to each function below. Plot the graph of the inverse transform.

(a) $\frac{1}{(s^2+1)^3}$

(b) $\frac{s^3-3s^2+1}{s^4+2s^3-3s^2+s+1}$

(c) $\frac{1}{s^6-s^4+3}$

(d) $\frac{s^4+6}{s^8-3s^5+s^2-1}$

3. Each transform below depends on a parameter. Determine its inverse Laplace transform using the Taylor series method, then plot the graphs of each inverse transform on a single screen as the parameter varies as specified. When using *Mathematica* or *MATLAB*, take advantage of their abilities to change the style of the plot according to parameter values.

(a) $\frac{1}{s^2+a^2}$; a varies between 1 and 3 at increments of 0.5

(b) $\frac{1}{s^2+cs+1}$; c varies between 0.1 and 0.5 at increments of 0.1

(c) $\frac{2s-1}{s^4+cs^2-3}$; c varies between 1 and 3 at increments of 0.5

(d) $\frac{s^3+3s^2-2s+1}{s^7+6s^5-4s^4+c}$; c varies between 1 and 10 at increments of 2

4. Apply the Laplace transform method to each system. Use the Taylor series method to graph the components of the solution on a single screen.

(a) $x' + 3x - y = 0$, $y' - 2x + 2y = 0$; $x(0) = 1$, $y(0) = -1$

(b) $x' - y' + y = 0$, $y' - x = \sin t$; $x(0) = 0$, $y(0) = -0$

(c) $x'' + x + 2y = \sin t$, $y' - 2x = 0$; $x(0) = 0$, $x'(0) = 0$, $y(0) = 0$

(d) $x'' + x' - x + 2y = t\cos 2t$, $y'' + y' - 2x = 0$; $x(0) = 0$, $x'(0) = 0$, $y(0) = 0$, $y'(0) = 0$

5. Consider the mass-spring system whose system of equations is given by

$$\begin{cases} x_1'' + (k_1 + 2)x_1 - 2x_2 & = 0, \\ x_2'' - 2x_1 + 3x_2 - x_3 & = 0, \\ x_3'' - x_2 + 4x_3 & = 0 \end{cases}$$

subject to the initial data

$$x_1(0) = 1, \ x_1'(0) = x_2(0) = x_2'(0) = x_3(0) = x_3'(0) = 0.$$

Here k_1 is the spring constant in the first spring. Apply the Taylor series method to this problem and plot the graphs of x_1 as k_1 varies between 0.1 and 0.5 at increments of 0.1. What is the effect of this parameter on x_1? For instance, does the amplitude of x_1 increase with k_1? How about its frequency? How does k_1 influence the behavior of x_2? and x_3?

5.6 The Heaviside Function

In many applications the physical setting one wishes to model has features in it that are best modeled by discontinuous functions. An electrical switch being turned on from the off position or a solid object moving in a fluid of certain density suddenly entering a fluid region of different density are just two examples of physical applications that could require use of discontinuous functions. In this section we will outline how one is able to solve differential equations whose forcing terms are discontinuous using the Laplace transform method. We point out that the existence and uniqueness theorem we stated in Section 3.3 does not apply to this new setting because the functions in that theorem were assumed to be at least continuous. The hypotheses of that result can be weakened however to accommodate our present situation, although we will not take this issue up in this text.

We begin with a definition.

Definition 5.6.1
We define the **Heaviside** *or the* **unit step** *function $u(t)$ by*

$$u(t) = \begin{cases} 1 & \text{if} \quad t \geq 0, \\ 0 & \text{otherwise.} \end{cases} \tag{5.58}$$

The unit step function $u(t)$ is discontinuous at $t = 0$ and experiences a jump at that point from the value 0 to 1. Similarly, the function $u(t - c)$ is zero until t reaches c; it then jumps to the value 1 and remains at that value for all t larger then c.

Example 5.6.1
Consider the function f defined by

$$f(t) = u(t - 2) + 6u(t - 4). \tag{5.59}$$

This function is zero until t reaches 2; then it takes on the value 1 until t reaches 4, at which point f experiences another jump and takes on the value 7 (see Figure 5.6a). In other words,

$$f(t) = \begin{cases} 0 & \text{for} \quad t < 2, \\ 1 & \text{for} \quad 2 \leq t < 4, \\ 7 & \text{for} \quad 4 \leq t. \end{cases} \tag{5.60}$$

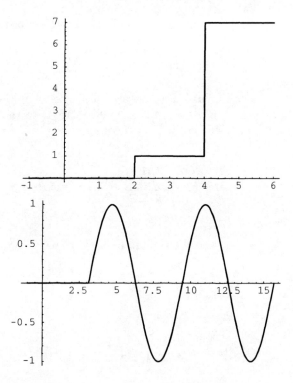

Figure 5.6: The graphs of $u(t-2) + 3u(t-4)$ and $u(t-\pi)\sin(t-\pi)$.

Similarly, $g(t) = u(t-\pi)\sin(t-\pi)$ is zero until t reaches π and then it evolves as the sine function shifted to the right by π units, that is,

$$g(t) = \begin{cases} 0 & \text{for} \quad t < \pi, \\ \sin(t-\pi) & \text{for} \quad \pi \le t. \end{cases} \tag{5.61}$$

(See Figure 5.6b.)

Example 5.6.2

The Heaviside function $u(t)$ is an internal function to *Mathematica* and is denoted by `UnitStep[t]`. This function is part of the `LaplaceTransform` package. After accessing `Calculus'LaplaceTransform'`, one draws the graph that appears in Figure 5.6a via

```
Plot[UnitStep[t-2]+6 UnitStep[t-4], {t, -1, 6}]
```

The Heaviside function is also available in *MATLAB* as `Heaviside(t-c)`. One way to get the graph in Figure 5.6a in *MATLAB* is as follows:

```
f = 'Heaviside(t-2) + 6*Heaviside(t-4)'
ezplot(f)
```

Next we compute the Laplace transform of $u(t - c)$ when $c > 0$:

$$L[u(t - c)] = \int_0^\infty e^{-st} u(t - c) \, dt = \int_c^\infty e^{-st} \, dt = \frac{e^{-cs}}{s}. \qquad (5.62)$$

In a similar fashion we are able to compute the Laplace transform of a translated function $u(t - c)f(t - c)$ and relate it to the Laplace transform of f. We state the result in the next theorem.

Theorem 5.6.1

Let f be a function with Laplace transform F. Let $c > 0$. Then

$$L[u(t - c)f(t - c)] = e^{-cs} L[f] \qquad (5.63)$$

for all s in the domain of $L[f]$. Consequently,

$$L^{-1}[e^{-cs} F(s)] = u(t - c)f(t - c). \qquad (5.64)$$

Proof: We begin with the left-hand side of (5.63):

$$L[u(t - c)f(t - c)] = \int_0^\infty e^{-st} u(t - c)f(t - c) \, dt = \int_c^\infty e^{-st} f(t - c) \, dt. \qquad (5.65)$$

The last integral is equal to the right-hand side of (5.64) after the change of variables $w = t - c$; note that $dw = dt$ and $\int_c^\infty e^{-st} f(t-c) \, dt = e^{-cs} \int_0^\infty e^{-sw} f(w) \, dw$. The latter integral is the Laplace transform of f, which completes the proof.

Example 5.6.2

According to the preceding theorem

$$L[u(t - \pi) \sin(t - \pi)] = e^{-\pi s} L[\sin t] = \frac{e^{-\pi s}}{1 + s^2}. \qquad (5.66)$$

Similarly,

$$L^{-1}\left[\frac{e^{-2s}}{s^2 + 3s + 2}\right] = L^{-1}\left[\frac{e^{-2s}}{s + 1} - \frac{e^{-2s}}{s + 2}\right] = u(t - 2)(e^{-t+1} - e^{-2t+4}). \qquad (5.67)$$

When we enter

```
InverseLaplaceTransform[Exp[-2s]/(s^2 + 3s + 2), s, t]
```

to *Mathematica*, we receive the output

```
      4 - 2 t     2 - t
(-E          + E       ) UnitStep[-2 + t]
```

MATLAB treats this problem in a similar fashion:

```
maple('invlaplace(exp(-2*s)/(s^2 +3*s+2),s,t)'
```

results in

```
Heaviside(t-2)*(exp(-2*t+4)+exp(-t+2))
```

A typical problem that involves an ODE with a discontinuous nonhomogeneity is considered in the following example.

Example 5.6.3
Consider the differential equation

$$x'' + x = 1 - u(t-2), \quad x(0) = x'(0) = 0. \tag{5.68}$$

This equation may be thought of as modeling a physical phenomenon whose internal mechanism is represented by the differential operator $x'' + x$ and the external forcing term by $1 - u(t-2)$. Note that the forcing term shuts off at $t = 2$.

We begin by taking the Laplace transform of both sides of (5.68) as follows

$$(s^2 + 1)L[x] = \frac{1}{s} - \frac{e^{-2s}}{s}. \tag{5.69}$$

This expression leads to

$$x(t) = L^{-1}\left[\frac{1 - e^{-2s}}{s(s^2 + 1)}\right]. \tag{5.70}$$

The inverse Laplace transform may either be determined by partial fractions ($\frac{1}{s(s^2+1)} = \frac{A}{s} + \frac{Bs+C}{s^2+1}$) or by using *Mathematica* or *MATLAB*. In either case the answer is

$$x(t) = 1 - \cos t - (1 - \cos(2-t))u(t-2). \tag{5.71}$$

Problems

1. Verify the computations in Example 5.6.3. Draw the graph of the solution for t in the interval $(0,6)$.

2. Draw the graphs of the following functions. Compare your graph with the one from *Mathematica* or *MATLAB* (recall that the package `Calculus'LaplaceTransform'` must be available to *Mathematica* before it can access the `UnitStep` function):

   (a) $u(t-1) + 2u(t-4) + 3u(t-7)$

   (b) $u(t-2)\sin t$

   (c) $u(t-2)\sin(t-2)$

   (d) $u(t-1)\sin(t-1) + 3u(t-3)\cos(t-3)$

   (e) $\sum_{i=1}^{10}(-1)^i u(t-i)$ (Hint: Use **Sum** of *Mathematica* or **sum** of *MATLAB*.)

3. Find the Laplace transform of each function in the previous problem. Check your answer in *Mathematica* or *MATLAB*.

4. Find the inverse Laplace transform of the following functions. Verify each result in *Mathematica* or *MATLAB*.

   (a) $F(s) = \frac{e^{-3s}}{s^2+1}$

   (b) $F(s) = \frac{e^{-2s}}{s^2+6s+9}$

   (c) $F(s) = \frac{e^{-s}}{s^3+1}$

   (d) $F(s) = \frac{s e^{-2s}}{s(s^2+4s+4)} - \frac{e^{-5s}}{s^2+\pi^2}$. Draw the graph of f when $t \in (0,10)$

5. Find the solution to the following initial value problems.

   (a) $y'' + 4y = u(t-2)$, $y(0) = y'(0) = 0$

   (b) $y'' + 4y = 1 - u(t-2)$, $y(0) = y'(0) = 0$. Draw the graphs of the solutions to this equation and the previous problem on the same screen.

   (c) $x'' + 4x = \sin 2t$, $x(0) = x'(0) = 0$. Replace the forcing term with $(1 - u(t - 2\pi))\sin 2t$, draw the solutions of the two equations, and compare their behavior for $t > 2\pi$.

   (d) $x'' + 4x' + 4x = e^{-2t}\sin 2t$, $x(0) = x'(0) = 0$. Replace the forcing term with $(1-u(t-2\pi))\sin 2t$, draw the solutions of the two equations, and compare their behavior for $t > 2\pi$.

6. Use the definition of Laplace transform and prove the identity

$$L[e^{at}f(t)] = F(s-a), \quad \text{where } F(s) = L[f]. \qquad (5.72)$$

   Use this identity and find the inverse Laplace transform of the following functions.

   (a) $\frac{1}{(s-1)^2+1}$

   (b) $\frac{1}{(s-3)^2+4}$

   (c) $\frac{s-2}{(s-1)^2+1}$

   (d) $\frac{1}{(s-3)^4}$

7. (**Transform of periodic functions**) Let f be a periodic function with period ω, that is, $f(t+\omega) = f(t)$ for all t (see Figure 5.7). Show that

$$L[f] = \frac{\int_0^\omega e^{-st}f(t)dt}{1 - e^{-s\omega}}. \qquad (5.73)$$

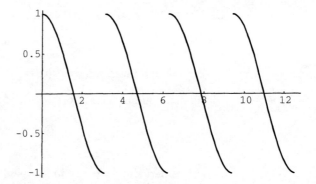

Figure 5.7: A periodic function.

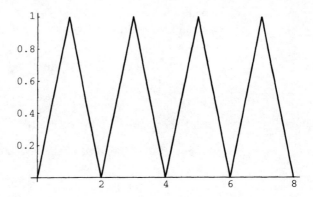

Figure 5.8: The graph of the saw-tooth function.

(Hint: Write $\int_0^\infty e^{-st} f(t)\, dt$ as $\int_0^\omega e^{-st} f(t)\, dt$ + $\int_\omega^{2\omega} e^{-st} f(t)\, dt$ + $\int_{2\omega}^{3\omega} e^{-st} f(t)\, dt$ + $\cdots$. Change variables in each integral and convert the limits of integration to the interval $(0, \omega)$. Use the perodicity of f and recall the geometric series expansion $1 + z + z^2 + z^3 + \cdots = \frac{1}{1-z}$.)

8. Use the previous result and find the Laplace transform of the saw-tooth function (see Figure 5.8)

$$f(t) = t \quad \text{for} \quad 0 \le t < 1, \quad f(t) = 2 - t \quad \text{for} \quad 1 \le t < 2, \qquad (5.74)$$

given that f is periodic with period 2.

9. (a) The internal function Floor in *Mathematica* returns the greatest integer less than or equal to t. Define f by

   f[a_, b_, t_] = a t - a/b [b t];

   Show that f is periodic with period 1/b. Plot the graph of f[1, 1, t], f[1, 1/2, t] and f[1, 1/3, t].

(b) Use NDSolve to solve the initial value problem

$$x'' + x = f(t), \quad x(0) = x'(0) = 0.$$

Plot the graph of x for $t \in (0, 10)$.

10. (a) Find the Laplace transform of the function g defined by

$$g(t) = \begin{cases} t & \text{when } 0 \leq t < 1, \\ 2 - t & \text{when } 1 \leq t < 2, \end{cases} \qquad (5.75)$$

and is periodic with period 2 otherwise.

(b) Define g in *Mathematica* by

g[a_, b_, t_] = Abs[f[a, b, t] - 1];

where f is defined in Problem 9. Show that g[1,1/2,t-1] is the same as g defined above. Plot the graph of g[1, 1, t], g[1, 1/2, t] and g[1, 2, t].

(c) Use NDSolve to solve the initial value problem

$$x'' + x = g(t), \quad x(0) = x'(0) = 0.$$

Plot the graph of x for $t \in (0, 10)$.

11. Find the Laplace transform of the **square wave function** h where

$$h(t) = \begin{cases} 1 & \text{for } 0 \leq t < 1, \\ -1 & \text{for } 1 \leq t < 2, \end{cases} \qquad (5.76)$$

and is periodic with period 2 otherwise.

(a) Define h[a, b, t] in *Mathematica* by

h[a, b, t] = Sign[f[a, b, t]-1];

Show that h[1, 1/2, t-1] is the same as the square wave function.

(b) Use NDSolve to solve the initial value problem

$$x'' + x = h(t), \quad x(0) = x'(0) = 0.$$

Plot the graph of x for $t \in (0, 10)$.

12. Find the Laplace transform of each of the following 2π periodic functions. Draw the graph of this function and its Laplace transform on two different sections of the same screen.

(a) $f(t) = \sin \frac{t}{2}$ for $t \in (0, 2\pi)$ and 2π periodic otherwise

(b) $f(t) = |\cos \frac{t}{2}|$ for $t \in (0, 2\pi)$

(c) $f(t) = t$ if $t \in (0, \pi)$ and $f(t) = 2\pi - t$ for $t \in (\pi, 2\pi)$

13. As we have seen previously, the differential equation

$$my'' + cy' + ky = f(t)$$

models the deviation, $y(t)$, of a mass m from its equilibrium under the restoring force of a spring with spring constant k, the resistive force of a dashpot with damping constant c, and an external force f. Solve each initial value problem below and plot the graph of y.

(a) $m = 4$, $c = 0$, $k = 3$, $f(t) = u(t-2)$, $y(0) = y'(0) = 0$

(b) $m = 1$, $c = 0$, $k = 3$, $f(t) = u(t-2) - u(t-4)$, $y(0) = y'(0) = 0$

(c) $m = 4$, $c = 0.1$, $k = 3$, $f(t) = u(t-2) - u(t-4)$, $y(0) = y'(0) = 0$

(d) $m = 1$, $c = 4$, $k = 3$, $f(t) = u(t-2)\sin(t-2)$, $y(0) = y'(0) = 0$

(e) $m = 1$, $c = 0.5$, $k = 0.4$, $f(t) = u(t-\pi)\sin 4(t-\pi)$, $y(0) = y'(0) = 0$

(f) $m = 1$, $c = 0$, $k = 9$, $f(t) = u(t-\pi)\sin 3(t-\pi)$, $y(0) = y'(0) = 0$. Compare the solution of this problem with the solution of the initial value problem

$$y'' + 9y = \sin 3t, \quad y(0) = y'(0) = 0.$$

14. Solve the initial value problem

$$y'' + y = f(t), \qquad y(0) = y'(0) = 0$$

where f is the square wave function

$$f(t) = \begin{cases} 0 & \text{if } 0 \le t < \pi, \\ 1 & \text{if } \pi \le t < 2\pi, \end{cases}$$

and is periodic with period 2π. Plot the graph of y.

5.7 The Dirac Delta Function

We now take the formalism of the previous section one step further and consider differential equations with nonhomogeneities that influence the evolution of the physical phenomenon only over a very short interval of time or over a small region of space. Examples of this type of forcing term arise in circuit theory when a circuit is subjected to a rather large amount of current for only a short period of time and in mechanics when one is interested in understanding the deflection of a beam, say, when the beam is acted upon by a point force. The traditional approach to such problems is to view an impulsive force as the limit of classical functions whose supports (that is, the sets over which the functions are nonzero) shrink to a point, while the average of the functions (in the sense of integration over the domain of definition) remains finite.

As the ensuing discussion will show, the limiting process alluded to previously does not lead to a convincing mathematical object in the classical sense.

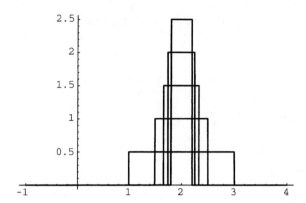

Figure 5.9: An approximating sequence to the Dirac delta function.

In fact, the limiting object is, in general, not a function in the classical sense since such a function turns out to be zero except at a single distinguished point at which it takes on the value of infinity and all of this happens in such a way that the integral of the function is unity! Despite this mathematical shortcoming, these impulse functions have tremendous operational value in mathematical physics. In honor of Paul A. M. Dirac, who popularized the use of impulse functions, these functions are often called the **Dirac delta** functions. Eventually a concerted effort was made to make mathematical sense of the Dirac delta function that culminated in the work of several mathematicians in the 1940s and 1950s, including that of Laurent Schwartz (and his treatise on the theory of *distributions*), which resulted in putting the concept of an impulse "function" on the right footing.

In this text we will confine ourselves to the traditional treatment of the delta function and refer the reader to texts on distribution theory and *generalized functions* for a rigorous mathematical introduction to such functions. Our goal in this section is to introduce the delta function, determine its Laplace transform, and understand its effect as a forcing term on differential equations.

Consider a sequence of nonnegative functions $f_n(t)$ such that each f_n has unit strength, that is,

$$\int_{-\infty}^{\infty} f_n(t)\, dt = 1, \tag{5.77}$$

and such that it is "concentrated" at the point t_0; that is, f_n is zero outside of a neighborhood $(t_0 - \epsilon_n, t_0 + \epsilon_n)$, where ϵ_n is a small number. For the sake of concreteness, let f_n be defined by

$$f_n(t) = \begin{cases} \frac{n}{2} & \text{when } t \in (t_0 - \frac{1}{n}, t_0 + \frac{1}{n}), \\ 0 & \text{otherwise} \end{cases} \tag{5.78}$$

(cf. Figure 5.9). It is easy to see that the integral of each f_n over the entire real line is one, independent of the index n. It is also evident that as n gets large, the peak in f_n gets large and its support (the interval $(t_0 - \frac{1}{n}, t_0 + \frac{1}{n})$) gets small. In the limit we expect that f_n has converged to an object whose integral is one, its pointwise value at each $t \neq t_0$ is zero, and its value at t_0 is infinite. This limiting object (which is clearly not an example of an elementary function we have encountered in calculus) is the **Dirac delta** function and is denoted by

$$\delta(t - t_0) \tag{5.79}$$

This function has several interesting properties that we summarize in the following theorem.

Theorem 5.7.1
The Dirac delta function has the following operational properties:

1. $\int_{-\infty}^{\infty} \delta(t - t_0) \, dt = 1.$

2. $\int_{-\infty}^{\infty} \delta(t - t_0) f(t) \, dt = f(t_0)$ *for any continuous function f.*

3. *Let $t_0 > 0$. Then $L[\delta(t - t_0)] = e^{-st_0}$. In particular, $L[\delta(t)] = 1$.*

We now give a sense of why these properties should hold. Property 1 was derived as the consequence of the limiting process that was described previously.

Let us consider property 2. Recalling the definition of the delta function and functions f_n, in place of the integral on the left-hand side of property 2 consider its approximation by $\int_{-\infty}^{\infty} f_n(t) f(t) \, dt$. Note that

$$\int_{-\infty}^{\infty} f_n(t) f(t) \, dt = \int_{t_0 - \frac{1}{n}}^{t_0 + \frac{1}{n}} \frac{n}{2} f(t) \, dt. \tag{5.80}$$

From the mean value for integrals (that $\int_a^b g(t) \, dt = g(\xi)(b - a)$, for some ξ between a and b) we have

$$\int_{t_0 - \frac{1}{n}}^{t_0 + \frac{1}{n}} \frac{n}{2} f(t) \, dt = f(\xi) \tag{5.81}$$

for some ξ between $t_0 - \frac{1}{n}$ and $t_0 + \frac{1}{n}$. Clearly as n approaches infinity the interval $(t_0 - \frac{1}{n}, t_0 + \frac{1}{n})$ shrinks to t_0 and, thus, ξ approaches t_0. Hence, the integral on the right-hand side of (5.80) approaches $f(t_0)$.

The analysis that leads to property 3 is similar. Again, we approximate the Laplace transform of the delta function by the Laplace transform of f_n. The latter transform is

$$L[f_n(t)] = \int_0^{\infty} s^{-st} f_n(t) dt = \int_{t_0 - \frac{1}{n}}^{t_0 + \frac{1}{n}} \frac{n}{2} e^{-st} \, dt = -\frac{n}{2s} e^{-st} \Big|_{t_0 - \frac{1}{n}}^{t_0 + \frac{1}{n}}. \tag{5.82}$$

Note that the right-hand side of (5.82) is equivalent to

$$e^{-st_0} \frac{\exp(s/n) - \exp(-s/n)}{2s/n} = e^{-st_0} \frac{\sinh s/n}{s/n}. \tag{5.83}$$

We apply the L'Hôpital's rule to the quotient $\frac{\sinh z}{z}$ as z approaches zero (where $z = \frac{1}{n}$) and get unity. Therefore, the limit of $L[f_n]$ is e^{-st_0} as n approaches infinity.

The next example shows the usage of the delta function as the forcing term in a differential equation.

Example 5.7.1
Consider the differential equation

$$x'' + 4x = \delta(t - 3), \quad x(0) = x'(0) = 0. \tag{5.84}$$

This equation models the mechanical displacement of a spring, say, with a unit mass and a restoring force that is four times its displacement. The system is idle until $t = 3$, at which time a unit impulse acts upon it. To get the response of the system for later times we start by taking the Laplace transform of both sides of (5.84) and find that $L[x]$ is

$$L[x] = \frac{e^{-3s}}{s^2 + 4}. \tag{5.85}$$

Note that, because of the term e^{-3s} in the numerator of (5.85), the inverse Laplace transform of $L[x]$ (that is, $x(t)$) will be given in terms of the Heaviside function $u(t - 3)$. In fact,

$$x(t) = u(t - 3)\sin(t - 3); \tag{5.86}$$

that is, the response of the system is a delayed sine function.

The Dirac delta function is denoted by `DiracDelta[t]` in *Mathematica* and is part of the `LaplaceTransform` package. This function is accessed by `Dirac(t)` in *MATLAB*.

Problems

1. Find the solution to the following initial value problems. Check all of your computations in *Mathematica* or *MATLAB*.

   (a) $x'' + x' + x = \delta(t - 1), \quad x(0) = x'(0) = 0$

   (b) $x'' + x' + x = \delta(t - 1), \quad x(0) = 1, x'(0) = 0$

(c) $y'' + y = \delta(t-1) + \delta(t-2), \quad y'(0) = y'(1) = 0$

(d) $y'' + y = 3\delta(t-2), \quad y(0) = y'(0) = 0$

2. Apply property 2 to show that $L[\delta(t-c)f(t)]$ is $e^{-cs}f(c)$. Use this fact and find the solution to

$$x'' + 2x' + 2x = \delta(t-\pi)\cos t, \quad x(0) = x'(0) = 0. \qquad (5.87)$$

3. In property 3 we derived the Laplace transform of the Dirac delta function as the limit of the Laplace transforms of the f_ns. Perform this limiting process in *Mathematica* or *MATLAB*. (Hint: Calculate the Laplace transform of f_n analytically and define the result (for instance, in terms of (5.82)) as a function (say `g[s_, n_, t0_]` in *Mathematica*). Then use `Limit[g[s, n, t0], {n->Infinity}]`. The approach in *MATLAB* is very similar.)

4. In this problem we seek a different sequential approximation to the Dirac delta function $f_n(t)$, one for which each f_n is continuous. Consider

$$f_n(t) = \begin{cases} A\cos(t-c) & \text{for } |t-c| \le \frac{\pi}{2n}, \\ 0 & \text{otherwise.} \end{cases} \qquad (5.88)$$

Determine A so that $\int_{-\infty}^{\infty} f_n(t)dt = 1$. With this value of A show that $\lim_{n\to\infty} f_n(c) = \infty$ while $\lim_{n\to\infty} f_n(t) = 0$ for any $t \ne c$. Use the `Integrate` command of *Mathematica* and compute the Laplace Transform of each f_n. Pass to the limit in $L[f_n]$ as n approaches infinity. What should the answer be?

5. Consider the mass-spring system with $m = 1$ and $k = 4$. Suppose this system is subjected to an external force of the form

$$f(t) = \delta(t-1) + \delta(t-3) + \delta(t-5).$$

What is the physical interpretation of f?

(a) Solve the differential equation

$$mx'' + kx = f(t), \quad x(0) = x'(0) = 0, \qquad (5.89)$$

with the preceding data and draw the graph of the displacement $x(t)$ as time evolves. Compare your graph to Figure 5.10.

(b) Draw the graph of the solution to the unforced equation in (5.89). Combine the two graphs and compare with Figure 5.11. What will happen to the effect of f as time approaches infinity?

(c) Find the solution to the prior equation using `DSolve` of *Mathematica* or `dsolve` of *MATLAB*. Compare and contrast your analytic solution with the one you get from *Mathematica* or *MATLAB*.

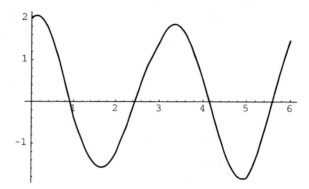

Figure 5.10: The solution to $x'' + 4x = \delta(t-1) + \delta(t-3) + \delta(t-5)$ with $x(0) = 2$ and $x'(0) = 1$.

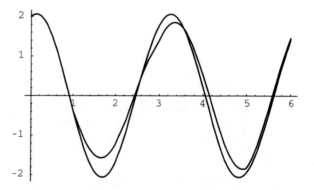

Figure 5.11: The graph in the previous figure combined with the solution to the unforced problem.

(d) Write a *Mathematica* or *MATLAB* program that begins with computing the Laplace transform of (5.89), determine the Laplace transform
of x, and after finding the inverse Laplace transform of x, plot its
graph.

6. An RLC electrical circuit consisting of $R = 0.4$ ohms, $L = 1$ henry, and
$C = 0.01$ farad is driven by an electromotive force $E(t)$. Suppose that
there is no current present in the circuit at time zero.

(a) Find the current in the circuit and the charge in the capacitor if

$$E'(t) = \delta(t-1) - 2\delta(t-3).$$

Draw the graphs of each quantity in the interval $(0, 10)$.

(b) Repeat the preceding problem for $E'(t) = u(t-1) - 2u(t-2)$.

7. Find the solution to the following initial value problems. Plot the graphs of x and y on a single screen.

(a) $2x' = 3x - y$, $y' = x + y + \delta(t - 2)$; $x(0) = 0$, $y(0) = 0$

(b) $x' = y$, $y' = -2x + y + 2\delta(t - 2) - 3\delta(t - 4)$; $x(0) = 0$, $y(0) = 0$

(c) $x' = x + 4y$, $y' = x + y + \delta(t - 2)\sin(t - 2)$; $x(0) = 0$, $y(0) = 0$.

5.8 Convolution of Functions

The convolution h of two functions f and g, denoted by $f * g$, is a special way of generating the "product" of f and g so that the Laplace transform of h becomes the ordinary product of the Laplace transforms of f and g. As the reader can verify with a simple example, the function h cannot be the pointwise product of f and g; that is, it is not true in general that $L[fg] = L[f]L[g]$ (try $f(t) = g(t) = t$). The purpose of the definition of convolution is to replace the pointwise product fg by a new concept of multiplication of function, denoted by $f * g$, so as to render the latter identity for Laplace transforms valid for the class of functions f and g that we encounter in applications.

Definition 5.8.1
*Let f and g be two functions with Laplace transforms F and G defined over a common domain $s > a$. Define the **convolution** of f and g by*

$$h(t) = (f * g)(t) = \int_0^t f(t - \tau)g(\tau)d\tau. \tag{5.90}$$

Example 5.8.1
Let $f(t) = t$ and $g(t) = t^2$. Then

$$h(t) = (f * g)(t) = \int_0^t (t - \tau)\tau^2 \, d\tau = \frac{t^4}{12}. \tag{5.91}$$

The convolution of the preceding two functions is a simple integration in *Mathematica* or *MATLAB*. For example, we use `Integrate` of *Mathematica* to get the desired result:

```
h[t_] = Integrate[(t - u)*u^2, {u, 0, t}]
```

Note in particular that $L[h] = 2/s^5 = L[t]L[t^2]$.
Similarly the convolution of $\sin t$ with $\cos 2t$ is

$$h(t) = \sin t * \cos 2t = \int_0^t \sin(t - \tau)\cos 2\tau \, d\tau = \frac{1}{3}(\cos t - \cos 2t). \tag{5.92}$$

We note that the identity element for the convolution operation (that is, the function g such that $f * g = f$ for all f) is not the function $g(t) \equiv 1$, as can easily be seen from the following example in which $f(t) = t$:

$$t * 1 = \int_0^t t - \tau \, d\tau = \frac{t^2}{2},$$ (5.93)

which is not $f(t)$. However, if we let $g(t) = \delta(t)$, the Dirac delta function, then

$$(f * \delta)(t) = \int_0^t f(t - \tau)\delta(\tau) \, d\tau = f(t),$$ (5.94)

by property 2 of Section 5.7.

We now state the main theorem for the convolution of two functions.

Theorem 5.8.1
Let f and g be as in Definition 5.8.1. Then

1. *The convolution operation is commutative, that is,*

$$f * g = g * f.$$ (5.95)

2. *The Laplace transform of the convolution of two functions is the product of their Laplace transforms*

$$L[f * g] = L[f]L[g],$$ (5.96)

with the consequence that

$$f * g = L^{-1}[L[f]L[g]].$$ (5.97)

The proof of (5.95) is left to the reader (see Problem 3). We now give a formal proof of (5.96). The proof we present here is formal because the rigorous proof requires the ability to exchange the order of a double integration in the calculations that follow. The theorem that enables us to perform this task is Fubini's theorem whose development is beyond the scope of this text. Still, the formal proof we will present is valuable because it indicates why we anticipate that (5.96) is true.

Note that

$$L[f] = \int_0^\infty e^{-st} f(t) \, dt \quad \text{and} \quad L[g] = \int_0^\infty e^{-su} g(u) \, du,$$ (5.98)

so that

$$L[f]L[g] = \left(\int_0^\infty e^{-st} f(t) \, dt \right) \left(\int_0^\infty e^{-su} g(u) \, du \right).$$ (5.99)

We combine the preceding interated integrals into the double integral

$$\int_0^\infty \int_0^\infty e^{-s(t+u)} f(t)g(u) \, dt \, du$$ (5.100)

and make the change of variables $w = t + u$ in the inner integral. With the new variables w and u, the expression in (5.100) becomes

$$\int_0^\infty \int_u^\infty e^{-sw} f(w-u)g(u)\, dw\, du. \tag{5.101}$$

We now exchange the order of integrations. The reader should convince himself (it may be helpful to draw a picture of the region of integration) that the region of integration in (5.101), which is parametrized as

$$u < w < \infty, \quad 0 < u < \infty, \tag{5.102}$$

can equally well be parametrized as

$$0 < u < w, \quad 0 < w < \infty. \tag{5.103}$$

Thus, the double integral in (5.101) is equivalent to

$$\int_0^\infty \int_0^w e^{-sw} f(w-u)g(u)\, du\, dw. \tag{5.104}$$

Note that e^{-sw} in (5.104) is independent of u and may be moved to the outside of the inner integral. What remains in the inner integral is the convolution of f and g. The outer integral is precisely the Laplace transform of the inner integral. Therefore, going back to (5.99), we have shown that $L[f]L[g] = L[f * g]$.

One of the applications of this theorem is in computing inverse Laplace transforms when a rather complicated function $F(s)$ is written as the product of two functions $G(s)$and $H(s)$, each of which has a simple inverse Laplace transform. The inverse Laplace transform of F is then the convolution of the inverse Laplace transforms of G and H. The next example illustrates this idea.

Example 5.8.2
Consider $F(s) = \frac{1}{(1+s^2)^2}$. We write F as

$$F(s) = \frac{1}{1+s^2}\frac{1}{1+s^2} = L[\sin t]L[\sin t]. \tag{5.105}$$

Hence,

$$f(t) = \sin t * \sin t = \int_0^t \sin(t-\tau)\sin\tau\, d\tau = \frac{1}{2}(\sin t - t\cos t). \tag{5.106}$$

Problems

1. Do the following convolutions. Verify your answer in either *Mathematica* or *MATLAB*.

    (a) $t * t$

    (b) $1 * 1$

(c) $t * \sin t$

(d) $\sin t * \sin t$

(e) $t * e^t$

(f) $t * \cos \omega t$ where ω is a positive constant

(g) $e^{2t} * \sin t$

2. Find the following inverse Laplace transforms. Check the integration as well as the inverse Laplace transforms in *Mathematica* or *MATLAB*.

   (a) $F(s) = \frac{1}{s^2(s^2+1)}$

   (b) $F(s) = \frac{s-2}{s^2(s^2+1)}$

   (c) $F(s) = \frac{1}{s^4-1}$

3. Prove that the operation of convolution is commutative, that is, $f*g = g*f$. Similarly, show that this operation is associative, that is, $f * (g * h) = (f * g) * h$.

4. Use Laplace transforms and the convolution theorem to show that the solution of the differential equation

$$x'' + a^2 x = f(t)$$

with $x(0) = x'(0) = 0$ is

$$x(t) = \frac{1}{a} f * \sin(at).$$

5. Find a formula for the solution to the initial value problem

$$x'' + x' + 4x = f(t), \quad x(0) = x'(0) = 0.$$

6. Consider the following integro-differential equation

$$x(t) + \int_0^t a(t - \tau)x(\tau) \, d\tau = f(t)$$

where f and a are given functions. Show that

$$x(t) = L^{-1} \frac{L[f]}{1 + L[a]}.$$

7. Determine the solution to the following integro-differential equations.

   (a) $x(t) + \int_0^t e^{-(t-\tau)}x(\tau) \, d\tau = 1$

   (b) $x(t) + \int_0^t e^{-2(t-\tau)}x(\tau) \, d\tau = 3\sin t$

   (c) $x(t) + \int_0^t \sin(t - \tau)x(\tau) \, d\tau = 2 + t - t^2$

   (d) $y(t) + \int_0^t y(t - \tau)\cosh \tau \, d\tau = 1$

8. Consider the integro-differential equation

$$x'(t) + \int_0^t a(t-\tau)x(\tau)\, d\tau = f(t), \quad x(0) = x_0,$$

where f and a are given functions. Show that

$$x(t) = L^{-1}\frac{x_0 + L[f]}{s + L[a]}.$$

9. Determine the solution to the following integro-differential equations.

   (a) $x'(t) + \int_0^t e^{-(t-\tau)}x(\tau)\, d\tau = 1$, $x(0) = 0$

   (b) $x'(t) + \int_0^t e^{-2\tau}x(t-\tau)\, d\tau = \sin t$, $x(0) = 0$

   (c) $y'(t) + \int_0^t \sin(t-\tau)y(\tau)\, d\tau = \sin 2t$, $y(0) = -1$

   (d) $y' + t * y = 1$, $y(0) = 1$

References

[1] Churchill, *Operational Mathematics*, 3rd Edition, McGraw-Hill, New York, 1972.

Chapter 6

Applied Linear Algebra

6.1 Introduction

In this chapter we will review the elementary concepts of vectors and vector spaces. These concepts play a fundamental role in the mathematical modeling of physical problems, as well as in the description of the numerical approximation of the differential equations that arise from these problems.

Our goal in this text is to present a computational approach to the study of matrices as opposed to the traditional axiomatic approach. This is not to say that we intend to ignore the abstract foundation of this topic; We will still be concerned with working definitions of vector spaces, subspaces, rank, and nullity, among others. Our ultimate goal, however, is to show how matrices (and linear transformations) play an important role in describing deformation of moving bodies as well as in demonstrating how the algebra of matrices is a crucial tool in compressing and transferring information from one set of ideas to another. It is the latter property of matrices that has enabled users of mathematics to look through a vast quantity of information and see a clear path for simple ideas and algorithms.

As it will turn out, the primary mathematical objects we need to compute are solutions of simultaneous algebraic equations, eigenvalues and eigenvectors, and, on occasion, inverses of matrices. These operations have been the focus of study by numerical analysts as well as algebraists for many decades. Much of the experience gained from these studies has been used in *MATLAB* and *Mathematica* to develop internal functions whose outputs are precisely the quantities we need to manipulate in applications. The presence of these software packages has thus made the job of a physical modeler considerably simpler.

A main application of the topics we study in this chapter is in generating solutions of linear systems of differential equations, a topic that we will take up in Chapter 7. In that chapter we will use the numerical and symbolic facilities of *MATLAB* and *Mathematica* in combination with the mathematical tools we develop in this chapter. This will in turn provide us with the necessary tools to consider physical applications that have finer structures than those we have been studying so far.

6.2 Vectors

Here we review the basic definitions and operations of vectors in two- and three-dimensional spaces.

Geometrically, a **vector** is a directed line segment. The length of the line segment is often referred to as the length or the magnitude of the vector. Two vectors are equal if their lengths and directions are the same. Typical examples of vectors are the velocity of a fluid particle or the gravitational force. The magnitude or the length of these vectors often have physical significance as well. For instance, the magnitude of the velocity vector is its speed.

Scalars are quantities for which direction is not significant, and a single number would suffice in their quantification. Temperature, speed, pressure, and voltage are examples of scalar physical quantities.

We will denote vectors by boldface letters. We will almost always denote a velocity vector by $\mathbf{v}$ and its speed, a scalar, by v. Pressure, a scalar quantity, will often be denoted by p. The symbol $|\mathbf{a}|$ denotes the magnitude of the vector $\mathbf{a}$. A vector that has length one is called a **unit vector**.

In two- and three-dimensional spaces we introduce coordinate systems with axes consisting of mutually perpendicular straight lines intersecting at a common point, the origin of the coordinate system. Such a coordinate system will be referred to as **rectangular** or **cartesian**. A vector $\mathbf{a}$ is a line segment that connects two points P and Q in this geometric setting. Assuming that $\mathbf{a}$ is directed from P toward Q and that P and Q have coordinates $P = (p_1, p_2, p_3)$ and $Q = (q_1, q_2, q_3)$ in the three-dimensional space, $\mathbf{a}$ is equivalent to

$$\mathbf{a} = \langle q_1 - p_1, q_2 - p_2, q_3 - p_3 \rangle \equiv \langle a_1, a_2, a_3 \rangle. \tag{6.1}$$

Its magnitude is the distance between P and Q, which is given by

$$\|\mathbf{a}\| = \sqrt{a_1^2 + a_2^2 + a_3^2}. \tag{6.2}$$

The scalars a_i in (6.1) are called the **components** of the vector $\mathbf{a}$. It should be clear that two vectors are equal if their respective components are equal.

The sum of two vectors $\mathbf{a}$ and $\mathbf{b}$ is another vector $\mathbf{c}$ whose components are the sum of the respective components of $\mathbf{a}$ and $\mathbf{b}$: If

$$\mathbf{a} = \langle a_1, a_2, a_3 \rangle \quad \text{and} \quad \mathbf{b} = \langle b_1, b_2, b_3 \rangle;$$

then

$$\mathbf{c} = \mathbf{a} + \mathbf{b} = \langle a_1 + b_1, a_2 + b_2, a_3 + b_3 \rangle.$$

Vector addition satisfies the usual properties of commutativity and associativity. With the zero vector $\mathbf{0} = \langle 0, 0, 0 \rangle$, we have

$$\mathbf{a} + \mathbf{0} = \mathbf{a}.$$

The vector $-\mathbf{a}$ is defined as the vector with the same magnitude as $\mathbf{a}$ but with opposite direction. Given a scalar quantity α, the **scalar product** of α and $\mathbf{a}$ is denoted by $\alpha\mathbf{a}$, which is the vector with components

$$\alpha\mathbf{a} = \langle \alpha a_1, \alpha a_2, \alpha a_3 \rangle.$$

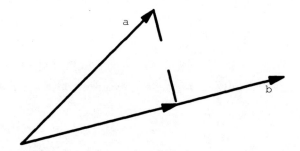

Figure 6.1: Dot product and projections of vectors.

Note that
$$|\alpha \mathbf{a}| = |\alpha| \, |\mathbf{a}|.$$

We associate three unit vectors $\mathbf{i}$, $\mathbf{j}$, and $\mathbf{k}$ with the axes of a rectangular coordinate system. These vectors point in the directions of the coordinates axes that define the three-dimensional space, each vector originating from $(0, 0, 0)$ and ending up a unit distance away along an axis, that is, $\mathbf{i} = \langle 1, 0, 0 \rangle$, $\mathbf{j} = \langle 0, 1, 0 \rangle$, and $\mathbf{k} = \langle 0, 0, 1 \rangle$. In terms of these unit vectors, any vector $\mathbf{a} = \langle a_1, a_2, a_3 \rangle$ can be written as
$$\mathbf{a} = a_1 \mathbf{i} + a_2 \mathbf{j} + a_3 \mathbf{k}.$$

The **dot product** or **inner product** of two vectors $\mathbf{a}$ and $\mathbf{b}$ is denoted by $\mathbf{a} \cdot \mathbf{b}$ and defined by
$$\mathbf{a} \cdot \mathbf{b} = \|\mathbf{a}\| \, \|\mathbf{b}\| \cos \gamma,$$
where γ is the angle between the vectors in the plane spanned by the two vectors. When either $\mathbf{a}$ or $\mathbf{b}$ is the zero vector, the inner product is defined to be zero. Two vector are **orthogonal** if their inner product is zero.

The inner product of two vectors has a simple and useful geometric interpretation. Let $\mathbf{b}$ be a unit vector. Then by (6.2) $\mathbf{a} \cdot \mathbf{b}$ is the magnitude of the vector one gets by projecting $\mathbf{a}$ in the direction of $\mathbf{b}$ (cf. Figure 6.1). The formula for this projection vector is
$$\operatorname{proj}_{\mathbf{b}} \mathbf{a} = \frac{\mathbf{a} \cdot \mathbf{b}}{\mathbf{b} \cdot \mathbf{b}} \mathbf{b}. \qquad (6.3)$$

A project at the end of this chapter outlines the program in *Mathematica* that leads to Figure 6.1.

It is not difficult to show that when $\mathbf{a} = \langle a_1, a_2, a_3 \rangle$ and $\mathbf{b} = \langle b_1, b_2, b_3 \rangle$, the preceding definition of the dot product is equivalent to
$$\mathbf{a} \cdot \mathbf{b} = \sum_{i=1}^{3} a_i b_i \qquad (6.4)$$

(see Problem 4(f)).

A *right-handed cartesian* coordinate system is one in which the vectors $\mathbf{i}$, $\mathbf{j}$, and $\mathbf{k}$ satisfy the right-hand rule; that is, if the right hand's palm is curved from $\mathbf{i}$ toward $\mathbf{j}$, then the thumb points in the direction of $\mathbf{k}$. We assume, unless otherwise stated, that the coordinate system is right-handed.

The **cross product** of two vectors **a** and **b** is defined as a vector **c** whose direction is perpendicular to both **a** and **b** and given by the right-hand rule and whose magnitude is

$$||\mathbf{c}|| = ||\mathbf{a}|| \, ||\mathbf{b}|| \sin \theta, \tag{6.5}$$

where θ is the angle between **a** and **b**. If $\mathbf{a} = \langle a_1, a_2, a_3 \rangle$ and $\mathbf{b} = \langle b_1, b_2, b_3 \rangle$, then **c** is given by

$$\mathbf{c} = \langle a_2 b_3 - a_3 b_2, a_3 b_1 - a_1 b_3, a_1 b_2 - a_2 b_1 \rangle. \tag{6.6}$$

Symbolically, the preceding formula is equivalent to computing the determinant of the 3×3 matrix

$$\begin{bmatrix} \mathbf{i} & \mathbf{j} & \mathbf{k} \\ a_1 & a_2 & a_3 \\ b_1 & b_2 & b_3 \end{bmatrix}. \tag{6.7}$$

The operation of cross product comes up naturally in determining the moment of a force about a point as well as in computing the normal vector to the plane passing through three points. We will see many instances in the text where this operation gives a concise mathematical representation of a physical phenomenon.

The **scalar triple product** of three vectors **a**, **b**, and **c** is denoted by

$$\mathbf{a} \cdot (\mathbf{b} \times \mathbf{c}).$$

In a right-handed coordinate system the preceding formula is equivalent to

$$\det \begin{bmatrix} a_1 & a_2 & a_3 \\ b_1 & b_2 & b_3 \\ c_1 & c_2 & c_3 \end{bmatrix}. \tag{6.8}$$

Geometrically, the triple scalar product gives us the (signed) volume of the parallelepiped constructed on the three vectors. In particular, when three points P, Q, and S are coplanar, the triple scalar product of the three vectors that pass through these points must be zero.

Vectors in *Mathematica* are lists. For example, the vector $\mathbf{a} = \langle 1, 2, 3 \rangle$ is entered as

```
a = {1, 2, 3}
```

The dot product of two vectors is determined in *Mathematica* by inserting a dot between the two vectors. For instance, if $\mathbf{a} = \langle -2, 3, 1 \rangle$ and $\mathbf{b} = \langle 1, -1, 1 \rangle$, we compute $\mathbf{a} \cdot \mathbf{b}$ as

```
a = {-2, 3, 1};   b = {1, -1, 1};
a . b
```

To determine the cross product of these two vectors we must first input an external package called `VectorAnalysis` (which is in a subdirectory called `Calculus`) by

```
<<Calculus`VectorAnalysis`
```

and then use `CrossProduct`:

```
CrossProduct[a, b]
```

Remark 6.2.1: As we have remarked previously, one must enter a package such as `VectorAnalysis` as a first input statement into *Mathematica* to avoid the risk of duplicate use of variable names.

In *MATLAB* we enter the vector $\mathbf{a} = \langle 1, -2, 4 \rangle$ as

```
a = [1 -1 4]
```

One finds the inner product of $\mathbf{a} = \langle 10 - 3 \rangle$ and $\mathbf{b} = \langle 0 - 12 \rangle$ via

```
a = [1 0 -3]; b = [0 -1 2];
a*b'
```

The cross product is not an internal function to *MATLAB*, but one can write a simple M-file to define such an operation.

Problems

1. Determine the components of a vector with initial point P and terminal point Q. In each case find the magnitude of the vector and draw its graph.

   (a) $P = (1, 0, -1)$, $Q = (3-, 1, 2)$
   (b) $P = (1, 0, 0)$, $Q = (0, 0, 1)$
   (c) $P = (\frac{1}{4}, \sqrt{2}, 0)$, $Q = (-1, \frac{3}{4}, 3)$.

2. Find a two-dimensional vector whose magnitude and angle with the positive x-axis are given.

   (a) $\|\mathbf{a}\| = 3$, $\theta = 30$ degrees
   (b) $\|\mathbf{a}\| = 1.3$, $\theta = 120$ degrees.

3. Find a unit vector perpendicular to the given vector.

   (a) $\mathbf{a} = 2\mathbf{i} - 3\mathbf{j}$
   (b) $\mathbf{a} = -2\mathbf{i} + 4\mathbf{j} + 3\mathbf{k}$
   (c) $\mathbf{a} = \mathbf{i} + \mathbf{j}$
   (d) $\mathbf{a} = \mathbf{i} - b\mathbf{j}$
   (e) $\mathbf{a} = a\mathbf{i} + b\mathbf{j} + c\mathbf{k}$.

4. Use *Mathematica* or *MATLAB* and prove the following statements. In each case input arbitrary vectors to these programs (i.e., do not verify the statements and assertions for specific vectors). In the case of identities it is often simpler to bring all terms to one side of the equality and write the identity as `statement = 0`. Then use the simplifying operations of the symbolic manipulator (such as `Expand` or `Factor`) on `statement`.

   (a) Show that $\|\mathbf{a}\|^2 = \mathbf{a} \cdot \mathbf{a}$.
   (b) Show that $\mathbf{a} \cdot \mathbf{b} = \mathbf{b} \cdot \mathbf{a}$ for all three-dimensional vectors $\mathbf{a}$ and $\mathbf{b}$.

(c) Show that $(\alpha \mathbf{a} + \beta \mathbf{b}) \cdot \mathbf{c} = \alpha \mathbf{a} \cdot \mathbf{c} + \beta \mathbf{b} \cdot \mathbf{c}$ for arbitrary scalars α and β and arbitrary vectors $\mathbf{a}$, $\mathbf{b}$, and $\mathbf{c}$.

(d) Prove the parallelogram equality

$$\|\mathbf{a} + \mathbf{b}\|^2 + \|\mathbf{a} - \mathbf{b}\|^2 = 2(\|\mathbf{a}\|^2 + \|\mathbf{b}\|^2).$$

(e) Prove that $\mathbf{a} \cdot \mathbf{b} = a_1 b_1 + a_2 b_2 + a_3 b_3$. (Hint: First determine the dot products of $\mathbf{i}$, $\mathbf{j}$, and $\mathbf{k}$).

5. Write an M-file in *MATLAB* that takes two vectors as its input and computes their cross product.

6. Determine the dot product of these vectors. Check your results in *Mathematica*.

(a) $\mathbf{a} = \langle 12, -3 \rangle$ and $\mathbf{b} = \langle 1, -1 \rangle$

(b) $\mathbf{a} = \langle 2, 0, -2 \rangle$ and $\mathbf{b} = \langle 1, -1, 3 \rangle$

(c) $\mathbf{a} = \langle 1, a, a^2 \rangle$ and $\mathbf{b} = \langle -a, \frac{1}{a}, a^3 \rangle$

7. Determine the cross product of the vectors in Problem 6. Use *Mathematica* to check your result in each case.

8. Let $\mathbf{a} = \langle 1, c, c^2 \rangle$ and $\mathbf{b} = \langle 1, 1, c \rangle$. Find all values of c for which $\mathbf{a}$ and $\mathbf{b}$ are perpendicular. (Hint: Use `Solve` in *Mathematica*.)

9. Let $\mathbf{a} = \langle 1, c, c^2 \rangle$ and $\mathbf{b} = \langle 1, 1, c \rangle$. Find all values of c for which $\mathbf{a}$ and $\mathbf{b}$ are parallel.

10. Prove the following results. When appropriate use *Mathematica* or *MATLAB* as an aid in your proofs.

(a) Show that if two vectors are parallel, their cross product is zero. Is the converse true?

(b) Show that $\mathbf{a} \times \mathbf{b} = -\mathbf{b} \times \mathbf{a}$. What does this statement say about the commutativity of the cross product?

(c) Show that $\mathbf{a} \times (\mathbf{b} + \mathbf{c}) = (\mathbf{a} \times \mathbf{b}) + (\mathbf{a} \times \mathbf{c})$ for any three vectors $\mathbf{a}$, $\mathbf{b}$, and $\mathbf{c}$.

(d) Verify whether the following statement is valid for all vectors $\mathbf{a}$, $\mathbf{b}$, and $\mathbf{c}$.

$$\mathbf{a} \times (\mathbf{b} \times \mathbf{c}) = (\mathbf{a} \times \mathbf{b}) \times \mathbf{c}.$$

11. Any plane in a three-dimensional space can be represented as

$$ax + by + cz = d,$$

where a, b, c, and d are constants. Note that

$$d = ax_0 + by_0 + cz_0$$

where (x_0, y_0, z_0) is a fixed but arbitrary point in the plane. Let (x, y, z) be any point in the plane. What is the geometric relationship between the vector $\langle x - x_0, y - y_0, z - z_0 \rangle$ and the plane? Show that the vector $\langle a, b, c \rangle$ is perpendicular to the plane.

12. Use the result of the above problem and find unit normals to the planes $x - y + z = 1$ and $2x - 4y + 3z = 12$.

13. Are the planes $x - y + z = 0$ and $2x + y - z = 1$ orthogonal?

14. Find all a's for which the plane $x + y + z = 1$ is orthogonal to $ax + (2a^2 - 3)y + (a^3 + 4a)z = 2$. Is there an a for which these planes are parallel?

15. Determine proj $_\mathbf{b}$ $\mathbf{a}$ with

   (a) $\mathbf{a} = \langle 1, -1 \rangle$, $\mathbf{b} = \langle 2, 2 \rangle$

   (b) $\mathbf{a} = \langle 1, 0 \rangle$, $\mathbf{b} = \langle 1, 1 \rangle$

   (c) $\mathbf{a} = \langle 1, 0, -1 \rangle$, $\mathbf{b} = \langle 1, -2, 2 \rangle$

   (d) $\mathbf{a} = \langle a, -a, 0 \rangle$, $\mathbf{b} = \langle 1, 1, 1 \rangle$, where a is a fixed real number

   (e) $\mathbf{a} = \langle x, y, z \rangle$, $\mathbf{b} = \langle 1, 1, 1 \rangle$, where x, y, and z are real numbers

16. Determine all vectors $\langle x, y, z \rangle$ that are orthogonal to $\langle -2, 3, 1 \rangle$.

17. Determine all vectors $\langle x, y, z \rangle$ that are orthogonal to $\langle 1, 1, 1 \rangle$ and $\langle -2, 2, 3 \rangle$.

18. Find a unit vector that is orthogonal to $\langle a, -a \rangle$.

19. Find a unit vector that is orthogonal to $\langle \cos \theta, \sin \theta \rangle$.

20. Prove the following identities. Check your computations in *Mathematica* or *MATLAB*.

   (a) $\|\mathbf{a} \times \mathbf{b}\| = \sqrt{\|\mathbf{a}\|^2 \|\mathbf{b}\|^2 - (\mathbf{a} \cdot \mathbf{b})^2}$

   (b) $\mathbf{a} \times (\mathbf{b} \times \mathbf{c}) = (\mathbf{a} \cdot \mathbf{c})\mathbf{b} - (\mathbf{a} \cdot \mathbf{b})\mathbf{c}$

   (c) $(\mathbf{a} \times \mathbf{b}) \times (\mathbf{c} \times \mathbf{d}) = (\mathbf{a} \cdot (\mathbf{b} \times \mathbf{d}))\mathbf{c} - (\mathbf{a} \cdot (\mathbf{b} \times \mathbf{c}))\mathbf{d}$

   (d) (Lagrange's identity) $(\mathbf{a} \times \mathbf{b}) \cdot (\mathbf{c} \times \mathbf{d}) = (\mathbf{a} \cdot \mathbf{c})(\mathbf{b} \cdot \mathbf{d}) - (\mathbf{a} \cdot \mathbf{d})(\mathbf{b} \cdot \mathbf{c})$.

21. Let B be a rigid body rotating about its center of mass with angular velocity $\mathbf{w} = w\mathbf{e}$, where $\mathbf{e}$ is a unit vector in the direction of the axis of rotation. Let O be a fixed point on the axis of rotation. Let $\mathbf{r}$ be the position vector from O to an arbitrary point P on B. Show that the velocity $\mathbf{v}$ at P is given by

$$\mathbf{v} = \mathbf{w} \times \mathbf{r}.$$

6.3 Vector Space E^n

We continue with our study of vectors and vector operations and extend some of the basic definitions and concepts of the previous section to higher dimensional spaces.

We define the set R^n as the set of all n-tuples $(a_1, a_2, \ldots, a_n)$, where each a_i belongs to R, the set of real numbers. We will often refer to each n-tuple of R^n as a *point* or an *element* of that set. We know from experience with R, R^2, and R^3 that we can add any two elements of these sets and again get an element of the same set. For example, given the two elements $a = (a_1, a_2)$ and $b = (b_1, b_2)$ from R^2, their sum $a + b$ is the element $(a_1 + b_1, a_2 + b_2)$, which belongs to in R^2. Similarly, we can multiply an element of each of these sets by any *scalar* (i.e., an element of R) and end up with yet another element of the same set; with $\alpha \in R$, $\alpha a = \langle \alpha a_a, \alpha a_2 \rangle$.

The two *operations* of addition and scalar multiplication are at the heart of the abstract notion of a **vector space**. These operations satisfy certain abstract properties (such as commutativity and associativity, among others) with which we are familiar from elementary algebra. The set R^2 is certainly endowed with such a structure. As we will see shortly, these operations and properties are shared by many other sets in mathematics for which we now give the following working definition.

Definition 6.3.1 (Vector Space)
*A set X together with the operations of addition and scalar multiplication is called a **Vector Space** if the following properties hold.*

1. *The zero element is in X; that is, there is an element $\mathbf{z} \in X$ such that*

$$\mathbf{a} + \mathbf{z} = \mathbf{a}$$

   for every element $\mathbf{a}$ in X.

2. *If $\mathbf{a}$ and $\mathbf{b}$ belong to X, then so does $\mathbf{a} + \mathbf{b}$.*

3. *If $\mathbf{a}$ belongs to X and α is a scalar (i.e., α belongs to the set of real numbers), then so does $\alpha\mathbf{a}$.*

Elements of a vector space are called *vectors*.

Example 6.3.1
Let $X = R$. Then clearly this set forms a vector space as the reader can easily verify. Similarly, R^2 and R^3 are vector spaces.

Example 6.3.2 (Vector Space E^3)
Let $X = \{\langle a_1, a_2, a_3 \rangle | a_i \in R\}$. Here the vector $\langle a_1, a_2, a_3 \rangle$ denotes the usual three-dimensional vector originating from the origin and ending up at the point (a_1, a_2, a_3), and the a_is are called its *components*. The zero vector $\mathbf{z}$ of this

vector space is $\langle 0, 0, 0 \rangle$. Addition of the two vectors $\langle a_1, b_1, c_1 \rangle$ and $\langle a_2, b_2, c_2 \rangle$ is performed in the usual sense of adding components, namely, $\langle a_1 + a_2, b_1 + b_2, c_1 + c_2 \rangle$. Clearly the sum belongs to X. Similarly, scalar multiplication is defined as usual; if $\alpha \in R$ and $\mathbf{x} = \langle a, b, c \rangle \in X$, then $\alpha \mathbf{x} \equiv \langle \alpha a, \alpha b, \alpha c \rangle$ is again in X. So X together with the usual concepts of addition and scalar multiplication forms a vector space. This vector space is usually denoted by E^3 and is called the three-dimensional *Euclidean* space (the concept of dimension will be introduced in the next section).

Example 6.3.3 (Vector Space E^n)

The vector space E^n, where n is any positive integer, is defined in the same manner. The underlying set consists of vectors of the form $\langle a_1, a_2, \ldots, a_n \rangle$. Vector addition and scalar multiplication are defined for E^n analogously to E^3.

For obvious reasons we do not attempt to draw these vectors when $n > 3$. Notwithstanding this shortcoming, this vector space plays an extremely important role in applied mathematics. We will see direct applications of E^n when we study approximate solutions of systems of differential equations, which require the manipulation of vectors in E^n where n could be as large as $k10^p$; here k is an integer and p may be 2 or 3 for the type of applications we will encounter in this text. The estimate on n could be much larger for more real-world models.

Vector addition and scalar multiplication are defined for E^n analogously to E^3.

Example 6.3.4

Let X be the set defined by

$$X = \{ \langle a, a^2 \rangle | \, a \in R \}. \tag{6.9}$$

Let us see that this set does not form a vector space under the standard operations of addition and scalar multiplication of vectors. Property 1 holds for X since the zero vector belongs to X when $a = 0$. But properties 2 and 3 both fail: Let $\mathbf{x} = \langle a_1, a_1^2 \rangle$ and $\mathbf{y} = \langle a_2, a_2^2 \rangle$. The sum of $\mathbf{x}$ and $\mathbf{y}$ is the vector $\langle a_1 + a_2, a_1^2 + a_2^2 \rangle$, which is not in X since, in general, $(a_1 + a_2)^2 \neq a_1^2 + a_2^2$. Property 3 fails to hold in this case by a similar reasoning.

We called Definition 6.3.1 a "working" definition because it contains the essence of what we need to test in order to determine whether a set and a collection of operations form a vector space. The complete definition of a vector space requires a more careful listing of the properties of addition and scalar multiplication, which we now state. Most of the properties listed in the following definition are familiar to the reader. As one will note in almost every example of a vector space we encounter, verifying the validity of these properties is nearly trivial. The properties in the original definition still constitute the main relations with which we need to be concerned when identifying vector spaces.

Definition 6.3.2 (General Vector Space)

A set X and the two operations of addition and scalar multiplication form a vector space, if, in addition to the properties stated in Definition 6.3.1, the following conditions hold.

1. *Commutativity of Vector Addition:* $\mathbf{a} + \mathbf{b} = \mathbf{b} + \mathbf{a}$.

2. *Associativity of Vector Addition:* $(\mathbf{a} + \mathbf{b}) + \mathbf{c} = \mathbf{a} + (\mathbf{b} + \mathbf{c})$.

3. For every element $\mathbf{a} \in X$, there is an element $-\mathbf{a} \in X$ such that $\mathbf{a} + (-\mathbf{a}) = \mathbf{0}$.

4. *Distributivity of Scalar Mulitiplication:* $\alpha(\mathbf{a} + \mathbf{b}) = \alpha\mathbf{a} + \alpha\mathbf{b}$, and $(\alpha + \beta)\mathbf{a} = \alpha\mathbf{a} + \beta\mathbf{b}$.

5. $1\mathbf{a} = \mathbf{a}$ for all $\mathbf{a} \in X$.

So far we have looked at examples of vector spaces whose elements have been vectors in the usual sense. We will now consider examples of vector spaces whose elements are from special classes of elementary functions such as polynomials or trigonometric functions. Such spaces arise naturally in approximation theory. In particular, we will come across several examples in later chapters that demonstrate how various vector spaces enter into the statements of theorems.

Example 6.3.5

Let X be defined by

$$X = \{ax + by \mid a, b \in R\}, \tag{6.10}$$

where x and y are independent variables. Here, the term "vector" refers to a typical element of X, such as $2x - 3y$. By addition and scalar multiplication of vectors we mean the natural way one can add elements of X and multiply an element by $\alpha \in R$. The reader can show that all three properties of Definitions 6.3.1 and 6.3.2 are satisfied in this case.

Example 6.3.6

Let X be the set of all functions of the form

$$X = \{a \sin x + b \sin 2x + c \sin 3x \mid a, b, c \in R\}. \tag{6.11}$$

A typical vector in X is a function f, such as $f(x) = \sin x - 2 \sin 2x + 9 \sin 3x$, which shows the dependence of f on all of the three scales of oscillations present in X. We define addition and scalar multiplication in the usual way we add functions and multiply a function by a scalar. These operations are commutative, associative, and distributive. The zero element of X is the identically zero function, which belongs to X with $a = b = c = 0$. The sum of two elements of X belong to it since if $f_1(x) = a_1 \sin x + b_1 \sin 2x + c_1 \sin 3x$ and

$f_2(x) = a_2 \sin x + b_2 \sin 2x + c_2 \sin 3x$, it follows that

$$(f_1 + f_2)(x) = f_1(x) + f_2(x)$$
$$= (a_1 + a_2) \sin x + (b_1 + b_2) \sin 2x + (c_1 + c_2) \sin 3x. \tag{6.12}$$

Hence, property 2 holds in this case. That property 3 holds follows in the same manner.

There is a special relationship between the vector spaces E^2 and E^3. In a sense any vector in E^2 also is a vector in E^3. To be precise, corresponding to any vector $\langle a, b \rangle$ in E^2 is the unique vector $\langle a, b, 0 \rangle$ in E^3. The set $Y = \{\langle a, b, 0 \rangle | a, b \in R\}$ forms a vector space under the same addition and scalar multiplication inherited from E^3. Note that

$$\langle a_1, b_1, 0 \rangle + \langle a_2, b_2, 0 \rangle = \langle a_1 + a_2, b_1 + b_2, 0 \rangle, \tag{6.13}$$

which is clearly in Y. Similarly, the other properties of the definition of a vector space hold for Y. The space Y is called a subspace of E^3.

Definition 6.3.3 (Subspace of a Vector Space)

*A set Y is called a **subspace** of a vector space X if $Y \subset X$ and if all of the properties of a vector space hold for Y.*

Example 6.3.7

Let $X = E^3$. Let $Y = \{\langle a, b, 0 \rangle | a, b \in R\}$. As shown previously this is a subspace of E^3. If we equip E^3 with coordinates x, y, and z, then Y can be identified with the xy-plane in R^2. Note that this plane passes through the origin. In fact, every plane that passes through the origin defines a subspace of E^3, which is the subject of our next example.

Example 6.3.8

Let $X = E^3$. Let

$$Y = \{\langle x, y, z \rangle | 2x - 3y + z = 0\}. \tag{6.14}$$

To see that Y is a vector space, we need to show that the three properties of a vector space hold for (6.14). First, the vector $\langle 0, 0, 0 \rangle$ satisfies $2x - 3y + z = 0$ and therefore the zero vector is in Y. Next, let vectors $\mathbf{a}$ and $\mathbf{b}$ belong to Y. We need to show that $\mathbf{a} + \mathbf{b}$ is a vector in Y. Since $\mathbf{a} \in Y$, it is defined by

$$\mathbf{a} = \langle a_1, a_2, a_3 \rangle \quad \text{such that} \quad 2a_1 - 3a_2 + a_3 = 0. \tag{6.15}$$

Similarly the vector $\mathbf{b}$ must have the form

$$\mathbf{b} = \langle b_1, b_2, b_3 \rangle \quad \text{such that} \quad 2b_1 - 3b_2 + b_3 = 0. \tag{6.16}$$

To show that $\mathbf{a} + \mathbf{b}$, whose components are given by

$$\mathbf{a} + \mathbf{b} = \langle a_1 + b_1, a_2 + b_2, a_3 + b_3 \rangle, \qquad (6.17)$$

belongs to Y, we need to show

$$2(a_1 + b_1) - 3(a_2 + b_2) + (a_3 + b_3) = 0. \qquad (6.18)$$

But

$$2(a_1 + b_1) - 3(a_2 + b_2) + (a_3 + b_3) = (2a_1 - 3a_2 + a_3) + (2b_1 - 3b_2 + b_3)$$
$$= 0 + 0 = 0, \qquad (6.19)$$

by (6.15) and (6.16). To show that the scalar multiplication property holds is very similar and is left to the reader.

Example 6.3.9

The set Y in the preceding example is, of course, a plane passing through the origin of E^3 with normal $\langle 2, -3, 1 \rangle$. There is nothing special about the numbers 2, -3, and 1. Any three real numbers replacing the latter would yield a plane subspace of E^3. In a similar fashion any straight line in R^3 passing through the origin gives rise to a subspace of E^3. For instance, let Y be defined by

$$Y = \{a\langle 1, -2, 3 \rangle | a \in R\}; \qquad (6.20)$$

that is, vectors in Y are constant multiples of the vector $\langle 1, -2, 3 \rangle$. Clearly the zero vector belongs to Y (let $a = 0$). We leave it to the reader to show that the other two properties of the definition of a vector space hold for all vectors of Y.

Example 6.3.10

The vector space $Y = \langle 0, 0, 0 \rangle$ is a subspace of E^3 and is called the *trivial* subspace. Here Y consists of a single vector and it is rather simple to go through the properties of the definition of a vector space and convince oneself that Y is indeed a vector space. In the other extreme, the vector space E^3 is a subspace of itself. Examples 6.3.6 to 6.3.9 complete the set of all subspaces of E^3. The three-dimensional Euclidean space has four different types of subspaces; E^3 itself, all planes passing through the origin, all lines passing through the origin, and the trivial subspace.

To get a sense for how a subspace such as that defined by (6.14) is generated by its elements, we will use *Mathematica* and construct a partial image of this subspace from its constituents, that is, the vectors that fill this region. Of course, we know that the image of this subspace is a plane with a particular normal, and *Mathematica* is capable of drawing a plane once one gives the software the normal vector of the plane. However, it is instructive to see how a subspace is determined by vectors that define it. The following command generates Figure 6.2.

```
Show[Graphics3D[Table[Line{{0, 0, 0}, {x, y, -2 x + 3 y}}],
    {x, -1, 1, 0.1}, {y, -1, 1, 0.1}]]]
```

Figure 6.2: The subspace $z = -2x + 3y$.

We end this section by emphasizing that Examples 6.3.5 and 6.3.6 give the term vector a slightly different texture than the concept we have carried with us from calculus. It would be fair to say that in most instances this term will still refer to an ordinary vector and the geometric interpretation with which we are familiar. In some occasions, however, which will be clear from the context, the term vector will refer to an object such as a trigonometric polynomial.

Problems

1. Show that the following sets form subspaces of E^2.

    (a) $\{\langle x, y \rangle \,|\, 2x - y = 0\}$

    (b) $\{\langle x, y \rangle \,|\, x + y = 0\}$

    (c) $\{\langle x, y \rangle \,|\, x = y\}$

    (d) $\{\langle x, y \rangle \,|\, 3x - 2y = 0\}$

2. Consider the set

$$A = \{\langle x, y \rangle \,|\, 2x - 3y = 1\}.$$

Does A form a subspace of E^2?

3. Show that the following sets form subspaces of E^3.

   (a) $\{\langle x, y, z\rangle | \, 2x - y + 2z = 0\}$

   (b) $\{\langle x, y, z\rangle | \, x + y = 0\}$

   (c) $\{\langle x, y, z\rangle | \, z - y = x\}$

   (d) $\{\langle x, y, z\rangle | \, ax + by = z\}$, where a and b are fixed constants

   (e) $\{\langle x, y, z\rangle | \, (\sin t)x + (\cos t)y = 0\}$, where t is a fixed constant

   (f) $\{\langle x, y, z\rangle | \, x = y = z\}$

   (g) $\{\langle x, y, z\rangle | \, x = 2y = z\}$.

4. Consider the set

   $$A = \{\langle x, y, z\rangle | \, (a^2 - 1)x + (a^2 + 1)y + z = a^2 - 3a + 2\}.$$

   For which values of a is A a subspace of E^3? Plot the graph of A.

5. Show that $A = \{a\langle 1, 1, 1, 1\rangle | a \in R\}$ forms a subspace of E^4.

6. Show that the following sets are subspaces of E^3. Describe the geometric object (such as line or plane) that each set represents. Use *Mathematica* and draw the graph of each subspace.

   (a) $S = \{\langle a, b, 0\rangle | a, b \in R\}$

   (b) $S = \{\langle a, a - 2b, b\rangle | a, b \in R\}$

7. Write down all subspaces of the vector space E^4. Give their description in terms of mathematical equations and identities.

8. Show that the intersection of any two subspaces of E^3 is a subspace of that vector space.

9. Show that the set

   $$X = \{a_1 \cos x + a_2 \cos 2x + \cdots + a_n \cos nx | \, a_i \in R\}$$

   forms a vector space with the usual addition and scalar multiplication of functions.

10. Show that the set of second-order polynomials

    $$Y = \{a + bt + ct^2 | \, a, b, c \in R\}$$

    with t an independent variable, forms a vector space.

11. Show that the set of trigonometric expressions defined by

    $$Z = \{a \sin x + b \sin 3x | \, a, b \in R\}$$

    forms a subspace of the vector space in Example 6.3.6.

12. Show that the set

$$X = \{ax^2 + bxy + cy^2 \,|\, a, b, c \in R\}$$

forms a vector space under the usual definition of addition and scalar multiplication.

13. Show that the set
$$Y = \{ax^2 + cy^2 \,|\, a, c \in R\}$$
forms a subspace of X defined in the previous problem.

14. Show that a set of all solutions of the second-order differential equation

$$x'' + 3x' + 2x = 0$$

forms a vector space under the usual definitions of function addition and scalar multiplication.

15. Show that the set
$$Y = \{ce^{-t} \,|\, c \in R\}$$
forms a subspace of the space X defined in the previous problem.

16. Consider the set of all solutions of the nonhomogeneous equation

$$x'' + 3x' + 2x = 1.$$

Does this set form a vector space under the usual definitions of additions and multiplication?

6.4 Linear Independence, Span, and Basis

We have seen in calculus and physics that the vectors $\mathbf{i} = \langle 1, 0, 0 \rangle$, $\mathbf{j} = \langle 0, 1, 0 \rangle$, and $\mathbf{k} = \langle 0, 0, 1 \rangle$ play a special role in E^3, the three-dimensional Euclidean vector space; any vector $\mathbf{a} = \langle a_1, a_2, a_3 \rangle$ can be written as a linear combination of these vectors, that is,

$$\mathbf{a} = a_1\mathbf{i} + a_2\mathbf{j} + a_3\mathbf{k}. \tag{6.21}$$

We called the scalars a_1, a_2, and a_3 the *coordinates* or the *components* of the vector $\mathbf{a}$.

The vectors $\{\mathbf{i}, \mathbf{j}, \mathbf{k}\}$ are by no means unique in having the property that all other vectors of E^3 can be written in terms of them. In fact, there are infinitely many such candidates and their proper definition is the subject of this section.

Definition 6.4.1 (Linear Independence)
Let X be a vector space. We say a set of vectors $\{\mathbf{a}_1, \mathbf{a}_2, \ldots, \mathbf{a}_n\}$ belonging to X is linearly independent if the only way the zero vector can be written as a linear combination of these vectors is with zero coefficients, that is,

$$\text{if} \quad \mathbf{0} = c_1\mathbf{a}_1 + c_2\mathbf{a}_2 + \cdots + c_n\mathbf{a}_n \quad \text{then} \quad c_1 = c_2 = \cdots = c_n = 0. \tag{6.22}$$

Remark 6.4.1: Note the similarity between this definition and that given for the linear independence of functions in Chapter 4.

Examples 6.4.1

Let $X = E^3$. Let $\mathbf{a}_1 = \langle 1, 3, 2 \rangle$ and $\mathbf{a}_2 = \langle -1, 2, 4 \rangle$. To check whether these vectors are linearly independent, we apply Definition 6.4.1. Let c_1 and c_2 be such that

$$\mathbf{0} = c_1 \mathbf{a}_1 + c_2 \mathbf{a}_2 = \langle c_1 - c_2, 3c_1 + 2c_2, 2c_1 + 4c_2 \rangle. \qquad (6.23)$$

Equation (6.23) leads to three equations in the two unknowns c_1 and c_2:

$$\begin{cases} c_1 - c_2 = 0, \\ 3c_1 + 2c_2 = 0, \\ 2c_1 + 4c_2 = 0. \end{cases} \qquad (6.24)$$

Vectors $\mathbf{a}_1$ and $\mathbf{a}_2$ are linearly independent if we could show that $c_1 = c_2 = 0$ is the *only* solution to (6.24). We eliminate c_2 between (6.24a) and (6.24b) and obtain the equation $5c_1 = 0$, which implies $c_1 = 0$. That $c_2 = 0$ follows from (6.24a). Therefore, $c_1 = c_2 = 0$ is the only solution to (6.24) and $\mathbf{a}_1$ and $\mathbf{a}_2$ are linearly independent.

Example 6.4.2 (*Mathematica*)

These computations could have been carried out in a symbolic manipulator. The `Solve` command of *Mathematica*, in particular, is designed to find the solution to a collection of equations that include systems of the type (6.24). Here is one way of solving the algebraic system (6.24) is:

```
eqns = {c1 - c2 == 0,3 c1 + 2 c2 == 0, 2 c1 + 4 c2 == 0};
Solve[eqns, {c1, c2}]
```

Mathematica returns

```
{{c1 -> 0, c2 -> 0}}
```

A slightly different way of entering this information to *Mathematica* is to break the preceding input statements into several smaller statements in the following manner:

```
vars = {c1, c2};
lhs = {c1 - c2, 3 c1 + 2 c2, 2 c1 + 4 c2};
Solve[lhs == 0, vars]
```

Yet another way of getting the same output is

```
a1 = {1, 3, 2}; a2 = {-1, 2, 4};
Solve[c1 a1 + c2 a2 == 0, {c1, c2}]
```

Example 6.4.3 (*MATLAB*)

The following is the syntax in *MATLAB* for carrying out the computations in Example 6.4.1:

```
syms c1 c2
[c1 c2]=solve(c1 - c2, 3*c1 + 2*c2, 2*c1 + 4*c2)
```

or

```
[c1 c2]=solve('c1 - c2 = 0', '3*c1 + 2*c2 = 0', '2*c1 + 4*c2 = 0')
```

Its output concurs with the outcome in Examples 6.4.1 and 6.4.2.

Example 6.4.4

Consider the vectors $\mathbf{a}_1 = \langle 1, -1, 1 \rangle$ and $\mathbf{a}_2 = \langle 2, -2, 2 \rangle$ in E^3. In this case, just by inspection, we observe the following relation between $\mathbf{a}_1$ and $\mathbf{a}_2$:

$$\mathbf{0} = 2\mathbf{a}_1 - \mathbf{a}_2, \tag{6.25}$$

which states that (6.22) does not hold; therefore, $\mathbf{a}_1$ and $\mathbf{a}_2$ are linearly dependent (which is clear since $\mathbf{a}_2$ is twice the vector $\mathbf{a}_1$).

It is interesting to look at the way *Mathematica* arrives at the same conclusion. Following the definition of linear independence, we need to find all solutions of the equation $c_1\mathbf{a}_1 + c_2\mathbf{a}_2 = \mathbf{0}$. This relation is given to *Mathematica* via the command

```
Solve[{c1 + 2 c2 == 0, - c1 - 2 c2 == 0, c1 + 2 c2 == 0}, {c1, c2}]
```

Mathematica responds with

```
{{c1 = -2 c2}}
```

which is its way of saying that c_1 and c_2 are related to each other in a nontrivial fashion. In particular, they do not have to be zero necessarily (for instance, we can choose $c_2 = 1$ and $c_1 = -2$). We get a similar result from *MATLAB*. The input

```
syms c1 c2
[c1 c2]=solve(c1 + 2*c2,-c1 - 2*c2,c1 + 2*c2)
```

results in the output

```
c1 = - 2*c2,  c2 = c2
```

which points out that $\mathbf{a}_1$ and $\mathbf{a}_2$ are linearly dependent.

Example 6.4.5

Consider the three vectors $\mathbf{a}_1 = \langle 1, 1, 0 \rangle$, $\mathbf{a}_2 = \langle 1, 0, 1 \rangle$, and $\mathbf{a}_3 = \langle 0, 1, 1 \rangle$. To check whether these vectors are linearly independent, we follow Definition 6.4.1. Let c_1, c_2, and c_3 be such that

$$\mathbf{0} = c_1\mathbf{a}_1 + c_2\mathbf{a}_2 + c_3\mathbf{a}_3, \tag{6.26}$$

or, equivalently, c_1, c_2, and c_3 satisfy the system of algebraic equations

$$\begin{cases} c_1 + c_2 = 0, \\ c_1 + c_3 = 0, \\ c_2 + c_3 = 0. \end{cases} \tag{6.27}$$

Subtract (6.27a) from (6.27b) to get $-c_2 + c_3 = 0$. Combine this equation with (6.27c) and obtain the following system of two equations in the unknowns c_2 and c_3:

$$\begin{cases} -c_2 + c_3 = 0, \\ c_2 + c_3 = 0. \end{cases} \tag{6.28}$$

This system has the unique solution $c_2 = c_3 = 0$. It then follows from (6.27a) that $c_1 = 0$. Therefore, all coefficients c_i's in (6.26) turn out to be zero, which implies that $\mathbf{a}_1$, $\mathbf{a}_2$, and $\mathbf{a}_3$ are linearly independent. We leave it to the reader to examine this example in either *Mathematica* or *MATLAB*.

Example 6.4.6

Let X be the vector space in Example 6.3.6. We recall that vectors in this vector space are functions of x defined on the real line and written as a linear combination of $\sin x$, $\sin 2x$, and $\sin 3x$. Let $\{\mathbf{a}_1, \mathbf{a}_2, \mathbf{a}_3\}$ be the vectors

$$\mathbf{a}_1 = 2\sin x + 3\sin 2x - \sin 3x, \qquad \mathbf{a}_2 = -3\sin x + \sin 2x + 2\sin 3x,$$
$$\mathbf{a}_3 = \sin x + 7\sin 2x. \tag{6.29}$$

To check for the linear independence of these vectors, we form the expression in (6.22) and collect the coefficients of $\sin x$, $\sin 2x$, and $\sin 3x$:

$$(2c_1 - 3c_2 + c_3)\sin x + (3c_1 + c_2 + 7c_3)\sin 2x + (-c_1 + 2c_2)\sin 3x = 0, \quad (6.30)$$

where the equality must hold for *every* value of x on the real line. The only way that the preceding identity can hold is if the respective coefficient of $\sin ix$ vanish. Here we use the concept of linear independence of $\sin ix$ as a set of functions, which we encountered in Chapter 4 (in one of the problems that follows the reader is asked to use (6.30) directly to reach the same conclusion). We get the system

$$\begin{cases} 2c_1 - 3c_2 + c_3 = 0, \\ 3c_1 + c_2 + 7c_3 = 0, \\ -c_1 + 2c_2 = 0. \end{cases} \tag{6.31}$$

We have reached the same stage as in all of the prior examples, namely, the problem of determining the linear independence of the $\mathbf{a}_i$s is reduced to solving a system of linear algebraic equations. We begin with the first equation in (6.30) and eliminate c_1 from the other two equations:

$$\begin{cases} 2c_1 - 3c_2 + c_3 = 0, \\ \frac{11}{2}c_2 + \frac{11}{2}c_3 = 0, \\ \frac{1}{2}c_2 + \frac{1}{2}c_3 = 0. \end{cases} \tag{6.32}$$

The last two equations in (6.32) are the same. We eliminate the third from the set, let c_3 be arbitrary, and solve for c_1 and c_2 from the first two equations. We get

$$c_1 = -2c_3 \quad \text{and} \quad c_2 = -c_3. \tag{6.33}$$

Since we can choose c_3 to be *any* nonzero number, we conclude that the three vectors a_is are linearly dependent. The reader can verify that, in fact, $\mathbf{a}_3 = 2\mathbf{a}_1 + \mathbf{a}_2$. Again, this conclusion may be verified in either *Mathematica* or *MATLAB*.

The following program in *Mathematica* shows how one could start with the original vectors a_is in the preceding example, use the dot product and the Coefficient commands of *Mathematica*, construct the system in (6.31), and solve it using the Solve command:

```
terms = {Sin[x], Sin[2 x], Sin[3 x]};
vectors = {2 Sin[x] + 3 Sin[2 x] - Sin[3 x],
           -3 Sin[x] + Sin[2 x] + 2 Sin[3 x],
           Sin[x] + 7 Sin[2 x]};
unknowns = {c1, c2, c3};
lhs = unknowns . vectors;
eqns = Table[Coefficient[lhs, terms[[i]], 1], {i, 3}];
Solve[eqns == 0, unknowns]
```

The output of this program is

```
{{c1 ->-2 c3,c2 -> -c3}}
```

which is equivalent to (6.33).

Remark 6.4.3: The usage of the Coefficient function in the preceding program should be noted. This command is very handy when we need to extract the coefficients of terms such as $\sin ix$ in a symbolic expression. This tool will become particularly important when we encounter Fourier series in a later chapter.

Definition 6.4.2 (Span)
Let $\{a_1, a_2, \ldots, a_n\}$ be a set of vectors that belong to a vector space X. The span of the set of vectors $\{a_1, a_2, \cdots, a_n\}$ is the vector space

$$\text{span}\{\mathbf{a}_1, \mathbf{a}_2, \mathbf{a}_n\} \equiv \{c_1\mathbf{a}_1 + c_2\mathbf{a}_2 + \cdots + c_n\mathbf{a}_n | c_1, c_2, \ldots, c_n \in R\}. \quad (6.34)$$

The definition states that the span of a set of vectors is the set of *all* vectors that can be written as a linear combination of these vectors. In this definition we asserted that the span of a set of vectors is a vector space, although this statement should be proved. Its proof is straightforward and we leave it to the reader to provide it (see also Problem 12).

Example 6.4.7
The span of $\langle 1,0 \rangle$ and $\langle 0,1 \rangle$ is the entire vector space E^2 since these vectors are linearly independent, and any vector in E^2 can be written as $c_1\langle 1,0 \rangle + c_2\langle 0,1 \rangle$. Geometrically, the span of these vectors is the plane that passes through them.

Example 6.4.8

Let us determine the span of $\mathbf{a}_1 = \langle 1, 0, 0 \rangle$, $\mathbf{a}_2 = \langle 0, 1, 0 \rangle$, and $\mathbf{a}_3 = \langle 1, 1, 0 \rangle$. First note that $\mathbf{a}_3 = \mathbf{a}_1 + \mathbf{a}_2$; consequently, these vectors are linearly dependent; and that $\mathbf{a}_3$ is already in the span of $\mathbf{a}_1$ and $\mathbf{a}_2$. We can therefore discard $\mathbf{a}_3$ from the list of vectors whose span we intend to determine. The two remaining vectors, $\mathbf{a}_1$ and $\mathbf{a}_2$, are linearly independent; and their span consists of all vectors of the form $\langle c_1, c_2, 0 \rangle$. Geometrically, this vector space is the xy-plane in R^3.

Definition 6.4.3 (Basis)

A set of vectors $\{\mathbf{a}_1, \mathbf{a}_2, \ldots, \mathbf{a}_n\}$ forms a basis for a vector space X if

1. *these vectors are linearly independent;*

2. *they span X.*

Definition 6.4.4 (Dimension)

The dimension of a vector space is the number of vectors in any basis of that vector space.

In order for the latter definition to make sense we need to prove that all bases of a vector space have the same number of vectors in them. This statement is true, and its proof is left as an exercise.

It follows from the preceding definitions that E^3 is a three-dimensional vector space because every basis of this vector space has three vectors in it. We will see in one of the exercises that any four vectors in E^3 are necessarily linearly dependent.

Throughout this text the term **natural basis** will refer to the set of unit vectors (vectors with magnitude one) whose directions are in the same directions as the coordinate axes of a set of coordinates we have chosen to represent the geometry of the physical model at hand. For instance, the natural basis for E^3 in rectangular coordinates consists of the vectors $\mathbf{i}$, $\mathbf{j}$, and $\mathbf{k}$.

Example 6.4.9

Let $S = \{\langle a, b, a + b \rangle\} | a, b \in R\}$. It will be left to the reader to show that S is a vector space. We note that the definition of S contains two parameters (or, two "degrees of freedom"), namely, a and b. Hence, to find a basis for this vector space, we let the values of a and b alternately toggle between 0 and 1. Let $a = 1$ and $b = 0$ and obtain the vector $\mathbf{a}_1 = \langle 1, 0, 1 \rangle$. Next, let $a = 0$ and $b = 1$ and get $\mathbf{a}_2 = \langle 0, 1, 1 \rangle$. The set of vectors $\{\mathbf{a}_1, \mathbf{a}_2\}$ forms a basis for S. To prove this, we need to show that $\mathbf{a}_1$ and $\mathbf{a}_2$ are linearly independent and that they span S. First, we apply the definition of linear independence to $\mathbf{a}_1$ and $\mathbf{a}_2$. Let c_1 and c_2 be such that the linear combination

$$c_1 \mathbf{a}_1 + c_2 \mathbf{a}_2 = c_1 \langle 1, 0, 1 \rangle + c_2 \langle 0, 1, 1 \rangle \tag{6.35}$$

equals the zero vector. We need to show that $c_1 = c_2 = 0$, which we leave to the reader.

To see that $\mathbf{a}_1$ and $\mathbf{a}_2$ span S, we need to show that any vector in S can be written as a linear combination of $\mathbf{a}_1$ and $\mathbf{a}_2$. A typical vector in S has the form $\langle a, b, a + b \rangle$. Note that

$$\langle a, b, a + b \rangle = a\langle 1, 0, 1 \rangle + b\langle 0, 1, 1 \rangle, \tag{6.36}$$

which proves our claim.

In summary, $\mathbf{a}_1$ and $\mathbf{a}_2$ are linearly independent and they span S, so the set $\{\mathbf{a}_1, \mathbf{a}_2\}$ forms a basis for S. Since there are two vectors in this basis, the dimension of S is two. Since each vector of S lives in the three-dimensional vector space E^3, S is a two-dimensional subspace of E^3. It has the geometrical interpretation of a plane. By looking at the relationship among the x, y, and z coordinates we see that the equation of this plane is

$$z = x + y. \tag{6.37}$$

Example 6.4.10

Consider the set $Y = \{a \sin x + b \sin 3x \,|\, a, b \in R\}$. We leave it to the reader to show that Y defines a vector space under the usual addition and scalar multiplication. We can use the same strategy we used in the previous example and construct a basis for Y. Let $a = 0$ and $b = 1$ to get the element $\mathbf{a}_1 = \sin x$, while $a = 0$ and $b = 0$ gives us the element $\mathbf{a}_2 = \sin 3x$. These two elements are linearly independent and span Y, as you can readily show. Hence, the dimension of Y is two.

The reader may have already noted that the concept of the component of a vector is basis dependent. Until now when we wrote $\langle a, b \rangle$ for a typical vector in E^2 we meant a vector whose component in the direction of $\mathbf{i}$ is a and in the direction of $\mathbf{j}$ is b. In other words, we were tacitly assuming that the basis in which the vectors were being represented was the natural basis. Now that we have discovered that a vector space has many different bases we need to be a bit careful and describe what exactly we mean by the notation $\langle a, b \rangle$. Traditionally the latter notation is used to denote a vector and its components in any specified basis; and when bases are changed in mid-application, still the same set of delimiters $\langle , \rangle$ is often used to denote vectors in the new basis. The practice we choose in the entire text is that, if the basis used is different from the natural basis, we will use two sets of brackets $\langle\langle , \rangle\rangle$ to denote the components of the vector. Perhaps the best way to avoid any form of confusion is to represent a vector as a linear combination of the basis in terms of which the vector is expanded. The following example addresses these comments.

Example 6.4.11

Let $X = E^2$. Let $\mathbf{a} = \langle 1, 1 \rangle$ in the natural basis, that is,

$$\mathbf{a} = \mathbf{i} + \mathbf{j}. \tag{6.38}$$

Now consider a different basis, namely, the vectors $\mathbf{e}_1 = \langle 1, 2 \rangle$, $\mathbf{e}_2 = \langle -2, 1 \rangle$ (we leave it to the reader to show that these vectors form a basis for E^2). To represent $\mathbf{a}$ in terms of $\{\mathbf{e}_1, \mathbf{e}_2\}$, we need to find constants α and β so that

$$\mathbf{a} = \alpha\mathbf{e}_1 + \beta\mathbf{e}_2, \tag{6.39}$$

which is equivalent to

$$\langle 1, 1 \rangle = \alpha\langle 1, 2 \rangle + \beta\langle -2, 1 \rangle. \tag{6.40}$$

We obtain the following two equations in α and β:

$$\begin{cases} \alpha - 2\beta = 1, \\ 2\alpha + \beta = 1, \end{cases} \tag{6.41}$$

whose solution pair is $(\alpha, \beta) = (\frac{3}{5}, -\frac{1}{5})$. Thus, we have two different ways to represent the vector $\mathbf{a}$: one given by (6.38) and the other by

$$\mathbf{a} = \left\langle \left\langle \frac{3}{5}, -\frac{1}{5} \right\rangle \right\rangle = \frac{3}{5}\mathbf{e}_1 - \frac{1}{5}\mathbf{e}_2. \tag{6.42}$$

Problems

1. Check whether the following set of vectors is linearly independent. Verify your answers in *Mathematica* or *MATLAB*.

   (a) $\langle 1, -1 \rangle$, $\langle 1, 1 \rangle$
   (b) $\langle 0, -1 \rangle$, $\langle -1, 0 \rangle$
   (c) $\langle 1, -1 \rangle$, $\langle -2, 2 \rangle$
   (d) $\langle 1, -1, 1 \rangle$, $\langle 1, 1, 0 \rangle$, $\langle 1, 1, 2 \rangle$
   (e) $\langle 1, -1, 1 \rangle$, $\langle 1, 1, 0 \rangle$, $\langle 1, 1, 2 \rangle$
   (f) $\langle a, 0, a \rangle$, $\langle a, a, 0 \rangle$, $\langle 1, 1, 1 \rangle$
   (g) $\langle b, 0, 1 \rangle$, $\langle 0, 1, a \rangle$, $\langle a, b, a \rangle$

2. Check whether the set of vectors $\mathbf{a}_1 = \langle 1, 0, 0 \rangle$, $\mathbf{a}_2 = \langle 1, 1, 0 \rangle$, $\mathbf{a}_3 = \langle 1, 1, 1 \rangle$ is linearly independent in E^3. What is the span of these vectors?

3. Check whether the set of vectors $\mathbf{a}_1 = \langle 1, 1, 1 \rangle$, $\mathbf{a}_2 = \langle 0, -1, 3 \rangle$, $\mathbf{a}_3 = \langle 2, 1, 5 \rangle$ is linearly independent in E^3. What is the span of these vectors? Use *Mathematica* or *MATLAB* and verify your claim.

4. Consider the set of vectors $\langle 1, 0, 0, 1 \rangle$, $\langle 1, 0, 1, 0 \rangle$, $\langle 1, 1, 0, 0 \rangle$, $\langle 0, 1, 1, 0 \rangle$. Are they linearly independent in E^4? What is their span?

5. Consider the set of vectors $\langle a, 1 \rangle$, $\langle 1, a \rangle$.

   (a) Find the set of all values of a for which the two vectors are linearly dependent in E^2.

   (b) Use the `Solve` (as in Example 6.4.2) command of *Mathematica* and redo this problem. Does the answer make sense? Use the `Reduce` and `Eliminate` commands for the same problem. Do you get a different amount of information from these commands?

(c) Use the `solve` command of *MATLAB* (as in Example 6.4.3) and find the range of a for which the two vectors are linearly independent.

6. Consider the set of vectors $\langle 1, a, 1 \rangle, \langle b, 0, b \rangle, \langle a, b - 1, a \rangle$. Find the set of all values a and b for which the set of vectors is linearly dependent in E^3. How can you solve this problem in *Mathematica* or *MATLAB*?

7. Show that the set of *any* three vectors in E^2 must be linearly dependent.

8. Show that the set of *any* four vectors in E^3 must be linearly dependent.

9. Let $S = \{\mathbf{a}_1, \mathbf{a}_2, \dots, \mathbf{a}_m\}$ be a set of vectors in E^n. Suppose that $m > n$. Prove that the vectors in S are linearly dependent. (Hint: Let $T = \{\mathbf{e}_1, \mathbf{e}_2, \dots, \mathbf{e}_n\}$ form the natural basis for E^n. Write each vector $\mathbf{a}_i$ from S in terms of T and use the linear independence of T.)

10. Use the result of the previous problem and prove that the dimension of any vector space is unique.

11. Let $S = \{\langle a, a - 2b, b \rangle | a, b \in R\}$. Show that S is a vector space. Find a basis for this vector space. What is its dimension? Draw the graph of this vector space in R^3.

12. Let $\{\mathbf{a}_1, \mathbf{a}_2, \dots, \mathbf{a}_n\}$ be a set of vectors from a vector space X. Show that the span of these vectors is a subspace of X.

13. Complete Example 6.4.10.

14. Let $X = \{a_1 \cos x + a_2 \cos 3x + a_3 \cos 5x | a_i \in R\}$.

    (a) Show that X is a vector space.

    (b) Find a basis for X. What is the dimension of X?

15. Let $X = \{\sum_{i=0}^{n} a_i t^i | a_i \in R\}$.

    (a) Show that X is a vector space.

    (b) Find a basis for X. What is its dimension?

16. Show that if $c_1 \sin x + c_2 \sin 2x + c_3 \sin 3x \equiv 0$, then $c_1 = c_2 = c_3 = 0$. (Hint: Evaluate this expression at $x = \frac{\pi}{4}, \frac{\pi}{2}$, and $\frac{3\pi}{4}$ and find the solution to the simultaneous equations you obtain.)

17. Show that the set of vectors $\{\langle 1, -1 \rangle, \langle 3, 3 \rangle\}$ forms a basis for E^2. Find the components of the vector $\langle 2, -1 \rangle$ in this basis.

18. Show that the set of vectors $\{\langle 1, -1, 0 \rangle, \langle 1, 3, 3 \rangle, \langle 0, 0, 1 \rangle\}$ forms a basis for E^3. Find the components of the vector $\langle 2, -1, 3 \rangle$ in this basis.

19. Show that the set of vectors $\{\langle 0, 0, 1 \rangle, \langle 0, 1, 1 \rangle, \langle 1, 1, 1 \rangle\}$ forms a basis for E^3. Find the components of the vector $\langle -1, 2, 1 \rangle$ in this basis.

6.5 Linear Transformations

In this section we will be concerned with properties of linear functions which have vector spaces for their domain and range. Such functions are called linear transformations. These transformations appear naturally in many disciplines in science. For us such a map is the idealization of a deformation of a material body, be it a fluid or a solid, under the action of internal and external forces. Understanding what a transformation does to its domain is crucial in several areas of applied science, most notably in solid mechanics, where locating the directions of extreme strain in a deformation plays an important role in the design of structures.

Definition 6.5.1 (Linear Transformations)
Let X and Y be two vector spaces. A linear transformation T with domain X and range in Y is a map that satisfies:

1. *$T(\mathbf{x} + \mathbf{y}) = T(\mathbf{x}) + T(\mathbf{y})$ for every $\mathbf{x}, \mathbf{y} \in X$;*

2. *$T(\alpha\mathbf{x}) = \alpha T(\mathbf{x})$ for every $\alpha \in R$ and $\mathbf{x} \in X$.*

Linear transformations reduce to the class of linear functions $f(x) = ax$ when the underlying vector spaces X and Y are restricted to be one-dimensional. When this restriction is removed, the structure of T becomes considerably more complicated and richer. Whereas in the one-dimensional case graphing f is a first step toward understanding the behavior of such a function, we do not expect to draw the graphs of linear transformations in higher dimensions because the sum of the dimensions of the domain and range of such transformations is typically larger than three. In what follows we give a few examples of linear transformations and, in lieu of graphing the entire deformation, discuss ways of graphing subsets of the domain and their images. We will also discuss in the problems that follow how one can use *Mathematica* or *MATLAB* and determine the directions in the domain which experience extreme amounts of stretching or compression.

Example 6.5.1
Let $X = Y = E^2$. We choose the natural basis (i.e., the set $\{\mathbf{i}, \mathbf{j}\}$) for both the domain and the range. Let T be defined by

$$T(\langle a, b \rangle) = \langle -b, a \rangle. \tag{6.43}$$

This transformation rotates every vector $\langle a, b \rangle \in E^2$ through an angle of $\frac{\pi}{2}$. To see that T is linear, we need to verify that the two properties of Definition 6.5.1 hold. Let $\mathbf{x} = \langle a_1, b_1 \rangle$ and $\mathbf{y} = \langle a_2, b_2 \rangle$. Then $T(\mathbf{x}) = \langle -b_1, a_1 \rangle$ and $T(\mathbf{y}) = \langle -b_2, a_2 \rangle$, so

$$T(\mathbf{x}) + T(\mathbf{y}) = \langle -b_1 - b_2, a_1 + a_2 \rangle. \tag{6.44}$$

But $T(\mathbf{x} + \mathbf{y}) = T(\langle a_1 + a_2, b_1 + b_2 \rangle) = \langle -b_1 - b_2, a_1 + a_2 \rangle$, which is the same as (6.44). Therefore, the first part of Definition 6.5.1 is satisfied. The second part of the definition follows in the same fashion.

Example 6.5.2

Let X and Y be as in the previous example. Let $T(\langle a, b \rangle) = \langle a^2, b \rangle$. Is T a linear transformation? The answer is no since Definition 6.5.1 is not satisfied in this case: For instance, $T(\alpha \langle a, b \rangle) = \langle \alpha^2 a, \alpha b \rangle$, which is not equal to $\alpha T(\langle a, b \rangle) = \langle \alpha a, \alpha b \rangle$.

Geometrically, a linear transformation maps a vector $\mathbf{a}$ to its image $T(\mathbf{a})$. In doing so T may stretch or compress the vector $\mathbf{a}$ or may rotate it. In a sense the linear transformation T is deforming the domain and creating the range. A good way to make sense of how T deforms a region is to draw the image of regions in the domain for which we have a good geometric understanding and see to which geometric regions in the range they get mapped. For instance, for two-dimensional problems, with $X = Y = E^2$, we often consider the image of the set of all vectors whose endpoints form a circle of radius one, say, and study the curve to which this circle is mapped. Recall that a set of all vectors $\mathbf{a} \in E^2$ of length one can be parametrized as

$$\mathbf{a}(\theta) = \cos \theta \mathbf{i} + \sin \theta \mathbf{j}, \quad \theta \in [0, 2\pi). \tag{6.45}$$

Therefore, the curve C defined by

$$C = \{\langle \cos \theta, \sin \theta \rangle | \theta \in [0, 2\pi)\} \tag{6.46}$$

is a circle of radius one. Its image under T is denoted by $T(C)$ and is defined by

$$T(C) = \{T(\mathbf{a}(\theta)) | \theta \in [0, 2\pi)\}, \tag{6.47}$$

where by $T(\mathbf{a}(\theta))$ we mean the endpoint of the vector $T(\mathbf{a}(\theta))$. The set $T(C)$ defines a curve in the range of T and its shape reveals quite a bit about the structure of the linear transformation. As we will see in the examples that follow, a linear transformation T maps a circle of radius one onto an ellipse whose principal axes are rotated with respect to the xy-axes of the domain. The shape of the ellipse gives an indication as to the extent the domain is being stretched or compressed in a specific direction. For linear transformations whose domains are E^3 we replace the circle by a sphere of radius one and consider the image of this sphere under T.

Example 6.5.3

Consider the linear transformation of Example 6.5.1, that is, $T(\langle a, b \rangle) = \langle -b, a \rangle$. This transformation takes the circle of radius one, C, and maps it to

$$T(C) = \{\langle -\sin \theta, \cos \theta \rangle | \theta \in [0, 2\pi)\}. \tag{6.48}$$

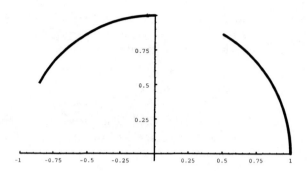

Figure 6.3: The curve C and its image $T(C)$ in *Mathematica*, when $T(\langle a, b \rangle) = \langle -b, a \rangle$.

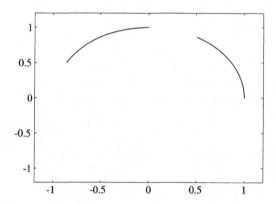

Figure 6.4: The curve C and its image $T(C)$ in *MATLAB*, when $T(\langle a, b \rangle) = \langle -b, a \rangle$.

Note that since $(-\sin \theta)^2 + (\cos \theta)^2 = 1$, the image $T(C)$ is also a circle of radius one; that is, T is a rigid rotation. It is instructive, however, to consider the image of a part of the circle C under this transformation. Let C_1 be the following segment of the circle:

$$C_1 = \left\{ \langle \cos \theta, \sin \theta \rangle | \, \theta \in \left[0, \frac{\pi}{3} \right) \right\}. \tag{6.49}$$

Then $T(C_1)$ is

$$T(C_1) = \left\{ \langle -\sin \theta, \cos \theta \rangle | \, \theta \in \left[0, \frac{\pi}{3} \right) \right\}. \tag{6.50}$$

The curve C_1 is a segment of the circle of arclength $\frac{\pi}{3}$, while $T(C_1)$ is that arc rotated clockwise through an angle of $\frac{\pi}{2}$. In the following problems we will use *Mathematica* and *MATLAB* and draw the graphs of C and $T(C)$ (see Figures 6.3 and 6.4).

Example 6.5.4

Let $T(\langle a, b \rangle) = \langle a - b, a + 2b \rangle$. Let C be as before. Then $T(C)$ is

$$T(C) = \{\langle \cos \theta - \sin \theta, \cos \theta + 2 \sin \theta \rangle | \theta \in [0, 2\pi)\}. \qquad (6.51)$$

Let $\langle x, y \rangle \in T(C)$ be a typical vector in the range of T. It follows from (6.51) that these vectors are parametrized as

$$\begin{cases} x = \cos \theta - \sin \theta, \\ y = \cos \theta + 2 \sin \theta. \end{cases} \qquad (6.52)$$

We find an equation for $T(C)$ by eliminating θ from the parametrization of $T(C)$. After solving (6.52) for $\sin \theta$ and $\cos \theta$, we find that

$$\sin \theta = -\frac{1}{3}(x - y) \quad \text{and} \quad \cos \theta = \frac{1}{3}(2x + y). \qquad (6.53)$$

Since $\sin^2 \theta + \cos^2 \theta = 1$, we are able to eliminate θ from the two previous equations and arrive at the following equation for x and y:

$$\frac{1}{9}(x - y)^2 + \frac{1}{9}(2x + y)^2 = 1 \qquad (6.54)$$

or, after some simplifying,

$$17x^2 + 6xy + 2y^2 = 9, \qquad (6.55)$$

which is the equation of an ellipse whose axes are rotated with respect to the x- and y-axes. (See Figure 6.5.)

Example 6.5.5 (*Mathematica*)

We now develop a *Mathematica* program to draw the graphs of C and $T(C)$. Let T be the linear transformation of Example 6.5.4. One available option is to use ListPlot to draw these graphs.

```
a = Table[{Cos[t], Sin[t]}, {t, 0, 2 Pi, 0.01}];
b = Table[{Cos[t]-Sin[t], Cos[t] + 2*Sin[t]}, {t, 0, 2*Pi, 0.01}];
c = ListPlot[a, PlotJoined -> True, AspectRatio -> 1];
d = ListPlot[b, PlotJoined -> True, AspectRatio -> 1];
Show[c, d]
```

Alternatively, the contents of b could be determined as follows:

```
T[{x_, y_}] = {x-y, x+2*y};
b=Table[T[a[[i]]], {i, Length[a]}];
```

We could also use ParametricPlot as follows to draw the preceding graphs.

```
a = ParametricPlot[{Cos[t], Sin[t]}, {t, 0, 2 Pi}];
b = ParametricPlot[{Cos[t] - Sin[t], Cos[t] + 2*Sin[t]},
            {t, 0, 2 Pi}];
Show[a, b]
```

See Figure 6.5 for the output.

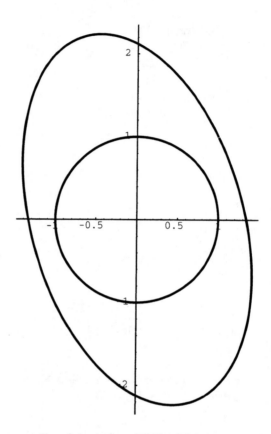

Figure 6.5: The curve C and its image $T(C)$ in *Mathematica*, when $T(\langle a, b \rangle) = \langle a - b, a + 2b \rangle$.

Example 6.5.6 (*MATLAB*)
Example 6.5.5 takes the following form in *MATLAB*:

```
t = 0:0.01:2*pi;
plot(cos(t)-sin(t), cos(t) + 2*sin(t));
hold on
plot(cos(t), sin(t));
hold off
```

Figure 6.6 shows the output of this program.

Problems

1. Generate the graphs in Examples 6.5.5 and 6.5.6.

2. Show that the following transformations are linear. Identify the domain and range of each T.

   (a) $T(\langle a, b \rangle) = \langle a - b, a + b \rangle$
   (b) $T(\langle a, b, c \rangle) = \langle a - b + 2c, a + c, b \rangle$

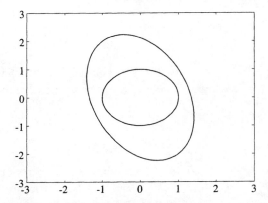

Figure 6.6: The curve C and its image $T(C)$ in *MATLAB*, when $T(\langle a, b \rangle) = \langle a - b, a + 2b \rangle$.

(c) $T(\langle a, b, c \rangle) = \langle a - b, b - c, c - a \rangle$

(d) $T(\langle a, b, c \rangle) = \langle a + b + c, a - b \rangle$

3. Show that the transformation T defined by

$$T(\langle x, y \rangle) = \langle x + y, x - y \rangle \tag{6.56}$$

is linear. Let C be a circle of radius one centered at the origin. Draw the graphs of C and $T(C)$ for this linear transformation.

4. Consider the transformation

$$T(\langle x, y \rangle) = \langle ax + by, cx + dy \rangle,$$

where a, b, c, and d are scalars. Show that T maps straight lines to straight lines.

5. Let C be a square with vertices at $(-1, -1)$, $(1, -1)$, $(1, 1)$, and $(-1, 1)$. Draw the image of C under T defined in the previous problem.

6. Use *Mathematica* or *MATLAB* to draw the graphs of the curves described in the preceding two problems.

7. Consider the transformation

$$T(\langle x, y \rangle) = \langle 2x - 5y, -5x + 3y \rangle. \tag{6.57}$$

(a) Show that T is a linear transformation.

(b) Let C be a circle of radius one centered at the origin. Use *Mathematica* or *MATLAB* and plot C and $T(C)$.

(c) Let $\|\mathbf{a}\|$ denote the magnitude of the vector $\mathbf{a} = \langle x, y \rangle$, that is,

$$\|\mathbf{a}\| = \sqrt{x^2 + y^2}. \tag{6.58}$$

Parametrize the circle C in the usual way:

$$C = \{\langle \cos t, \sin t \rangle \mid t \in [0, 2\pi)\}. \tag{6.59}$$

Define the function $f(t)$ to be the square of the magnitude of the image of a vector in the domain given by the parameter t, that is,

$$f(t) \equiv \|T(\langle \cos t, \sin t \rangle)\|^2. \tag{6.60}$$

Find a formula for f.

(d) In order to find which vectors in the domain are stretched and com- pressed the most by T you need to compute the derivative of f with respect to t and find its roots. Compute the derivative of f by hand and also by using *Mathematica*. Find the roots of the derivative using `FindRoot`.

(e) You will obtain two roots of f'. Denote them by t_1 and t_2. Find the vectors in the domain that correspond to these roots and plot them. Also plot their images under T. Combine the graphs of the images on the same screen. Which one of the two parameter values t_1 or t_2 corresponds to the direction of most stretching in $T(C)$?

6.6 Matrices

Matrices are representations of linear transformations. They will carry the min- imal amount of information necessary to distinguish one linear transformation from another.

Definition 6.6.1
A **matrix** *is an array of numbers. It has m rows and n columns. The dimension (or order) of a matrix refers to the number of rows and columns.*

Example 6.6.1
The array

$$\begin{bmatrix} 0 & -1 \\ 1 & 0 \end{bmatrix} \tag{6.61}$$

is an example of a matrix that has 2 rows and 2 columns.

The entries of a matrix are denoted by a_{ij} where i corresponds to the row and j to the column. For instance, the entry 0 in the first row and column of the preceding example is the a_{11} entry of that matrix. Similarly, -1 is the a_{12} entry. The preceding matrix is a 2×2 matrix.

We can multiply matrices by scalars and add matrices in the following nat- ural way. If A is a matrix, of any order, then αA is a matrix of the same order, every entry of which is α times the corresponding entry of A. So,

$$2 \begin{bmatrix} 1 & -1 \\ -1 & 3 \end{bmatrix} = \begin{bmatrix} 2 & -2 \\ -2 & 6 \end{bmatrix}.$$

Similarly, if A and B are two matrices of the same order, then they can be added together and the resulting matrix is a matrix of the same order as A and B, every entry of which is the sum of the corresponding entries of A and B. For example,

$$\begin{bmatrix} 1 & -1 \\ 2 & -2 \end{bmatrix} + \begin{bmatrix} -1 & 2 \\ 4 & 3 \end{bmatrix} = \begin{bmatrix} 0 & 1 \\ 6 & 1 \end{bmatrix}.$$

Matrices give rise to linear transformations by the special way that they multiply (or act on) vectors (matrix multiplication will be discussed in the next section). We will think of vectors in E^2, say, as matrices written as a column or as a row. For instance, the vector $\langle 3, 2 \rangle \in E^2$ can be represented by the 2×1 matrix

$$\begin{bmatrix} 3 \\ 2 \end{bmatrix} \tag{6.62}$$

or by the 1×2 matrix

$$\begin{bmatrix} 3 & 2 \end{bmatrix}. \tag{6.63}$$

The two matrices in (6.62) and (6.63) are quite different mathematical objects, although they may represent the same physical object or a geometric quantity. This discussion is generalized to any vector in E^2 or E^3 or in E^n.

Definition 6.6.2
Two matrices A and B are equal if they have the same order and if $a_{ij} = b_{ij}$ for all i and j.

Note that the two matrices (6.62) and (6.63) are not equal according to the previous definition since they do not have the same order.

The $n \times n$ matrix

$$I \equiv \begin{bmatrix} 1 & 0 & 0 & \dots & 0 \\ 0 & 1 & 0 & \dots & 0 \\ \dots & \dots & \dots & \dots & \dots \\ 0 & 0 & & & \\ \dots & 0 & 1 & & \end{bmatrix} \tag{6.64}$$

is called the **identity matrix**. The entries of the identity matrix satisfy the relations

$$a_{ii} = 1 \quad \text{for} \quad 1 \le i \le n \quad \text{and} \quad a_{ij} = 0 \quad \text{if} \quad i \ne j. \tag{6.65}$$

The **zero matrix** Z is a matrix every entry of which is zero. It acts as the additive zero in the algebra of matrices, that is $A + Z = A$, for every matrix for which $A + Z$ is allowable.

The **transpose** of a matrix A, denoted by A^{T}, is a matrix we obtain from A by exchanging the rows of A with its columns. Putting this definition in terms of the entries of A, the ij entry of A^{T} is the ji entry of A. For example, the transpose of the 3×2 matrix

$$\begin{bmatrix} 1 & 2 \\ -1 & 3 \\ 3 & 2 \end{bmatrix} \tag{6.66}$$

is the 2×3 matrix

$$\begin{bmatrix} 1 & -1 & 3 \\ 2 & 3 & 2 \end{bmatrix}. \tag{6.67}$$

A matrix is said to be a **square** matrix if it has the same number of rows and columns. A square matrix is called **symmetric** if $A = A^{\mathrm{T}}$, or $a_{ij} = a_{ji}$ for all i and j; and it is called **anti-symmetric** if $A^{\mathrm{T}} = -A$. For instance, the matrix

$$\begin{bmatrix} 1 & 2 \\ 2 & 3 \end{bmatrix} \tag{6.68}$$

is symmetric, while

$$\begin{bmatrix} 0 & -1 \\ 1 & 0 \end{bmatrix} \tag{6.69}$$

is anti-symmetric.

Mathematica and *MATLAB* are both capable of manipulating matrices. One introduces a matrix $A = \begin{bmatrix} a & b \\ c & d \end{bmatrix}$, say, to *Mathematica* by giving the command

```
A = {{a, b}, {c, d}}
```

Two matrices A and B can be added by entering

```
C = A + B;
```

Matrices are introduced to *MATLAB* in a similar fashion. Matrix A is given to *MATLAB* by

```
A = [a b;c d];
```

if the numerical values for a through d are known, otherwise the matrix should be entered symbolically:

```
A = sym('[a b; c d]');
```

Thus, both of these packages are capable of treating the variables a, b, c, d as abstract "symbols" or as variables with numerical values attached to them.

The identity and zero matrices are also available to *Mathematica* and *MATLAB*. In *Mathematica* the internal function `IdentityMatrix[n]` defines an $n \times n$ identity matrix, while `Table[0,n,n]` defines an $n \times n$ zero matrix. The command `MatrixForm[A]` writes A in its matrix form as opposed to the list form in which every vector or matrix is represented in *Mathematica*.

In *MATLAB* the operation

A = eye(n);

assigns the $n \times n$ identity matrix to A and

B = zeros(A);

creates a zero matrix that is the same size as A and assigns it to B.

Problems

1. Is it possible to add the matrices

$$\begin{bmatrix} 1 & 0 \\ 0 & 1 \end{bmatrix} \quad \text{and} \quad [\ 1 \ \ 0 \ ]?$$

If not, why not? If yes, what is the result?

2. Check to see if the following matrix operations are allowable. Give each problem to *Mathematica* or *MATLAB* and verify your answers.

(a) $A = \begin{bmatrix} 2 & -2 & -1 \\ 3 & -1 & 2 \end{bmatrix}$. Determine $2A$, $A + A$, A^{T}, and $A + A^{\mathrm{T}}$.

(b) $A = \begin{bmatrix} -2 & 1 & 0 & 0 \\ 1 & -2 & 0 & 0 \\ 0 & 1 & -2 & 1 \\ 0 & 0 & 1 & -2 \end{bmatrix}$. Determine $-3A$, $A + A + A$, A^{T}, $(A^{\mathrm{T}})^{\mathrm{T}}$

and $\sum_{i=1}^{10} \frac{1}{i!} A^i$.

3. Show that if A is an $n \times n$ symmetric matrix, then

$$a_{ij} = a_{ji} \tag{6.70}$$

for all i and j.

4. Show that if A is an anti-symmetric matrix, then

$$a_{ii} = 0 \tag{6.71}$$

for all i.

5. Let A be an $n \times n$ matrix. Show that there are two matrices B and C such that B is symmetric, C is anti-symmetric, and

$$A = B + C. \tag{6.72}$$

6. Referring to the previous problem, find the symmetric and anti-symmetric parts of each matrix that follows.

(a) $\begin{bmatrix} 1 & 2 \\ 3 & 4 \end{bmatrix}$

(b) $\begin{bmatrix} 1 & -1 \\ 3 & 0 \end{bmatrix}$

(c) $\begin{bmatrix} 1 & 2 \\ 1 & 1 \end{bmatrix}$

(d) $\begin{bmatrix} 1 & 2 & 3 \\ -1 & 0 & 1 \\ 1 & -1 & 2 \end{bmatrix}$

(e) $\begin{bmatrix} 0 & 1 & 0 \\ 0 & 0 & 1 \\ a & b & c \end{bmatrix}$

(f) $\begin{bmatrix} 0 & 1 & 0 & 1 \\ 1 & 0 & 1 & 0 \\ -1 & 2 & 0 & 2 \\ 3 & 0 & 2 & 4 \end{bmatrix}$

(g) $\begin{bmatrix} \cos t & \sin t \\ -\sin t & \cos t \end{bmatrix}$

(h) $\begin{bmatrix} 1 & t & t^2 \\ t & t^2 & t^3 \\ t^2 & -t^2 & t^4 \end{bmatrix}$.

7. Show that matrix addition is commutative, that is, for any two matrices A and B with the same order, $A + B = B + A$.

8. Show that

$$(A + B)^{\mathrm{T}} = A^{\mathrm{T}} + B^{\mathrm{T}}, \quad (\alpha A)^{\mathrm{T}} = \alpha A^{\mathrm{T}}, \quad (A^{\mathrm{T}})^{\mathrm{T}},$$

where α is a scalar. Check the validity of each of these statements in *Mathematica* or *MATLAB*.

9. Let A be an arbitrary $n \times n$ matrix. Prove that $A + A^{\mathrm{T}}$ is symmetric. How would you prove this statement in *Mathematica* or *MATLAB*?

10. Is it true that the sum of two symmetric matrices is symmetric? Is it possible for the sum of two non-symmetric matrices to be symmetric?

11. Show that for any two scalars α and β and any two matrices A and B, having the same order, $\alpha(A + B) = \alpha A + \alpha B$ and $(\alpha + \beta)A = \alpha A + \beta B$.

12. Show that for any three matrices A, B, and C having the same order, we have $A + (B + C) = (A + B) + C$.

13. Consider the following set:

$$A = \left\{ \begin{bmatrix} a & 0 \\ 0 & b \end{bmatrix} \;\middle|\; a, b \in R \right\}. \tag{6.73}$$

Show that this set forms a vector space under the usual addition of matrices and scalar multiplication. What is the dimension of this vector space?

14. Let $S = \left\{ \begin{bmatrix} a & a+b \\ 0 & b \end{bmatrix} \middle| a, b \in R \right\}$. Show that S is a vector space under the usual addition and scalar multiplication of matrices. Find a basis for S. What is the dimension of S?

15. Consider the set

$$A = \left\{ \begin{bmatrix} a & 0 \\ 1 & b \end{bmatrix} \;\middle|\; a, b \in R \right\}. \tag{6.74}$$

Does this set form a vector space? Why?

6.7 Matrix Multiplication

The operation of matrix multiplication is motivated by the definition of linear transformation and by the way systems of linear simultaneous equations are formed. This operation is intended for two compatible matrices; the matrix A must have as many columns as B has rows.

Definition 6.7.1 (Matrix Multiplication)
Let A and B have dimensions $m \times p$ and $p \times n$, respectively. The matrix AB is defined as the $m \times n$ matrix whose ijth entry is given by the formula

$$(AB)_{ij} = \sum_{k=1}^{p} a_{ik} b_{kj}. \tag{6.75}$$

That is, to get the ijth entry of AB we take the dot product of the ith row of A with the jth column of B.

Example 6.7.1
Let A and B be defined by

$$A = \begin{bmatrix} 1 & 0 & -1 \\ -1 & 2 & 3 \end{bmatrix} \quad B = \begin{bmatrix} 1 & -1 & 2 & 0 \\ 2 & 3 & -1 & 2 \\ -2 & 4 & 0 & 0 \end{bmatrix}.$$

Since A is 2×3 and B is 3×4, the multiplication operation is permissible and the resultant matrix will be 2×4. Following the rule described in (6.75), we get the matrix

$$\begin{bmatrix} 3 & -5 & 2 & 0 \\ -3 & 19 & -4 & 4 \end{bmatrix}. \tag{6.76}$$

Example 6.7.2
Consider the matrices

$$A = \begin{bmatrix} 1 & 0 \\ 0 & 1 \end{bmatrix} \quad B = \begin{bmatrix} 1 & 1 & 1 \end{bmatrix}.$$

These two matrices cannot be multiplied since A is 2×2 and B is 1×3: In order to multiply A by B, we need the number of columns of A to be the same as the number of rows of B. Similarly, the product BA does not exist.

Example 6.7.3
Matrix multiplication is not commutative. That is, in general, given two matrices A and B that have compatible rows and columns so that both AB and BA can be formed, the resulting matrices may not be the same. We point out that two matrices A and B are equal if they have the same order and if $a_{ij} = b_{ij}$ for all i and j. To see an example of noncommutativity of matrix multiplication, let

$$A = \begin{bmatrix} 1 & 2 \\ 3 & 4 \end{bmatrix} \quad \text{and} \quad B = \begin{bmatrix} 0 & 1 \\ 1 & 0 \end{bmatrix}.$$

Then

$$AB = \begin{bmatrix} 2 & 1 \\ 4 & 3 \end{bmatrix}, \quad \text{while} \quad BA = \begin{bmatrix} 3 & 4 \\ 1 & 2 \end{bmatrix}.$$

Every matrix gives rise to a linear transformation in the following way; let A be a 2×2 matrix, say. Define a linear transformation T from E^2 into E^2 by the rule

$$T(\mathbf{x}) = A\mathbf{x}, \tag{6.77}$$

where the vector $\mathbf{x}$ on the right-hand side of (6.77) is written as a 2×1 matrix in some basis of E^2. Therefore, the action of the linear transformation T on $\mathbf{x}$ is defined by the matrix multiplication of A (a fixed matrix) and the matrix representation of $\mathbf{x}$. More importantly, the converse of the preceding procedure holds true. That is, **every linear transformation T can be represented by a matrix A via** (6.77). We will demonstrate this by an example. Let T be the linear transformation

$$T(\langle a, b \rangle) = \langle -b, a \rangle. \tag{6.78}$$

The vectors in E^2 are written in the natural basis $\{\mathbf{i}, \mathbf{j}\}$. To find the matrix A in this case we first determine what T does to the basis vectors $\mathbf{i}$ and $\mathbf{j}$. With $\langle a, b \rangle = \mathbf{i}$, we have

$$T(\mathbf{i}) = T(\langle 1, 0 \rangle) = \langle 0, 1 \rangle. \tag{6.79}$$

The first column of the matrix A will then be

$$\begin{bmatrix} 0 \\ 1 \end{bmatrix}.$$

Next
$$T(\mathbf{j}) = T(\langle 0, 1 \rangle) = \langle -1, 0 \rangle. \tag{6.80}$$

The second column of A will be
$$\begin{bmatrix} -1 \\ 0 \end{bmatrix}.$$

Thus, A takes the form
$$A = \begin{bmatrix} 0 & -1 \\ 1 & 0 \end{bmatrix}.$$

Note that
$$T(\langle a, b \rangle) = A \begin{bmatrix} a \\ b \end{bmatrix} = \begin{bmatrix} -b \\ a \end{bmatrix},$$

as required by (6.78).

We described the preceding procedure in terms of a 2×2 matrix. It should be clear to the reader that A could be any $m \times n$ matrix; T any linear transformation from E^m to E^n; and as we will see in the following examples, the domain and range of T could even be the abstract vector spaces we encountered in Section 6.3.

Example 6.7.4
Let A be the 2×3 matrix
$$A = \begin{bmatrix} 2 & -3 & 0 \\ 1 & 1 & 1 \end{bmatrix}. \tag{6.81}$$

We define a linear transformation T with domain E^3 and range in E^2 by the relation
$$T(\mathbf{x}) = A\mathbf{x}. \tag{6.82}$$

For instance,
$$T(\langle 1, -1, 1 \rangle) = \begin{bmatrix} 2 & -3 & 0 \\ 1 & 1 & 1 \end{bmatrix} \begin{bmatrix} 1 \\ -1 \\ 1 \end{bmatrix} = \begin{bmatrix} 5 \\ 1 \end{bmatrix}. \tag{6.83}$$

Note that $T(\mathbf{x}) \in E^2$. Similarly, given a linear transformation
$$S(\langle a, b \rangle) = \langle a - b, a + b, a + 2b \rangle, \tag{6.84}$$

we are able to find a 2×3 matrix B to represent T by determining the action of T on a basis of E^2. Let $\{\mathbf{i}, \mathbf{j}\}$ be the basis for the domain, and $\{\mathbf{i}, \mathbf{j}, \mathbf{k}\}$ the basis for E^3. It follows from (6.84) that
$$T(\mathbf{i}) = \langle 1, 1, 1 \rangle, \tag{6.85}$$

which is the first column of B, and
$$T(\mathbf{j}) = \langle -1, 1, -2 \rangle, \tag{6.86}$$

which is the second column of B. Thus, T is represented by

$$B = \begin{bmatrix} 1 & -1 \\ 1 & 1 \\ 1 & -2 \end{bmatrix}. \tag{6.87}$$

Remark 6.7.1: We emphasize that had we used different bases to represent the vectors in the domain and range of T we would obtain a different matrix B to represent T. We will see later that different matrix representations of the same linear transformation are related in a special way.

Example 6.7.5
Let T be the linear transformation

$$T(a \sin x + b \sin 2x)) = (a + b) \sin x + (a - b) \sin 2x. \tag{6.88}$$

The domain and range of this T are the vector space X we obtain from the span of $\sin x$ and $\sin 2x$. We could get a matrix representation of T if we generalize our point of view slightly and think of a typical vector in the domain of T, which is $a \sin x + b \sin 2x$, as $\langle a, b \rangle$. What we are doing at this stage is identifying the vector space X with the vector space E^2. With this identification the basis vectors in X (i.e., $\sin x$ and $\sin 2x$) are nothing but the natural basis in E^2. The action of T on the basis vectors is found as follows

$$T(\mathbf{i}) = \begin{bmatrix} 1 \\ 1 \end{bmatrix} \quad \text{and} \quad T(\mathbf{j}) = \begin{bmatrix} 1 \\ -1 \end{bmatrix}. \tag{6.89}$$

Hence, the matrix representation of T is

$$\begin{bmatrix} 1 & 1 \\ 1 & -1 \end{bmatrix}. \tag{6.90}$$

For instance, to understand to which vector T maps $-3 \sin x + 5 \sin 2x$, we can either use the definition of T and get

$$T(-3 \sin x + 5 \sin 2x) = 2 \sin x - 8 \sin 2x$$

or determine the same result by multiplying the matrix representation of T by the vector

$$\begin{bmatrix} -3 \\ 5 \end{bmatrix}. \tag{6.91}$$

Example 6.7.6
We multiply the two matrices

$$A = \begin{bmatrix} a & b & c \\ e & f & g \end{bmatrix} \quad \text{and} \quad B = \begin{bmatrix} 1 & 0 \\ 0 & 1 \\ 1 & 1 \end{bmatrix},$$

in *Mathematica* as follows:

```
A = {{a,b,c}, {e,f,g}};
B={{1,0}, {0,1}, {1,1}};
A.B
```

The output is

```
{{a + c, b + c}, {d + f, e + f}}
```

When we apply `MatrixForm` to this output we get

```
//MatrixForm= a + c    b + c

             d + f   e + f
```

The operation BA is permissible, although its result is different from AB; in fact,

```
B.A
```

returns

```
{{a, b, c}, {d, e, f}, {a + d, b + e, c + f}}
```

The operation AA (or A^2) is not permissible since A is not a square matrix. When we try `A.A`, we receive the error message

```
Dot::dotsh:
   Tensors {{a, b, c}, {d, e, f}} and
      {{a, b, c}, {d, e, f}} have incompatible shapes.
```

Returning to *MATLAB*, after introducing A and B by

```
A = sym('[a b c; e f g]');
B = sym('[1 0; 0 1; 1 1]');
```

we multiply A by B by applying `symmul`:

```
A*B
```

whose output is

```
ans =

[a+c, b+c]
[e+g, f+g]
```

Similarly, the product BA is obtained by executing `B*A`, which results in

```
ans =

[ a,   b,   c]
[ e,   f,   g]
[a+e, b+f, c+g]
```

When we attempt to multiply A by itself, which is not permissible, we receive the following error message:

```
??? Error using ==> sym/mtimes
Inner matrix dimensions must agree.
```

Problems

1. Let
$$A = \begin{bmatrix} 1 & -1 \\ 2 & 3 \end{bmatrix} \quad \text{and} \quad B = \begin{bmatrix} 1 & -1 & 2 \\ -3 & 4 & 3 \end{bmatrix}.$$
Find AB and BA.

2. Write the following linear transformations in matrix form.

   (a) $T_1(\langle a, b \rangle) = \langle a - b, a + b \rangle$

   (b) $T_2(\langle a, b, c \rangle) = \langle a - 2b + c, a - c, b + 3c \rangle$

3. Let
$$A = \begin{bmatrix} 1 & 2 & 3 \\ -1 & 0 & 1 \\ 1 & 1 & 2 \end{bmatrix}, \quad B = \begin{bmatrix} 0 & 1 & 1 \\ 0 & 0 & 1 \\ 1 & 2 & 3 \end{bmatrix}. \tag{6.92}$$

   Do the following operations. Check your answers using *Mathematica* or *MATLAB*.

   (a) AB

   (b) BA

   (c) A^2

   (d) ABA

   (e) A^2B

   (f) BA^2

   (g) $A^2 - B^2$

   (h) $(A + B)(A - B)$

   (i) $(A + B)^2$

   (j) $A^2 + 2AB + B^2$

   (k) $(A + B)^3$

   (l) $A^3 + 3A^2B + 3AB^2 + B^3$

   Are (i) and (j) the same? How about (k) and (l)? Why?

4. Write the following linear transformations in matrix form.

   (a) $T_1(\langle a, b \rangle) = \langle a - b, a + b \rangle$

   (b) $T_2(\langle a, b, c \rangle) = \langle a - 2b + c, a - c, b + 3c \rangle$

For each linear transformation find $T_i(T_i(\mathbf{x}))$, where $\mathbf{x}$ is an arbitrary vector in the domain. Draw the vectors $\mathbf{x}$, $T_1(\mathbf{x})$, and $T_1(T_1(\mathbf{x}))$. Next compute $A_i^2\mathbf{x}$ where A_i is the matrix representation of T_i (choose the natural basis for the domain and range). How do $T_i(T_i(\mathbf{x}))$ and $A_i^2\mathbf{x}$ compare in each case?

5. Give a geometric interpretation of a linear transformation whose matrix is given. For a geometric interpretation you may wish to illustrate the effect of the linear transformation on a circle in the domain or, in general, what happens to a typical vector in the domain.

(a) $\begin{bmatrix} 2 & 0 \\ 0 & 2 \end{bmatrix}$

(b) $\begin{bmatrix} -2 & 0 \\ 0 & 2 \end{bmatrix}$

(c) $\begin{bmatrix} 2 & 0 \\ 0 & 3 \end{bmatrix}$

(d) $\begin{bmatrix} 2 & 0 \\ 0 & 0 \end{bmatrix}$

(e) $\begin{bmatrix} 0 & 0 \\ 1 & 0 \end{bmatrix}$

(f) $\begin{bmatrix} 0 & 1 \\ -1 & 0 \end{bmatrix}$

(g) $\begin{bmatrix} 0 & -1 \\ -10 & 0 \end{bmatrix}$

6. Show that the matrix

$$A = \begin{bmatrix} \cos\theta & -\sin\theta \\ \sin\theta & \cos\theta \end{bmatrix}$$

defines a linear transformation that rotates a vector in the domain by the angle θ. Use a specific θ and verify this fact in *Mathematica* or *MATLAB*. What is the geometric interpretation of A^2? What about A^n, where n is a positive integer?

7. Show that for any three matrices for which the following operations are allowed we have

$$A(B + C) = AB + AC, \tag{6.93}$$

that is, matrix addition and multiplication are associative. (Hint: Find the typical ijth entry of $A(B + C)$ and compare it with the ijth entry of $AB + AC$.)

8. Show that $A(BC) = (AB)C$ whenever these operations make sense.

9. Show that $\alpha(AB) = (\alpha A)B$, where α is a scalar and A and B are matrices for which the preceding operations make sense.

10. Show that if A and B are symmetric, then AB is symmetric if and only if A and B commute, that is, $AB = BA$.

11. Recall from calculus that the exponential function e^x has the Taylor series expansion

$$e^x = 1 + x + \frac{x^2}{2!} + \frac{x^3}{3!} + \cdots + \frac{x^n}{n!} + \cdots = \sum_{n=0}^{\infty} \frac{x^n}{n!}, \qquad (6.94)$$

where, by definition, $0! = 1$. The series on the right-hand side of (6.94) can be used to define e^A, the exponential of a matrix A, as

$$e^A \equiv I + A + \frac{1}{2!} A^2 + \frac{1}{3!} A^3 + \cdots + \frac{1}{n!} A^n + \cdots. \qquad (6.95)$$

We approximate e^A by truncating the infinite series after N terms. Let $S(A, N)$ be the Nth partial sum of (6.95)

$$S(A, N) = \sum_{n=0}^{N} \frac{1}{n!} A^n, \qquad (6.96)$$

where, by definition, $A^0 \equiv I$, the identity matrix.

Write a *Mathematica* or a *MATLAB* program to compute the function $S(A, N)$ once A and N are known to it. For the *Mathematica* program you need to use the built-in function `Sum`, while the *MATLAB* program will require a `For` loop. Let

$$A = \begin{bmatrix} 0 & 1 \\ -1 & 0 \end{bmatrix}. \qquad (6.97)$$

Use your program with $N = 5, 10, 20,$ and 50; and compute the approximation to e^A. Both *Mathematica* and *MATLAB* have built-in functions that compute e^A (`MatrixExp` in *Mathematica* and `expm` in *MATLAB*). Compare your answers with these built-in functions as N varies from 5 to 50.

6.8 Systems of Algebraic Equations

Historically, matrix notation became prevalent in mathematics because it provided a concise and easily readable way of representing systems of algebraic and differential equations. In this section we will study systems of algebraic equations in the context of matrix algebra.

We begin with an example. Consider the following system of two equations in two unknowns:

$$2x + 3y = 1, \quad x - y = 3. \qquad (6.98)$$

This system is equivalent to

$$\begin{bmatrix} 2 & 3 \\ 1 & -1 \end{bmatrix} \begin{bmatrix} x \\ y \end{bmatrix} = \begin{bmatrix} 1 \\ 3 \end{bmatrix}. \qquad (6.99)$$

Let A stand for the matrix of coefficients $\begin{bmatrix} 2 & 3 \\ 1 & -1 \end{bmatrix}$, $\mathbf{x}$ be the vector of unknowns $\begin{bmatrix} x \\ y \end{bmatrix}$, and $\mathbf{b}$ be $\begin{bmatrix} 1 \\ 3 \end{bmatrix}$. We can then write (6.99) as

$$A\mathbf{x} = \mathbf{b}. \tag{6.100}$$

System (6.100) contains all of the information in the simultaneous equation (6.98). In analogy with the single equation $ax = b$ whose solution is $x = a^{-1}b$, we seek a matrix A^{-1} and write the solution to (6.100) in the form

$$\mathbf{x} = A^{-1}\mathbf{b}. \tag{6.101}$$

The matrix A^{-1} is unknown at this point but, much like the definition of a^{-1} when a is a scalar, it must satisfy the relation

$$A^{-1}A = I, \tag{6.102}$$

where I is the *Identity* matrix,

$$I = \begin{bmatrix} 1 & 0 \\ 0 & 1 \end{bmatrix}. \tag{6.103}$$

This discussion motivates the following definition.

Definition 6.8.1 (Inverse of a Matrix)
The inverse of a square matrix A is another square matrix, denoted by A^{-1}, with the property that

$$AA^{-1} = A^{-1}A = I. \tag{6.104}$$

A square matrix that has an inverse is called a **nonsingular** *matrix. Otherwise, it is called a* **singular** *matrix.*

Once we find A^{-1} we multiply (6.100) on the left by A^{-1} to get

$$A^{-1}A\mathbf{x} = A^{-1}\mathbf{b}. \tag{6.105}$$

Since $A^{-1}A = I$ and $I\mathbf{x} = \mathbf{x}$, we arrive at (6.101).

It is not difficult to see from (6.104) that A and A^{-1} must have the same dimensions. For our particular problem in (6.98) the matrix A^{-1} should be 2×2. Let l, m, n, and p denote its entries:

$$A^{-1} = \begin{bmatrix} l & m \\ n & p \end{bmatrix}. \tag{6.106}$$

The product AA^{-1} equals

$$\begin{bmatrix} 2l + m & 3l - m \\ 2n + p & 3n - p \end{bmatrix}.$$

Since $A^{-1}A = I$, we need to determine l, m, n, and p so that the four equations

$$2l + m = 1, \quad 3l - m = 0, \quad 2n + p = 0, \quad 3n - p = 1 \qquad (6.107)$$

hold.

The system of equations in (6.107) is partially uncoupled. The equations for l and m do not require information about n and p. We solve these pairs of equations separately and arrive at the solution $l = \frac{1}{5}$, $m = \frac{3}{5}$, $n = \frac{1}{5}$, and $p = -\frac{2}{5}$. Therefore,

$$A^{-1} = \frac{1}{5} \begin{bmatrix} 1 & 3 \\ 1 & -2 \end{bmatrix}. \qquad (6.108)$$

Hence, the solution to (6.98) is

$$\mathbf{x} = \begin{bmatrix} x \\ y \end{bmatrix} = A^{-1} \begin{bmatrix} 1 \\ 3 \end{bmatrix} = \begin{bmatrix} 2 \\ -1 \end{bmatrix}. \qquad (6.109)$$

Remark 6.8.1: It may seem that we have done a lot more work than was necessary to find the solution to system (6.98). That is true to a point. By finding the inverse of the matrix A, however, we have accomplished a bit more than just determining the solution to one particular system. As along as A remains the same, we have ostensibly solved all systems of linear equations of the form (6.100) no matter what the vector $\mathbf{b}$ is; given any vector $\mathbf{b}$, the solution to (6.98) will simply be computed from (6.101) by a single matrix multiplication.

Remark 6.8.2: As we discussed earlier, Solve in *Mathematica* and solve in *MATLAB* are capable of obtaining the solution to the system in (6.98).

The number 5 that appeared in the denominator of each entry of A^{-1} is called the determinant of the matrix A. In general, we have the following definition:

Definition 6.8.2 (Determinant of a Square Matrix)

Let A be the 2×2 matrix $\begin{bmatrix} a & b \\ c & d \end{bmatrix}$. *The* **determinant** *of A, denoted by* $\det(A)$ *or* $|A|$, *is*

$$\det A = ad - bc. \qquad (6.110)$$

Using this notation, the inverse of the matrix $A = \begin{bmatrix} a & b \\ c & d \end{bmatrix}$ is

$$A^{-1} = \frac{1}{\det A} \begin{bmatrix} d & -b \\ -c & a \end{bmatrix}, \qquad (6.111)$$

as the reader can verify directly.

The inverse of a 3×3 matrix can be found in a similar fashion. We may follow the strategy we used earlier and apply the definition of A^{-1} directly. Since A is now a 3×3 matrix, the matrix A^{-1} introduces nine unknown entries. The identity $A^{-1}A = I$ will lead to a simultaneous system of nine algebraic equations in these unknowns. As in the previous example, we can solve these equations and obtain a general formula for computing the inverse of A. Although this process is straightforward, it is rather tedious and uses a larger than necessary number of algebraic operations, especially when n is large. Instead of pursuing this path, in the next section we will present an alternate algorithm which is capable of computing the inverse of any $n \times n$ matrix, while employing fewer algebraic operations.

Not every matrix is invertible, as some of the exercises will demonstrate. In fact, it is clear from (6.111) that something goes wrong when $\det A = 0$. If a matrix A does have an inverse, however, that inverse is unique. As is the case with any uniqueness property, the fact that the inverse of a matrix is unique allows us to use different algorithms to determine A^{-1}, while being confident that all such algorithms lead to the same conclusion.

Example 6.8.1 (Inverse of Matrices in *Mathematica* and *MATLAB*)
Both *Mathematica* and *MATLAB* are capable of determining the inverse of a matrix, when it exists, symbolically and analytically. The internal functions `Inverse` in *Mathematica* and `inv` in *MATLAB* do the job. Also, `Det` and `det` functions compute the determinant of a matrix in *Mathematica* and *MATLAB*, respectively. For instance, let

$$A = \begin{bmatrix} 3.1 & \sqrt{3} \\ -e^2 & 1 \end{bmatrix}. \tag{6.112}$$

Then

```
A = {{3.1, -Sqrt[3]}, {-Exp[2], 1}};
Det[A]
```

leads to the determinant of A, while

```
Inverse[A]
```

yields

```
{{0.0629001, 0.108946}, {-0.464773, 0.19499}}
```

Note that once one entry of A is specified in floating-point arithmetic the remaining entries are converted to that form. On the other hand, the preceding calculations with the matrix

$$A = \begin{bmatrix} \frac{31}{10} & \sqrt{3} \\ -e^2 & 1 \end{bmatrix}$$

lead to the following inverse:

```
           10                    10 Sqrt[3]
{{------------------- ,        ------------------} ,
                     2                           2
  31 + 10 Sqrt[3] E            31 + 10 Sqrt[3] E


              2
       -10 E                        31
  {------------------- ,        ------------------}}
                     2                           2
  31 + 10 Sqrt[3] E            31 + 10 Sqrt[3] E
```

The following series of commands in *MATLAB*

```
A = [3.1 -sqrt(3); -exp(2) 1];
det(A)
inv(A)
```

gives the expected output in floating-point arithmetic. To carry the preceding computations in exact arithmetic, we must declare the matrix A symbolically as

```
A = sym('[31/10 sqrt(2); -exp(2) 1]')
```

and follow up with computing the determinant and the inverse as before.

Problems

1. Write the following systems of equations in the matrix form $A\mathbf{x} = \mathbf{b}$.

   (a)
   $$\begin{cases} 2x + 3y - 3z = 1, \\ x - y + z = 0, \\ -3x + 7y + 2z = -1; \end{cases}$$

   (b)
   $$\begin{cases} ax + by + cz = d_1, \\ bx + cy + az = d_2, \\ cx + ay + bz = d_3; \end{cases}$$

   (c)
   $$\begin{cases} 4x_1 - x_2 = 0.1, \\ -x_1 + 4x_2 - x_3 = 0.2, \\ -x_2 + 4x_3 - x_4 = -0.2, \\ -x_3 + 4x_4 - x_5 = 0.1, \\ -x_4 + 4x_5 = 0.2 \end{cases}$$

2. Derive formula (6.111) by following the procedure described previously, starting with (6.104).

3. Solve each algebraic system by writing the system in the form $A\mathbf{x} = \mathbf{b}$ and by computing the inverse of A. Check your calculations in either *Mathematica* or *MATLAB*. In particular, compare the answers from the inverse matrix method to that obtained by applying Solve or solve.

(a)
$$\begin{cases} 2x - 3y = 12, \\ x + y \quad = 9 \end{cases}$$

(b)
$$\begin{cases} 2x + 3y = 1, \\ x - y \quad = -2 \end{cases}$$

(c)
$$\begin{cases} \frac{1}{2}x - 2y = \frac{1}{4}, \\ 3x - \frac{1}{3}y = -\frac{1}{2} \end{cases}$$

(d)
$$\begin{cases} ax - y = 1, \\ x - ay = -2 \end{cases}$$

(e)
$$\begin{cases} 2a - 3b = -1, \\ 3a - 4b = 5 \end{cases}$$

(f)
$$\begin{cases} aw + bz = 0, \\ cw + dz = 1 \end{cases}$$

4. Consider the system
$$\begin{cases} 2x - 3y = 10, \\ 4x - 6y = 2. \end{cases} \tag{6.113}$$
Solve this system as in Problem 3. What goes wrong? How many solutions does this problem have? Draw the graph of each equation in the xy-plane, and explain the relation between the graphs of the equations in (6.113) and the solvability of this system. What is the response of Solve or solve when applied to this problem?

5. Show directly that the matrix $A = \begin{bmatrix} 0 & 1 \\ 0 & 0 \end{bmatrix}$ does not have an inverse.

(Hint: Suppose, by way of contradiction, that $B = \begin{bmatrix} a & b \\ c & d \end{bmatrix}$ is the inverse of A. Compute $AB = I$ and arrive at a contradiction). What is *Mathematica* or *MATLAB*'s response when you attempt to determine the inverse of A?

6. Show directly that the following matrices do not possess an inverse for any a or b.

$$\begin{bmatrix} a & a \\ a & a \end{bmatrix} \qquad \begin{bmatrix} a & b \\ a & b \end{bmatrix} \qquad \begin{bmatrix} b & a \\ b & a \end{bmatrix} \qquad \begin{bmatrix} a & -b \\ a & -b \end{bmatrix}$$

7. Show that if the inverse of a square matrix A exists, then it must be unique. (Hint: Suppose that there are two inverses B_1 and B_2 for A. Then use $AB_1 = B_1 A = AB_2 = B_2 A = I$ and the associativity of matrix multiplication to show $B_1 = B_2$.)

8. Let A be the general 3×3 matrix

$$A = \begin{bmatrix} a & b & c \\ d & e & f \\ g & h & i \end{bmatrix}.$$

Find the inverse of A by generalizing to 3×3 matrices the ideas that led to (6.108). Verify the final result in *Mathematica* or *MATLAB*.

9. Let $\mathbf{e}_1 = \begin{bmatrix} a \\ b \end{bmatrix}$ and $\mathbf{e}_2 = \begin{bmatrix} c \\ d \end{bmatrix}$ be representing two vectors in E^2. Suppose that $\mathbf{e}_1$ and $\mathbf{e}_2$ are linearly dependent. Show that the determinant of the matrix one constructs from $\mathbf{e}_1$ and $\mathbf{e}_2$ is zero.

6.9 Computing det A and A^{-1} Using Cofactors

In this section we will present an algorithm for computing the determinant and the inverse of an $n \times n$ matrix using the method of cofactors. This algorithm will give us a working definition of the determinant of a square matrix of any size, while providing a theoretical basis for establishing facts about this operation.

Definition 6.9.1 (Cofactor of a Matrix)
Let A be an $n \times n$ matrix. The cofactor of the a_{ij} entry of A, denoted by A_{ij}, is a scalar obtained in the following way.

1. *Construct an $(n-1) \times (n-1)$ matrix from A by deleting its ith row and jth column.*

2. *A_{ij} is $(-1)^{i+j}$ times the determinant of the resulting $(n-1) \times (n-1)$ matrix.*

Example 6.9.1
   Let

$$A = \begin{bmatrix} 1 & 0 & 1 \\ 2 & -1 & 1 \\ 3 & 2 & 1 \end{bmatrix}.$$

Then A_{11} is the determinant of the 2×2 matrix

$$\begin{bmatrix} -1 & 1 \\ 2 & 1 \end{bmatrix},$$

which is -3. Similarly, $A_{12} = 1$, $A_{22} = 2$, and so on.

Definition 6.9.2 (Determinant of $n \times n$ Matrices)
The determinant of a matrix A is defined by

$$\det A = \sum_{j=1}^{n} a_{ij} A_{ij}. \tag{6.114}$$

The index i in the preceding definition is kept fixed; we choose a fixed row of A, compute all cofactors of that row only, and then use (6.114). We can also write the previous definition using any column of A, that is,

$$\det A = \sum_{i=1}^{n} a_{ij} A_{ij}. \qquad (6.115)$$

Equations (6.114) and (6.115) lead to the same value for the determinant of a square matrix A. Although this fact is by no means obvious, the reader can verify it directly for general $n \times n$ matrices when n is small. We do not give a formal proof of this fact, but refer the reader to Problem 5 for more detail.

Example 6.9.2
Let A be the matrix defined in Example 6.9.1. We compute its determinant using its first row. Since $A_{11} = -3$, $A_{12} = 1$, and $A_{13} = 7$, we have

$$\det A = 1 \times (-3) + 0 \times (1) + 1 \times 7 = 4. \qquad (6.116)$$

Definition 6.9.2 is recursive in the sense that to compute the determinant of a 4×4 matrix, it is necessary to compute four determinants of 3×3 matrices (the cofactors). Although the number of algebraic operations in the cofactor method is quite manageable as long the size of the matix is small, the amount of algebra involved becomes prohibitive rather quickly for large-sized matrices. For this reason this method has more of a theoretical value than a numerical one. We will see a different method for computing determinants shortly.

Definition 6.9.3 (Adjugate of a Matrix)
The adjugate of an $n \times n$ matrix A, denoted by adj A, is the transpose of the matrix of cofactors of A.

Example 6.9.2
Returning to the previous example, the adjugate of the matrix A defined by

$$A = \begin{bmatrix} 1 & 0 & 1 \\ 2 & -1 & 1 \\ 3 & 2 & 1 \end{bmatrix}$$

is the matrix

$$\text{adj } A = \begin{bmatrix} -3 & 2 & 1 \\ 1 & -2 & 1 \\ 7 & -2 & -1 \end{bmatrix}. \qquad (6.117)$$

Definition 6.9.4 (Inverse of an $n \times n$ Matrix)
The inverse of an $n \times n$ matrix A, when it exists, is

$$A^{-1} = \frac{1}{\det A} \text{adj } A. \qquad (6.118)$$

Remark 6.9.1: Although we are not motivating the definition of the adj A, the reader should verify the validity of (6.118) for general 2×2, 3×3, and 4×4 matrices, to get a sense of the computations involved in the derivation of (6.118). Since the inverse of a matrix, when it exists, is unique (see Problem 7 of Section 6.8), the two methods described here and in the previous section lead to the same inverse of A.

Example 6.9.3
To find the inverse of the matrix A in the previous example we need to multiply (6.117) by $\frac{1}{4}$.

It is clear from the definition of the determinant of a matrix that two properties need to hold for a matrix to have an inverse. First, the matrix must be *square*; and second, the determinant of A must not be zero. This second property is so significant that we state it as a lemma.

Lemma 6.9.1
A square matrix has an inverse if and only if its determinant is nonzero.

The necessary and sufficient condition for the existence of an inverse of a matrix also enables us determine whether a system of algebraic linear equations written in the form $A\mathbf{x} = \mathbf{b}$ has a unique solution.

Lemma 6.9.2
A system of linear algebraic equations $A\mathbf{x} = \mathbf{b}$, with A a square matrix, has a unique solution if and only if

$$\det A \neq 0 \quad \text{and in that case } \mathbf{x} = A^{-1}\mathbf{b}. \tag{6.119}$$

Example 6.9.4
Consider the system of equations

$$\begin{cases} ax + y - z & = -1, \\ 2x - ay + z & = 0, \\ x + y + az & = 2. \end{cases} \tag{6.120}$$

Here a is a real parameter. The problem is to find all as for which (6.120) fails to have a unique solution. According to Lemma 6.9.1, this system possesses a unique solution if the determinant of the matrix of coefficients

$$A = \begin{bmatrix} a & 1 & -1 \\ 2 & -a & 1 \\ 1 & 1 & a \end{bmatrix}$$

is nonzero. The determinant of A is the polynomial

$$-1 - 4a - a^3.$$

We enter Solve[-1 - 4 a - a^3 == 0, a] into *Mathematica* and obtain

{{a -> -0.246266}, {a -> 0.123133 + 2.01134 I},
        {a -> 0.123133 - 2.01134 I}}

as its roots. Thus, as long as $a \neq -0.246244$ and is real, the above system has a unique solution.

In the next section we will study the method of Gaussian elimination. This method has several important applications, among which is the resolution of the problem of obtaining information about the solutions of systems of algebraic equations whose matrix of coefficients is singular.

Problems

1. Find the determinant and inverse (if it exists) of each matrix given. In each case use *Mathematica* or *MATLAB* and verify your answers.

   (a) $\begin{bmatrix} 0 & 0 & 1 \\ 0 & 1 & 0 \\ 1 & 0 & 0 \end{bmatrix}$

   (b) $\begin{bmatrix} 0 & 1 & 0 & 1 \\ 1 & 0 & 1 & 0 \\ 0 & 0 & 1 & 1 \\ 1 & 0 & 0 & 1 \end{bmatrix}$

   (c) $\begin{bmatrix} a & a \\ -a & a \end{bmatrix}$

   (d) $\begin{bmatrix} a & a \\ -a & a \end{bmatrix}$

   (e) $\begin{bmatrix} a & a & a \\ -a & a & a \\ a & 0 & 0 \end{bmatrix}$

   (f) $\begin{bmatrix} 0 & 1 & 0 \\ 0 & 0 & 1 \\ a & b & c \end{bmatrix}$

   (g) $\begin{bmatrix} 4 & -1 & 0 & 0 \\ -1 & 4 & -1 & 0 \\ 0 & -1 & 4 & -1 \\ 0 & 0 & -1 & 4 \end{bmatrix}$

   (h) The 4×4 **Hilbert** matrix whose ijth entry is $\frac{1}{i+j-1}$

2. Prove that any square matrix that has a row or a column of only zeros has determinant equal to zero.

3. Consider the two matrices $\begin{bmatrix} a & b \\ c & d \end{bmatrix}$ and $\begin{bmatrix} c & d \\ a & b \end{bmatrix}$. How are their determinants related?

4. Let A, B, and C be 2×2 matrices. Is it true that if $AB = AC$, then necessarily $B = C$? If yes, prove your statement. If not, give a counterexample.

5. Let A be a general 2×2 matrix.

   (a) Apply the formula in (6.114) with $j = 1$ and $j = 2$. Do the answers differ?

   (b) Apply the formula in (6.115) with $i = 1$ and $i = 2$. Do the answers differ? How do these results differ from the values obtained from (6.114)?

6. Let A be a general 3×3 matrix. Compute the determinant of A using (6.114) and (6.115) for various values of i and j. Show that in all cases the answer is the unique value one obtains from *Mathematica* or *MATLAB*.

7. Let A be a 3×3 matrix. Use either *Mathematica* or *MATLAB* to compute the adj A symbolically and then verify (6.118).

8. Repeat the previous problem for an arbitrary 4×4 matrix.

9. Let A and B be any two 2×2 matrices. Show that $\det(AB) = \det(A) \det(B)$.

10. Let A and B be any 3×3 matrices. Verify that the identity in the previous problem holds for these matrices. How would you prove this statement on either *Mathematica* or *MATLAB*? (Hint: Define A and B symbolically to either software package. Compute the left-hand and right-hand sides of the identity and take the difference. Use the Simplify command of *Mathematica* or the simplify command of *MATLAB* on the difference.)

   Remark 6.9.3: The determinant of the product of two $n \times n$ matrices is equal to the product of the determinants. The proof of this result is rather difficult. It could be found in texts on advanced linear algebra such as [1].

11. Let A be a square matrix. Use the result from the previous problem to show that $\det(A^2) = (\det(A))^2$, $\det(A^3) = (\det(A))^3$, and in general, $\det(A^n) = (\det(A))^n$.

6.10 Gaussian Elimination, Row Operations, and Rank

In the previous section we discussed a method for finding the solution to a system of linear equations as long as A, the matrix of the coefficients, is nonsingular. In this section we will study a different technique that not only leads to the solution of

$$A\mathbf{x} = \mathbf{b}, \tag{6.121}$$

when A is nonsingular, but will also give us information about the set of all solutions of this equation when A is singular. This method, known as **Gaussian**

elimination, will also provide an alternative way of computing the determinant and inverse of a given matrix.

The idea behind Gaussian elimination is motivated by the way we solve a system of linear algebraic equations. As a concrete example, consider the system of two equations in the two unknowns x and y:

$$\begin{cases} x - y = 1, \\ x + y = 0, \end{cases} \tag{6.122}$$

which has the unique solution $x = \frac{1}{2}, y = -\frac{1}{2}$. The way we generally proceed to find this solution is by manipulating the equations algebraically: We multiply an equation by an appropriate scalar and add the result to the other, thus eliminating one unknown from the system. The resulting equation has a single unknown in it and is easy to solve. Of course this idea is by no means only applicable to systems of two equations; it works quite satisfactorily for general systems as well.

This process works because of the following properties that are enjoyed by (6.122) as well as by all systems of linear equations we will come across. If we multiply an equation in (6.122) by a nonzero constant, we do not alter its set of solutions. If we exchange the two equations in (6.122), we do not alter the final solution. Finally, if we multiply an equation by a constant, add the result to the remaining equation, and use the new equation as the second equation in the system, we do not alter the solution. Again, we emphasize that these operations involving the equations in (6.122) are not special to this system alone and hold for any system of linear algebraic equations.

It is clear that these three operations only affect the coefficients of the equations in a system, so it is reasonable to adopt these operations for matrices as well. In solving the system of equations $A\mathbf{x} = \mathbf{b}$, we first construct the **augmented matrix** $[A\,|\,\mathbf{b}]$, which is the matrix we obtain by appending the column vector $\mathbf{b}$ to A. The augmented matrix has all the information we need to obtain a solution to $A\mathbf{x} = \mathbf{b}$. Furthermore, the three operations discussed previously only alter the entries of this augmented matrix and not the solution vector $\mathbf{x}$.

Definition 6.10.1 (Row Operations)
Let $[A\,|\,\mathbf{b}]$ *be the augmented matrix obtained from the system of linear equations* $A\mathbf{x} = \mathbf{b}$. *We say* $[A\,|\,\mathbf{b}]$ *is equivalent to the matrix M if M is obtained from* $[A\,|\,\mathbf{b}]$ *through the following* **row operations***:*

1. *Exchanging two rows of* $[A\,|\,\mathbf{b}]$.

2. *Multiplying a row of* $[A\,|\,\mathbf{b}]$ *by a nonzero constant.*

3. *Multiplying a row of* $[A\,|\,\mathbf{b}]$ *by a constant and adding the result to another row.*

Definition 6.10.2
The diagonal of a matrix is the set of entries of the form a_{ii}.

Definition 6.10.3 (Upper Triangular Echelon Form)
    A matrix is in upper triangular echelon form if every entry of the matrix below the diagonal is zero.

The method of Gaussian elimination is to use these row operations and reduce $[A|\mathbf{b}]$ to an **upper triangular echelon form**. The goal of the algorithm is to use these operations in an iterative fashion and reduce the entries below the diagonal of $[A|\mathbf{b}]$ to zero. Once this is achieved, the solution to the equivalent system $A\mathbf{x} = \mathbf{b}$ is obtained by **backward substitution**. The next example illustrates this procedure.

Example 6.10.1
    Consider the system of three linear equations

$$\begin{cases} x + y + z = 6, \\ 2x - y + z = 3, \\ -x + 2y - z = 0, \end{cases} \tag{6.123}$$

whose unique solution is $x = 1, y = 2, z = 3$. The augmented matrix in this case is

$$[A|\mathbf{b}] = \begin{bmatrix} 1 & 1 & 1 & 6 \\ 2 & -1 & 1 & 3 \\ -1 & 2 & -1 & 0 \end{bmatrix}. \tag{6.124}$$

We begin with row one and eliminate the entries 2 and -1 in the second and the third rows of the first column. To eliminate 2 in the second row of the first column (the a_{12} entry of $[A|\mathbf{b}]$), we multiply the first row by -2 and add to the second row. This row operation, of course, alters all of the entries in the second row, including its first entry. To eliminate -1 in the third row in the first column, we add the first row to the third row. We obtain the equivalent matrix

$$\begin{bmatrix} 1 & 1 & 1 & 6 \\ 0 & -3 & -1 & -9 \\ 0 & 3 & 0 & 6 \end{bmatrix}. \tag{6.125}$$

To eliminate the entry 3 in the third row (the a_{32} entry), we use the second row. We add the second row to the third row and get

$$\begin{bmatrix} 1 & 1 & 1 & 6 \\ 0 & -3 & -1 & -9 \\ 0 & 0 & -1 & -3 \end{bmatrix}. \tag{6.126}$$

The matrix (6.126) is in upper triangular echelon form. Note that it is equivalent to the system of equations

$$\begin{cases} x + y + z = 6, \\ -3y - z = -9, \\ -z = -3. \end{cases} \tag{6.127}$$

It should now be clear how to solve for x, y, and z. First we solve the third equation to get $z = 3$. Then we proceed to the second equation and solve for y (now that z is known) to get $y = 2$. Finally we solve the first equation (now that y and z are known) to get $x = 1$. We will refer to this procedure as backward substitution.

In this example A is nonsingular and we would have been able to find the unique solution to (6.127) by the method described in Section 6.8. There are, however, two major advantages to the method of Gaussian elimination:

1. This method applies equally well to singular problems as well as to systems of equations of the form (6.121) for which A is not square.

2. This algorithm is suited very well to numerical computations and generally uses fewer arithmetic operations than the inverse operation does.

Example 6.10.2

Consider the system of linear equations

$$\begin{cases} y - z = 3, \\ 2x - y + 3z = -1, \\ 2x + y + z = 5. \end{cases} \tag{6.128}$$

The augmented matrix for this system is

$$\begin{bmatrix} 0 & 1 & -1 & 3 \\ 2 & -1 & 3 & -1 \\ 2 & 1 & 1 & 5 \end{bmatrix}. \tag{6.129}$$

Note that the 11-entry of this matrix is zero, which stops us from starting the Gaussian elimination procedure. We therefore exchange the first row with the second:

$$\begin{bmatrix} 2 & -1 & 3 & -1 \\ 0 & 1 & -1 & 3 \\ 2 & 1 & 1 & 5 \end{bmatrix}. \tag{6.130}$$

We now proceed as in the preceding example and reduce this matrix to its upper triangular echelon form:

$$\begin{bmatrix} 2 & -1 & 3 & -1 \\ 0 & 1 & -1 & 3 \\ 0 & 0 & 0 & 0 \end{bmatrix}. \tag{6.131}$$

The last row in this matrix is identically zero. Thus, we are left with two equations in the three unknowns x, y, and z. We let

$$z = c,$$

where c is an arbitrary constant. Next we apply the backward substitution to solve for x an y in terms of z. We obtain infinitely many solutions parametrized by c:

$$x = 1 - c, \quad y = 3 + c, \quad z = c. \tag{6.132}$$

We reach the same conclusion when we give this problem to *Mathematica* and *MATLAB*. In *Mathematica* we use the function Solve:

```
lhs = {y - z, 2 x - y + 3 z, 2 x + y + z};
rhs = {3, -1, 5};
Solve[lhs == rhs, {x, y, z}]
```

which results in

```
{{x -> 1 - z, y -> 3 + z}}
```

which indicates that z is arbitrary. We can also use the RowReduce command of *Mathematica* and get the upper triangular echelon form of the augmented matrix:

```
aug = {{0, 1, -1, 3}, {2, -1, 3, -1}, {2, 1, 1, 5}};
RowReduce[aug]
```

The response is

```
{{1, 0, 1, 1}, {0, 1, -1, 3}, {0, 0, 0, 0}}
```

In *MATLAB* the function linsolve will do the job:

```
A = [0 1 -1; 2 -1 3; 2 1 1];
b = [3; -1; 5];
linsolve(A,b)
```

MATLAB responds with

```
Warning: Matrix is rank deficient; solution is not unique.

ans =

[1]
[3]
[0]
```

which is the solution for $c = 0$.

Let us take a close look at the general case when A is singular. Let A be an $n \times n$ matrix. When we reduce $[A|\mathbf{b}]$ by Gaussian elimination, we will end up with an upper triangular echelon matrix of the form

$$
\begin{bmatrix}
a'_{11} & a'_{12} & a'_{13} & a'_{14} & \cdots & a'_{1n} & b'_1 \\
0 & a'_{22} & a'_{23} & a'_{24} & \cdots & a'_{2n} & b'_2 \\
\cdots & \cdots & \cdots & \cdots & \cdots & \cdots & \cdots \\
\cdots & \cdots & \cdots & \cdots & \cdots & \cdots & \cdots \\
0 & 0 & 0 & 0 & 0 & 0 & b'_n
\end{bmatrix}. \tag{6.133}
$$

Since A is singular, all entries in (at least) the last row are zero, except perhaps for b'_n which may or may not be zero. Whether (6.121) has a solution or not depends now on whether b'_n is zero. If $b'_n \neq 0$, then the last equation in (6.133) reduces to

$$0 = b'_n, \tag{6.134}$$

leading to a contradiction; therefore, the entire system (6.121) has no solution in this case. On the other hand, when $b'_n = 0$, the last equation can be eliminated entirely from the list of equations in (6.133): This system is now reduced to $n - 1$ equations in n unknowns. We arbitrarily choose one of the variables, say x_n, as parameter in terms of which all other variables, x_1, x_2, through x_{n-1}, can be solved. Thus, barring other redundancies, the system, in general, will have infinitely many solutions. Example 6.10.2 described one such system.

Example 6.10.3
Consider the system of equations

$$\begin{cases} x + y = 0, \\ x + y = 1. \end{cases} \tag{6.135}$$

Clearly this system has no solutions. The upper triangular echelon matrix of this system is

$$\begin{bmatrix} 1 & 1 & 0 \\ 0 & 0 & 1 \end{bmatrix}, \tag{6.136}$$

which states that the second row of this system is equivalent to $0 = 1$, thus leading us to conclude that the system has no solutions. When we give this system to either *Mathematica* or *MATLAB*, we are warned about the inconsistencies in the set of equations.

We will not summarize this discussion in terms of a theorem at this point because we still need to worry about redundancy in the reduced (i.e., the $(n - 1) \times n$) matrix described previously. This issue will be resolved once we introduce the concept of rank.

One of the properties of row operations is how they affect the value of the determinant of a matrix. We have the following lemma.

Lemma 6.10.3
Let A be a square matrix. Suppose that:

1. *B is obtained from A by exchanging two rows of A. Then $\det B = -\det A$.*

2. *B is obtained from A by multiplying a row of A by a constant α. Then $\det B = \alpha \det A$.*

3. *B is obtained from A by multiplying a row of A by a constant and adding it to another row of A. Then $\det B = \det A$.*

The reader can verify these properties for 2×2 and 3×3 matrices using the definition of determinant. Although the proof of the general case is not difficult, it will not be given here.

Lemma 6.10.4

The determinant of a matrix in upper triangular form is the product of its entries on the diagonal.

The proof of this result follows directly from the definition of determinant (compute the cofactors of the first row, say).

Lemmas (6.126) and (6.127) combine to give us a very useful way of computing determinants of matrices, as the next example demonstrates.

Example 6.10.4

We note that the matrix A given in Example 6.10.1 is equivalent to the upper triangular echelon form

$$\begin{bmatrix} 1 & 1 & 1 \\ 0 & -3 & -1 \\ 0 & 0 & -1 \end{bmatrix}, \tag{6.137}$$

where only the third row operation (multiplying a row by a scalar and adding the result to another row) is applied. Since this operation does not alter the value of a determinant, it follows that the determinant of A is 3, the product of the entries on the diagonal of the upper triangular echelon form.

Row operations can also be used to find the inverse of a matrix A if it exists. Since A and its inverse satisfy $AA^{-1} = I$, the problem of determining A^{-1} reduces to solving n linear systems of the form $A\mathbf{x} = \mathbf{b}_i$, where $\mathbf{b}_i$ is the i-th column of the identity matrix. We solve these systems simultaneously, by constructing the augmented matrix

$$[A|\,I],$$

and then apply row operations to reduce it to $[I|\,B]$. The matrix B will be the inverse of A.

Example 6.10.5

To find the inverse of the matrix A in Example 6.10.1 (which exists since $\det A = 3 \neq 0$), we consider the augmented matrix

$$\begin{bmatrix} 1 & 1 & 1 & 1 & 0 & 0 \\ 2 & -1 & 1 & 0 & 1 & 0 \\ -1 & 2 & -1 & 0 & 0 & 1 \end{bmatrix}. \tag{6.138}$$

We begin by multiplying the first row of (6.138) by -2 and adding to the second row, while at the same time adding the first row to the third row to get

$$\begin{bmatrix} 1 & 1 & 1 & 1 & 0 & 0 \\ 0 & -3 & -1 & -2 & 1 & 0 \\ 0 & 3 & 0 & 1 & 0 & 1 \end{bmatrix}. \tag{6.139}$$

Next we add the second row to the third row, and $\frac{1}{3}$ of the second row to the first row to get

$$\begin{bmatrix} 1 & 0 & \frac{2}{3} & \frac{1}{3} & \frac{1}{3} & 0 \\ 0 & -3 & -1 & -2 & 1 & 0 \\ 0 & 0 & -1 & -1 & 1 & 1 \end{bmatrix}. \tag{6.140}$$

Next we multiply the third row by -1 and add to the second row and multiply the third row by $\frac{2}{3}$ and add to the first row to get

$$\begin{bmatrix} 1 & 0 & 0 & -\frac{1}{3} & 1 & \frac{2}{3} \\ 0 & -3 & 0 & -1 & 0 & -1 \\ 0 & 0 & -1 & -1 & 1 & 1 \end{bmatrix}. \tag{6.141}$$

In order to make the left half part of (6.141) the identity matrix, we multiply the second row by $-\frac{1}{3}$ and the third row by -1. It follows from the right-half of the new matrix that

$$A^{-1} = \begin{bmatrix} -\frac{1}{3} & 1 & \frac{2}{3} \\ \frac{1}{3} & 0 & \frac{1}{3} \\ 1 & -1 & -1 \end{bmatrix}. \tag{6.142}$$

This result is, of course, verified by applying **Inverse** of *Mathematica* to A. It also turns out that these computations lead precisely to what one expects from the output of **RowReduce**. Here is a short program that begins with A and ends with producing the inverse of A:

```
A = {{1,1,1}, {2,-1,1}, {-1,2,-1}};
n = Length[A];
id = IdentityMatrix[n];
augmented = Table[Flatten[Append[A[[i]], id[[i]]]], {i, n}];
reduced = RowReduce[augmented];
inverseA = Table[Table[reduced[[i, j]], {j, n+1, 2n}], {i, n}];
```

The term **reduced** produces

```
           1      2            1      1
{{1, 0, 0, --, 1, -}, {0, 1, 0, -, 0, -},
           3      3            3      3

  {0, 0, 1, 1, -1, -1}}
```

which is equivalent to (6.141). The term `inverseA` contains the right-half of the matrix `reduced`.

The above *Mathematica* program may prove useful for computing the inverse of high order matrices since it relies on the Gaussian elimination algorithm, which is considered quite an efficient computational method. Note that only the first line of this program, the one that defines A to *Mathematica*, needs be changed when it is applied to a different matrix.

Another important application of the upper triangular echelon form has to do with the concept of the **rank** of a matrix. We will give a working definition of this concept here and develop its properties and its relation to the solvability of linear systems in terms of several theorems. As the preceding examples have shown, the number of nonzero rows of a matrix in its upper triangular echelon form measures the amount of redundancy in the information a matrix carries. Our next definition defines the rank of a matrix to be that number.

Definition 6.10.4 (Rank of a Matrix)
Let A be an $m \times n$ matrix. Let M be an upper triangular echelon form of the matrix A. Then the rank of A, denoted by $r(A)$, is the number of nonzero rows of M.

There is a direct relationship between the rank of a matrix A and the number of linearly independent rows in A. A matrix A is reduced to its upper triangular form by applying row operations. These are the same operations one performs on the rows of a matrix when testing for their linear independence. We therefore conclude that the rank of A and the number of linearly independent rows of A are the same. We state this fact as a corollary to Definition 6.10.4.

Corollary 6.10.1
Let A be an $m \times n$ matrix. The rank of A is equal to the maximum number of its linearly independent rows.

The next theorem demonstrates that we can replace columns for rows in the statement of the preceding corollary.

Theorem 6.10.1
Let A be an $m \times n$ matrix. The rank of A is equal to the maximum number of its linearly independent columns.

Proof: Let $\{\mathbf{r}_1, \mathbf{r}_2, \dots, \mathbf{r}_m\}$ denote the m rows of A. Suppose that A has rank $k \leq m$. Without loss of generality, let us assume that the first k rows of A are linearly independent. Since all rows of A can be written in terms of the first

k rows, there are constants c_{ji} such that

$$\mathbf{r}_j = c_{j1}\mathbf{r}_1 + c_{j2}\mathbf{r}_2 + \cdots + c_{jk}\mathbf{r}_k.$$

The typical ith entry of this vector is

$$a_{ji} = c_{j1}\mathbf{r}_{1i} + c_{j2}\mathbf{r}_{2i} + \cdots + c_{jk}\mathbf{r}_{ki}. \tag{6.143}$$

We now let i vary between 1 and m in (6.143). The left-hand side of (6.143) generates the jth column of A. The right-hand side of this equation is a linear combination (with constants c_{j1}, for example) of the k rows of A. Hence, we have shown that a typical column of A is a linear combination of the k rows of A. This proves that the maximum number of linearly independent columns of A also equals k.

The maximum number of linearly independent columns of A cannot be less than k. By way of contradiction, suppose that this number is, in fact, $k - 1$ (or less). Then, repeating the preceding argument for A^{T}, the transpose of A, we see that A^{T}s maximum number of linearly independent columns cannot exceed $k - 1$. But columns of A^{T} are rows of A, thus contradicting the assumption that the rank of A is k (see also Corollary 6.10.1). This completes the proof of the theorem.

We have thus shown that for an $m \times n$ matrix we have the following inequality that relates the rank of A and its dimension:

$$r(A) \leq \min(m, n). \tag{6.144}$$

A matrix is said to have *full rank* if equality holds in (6.144). Our discussions in this section and the examples we presented show that when the matrix A is square, the concept of full rank and the nonsingularity of A are equivalent. When strict inequality holds in (6.144), on the other hand, we acquire more information than just the fact that A is a singular matrix. We find precisely how many rows or columns of A are linearly independent. This information gives us quite a bit of insight into the structure of the mathematical model. In the context of an electrical network, for instance, the rank of A becomes related to the efficiency of the electrical loops used in the network, while in control theory the rank of the underlying matrices determines the optimality of certain design factors.

We next formulate a theorem that summarizes our discussion of the number of solutions to (6.121) when A is either singular or nonsquare.

Theorem 6.10.2
Consider the system of linear equations

$$A\mathbf{x} = \mathbf{b}, \tag{6.145}$$

where A is an $m \times n$ matrix. Then this system

1. *has a solution if* $r([A|\mathbf{b}]) = r(A)$—*infinitely many if A does not have full rank;*

2. *has no solutions if* $r(A) < r([A|\mathbf{b}])$.

Example 6.10.6

We now apply Theorem 6.10.2 with *Mathematica* and *MATLAB* to determine the number of solutions in the following system of equations:

$$\begin{cases} 2x + 3y - 3z + 4w - 5v = -1, \\ 3x - y + 2z + w = -2, \\ -x + y + z = 0, \\ 6x + y - 2z + 5w - 5v = -3, \\ 8y + 7z - 7w + 10v = 0. \end{cases} \tag{6.146}$$

The matrix of coefficients and the augmented matrix of (6.146) are

$$A = \begin{bmatrix} 2 & 3 & -3 & 4 & -5 \\ 3 & -1 & 2 & 1 & 0 \\ -1 & 1 & 1 & 0 & 0 \\ 6 & 1 & -2 & 4 & -5 \\ 0 & 8 & 7 & -7 & 10 \end{bmatrix},$$

$$[A|\mathbf{b}] = \begin{bmatrix} 2 & 3 & -3 & 4 & -5 & -1 \\ 3 & -1 & 2 & 1 & 0 & -2 \\ -1 & 1 & 1 & 0 & 0 & 0 \\ 6 & 1 & -2 & 5 & -5 & -3 \\ 0 & 8 & 7 & -7 & 10 & 0 \end{bmatrix}. \tag{6.147}$$

We use the **rank** command of *MATLAB* on these matrices and find that both have rank 3. It follows from Theorem 6.10.2 that this system has infinitely many solutions since the matrix of coefficients does not have full rank. Because the difference between the number of variables (5 in this case) and the rank of A is 2, we expect that two of the equations in (6.146) are redundant. After using Gaussian elimination on this system, we will end up with three equation in five unknowns. Hence, the solution set will have two free parameters in it.

Problems

1. Apply Gaussian elimination to solve the following systems. Use the results of this section on rank and determinant of matrices to determine the number of solutions each system possesses. Also use the *Mathematica* functions **RowReduce**, **Solve**, or **LinearSolve** or the *MATLAB* functions **rank** and **linsolve** to check each result.

$$\text{(a)} \begin{cases} y + z = 1, \\ x - 2y - z = 2, \\ x + 2y + z = 3 \end{cases}$$

(b) $\begin{cases} x + y + z \ = 0, \\ x - y + 2z = 2, \\ x + y + 4z = 2 \end{cases}$

(c) $\begin{cases} x - z = 1, \\ y + z = 3, \\ x + y = 5 \end{cases}$

(d) $\begin{cases} x + y + z + w = 0, \\ x - y + z + w = 0, \\ x + y - z + w = 0, \\ x + y + z - w = 0 \end{cases}$

(e) $\sum_{i=1}^{5} a_{ij} x_i = \frac{1}{j}, \ j = 1, \cdots, 5$, where $a_{ij} = \frac{1}{i+j}$

(f) $\begin{cases} x + y + z + u \quad\ = 0, \\ 2x - 4y - 3z + 2w = 3 \end{cases}$

(g) $\begin{cases} x - y + z - u \ \ = 1, \\ 2x - y + z + w = 0, \\ y - z + 3w \quad\ = -1 \end{cases}$

2. Let a and b be two parameters in R, the set of real numbers, and consider the system of equations

$$\begin{cases} 2x - 3y + z = 0, \\ x - y + 2z = 1, \\ 5x - 7y + az = b. \end{cases} \tag{6.148}$$

(a) Find a value for a so that the system does not have a unique solution.

(b) For this value of a, choose b so that the system does not have a solution.

(c) How would you solve part (a) of this problem in either *Mathematica* or *MATLAB*?

3. Consider the system of equations

$$\begin{cases} 3x + ay - z = 1, \\ 2x - 2ay + 3z = b, \\ ax - 2y + z = 2b, \end{cases} \tag{6.149}$$

where a and b are parameters. Find all values of a for which the system does not have a unique solution. You may need to use either *Mathematica* or *MATLAB* to find these values. For each value of a find all those values of b for which the system has infinitely many solutions, and find these solutions.

4. Find the rank of the following matrices.

(a) $\begin{bmatrix} 1 & 0 \\ 0 & 1 \end{bmatrix}$

(b) $\begin{bmatrix} 1 & 1 \\ 0 & 1 \end{bmatrix}$

(c) $\begin{bmatrix} 1 & 1 \\ 1 & 1 \end{bmatrix}$

(d) $\begin{bmatrix} 0 & 0 \\ 0 & 1 \end{bmatrix}$

(e) $\begin{bmatrix} 1 & 0 & 1 \\ 0 & 1 & 1 \\ -2 & 2 & 3 \end{bmatrix}$

(f) $\begin{bmatrix} 0 & 0 & 1 \\ 1 & 0 & 0 \\ 2 & 0 & 2 \end{bmatrix}$

(g) $\begin{bmatrix} a & a & a \\ 0 & a & 2a \\ -a & 2a & -a \end{bmatrix}$

(h) $\begin{bmatrix} 1 & 0 & 1 & 0 \\ 1 & 0 & 1 & 1 \\ 1 & -2 & 2 & 3 \end{bmatrix}$

(i) $\begin{bmatrix} 1 & -1 & 1 & -1 \\ 2 & 0 & 1 & 1 \\ 1 & 1 & 2 & 3 \\ 3 & -2 & -4 & 3 \end{bmatrix}$

(j) $\begin{bmatrix} 1 & a & -a & 0 \\ 0 & 0 & 0 & 1 \\ b & 1 & 1 & -b \\ 0 & 1 & 0 & 0 \end{bmatrix}$

5. Is it true that the rank of $A + B$ is equal to $r(A) + r(B)$?

6. Is it true that $r(A^2) = (r(A))^2$?

6.11 Inner Product of Vectors and Change of Coordinates

The setting of many applications often suggests a natural set of vectors as a basis or set of coordinates. The directions of extreme strain in a material that is undergoing a deformation is an example of such a setting. When these directions are not aligned with the preassigned coordinate system, we are confronted with the task of changing coordinates from an old basis to a new, and perhaps, more natural one. In this section we derive the necessary formulas for accomplishing a change of basis.

We first recall the geometric interpretation of the dot product of vectors. Given two vectors $\mathbf{a}$ and $\mathbf{b}$ in E^2 or E^3, the dot product of $\mathbf{a}$ and $\mathbf{b}$ is defined (see Section 6.2) as the scalar

$$\mathbf{a} \cdot \mathbf{b} = ||\mathbf{a}||\,||\mathbf{b}||\cos\gamma = \sum_{i=1}^{2 \text{ or } 3} a_i b_i, \tag{6.150}$$

where γ is the angle between $\mathbf{a}$ and $\mathbf{b}$. One of the primary applications of this operation is in determining the projection of one vector on another, that is,

$$\text{proj }_b\mathbf{a} = \frac{\mathbf{a} \cdot \mathbf{b}}{\mathbf{b} \cdot \mathbf{b}}\mathbf{b}. \tag{6.151}$$

We also recall the fact that two vectors are orthogonal if and only if their dot product is zero.

The definition of the dot product easily generalizes to any vector space of the form E^n.

Definition 6.11.1 (Inner (or Dot) Product)
The inner product (or dot product) of two vectors $\mathbf{a}$ and $\mathbf{b}$ in E^n is defined by

$$(\mathbf{a}, \mathbf{b}) = ||\mathbf{a}||\,||\mathbf{b}||\cos\gamma = \sum_{i=1}^{n} a_i b_i. \tag{6.152}$$

Two vectors in E^n are *orthogonal* if $\mathbf{a} \cdot \mathbf{b} = 0$. We also note that the operation of inner product is a linear operation, that is,

$$(\mathbf{a} + \mathbf{b}, \mathbf{c}) = (\mathbf{a}, \mathbf{c}) + (\mathbf{b}, \mathbf{c}), \tag{6.153}$$

and

$$(\alpha\mathbf{a}, \mathbf{b}) = \alpha(\mathbf{a}, \mathbf{b}) \tag{6.154}$$

for any scalar α.

One of the important applications of the concept of inner product is in changing bases. We give an outline of the role this operation plays in changing bases in E^3, leaving the natural generalization to E^n for $n \neq 3$ to the reader. Suppose that we have two bases for the vector space E^3: one the natural basis $\{\mathbf{i}, \mathbf{j}, \mathbf{k}\}$, and the other denoted by $\{\mathbf{a}_1, \mathbf{a}_2, \mathbf{a}_3\}$. Let us assume further that the second basis has the property that its vectors are mutually orthogonal, that is,

$$(\mathbf{a}_i, \mathbf{a}_j) = 0 \quad \text{if} \quad i \neq j. \tag{6.155}$$

Let $\mathbf{u} = \langle x, y, z \rangle$ be a vector written in the natural basis. What are the components of $\mathbf{u}$ in terms of the second basis $\{\mathbf{a}_1, \mathbf{a}_2, \mathbf{a}_3\}$? To answer this questions, we need to find c_1, c_2, and c_3 such that

$$\mathbf{u} = \langle x, y, z \rangle = c_1\mathbf{a}_1 + c_2\mathbf{a}_2 + c_3\mathbf{a}_3. \tag{6.156}$$

The scalars c_1, c_2, and c_3 in (6.156) are the magnitudes of the projections of the vector $\mathbf{u}$ in the directions of $\mathbf{a}_1$, $\mathbf{a}_2$, and $\mathbf{a}_3$. We use the definition of projection to find the c_is by taking the inner product of both sides of (6.156) with the vectors $\mathbf{a}_i$. For instance, to find c_1 we take the inner product of (6.156) with $\mathbf{a}_1$ and get

$$(\mathbf{u}, \mathbf{a}_1) = c_1(\mathbf{a}_1, \mathbf{a}_1) + c_2(\mathbf{a}_2, \mathbf{a}_1) + c_3(\mathbf{a}_3, \mathbf{a}_1), \tag{6.157}$$

where we used the linearity of the inner product. However, the vectors $\mathbf{a}_i$ are mutually orthogonal, so $(\mathbf{a}_1, \mathbf{a}_i) = 0$ if i is different from one. Thus,

$$c_1 = \frac{(\mathbf{u}, \mathbf{a}_1)}{(\mathbf{a}_1, \mathbf{a}_1)} \tag{6.158}$$

(compare this formula with (6.151)). The determination of c_2 and c_3 is similar.

Example 6.11.1
It is not difficult to show that the three vectors

$$\mathbf{a}_1 = \langle 1, 0, 1 \rangle, \quad \mathbf{a}_2 = \langle 0, 1, 0 \rangle, \quad \mathbf{a}_3 = \langle -1, 0, 1 \rangle,$$

are mutually orthogonal. Let

$$\mathbf{u} = \langle 1, 1, 1 \rangle$$

be a vector in E^3. We use the formulas in (6.158) in order to determine the components of $\mathbf{u}$ in the new basis. We need to find c_1, c_2, and c_3 so that

$$\langle 1, 1, 1 \rangle = c_1 \langle 1, 0, 1 \rangle + c_2 \langle 0, 1, 0 \rangle + c_3 \langle -1, 0, 1 \rangle. \tag{6.159}$$

It follows from (6.158) that $c_1 = 1$, $c_2 = 1$, and $c_3 = 0$. Therefore, the vector that has the representation $\langle 1, 1, 1 \rangle$ in the natural basis has the representation $\langle\langle 1, 1, 0 \rangle\rangle$ in the new basis (we recall our convection of Section 6.4, where double brackets $\langle\langle$ and $\rangle\rangle$ were used to denote vector components in a second basis of a vector space.)

We now describe this procedure in the notation of matrices. Again, we present the notation for E^3 and leave the natural generalization to E^n for n larger or smaller than 3 to an exercise.

Definition 6.11.2 (Orthonormal Vectors)
We call a set of vectors $\{\mathbf{a}_1, \mathbf{a}_2, \ldots, \mathbf{a}_n\}$ an **orthonormal** *basis for E^n if these vectors satisfy*

$$(\mathbf{a}_i, \mathbf{a}_j) = \begin{cases} 0 & \text{if } i \neq j, \\ 1 & \text{if } i = j. \end{cases} \tag{6.160}$$

Let $\{\mathbf{a}_1, \mathbf{a}_2, \mathbf{a}_3\}$ and $\{\mathbf{b}_1, \mathbf{b}_2, \mathbf{b}_3\}$ be two orthornormal bases for E^3. Since $\mathbf{b}_i$s form a basis for E^3, each $\mathbf{a}_j$ can be written as a linear combination of the

$\mathbf{b}_i$s; in other words, there exist scalars α_{ij} such that

$$\begin{cases} \mathbf{a}_1 = \alpha_{11}\mathbf{b}_1 + \alpha_{21}\mathbf{b}_2 + \alpha_{31}\mathbf{b}_3, \\ \mathbf{a}_2 = \alpha_{12}\mathbf{b}_1 + \alpha_{22}\mathbf{b}_2 + \alpha_{32}\mathbf{b}_3, \\ \mathbf{a}_3 = \alpha_{13}\mathbf{b}_1 + \alpha_{23}\mathbf{b}_2 + \alpha_{33}\mathbf{b}_3. \end{cases} \tag{6.161}$$

Since the $\mathbf{b}_i$s are orthonormal, we can readily find a relationship between the entries of the matrix A, defined by

$$A = [\alpha_{ij}], \tag{6.162}$$

and the two bases: Take the inner product of the first equation in (6.161) with $\mathbf{b}_1$ and obtain

$$\alpha_{11} = (\mathbf{b}_1, \mathbf{a}_1). \tag{6.163}$$

The inner product of the same equation with $\mathbf{b}_2$ yields

$$\alpha_{21} = (\mathbf{b}_2, \mathbf{a}_1).$$

In general,

$$\alpha_{ij} = (\mathbf{b}_i, \mathbf{a}_j). \tag{6.164}$$

Now, let $\mathbf{x}$ be a vector in E^3 with components x_1, x_2, x_3 in the $\mathbf{a}_i$ basis, and with components y_1, y_2, y_3 in the $\mathbf{b}_i$ basis; in other words, we have

$$\mathbf{x} = x_1\mathbf{a}_1 + x_2\mathbf{a}_2 + x_3\mathbf{a}_3 = y_1\mathbf{b}_1 + y_2\mathbf{b}_2 + y_3\mathbf{b}_3. \tag{6.165}$$

What is the relationship between the scalars x_i and y_j? To answer this question we substitute the expressions (6.161) in the first part of (6.165). We therefore write $\mathbf{x}$ in terms of the $\mathbf{b}_i$s in the form

$$\mathbf{x} = (\alpha_{11}x_1 + \alpha_{12}x_2 + \alpha_{13}x_3)\mathbf{b}_1 + (\alpha_{21}x_1 + \alpha_{22}x_2 + \alpha_{23}x_3)\mathbf{b}_2 + (\alpha_{31}x_1$$
$$+ \alpha_{32}x_2 + \alpha_{33}x_3)\mathbf{b}_3. \tag{6.166}$$

Comparing (6.166) with the second part of (6.165) and recalling that the $\mathbf{b}_i$s are linearly independent, we arrive at the following algebraic equations that relate the x_is with the y_js:

$$\begin{cases} y_1 = \alpha_{11}x_1 + \alpha_{12}x_2 + \alpha_{13}x_3, \\ y_2 = \alpha_{21}x_1 + \alpha_{22}x_2 + \alpha_{23}x_3, \\ y_3 = \alpha_{31}x_1 + \alpha_{32}x_2 + \alpha_{33}x_3. \end{cases} \tag{6.167}$$

System (6.167) is equivalent to the matrix equation

$$\mathbf{y} = A\mathbf{x} \quad \text{and} \quad \mathbf{x} = A^{-1}\mathbf{y}, \tag{6.168}$$

where A is defined in (6.162) and (6.164). Thus, once we know the two bases and the components of $\mathbf{x}$ in one of the bases, we can obtain the components of $\mathbf{x}$ in the other basis by applying one of the formulas in (6.168). We summarize these findings in the following theorem.

Theorem 6.11.1 (Change of Bases)
Let $\{\mathbf{a}_1, \mathbf{a}_2, \cdots, \mathbf{a}_n\}$ and $\{\mathbf{b}_1, \mathbf{b}_2, \cdots, \mathbf{b}_n\}$ be two sets of orthonormal bases for E^n. Let A be an $n \times n$ matrix with coefficients defined by

$$\alpha_{ij} = (\mathbf{b}_i, \mathbf{a}_j). \tag{6.169}$$

Let $\mathbf{x}$ be a vector in E^n with the following representations in terms of two sets of bases:

$$\mathbf{x} = x_1\mathbf{a}_1 + x_2\mathbf{a}_2 + \cdots + x_n\mathbf{a}_n = y_1\mathbf{a}_1 + y_2\mathbf{a}_2 + \cdots + y_n\mathbf{a}_n. \tag{6.170}$$

Then the components $\{x_1, x_2, \ldots, x_n\}$ and $\{y_1, y_2, \ldots, y_n\}$ are related by

$$\mathbf{y} = A\mathbf{x}. \tag{6.171}$$

Example 6.11.2
Let $\{\mathbf{i}, \mathbf{j}\}$ be the natural basis in E^2. Let

$$\{\mathbf{a}_1, \mathbf{a}_2\} = \left\{ \frac{1}{\sqrt{2}}(\mathbf{i} - \mathbf{j}), \frac{1}{\sqrt{2}}(\mathbf{i} + \mathbf{j}) \right\} \tag{6.172}$$

and

$$\{\mathbf{b}_1, \mathbf{b}_2\} = \left\{ \frac{1}{\sqrt{5}}(\sqrt{2}\mathbf{i} + \sqrt{3}\mathbf{j}), \frac{1}{\sqrt{5}}(\sqrt{3}\mathbf{i} - \sqrt{2}\mathbf{j}) \right\}. \tag{6.173}$$

In this case, the matrix A defined in (6.169) is

$$\frac{1}{\sqrt{10}} \begin{bmatrix} \sqrt{2} + \sqrt{3} & \sqrt{3} - \sqrt{2} \\ \sqrt{2} - \sqrt{3} & \sqrt{3} + \sqrt{2} \end{bmatrix}. \tag{6.174}$$

Let $\mathbf{x}$ be a vector in E^2 that has coordinates $\langle\langle 1, 1 \rangle\rangle$ in the $\mathbf{a}_i$ basis (i.e., $\mathbf{x} = \mathbf{a}_1 + \mathbf{a}_2$). Its coordinates in the $\mathbf{b}_i$ basis are given by

$$\begin{bmatrix} y_1 \\ y_2 \end{bmatrix} = \frac{1}{\sqrt{10}} \begin{bmatrix} \sqrt{2} + \sqrt{3} & \sqrt{3} - \sqrt{2} \\ \sqrt{2} - \sqrt{3} & \sqrt{3} + \sqrt{2} \end{bmatrix} \begin{bmatrix} 1 \\ 1 \end{bmatrix} = \frac{2}{\sqrt{10}} \begin{bmatrix} \sqrt{3} \\ \sqrt{2} \end{bmatrix}. \tag{6.175}$$

Problems

1. Show that each set of vectors forms a basis in E^2. Let $\mathbf{x} = \mathbf{i} + \mathbf{j}$. Find the components of $\mathbf{x}$ in the new basis.

   (a) $\{2\mathbf{i} - \mathbf{j}, 1/2\mathbf{i} + \mathbf{j}\}$
   (b) $\{\mathbf{i} - 2\mathbf{j}, \mathbf{i} + 1/2\mathbf{j}\}$
   (c) $\{\mathbf{i} + 3\mathbf{j}, 2\mathbf{i} - 2/3\mathbf{j}\}$
   (d) $\{\mathbf{i} + \frac{1}{2}\mathbf{j}, 2\mathbf{i} - 4\mathbf{j}\}$

2. Show that each set of vectors forms a basis in E^3. Let $\mathbf{x} = \mathbf{i} + 2\mathbf{j} - \mathbf{k}$. Find the components of $\mathbf{x}$ in the new basis.

    (a) $\{2\mathbf{i} - \mathbf{k}, \mathbf{j}, \mathbf{i} + 2\mathbf{k}\}$

    (b) $\{\mathbf{i} + \mathbf{j}, -\mathbf{i} + \mathbf{j} + \mathbf{k}, \mathbf{i} - \mathbf{j} + 2\mathbf{k}\}$

    (c) $\{\mathbf{i} - \mathbf{j}, 2\mathbf{i} + 2\mathbf{j} - 3\mathbf{k}, 3\mathbf{i} + 3\mathbf{j} + 4\mathbf{k}\}$

    (d) $\{3\mathbf{i} - 2\mathbf{k}, \mathbf{j}, 1/3\mathbf{i} + 1/2\mathbf{k}\}$

3. Show that the matrix A defined by (6.162) is nonsingular.

4. Show that $\langle 1, 2, 1 \rangle$, $\langle -1, 1, -1 \rangle$, and $\langle -3, 0, 3 \rangle$ form a mutually orthogonal basis for E^3. What are the components of the vector $3\mathbf{i} - 2\mathbf{j} + 3\mathbf{k}$ in this new basis?

5. Show by direct calculation that if A is any 2×2 matrix and $\mathbf{a}$ and $\mathbf{b}$ are any two vectors in E^2, then

$$(\mathbf{a}, A\mathbf{b}) = (A^T\mathbf{a}, \mathbf{b}), \qquad (6.176)$$

where A^T is the transpose of A.

6. Let $\mathbf{b}_1 = \alpha(\mathbf{i} + 2\mathbf{j})$. Choose α and a second vector $\mathbf{b}_2$ so that the set $\{\mathbf{b}_1, \mathbf{b}_2\}$ forms an orthonormal basis for E^2. With $\mathbf{a}_1 = \mathbf{i}$ and $\mathbf{a}_2 = \mathbf{j}$, find the matrix A that defines the change of coordinate systems. Let $\mathbf{x}$ have coordinates $\langle\langle 1, -1 \rangle\rangle$ in the $\mathbf{b}_i$ basis. Find its coordinates in the natural basis.

7. Consider the two sets of vectors $\{\langle 1, 0, 1 \rangle, \langle 0, 1, 0 \rangle, \langle -1, 0, 1 \rangle\}$ and $\{\langle 1, 1, 1 \rangle, \langle -1, 0, 1 \rangle, \langle 1, -2, -1 \rangle\}$, where each vector is written in terms of the natural basis of E^3. Show that the vectors in each set are mutually orthogonal. Normalize these vectors so that each set forms an orthonormal basis for E^3. Find the matrix A that defines the change of coordinates in these bases. Let $\mathbf{x}$ have coordinates $\langle\langle 1, 0, 2 \rangle\rangle$ in the first basis. Find its coordinates in the second basis.

8. Recall that the equations

$$\begin{aligned} x &= r \cos\theta, \\ y &= r \sin\theta, \qquad\qquad (6.177) \\ z &= z \end{aligned}$$

define the cylindrical coordinates at a point $(x, y, z) \in R^3$. The set of vectors $\{\cos\theta\,\mathbf{i} + \sin\theta\,\mathbf{j}, -\sin\theta\,\mathbf{i} + \cos\theta\,\mathbf{j}, \mathbf{k}\}$ forms an orthonormal basis for this system at the point (x, y, z). Similarly, the equations

$$\begin{aligned} x &= \rho\sin\phi\cos\theta, \\ y &= \rho\sin\phi\sin\theta, \qquad\qquad (6.178) \\ z &= \rho\cos\phi \end{aligned}$$

define the spherical coordinates of a point $(x, y, z) \in R^3$. In this case, the vectors $\langle \sin\phi\cos\theta, \sin\phi\sin\theta, \cos\phi \rangle$, $\langle \cos\phi\cos\theta, \cos\phi\sin\theta, -\sin\phi \rangle$, and $\langle -\sin\theta, \cos\theta, 0 \rangle$ form an orthonormal basis at the point (x, y, z). Find the matrix A that relates these two coordinate systems. Compute the inverse of this matrix. Check your calculations by introducing the formulas (6.161) through (6.168) to *Mathematica* or *MATLAB*.

6.12 Eigenvalues and Eigenvectors

In this section we take up the concept of eigenvalues and eigenvectors of matrices. This topic constitutes one of the most important topics in mathematical physics with applications ranging from solving systems of ordinary and partial differential equations to analyzing the stability and control of physical motions (such as an aircraft or a submarine) to fracture and failure in everyday appliances.

The definition of an eigenvector of a matrix is best understood when we recall that an $m \times n$ matrix A can be viewed as a representation of a linear transformation that maps vectors from the vector space E^n to the vector space E^m. This linear transformation is defined by

$$T(\mathbf{x}) = A\mathbf{x}, \quad \mathbf{x} \in E^n. \tag{6.179}$$

In general, this transformation maps a vector $\mathbf{x}$ in E^n to another vector $A\mathbf{x}$ in E^m. An interesting situation arises when the domain and the range of T are the same vector space E^n, or in other words A is an $n \times n$ matrix. We then ask the question: Is there any vector $\mathbf{x}$ in E^n that is mapped by T into a vector in the same direction as $\mathbf{x}$? Or equivalently, is there a direction in the vector space E^n that remains invariant under T? Such vectors or directions are called **eigenvectors**.

Definition 6.12.1 (Eigenvalue and Eigenvector)
Let A be an $n \times n$ matrix. A nonzero vector $\mathbf{x}$ is called an eigenvector of A corresponding to the eigenvalue λ if

$$A\mathbf{x} = \lambda\mathbf{x}. \tag{6.180}$$

Example 6.12.1
Consider the 2×2 matrix

$$A = \begin{bmatrix} 1 & -1 \\ -1 & 1 \end{bmatrix}.$$

The reader can show by direct computation that

$$\begin{bmatrix} 1 & -1 \\ -1 & 1 \end{bmatrix} \begin{bmatrix} 1 \\ 1 \end{bmatrix} = \begin{bmatrix} 0 \\ 0 \end{bmatrix}, \tag{6.181}$$

so 0 is an eigenvalue of this matrix corresponding to the eigenvector $\begin{bmatrix} 1 \\ 1 \end{bmatrix}$. Similarly,

$$\begin{bmatrix} 1 & -1 \\ -1 & 1 \end{bmatrix} \begin{bmatrix} 1 \\ -1 \end{bmatrix} = 2 \begin{bmatrix} 1 \\ -1 \end{bmatrix}, \tag{6.182}$$

so 2 is an eigenvalue of this matrix corresponding to the eigenvector $\begin{bmatrix} 1 \\ -1 \end{bmatrix}$.

Of course, we do not want to guess what values constitute as eigenvalues of a matrix. Going back to the definition of λ and $\mathbf{x}$ in (6.180), we note that these quantities must satisfy the system of linear equations $A\mathbf{x} = \lambda\mathbf{x}$. This system is in turn equivalent to the system

$$(A - \lambda I)\mathbf{x} = \mathbf{0}. \tag{6.183}$$

This expression defines a system of linear equations for the unknown $\mathbf{x}$. Clearly $\mathbf{x} = \mathbf{0}$ is a solution of (6.183). However, as stated in Definition 6.12.1, $\mathbf{x}$ is an eigenvector of A if $\mathbf{x}$ is a *nonzero* (or nontrivial) vector which satisfies (6.183). In order for such a vector to exist, the system in (6.183) must have multiple solutions. In that case the matrix $A - \lambda I$ must not have an inverse, since if B is the inverse of $A - \lambda \mathbf{I}$, (6.183) has the unique solution $\mathbf{x} = B\mathbf{0}$, which is the zero vector. We saw in Section 6.10 that the necessary and sufficient condition for a matrix A not to have an inverse is that its determinant must vanish. Thus, the condition for λ to be an eigenvalue of A is that the determinant of $A - \lambda I$ must be zero. We have proved the following result.

Lemma 6.12.1
λ is an eigenvalue of a matrix A if it is a root of the polynomial equation

$$\det(A - \lambda I) = 0. \tag{6.184}$$

The function

$$p(\lambda) = \det(A - \lambda I)$$

is called the **characteristic polynomial** of A and equation (6.184) its **characteristic equation**.

Example 6.12.2
Let A be the matrix defined in Example 6.12.1. In order for λ to be an eigenvalue of this matrix condition (6.184) must hold, that is,

$$\det \begin{bmatrix} 1 - \lambda & -1 \\ -1 & 1 - \lambda \end{bmatrix} = 0. \tag{6.185}$$

The characteristic polynomial of A is

$$p(\lambda) = \lambda^2 - 2\lambda. \tag{6.186}$$

This polynomial has the two roots 0 and 2. To find the eigenvector corresponding to 0, say, we need to go back to (6.183) and solve that system using Gaussian elimination. Starting with the augmented matrix

$$\begin{bmatrix} 1 & -1 & 0 \\ -1 & 1 & 0 \end{bmatrix}, \tag{6.187}$$

and applying one step of the Gaussian elimination, we end up with

$$\begin{bmatrix} 1 & -1 & 0 \\ 0 & 0 & 0 \end{bmatrix}. \tag{6.188}$$

Noting that the second row of (6.188) is the zero vector, we conclude that the components of

$$\mathbf{x} = \begin{bmatrix} x_1 \\ x_2 \end{bmatrix}$$

must satisfy only the first equation

$$x_1 - x_2 = 0. \tag{6.189}$$

Since we have one equation in two unknowns, we let one of the variables, say x_1, take any nonzero value. Let $x_1 = 1$. Then $x_2 = 1$ from (6.189). So the eigenvector corresponding to the eigenvalue 0 is

$$\begin{bmatrix} 1 \\ 1 \end{bmatrix}.$$

Similarly, when $\lambda = 2$ the system (6.183) becomes

$$\begin{bmatrix} -1 & -1 & 0 \\ -1 & -1 & 0 \end{bmatrix}. \tag{6.190}$$

Again one step of the Gaussian elimination results in

$$\begin{bmatrix} -1 & -1 & 0 \\ 0 & 0 & 0 \end{bmatrix}, \tag{6.191}$$

which leads to the eigenvector

$$\begin{bmatrix} 1 \\ -1 \end{bmatrix}.$$

Example 6.12.3
Consider the 3×3 matrix

$$A = \begin{bmatrix} 0 & 1 & 0 \\ 0 & 0 & 1 \\ 6 & -11 & 6 \end{bmatrix} \tag{6.192}$$

The charateristic polynomial of A is

$$p(\lambda) = \det(A - \lambda I) = -\lambda^3 + 6\lambda^2 - 11\lambda + 6,$$

with roots

$$\lambda_1 = 1, \quad \lambda_2 = 2, \quad \lambda_3 = 3.$$

Let us begin with $\lambda = 1$ and determine its associated eigenvector. The system of linear equations $A\mathbf{x} = \mathbf{x}$ leads to the augmented matrix

$$\begin{bmatrix} -1 & 1 & 0 & 0 \\ 0 & -1 & 1 & 0 \\ 6 & -11 & 5 & 0 \end{bmatrix}.$$

We apply two steps of the Gaussian elimination this matrix and obtain

$$\begin{bmatrix} -1 & 1 & 0 & 0 \\ 0 & -1 & 1 & 0 \\ 0 & 0 & 0 & 0 \end{bmatrix}.$$

Letting $z = 1$ in the first two equations that result from the first two rows of the above matrix, we find

$$x = y = 1.$$

Thus,

$$\mathbf{e}_1 = \begin{bmatrix} 1 \\ 1 \\ 1 \end{bmatrix}$$

is the eigenvector associated with $\lambda_1 = 1$. The augmented matrix associated with $\lambda_2 = 2$ is

$$\begin{bmatrix} -2 & 1 & 0 & 0 \\ 0 & -2 & 1 & 0 \\ 6 & -11 & 4 & 0 \end{bmatrix}.$$

Again, two steps of the Gaussian elimination lead to

$$\begin{bmatrix} -2 & 1 & 0 & 0 \\ 0 & -2 & 1 & 0 \\ 0 & 0 & 05 & 0 \end{bmatrix}.$$

Let $z = 1$. Then $x = \frac{1}{4}$ and $y = \frac{1}{2}$. Thus, the eigenvector associated with $\lambda_2 = 2$ is

$$\mathbf{e}_2 = \begin{bmatrix} \frac{1}{4} \\ \frac{1}{2} \\ 1 \end{bmatrix}.$$

Similarly, the eigenvector associated with $\lambda_3 = 3$ is

$$\mathbf{e}_3 = \begin{bmatrix} \frac{1}{9} \\ \frac{1}{3} \\ 1 \end{bmatrix}.$$

We note that any nonzero multiple of $\mathbf{e}_i$ is also an eigenvector associated with the eigenvalue λ_i. For instance,

$$9\mathbf{e}_3 = \begin{bmatrix} 1 \\ 3 \\ 9 \end{bmatrix}.$$

is an eigenvector associated with λ_3.

Example 6.12.4 (Eigenvalues in *Mathematica* and *MATLAB*)

In *Mathematica* the internal functions `Eigenvalues[A]` and `Eigenvectors[A]` determine the eigenvalues and eigenvectors of A, respectively. We can also get hold of the characteristic polynomial of a matrix by using its definition in *Mathematica*:

```
Det[A - l IdentityMatrix[A]]
```

The following *Mathematica* commands determine the eigenvalues and eigenvectors of the matrix in the preceding example:

```
a = {{1, -1}, {-1, 1}};
eigvals = Eigenvalues[a];
eigvecs = Eigenvectors[a];
```

The function `Eigensystem[a]` produces the eigenvalues and eigenvectors of `a` in one operation. We point out that *Mathematica* attempts to determine the eigenvalues and eigenvectors symbolically, unless we tell it to proceed with its numerical algorithms. To do so we need to replace the matrix `a` with `N[a]` when calling upon `Eigenvalues` and `Eigenvectors`.

MATLAB accomplishes the task of determining the eigenvalues and eigenvectors of a matrix in much the same way. Let B be the matrix

$$B = \begin{bmatrix} a & -a \\ -a & a \end{bmatrix}. \tag{6.193}$$

Here is how one proceeds to determine the eigenvalues and eigenvectors of A:

```
A = sym('[a -a; -a -a]');
[V, E] = eig(A);
```

The middle line indicates that the eigenvectors of A are stored in V and its eigenvalues in E. *MATLAB's* output will have two parameters in it denoted by `E(1)` and `E(2)` that refer to the eigenvalues in `E`.

The eigenvalues of a matrix A do not have to be real numbers even when all entries of A belong to R, the set of real numbers. Since eigenvalues are roots of a polynomial with real coefficients, it is easy to construct examples in which the eigenvalues have nonzero imaginary parts. In such cases, the eigenvectors will have complex numbers in them as well. Although a vector with complex entries does not have a geometric interpretation, these eigenvectors still play a crucial role in all aspects of physics, in general, and in wave motion, in particular.

Example 6.12.5 (Complex Eigenvalues and Eigenvectors)
Consider the matrix

$$A = \begin{bmatrix} 1 & 2 \\ -2 & 1 \end{bmatrix}.$$

Its characteristic polynomial is

$$p(\lambda) = \lambda^2 - 2\lambda + 5$$

with roots

$$\lambda_1 = 1 + 2i, \quad \lambda_2 = 1 - 2i.$$

Let us determine the eigenvector associated with λ_1. The system of linear equations $A\mathbf{x} = (1 + 2i)\mathbf{x}$ leads to the augmented matrix

$$\begin{bmatrix} -2i & 2 & 0 \\ -2 & -2i & 0 \end{bmatrix}.$$

One step of Gaussian elimination (multiply the first row by i and add to the second row) yields

$$\begin{bmatrix} -2i & 2 & 0 \\ 0 & 0 & 0 \end{bmatrix}.$$

Let $y = 1$. The first equation resulting from the above system gives us $x = -i$. Thus

$$\mathbf{e}_1 = \begin{bmatrix} -i \\ 1 \end{bmatrix},$$

is the eigenvector associated with the eigenvalue $1 + 2i$. Similarly,

$$\mathbf{e}_2 = \begin{bmatrix} i \\ 1 \end{bmatrix},$$

is the eigenvector associated with $1 - 2i$.

The characteristic polynomial, defined by (6.184), is an nth order polynomial in λ with coefficients that are functions of the entries of A. The fundamental theorem of algebra states that such a polynomial has n roots, real or complex, and counting multiplicity. When a root of this polynomial repeats, it is then possible that more than one (linearly independent) eigenvector is associated with that eigenvalue. In general, we expect that if an eigenvalue has multiplicity m, we will be able to find m linearly independent eigenvectors corresponding to it. This, however, is not always the case and on occasion there are fewer than m linearly independent eigenvectors available in a physical problem. We will see examples of all of these cases in the exercises at the end of this section.

Example 6.12.6
Consider the matrix

$$A = \begin{bmatrix} 0 & 1 \\ 0 & 0 \end{bmatrix}.$$

Both eigenvalues of A are 0. However, there is only one eigenvector associated with 0. To see this, we construct the augmented matrix for $A\mathbf{x} = \mathbf{0}$, which is

$$\begin{bmatrix} 0 & 1 & 0 \\ 0 & 0 & 0 \end{bmatrix}.$$

This matrix is already in upper echelon form. The equation resulting from the first row of the matrix is

$$y = 0,$$

implying that the variable x is arbitrary. We let $x = 1$. Thus the vector

$$\mathbf{e}_1 = \begin{bmatrix} 1 \\ 0 \end{bmatrix}$$

is the (only) eigenvector associated with $\lambda = 0$.

We say an $n \times n$ matrix A has a **complete** set of eigenvectors if it possesses n linearly independent eigenvectors. Let A be such a matrix with $\mathbf{a}_1$, $\mathbf{a}_2$, $\cdots$, $\mathbf{a}_n$ standing for its eigenvectors corresponding to the eigenvalues λ_1, λ_2, $\cdots$, λ_n. We emphasize that the eigenvalues do not have to be distinct. Being linearly independent, the set of eigenvectors forms a basis for E^n. Thus any vector $\mathbf{u} \in E^n$ can be written as a linear combination of these vectors as

$$\mathbf{u} = c_1 \mathbf{a}_1 + c_2 \mathbf{a}_2 + \cdots + c_n \mathbf{a}_n \qquad (6.194)$$

for some set of constants $\{c_1, \cdots, c_n\}$. Because each $\mathbf{a}_i$ is an eigenvector of A we can determine the action of A on $\mathbf{u}$ in the following way:

$$A\mathbf{u} = A(c_1 \mathbf{a}_1 + \cdots + c_n \mathbf{a}_n) = c_1 A\mathbf{a}_1 + \cdots + c_n A\mathbf{a}_n = c_1 \lambda_1 \mathbf{a}_1 + \cdots + c_n \lambda_n \mathbf{a}_n,$$
$$(6.195)$$

where we used the relation $A\mathbf{a}_i = \lambda_i \mathbf{a}_i$. So by knowing the directions along which A remains invariant we are able to find the action of A on *any* vector in E^n by (6.195).

Let us assume further that the eigenvectors of A are in fact mutually orthogonal, that is,

$$(\mathbf{a}_i, \mathbf{a}_j) = 0 \quad \text{if } i \neq j, \qquad (6.196)$$

and that they are **normalized**, that is,

$$(\mathbf{a}_i, \mathbf{a}_i) = 1 \quad \text{for every } i. \qquad (6.197)$$

Let us take the inner product of $A\mathbf{a}_i = \lambda_i \mathbf{a}_i$ with $\mathbf{a}_i$:

$$(\mathbf{a}_i, A\mathbf{a}_i) = \lambda_i, \qquad (6.198)$$

and take the inner product of the same expression with $\mathbf{a}_j$:

$$(\mathbf{a}_i, A\mathbf{a}_j) = 0 \quad \text{if } \quad i \neq j. \qquad (6.199)$$

We note that the right-hand sides of (6.198) and (6.199) are precisely the entries of the diagonal matrix

$$\text{diag}(\lambda_1, \lambda_2, \cdots, \lambda_n) \equiv \begin{bmatrix} \lambda_1 & 0 & 0 & \cdots & 0 \\ 0 & \lambda_2 & 0 & \cdots & 0 \\ \cdots & & & & \\ \cdots & & & & \\ 0 & 0 & \cdots & \cdots & \lambda_n \end{bmatrix}. \qquad (6.200)$$

Let Q stand for the matrix of eigenvectors, where each column of Q is an eigenvector of A. Then the left-hand sides of (6.198) and (6.199) are precisely the entries of the $Q^{\mathrm{T}}AQ$ (recall that the ijth entry of the product of two matrices A and B is the inner (dot) product of the ith row of A with the jth column of B). We summarize the above discussion in the following lemma.

Lemma 6.12.2 (Diagonalizable Matrices)

Let A be an $n \times n$ matrix with a complete set of n mutually orthogonal and normalized eigenvectors. Let Q stand for the matrix of eigenvectors. Then the matrix Q diagonalizes A, that is,

$$Q^{\mathrm{T}}AQ = \mathrm{diag}(\lambda_1, \cdots, \lambda_n). \tag{6.201}$$

Definition 6.12.3 (Unitary Matrices)

*An $n \times n$ real-valued matrix Q is called a **unitary** (or **orthogonal**) matrix if it satisfies*

$$Q^{-1} = Q^{\mathrm{T}}. \tag{6.202}$$

We now state two important lemmas concerning the diagonalization of special matrices.

Lemma 6.12.3

Let A be an $n \times n$ matrix with n distinct eigenvalues. Then the eigenvectors form a set of linearly independent vectors for E^n. Let Q be the matrix of eigenvectors of A. Then

$$Q^{-1}AQ = \mathrm{diag}(\lambda_1, \cdots, \lambda_n). \tag{6.203}$$

Proof: Let $T = \{\mathbf{a}_1, \mathbf{a}_2, \cdots, \mathbf{a}_n\}$ be the eigenvectors of A. By way of contradiction, suppose that these eigenvectors are not linearly independent. Let $\{\mathbf{a}_1, \mathbf{a}_2, \cdots, \mathbf{a}_m\}$ be the largest set of linearly independent vectors in T (note that $m > 0$ because eigenvectors are nonzero vectors). Then $\{\mathbf{a}_1, \mathbf{a}_2, \cdots, \mathbf{a}_m, \mathbf{a}_{m+1}\}$ must be linearly dependent. Let $c_1, c_2, \cdots, c_{m+1}$ be such that

$$c_1\mathbf{a}_1 + c_2\mathbf{a}_2 + \cdots + c_m\mathbf{a}_m + c_{m+1}\mathbf{a}_{m+1} = \mathbf{0}, \tag{6.204}$$

with at least one of c_is nonzero. Let A act on the preceding expression. Using the fact that $\mathbf{a}_i$s are the eigenvectors of A, we get

$$c_1\lambda_1\mathbf{a}_1 + c_2\lambda_2\mathbf{a}_2 + \cdots + c_m\lambda_m\mathbf{a}_m + c_{m+1}\lambda_{m+1}\mathbf{a}_{m+1} = \mathbf{0}. \tag{6.205}$$

Next multiply (6.204) by λ_{m+1} and subtract from (6.205). We obtain a linear combination of $\mathbf{a}_i$s with $i = 1, \cdots, m$ that is equal to $\mathbf{0}$. Since the latter vectors are linearly independent, their coefficients must be zero. But these coefficients are

$$c_1(\lambda_1 - \lambda_{m+1}), \cdots, c_m(\lambda_m - \lambda_{m+1}). \tag{6.206}$$

Since all of the eigenvalues are distinct, all the c_is in (6.206) must be zero. It then follows from (6.204) that $c_{m+1} = 0$, which is a contradiction. The derivation of (6.203) is now straightforward.

Lemma 6.12.4 (Properties of Symmetric Matrices)
Let A be an $n \times n$ real-valued symmetric matrix. Then,

1. All eigenvalues of A are real.

2. Eigenvectors corresponding to distinct eigenvalues are orthogonal.

3. $Q^{-1} = Q^{\mathrm{T}}$.

Proof: 1. Let $\mathbf{x}$ be an eigenvector of A corresponding to the eigenvalue λ. Then

$$A\mathbf{x} = \lambda\mathbf{x}. \tag{6.207}$$

We note that $A = \bar{A}$ since A is real-valued. Here $\bar{x}$ is the complex conjugate of x. Apply the complex conjugate operation to both sides of (6.207). We arrive at

$$A\bar{\mathbf{x}} = \bar{\lambda}\bar{\mathbf{x}}. \tag{6.208}$$

Next take the inner product of both sides of (6.208) with $\mathbf{x}$ to obtain

$$(A\bar{\mathbf{x}}, \mathbf{x}) = \bar{\lambda}(\bar{\mathbf{x}}, \mathbf{x}). \tag{6.209}$$

But $(A\mathbf{x}, \mathbf{y}) = (\mathbf{x}, \mathbf{y})$ since A is a symmetric matrix. Equation (6.209) now becomes

$$(\bar{\mathbf{x}}, A\mathbf{x}) = \bar{\lambda}(\bar{\mathbf{x}}, \mathbf{x}). \tag{6.210}$$

Equation (6.207) simplifies the left-hand side of (6.210). We have

$$\lambda(\bar{\mathbf{x}}, \mathbf{x}) = \bar{\lambda}(\bar{\mathbf{x}}, \mathbf{x}). \tag{6.211}$$

Since $\mathbf{x}$ is an eigenvector, the quantity $(\bar{\mathbf{x}}, \mathbf{x})$ is nonzero. Expression (6.211) now implies $\lambda = \bar{\lambda}$, which in turn states that λ is real.

2. In order to show that eigenvectors corresponding to distinct eigenvalues are orthogonal, we let $\mathbf{x}$ and $\mathbf{y}$ correspond to two eigenvalues λ and μ, which must satisfy

$$A\mathbf{x} = \lambda\mathbf{x}, \quad A\mathbf{y} = \mu\mathbf{y}. \tag{6.212}$$

We take the inner product of (6.212a) with $\mathbf{y}$ and (6.212b) with $\mathbf{x}$. We have

$$(A\mathbf{x}, \mathbf{y}) = \lambda(\mathbf{x}, \mathbf{y}), \quad (A\mathbf{y}, \mathbf{x}) = \mu(\mathbf{y}, \mathbf{x}). \tag{6.213}$$

Since A is symmetric, the left-hand sides of (6.213a) and (6.213b) are equal. Then $(\lambda - \mu)(\mathbf{x}, \mathbf{y}) = 0$, which implies $(\mathbf{x}, \mathbf{y}) = 0$ since λ and μ are distinct. Therefore, $\mathbf{x}$ and $\mathbf{y}$ are orthogonal.

We leave the proof of 3 to the reader.

We have not shown that the eigenvectors of a symmetric matrix A form a complete set of linearly independent vectors, although this statement is a fact. The proof of this assertion requires a careful study of the eigenvectors corresponding to those roots of the characteristic polynomial that may have multiplicity larger than one. We will not pursue the proof of this fact here because its proof is beyond the scope of this text, but allow ourselves the following important conclusion that follows from the completeness of the eigenvectors of symmetric matrices.

Lemma 6.12.5

Every symmetric matrix is diagonalizable. Moreover, the matrix Q of the eigenvectors of A is a unitary matrix.

We will come back to these results when we study systems of linear ODEs in the next chapter. We would like to caution the reader, however, that the class of diagonalizable matrices is larger than the special matrices we discussed. In particular, there are real-valued matrices that are neither symmetric nor have distinct eigenvalues but that are still diagonalizable.

We finish the section by stating the following theorem, whose proof is left to the reader.

Theorem 6.12.1 (Null Space)

*The set of all eigenvectors corresponding to an eigenvalue of A forms a vector space. The **Null Space** is the vector space corresponding to the zero eigenvalue.*

Remark 6.12.1: Note that the null space is empty when zero is not an eigenvalue of A.

Both *Mathematica* and *MATLAB* have internal functions capable of constructing the null space of a matrix.

Problems

1. Find all eigenvalues and eigenvectors of the following matrices. Check your result in either *Mathematica* or *MATLAB*.

   (a) $\begin{bmatrix} 1 & 1 \\ 1 & 0 \end{bmatrix}$

   (b) $\begin{bmatrix} 1 & 1 \\ 1 & 1 \end{bmatrix}$

   (c) $\begin{bmatrix} -1 & 1 \\ 1 & 3 \end{bmatrix}$

   (d) $\begin{bmatrix} a & b \\ c & d \end{bmatrix}$

(e) $\begin{bmatrix} a & a \\ -a & a \end{bmatrix}$

(f) $\begin{bmatrix} 0 & 1 & 0 \\ 0 & 0 & 1 \\ 6 & -11 & 6 \end{bmatrix}$

(g) $\begin{bmatrix} 0 & 1 & 0 \\ 0 & 0 & 1 \\ a^3 & -3a^2 & 3a \end{bmatrix}$

(h) $\begin{bmatrix} 2 & -1 & 0 \\ -1 & 2 & -1 \\ 0 & -1 & 2 \end{bmatrix}$

(i) $\begin{bmatrix} 0 & 1 & 0 \\ 1 & 0 & 0 \\ 0 & 0 & 1 \end{bmatrix}$

(j) $\begin{bmatrix} 1 & \frac{1}{2} & \frac{1}{3} \\ \frac{1}{2} & \frac{1}{3} & \frac{1}{4} \\ \frac{1}{3} & \frac{1}{4} & \frac{1}{6} \end{bmatrix}$

(k) $\begin{bmatrix} 0 & -1 & 0 & 0 \\ -1 & 0 & 0 & 0 \\ 0 & 0 & -1 & 0 \\ 0 & 0 & 0 & -1 \end{bmatrix}$

(l) $\begin{bmatrix} 2 & -1 & 0 & 0 \\ -1 & 2 & -1 & 0 \\ 0 & -1 & 2 & -1 \\ 0 & 0 & -1 & 2 \end{bmatrix}$

(m) $\begin{bmatrix} 0 & 1 & 0 & 0 & 0 \\ 0 & 0 & 1 & 0 & 0 \\ 0 & 0 & 0 & 1 & 0 \\ 0 & 0 & 0 & 0 & 1 \\ 108 & -216 & 171 & -67 & 13 \end{bmatrix}$ (Hint: Use *Mathematica* or *MAT-*

LAB to show that roots of the characteristic polynomial are 2, 2, 3, 3, and 3.)

2. Let $A = \begin{bmatrix} 0 & 1 \\ 1 & 0 \end{bmatrix}$. Construct the matrix Q and verify the conclusion of Lemma 6.12.2. What does Q^TQ equal? What does your result say about Q and Q^T?

3. Find all eigenvalues and eigenvectors of the matrix $\begin{bmatrix} 2 & 1 & 0 \\ 1 & 1 & 1 \\ 0 & 1 & -3 \end{bmatrix}$. Do the same as in Problem 2.

4. Prove Theorem 6.12.1.

5. Let $A = \begin{bmatrix} a & b \\ c & d \end{bmatrix}$. Show that the product of the eigenvalues of A equals the determinant of A. Similarly, show that the sum of the eigenvalues equals trace of A (the trace of a square matrix A, denoted by $\text{Tr}(A)$, is the sum of its entries on the diagonal). How would you use either *Mathematica* or *MATLAB* to prove this statement?

6. Let A be any 3×3 matrix. Prove the analogue of the previous problem for A.

7. Let λ be an eigenvalue of a matrix A. Let Q be a nonsingular matrix with the same dimensions as A. Show that λ is an eigenvalue of $Q^{-1}AQ$. (Hint: Use the fact that $\det(AB) = \det(A)\det(B)$.)

8. (**Cayley–Hamilton Theorem**) Let A be a 2×2 matrix. Let $p(\lambda)$ be its characteristic polynomial. Show that $p(A) = Z$, where Z is the 2×2 zero matrix (for instance, if $p(\lambda) = \lambda^2 - 2\lambda + 3$, by $p(A)$ we mean $A^2 - 2A + 3I$, where I is 2×2 identity matrix). How would you prove this result in either *Mathematica* ot *MATLAB*?

9. Use either *Mathematica* or *MATLAB* to prove the Cayley–Hamilton theorem for 3×3 matrices.

10. Let A be a 2×2 real-valued anti-symmetric matrix. Show that A has only pure imaginary eigenvalues.

11. Let A be a 3×3 real-valued anti-symmetric matrix. Show that

    (a) $\lambda = 0$ is an eigenvalue of A

    (b) A has two pure imaginary eigenvalues

12. Prove that the determinant of a unitary matrix is 1.

13. Prove that the eigenvalues of a unitary matrix have absolute value equal to one.

6.13 Project A—Dot Product and Projections

In this project we will carry out the details of the *Mathematica* program, whose output is Figure 6.1. This program's goal is to take two vectors **a** and **b** and graph the projection of one on the other.

Graphing vectors in R^2 versus R^3 are quite different in *Mathematica*. We first develop the projection program in R^2 and then remark on what other tools are needed to have an equivalent program in R^3. The following program draws the projection of the vector $\mathbf{a} = \langle 1, 1 \rangle$ on $\mathbf{b} = \langle 2, 0.5 \rangle$:

```
<<Graphics'Arrow'
a = {1, 1};
b = {2, 0.5};
c = Graphics[{Arrow[{0,0}, a], Text["a", {0.8, 0.9}]}];
d = Graphics[{Arrow[{0,0},b], Text["b", {1.9, 0.4}]}];
e = (a . b)/(b . b) b;
f = Graphics[Arrow[{0,0}, e]];
g = Graphics[{Dashing[{0.05, 0.05}], Line[{a, e}]}];
output = Show[c, d, f, g, AspectRatio->Automatic]
```

The first line of the program makes the subroutine **Arrow** available to *Mathematica*. This program allows the user to draw a vector once the coordinates of the tip and the origin of the vector are given to it. For example,

```
Graphics[{Arrow[{0, 0}, a]}];
```

draws an arrow that starts with $(0,0)$ and ends up at **a**. *Mathematica* generates the necessary data to graph the vector, but the output is not displayed on the screen. The **Show** command acting on the preceding output displays the graph of the vector. Note that the line that defines **e** is the place where our concept of projection is introduced to the program.

1. Create a file that contains the preceding instructions in it. Execute this program in *Mathematica* to produce Figure 6.1.

2. Alter the appropriate lines in the previous program and draw the projections of **a** on **b** for the following pairs.

    (a) $\mathbf{a} = \langle 2, 2 \rangle$, $\mathbf{b} = \langle -0.2, 1 \rangle$

    (b) $\mathbf{a} = \langle 2, 2 \rangle$, **b** a unit vector on the x-axis

    (c) $\mathbf{a} = \langle 2, 2 \rangle$, **b** a unit vector on the y-axis

3. In order to draw vectors in R^3, one must first introduce a subroutine to *Mathematica* (similar to **Arrow** above) that enables it to draw such vectors. The following program will effectively do the job of **Arrow** in R^3:

```
Needs["Graphics'PlotField3D'"];
arrow3D[point1:{x1_, y1_, z1_}, point2:{x2_, y2_, z2_}] :=
    Module[{dx,dy,dz},
        dx = x2 - x1; dy = y2 - y1; dz = z2 - z1;
        Graphics'PlotField3D'Private'vector3D[
            {x1,y1,z1}, {dx, dy, dz}, True]
    ]
```

An example that shows the usage of **arrow3D** is

```
Show[Graphics3D[arrow3D[{0, 0, 0}, {1, 1, 1}]]]
```

Modify the program in Problem 1 to include `arrow3D`. Graph vectors **a** and **b**, and the projection of vector **a** on vector **b** on the same screen.

(a) $\mathbf{a} = \langle 2, 2, 2 \rangle$, $\mathbf{b} = \langle -0.2, 1, 0 \rangle$

(b) $\mathbf{a} = \langle 2, 2, -2 \rangle$, $\mathbf{b} = \langle -0.2, 1, 0 \rangle$

(c) $\mathbf{a} = \langle 2, -2, -2 \rangle$, $\mathbf{b}$ a unit vector on the z-axis

6.14 Project B—An Application of Matrix Algebra to Boundary Value Problems

One of the main applications of matrix algebra appears in the process of discretizing ordinary and partial differential equations. In this project we touch upon this subject in the context of solving the **boundary value problem**

$$y'' + ay' + by = f(x), \quad y(x_l) = y_l, \, y(x_r) = y_r. \tag{6.214}$$

The parameters a and b are given constants, although they could very well be known functions of x. The algorithm we will outline should be contrasted with the shooting method we presented in Chapter 4.

The method we will use here is called the **finite difference method**. In this method the interval (x_l, x_r) on the x-axis is divided into small subintervals. On each subinterval the derivatives in (6.214) are replaced by finite differences. As we will see shortly, this technique gives us an approximation to the values of $y(x_i)$s, where $y(x_i)$ is the exact solution of (6.214) at x_i, an endpoint of a typical subinterval. Equation (6.214) is then replaced by a system of linear algebraic equations for the unknown vector that consists of those approximate values.

We begin the process by dividing the interval (x_l, x_r) into n equidistant subintervals $x_l, x_1, x_2, \ldots, x_{n-1}, x_r$ and define the step-size h by

$$h = \frac{x_r - x_l}{n}. \tag{6.215}$$

Denote the quantity $y(x_i)$ by y_i. Recall that by definition

$$y'(x_i) = \lim_{h \to 0} \frac{y(x_i + h) - y(x_i)}{h}. \tag{6.216}$$

Equation (6.216) suggests that we approximate $y'(x_i)$ by

$$\frac{y(x_i + h) - y(x_i)}{h} \tag{6.217}$$

when h is small. Since $x_i + h = x_{i+1}$ and $y(x_i) = y_i$, the expression in (6.217) states that $y'(x_i)$ can be approximated by the **forward difference** quotient

$$\frac{y_{i+1} - y_i}{h}. \tag{6.218}$$

We could have approximated $y'(x_i)$ equally well by the **centered difference quotient** (see Problem 1(b))

$$\frac{y(x_{i+1}) - y(x_{i-1})}{2h} = \frac{y_{i+1} - y_{i-1}}{2h}.$$ (6.219)

Because of the symmetry of (6.219) (and other more significant properties of centered differencing that will not be emphasized at this point) we will use (6.219) to approximate $y'(x_i)$. Similarly the second derivative y'' can be approximated (see Problem 3) by

$$\frac{y_{i+1} - 2y_i + y_{i-1}}{h^2}.$$ (6.220)

We can obtain (6.220) from (6.219) by applying the latter operation twice with $\frac{h}{2}$ replacing h.

The discretized version of our differential equation now becomes

$$\frac{1}{h^2}[y_{i+1} - 2y_i + y_{i-1}] + \frac{a}{2h}[y_{i+1} - y_{i-1}] + by_i = f_i, \quad i = 1, 2, 3, \ldots, n-1. \quad (6.221)$$

Equation (6.221) leads to a system of linear algebraic equations in the unknowns $y_1, y_2, \ldots, y_{n-1}$. This system is written in the form

$$\begin{aligned}
Dy_1 + Ey_2 &= f_1 - (\frac{1}{h^2} - \frac{a}{2h})y_l \\
Cy_1 + Dy_2 + Ey_3 &= f_2 \\
Cy_2 + Dy_3 + Ey_4 &= f_3 \\
&\vdots \\
Cy_{n-2} + Dy_{n-1} &= f_{n-1} - (\frac{1}{h^2} + \frac{a}{2h})y_r.
\end{aligned}$$ (6.222)

where $f_i = f(x_i)$, $C = \frac{1}{h}(\frac{1}{h} - \frac{a}{2})$, $D = -\frac{2}{h^2} + b$, and $E = \frac{1}{h}(\frac{1}{h} + \frac{a}{2})$. Equations (6.222), when viewed in matrix form, can be written as

$$A\mathbf{x} = \mathbf{b},$$ (6.223)

where A is the **tridiagonal** matrix

$$A = \begin{bmatrix}
D & E & 0 & 0 & 0 & \cdots & 0 \\
C & D & E & 0 & 0 & \cdots & 0 \\
0 & C & D & E & 0 & \cdots & 0 \\
\cdots & & & & & & \\
\cdots & & & & & & \\
0 & \cdots & \cdots & \cdots & C & D & E \\
0 & \cdots & 0 & 0 & 0 & C & D
\end{bmatrix},$$ (6.224)

$$\mathbf{x} = \begin{bmatrix} y_1 \\ y_2 \\ .. \\ .. \\ y_{n-1} \end{bmatrix},$$ (6.225)

and

$$\mathbf{b} = \begin{bmatrix} f_1 - (\frac{1}{h^2} - \frac{a}{2h})y_l \\ f_2 \\ f_3 \\ \cdots \\ \cdots \\ f_{n-1} - (\frac{1}{h^2} + \frac{a}{2h})y_r \end{bmatrix}. \tag{6.226}$$

Although from the point of view of computational efficiency it is **not** desirable to solve (6.224)–(6.226) by inverting the matrix A, because of the ease with which *Mathematica* inverts such a matrix, we will present the outline of how to enter this problem into *Mathematica* and draw the graph of the approximate solution.

Remark 6.14.1: One of the measures of the effectiveness and efficiency of a numerical algorithm is the cost of carrying out all of the algebraic operations (addition, subtraction, multiplication, and division) in executing the algorithm once. It is not difficult to show that the number of algebraic operations required to determine the inverse of an $n \times n$ matrix using the cofactor method is proportional to n^3 while the same count for the Gaussian elimination algorithm is only proportional to n^2. Thus, it is preferable to solve an equation such as $A\mathbf{x} = \mathbf{b}$ using the latter rather than the former. Moreover, there are iterative techniques, such as the **Jacobi** and **Gauss–Seidel** algorithms, that often improve upon the performance of Gaussian elimination.

For the sake of concreteness, let $a = 1$, $b = 1$, $x_l = 0$, $x_r = 1$, $y_l = 1$, $y_r = 2$, and $f(x) = x$. The following paragraph summarizes the *Mathematica* program that begins with the data and solves for $\mathbf{x}$ in (6.224)–(6.226):

```
xl=0; xr=1; a= 1; b=1; f[x_] = x; yl = 1; yr = 2;
n = 20; h = (xr-xl)/n; c = 1/h^2 - a/(2h); d = b - 2/h^2;
     e = 1/h^2 + a/(2h);
X = N[Table[xl + i*h, {i, n - 1}]];
aa[i_, j_] = If[i == j, d, If[i == j + 1, c,
     If[i == j - 1, e, 0]]];
A[n_] = N[Table[aa[i, j], {i, n - 1}, {j, n - 1}]];
B = f[X];
B[[1]] = B[[1]] - c*yl;
B[[n - 1]] = B[[n - 1]] - e*yr;
y = Inverse[A[n]].B;
solution = Prepend[y, yl]; solution = Append[solution, yr];
X = Prepend[X, xl]; X = Append[X, xr];
vector = Table[{X[[i]], solution[[i]]}, {i, n + 1}];
output = ListPlot[vector, PlotJoined -> True];
```

The output of this program is shown in Figure 6.7.

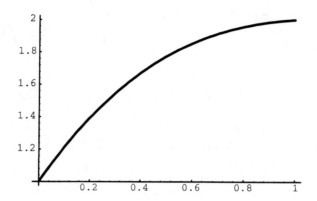

Figure 6.7: The output of the *Mathematica* program.

We now develop the preceding program in *MATLAB*. Since tridiagonal matrices arise naturally in the discretization of many physical problems, *MATLAB* has developed the facility of entering such a matrix directly into a program. The following *MATLAB* M-file is written for the same set of parameters used in the *Mathematica* program.

```
n=20; xl=0; xr=1; yl = 1; yr = 2;
%
a=1; b=1;
%
h=(xr-xl)/n; c=1/h*(1/h-a/2); d=-2/h^2+b; e=1/h*(1/h+a/2);
%
x=xl+h:h:xr-h;
%
vector1=ones(n-1,1);
B=x; B(1)=B(1)-c*yl; B(n-1)=B(n-1)-e*yr;
diagonal=d*vector1;
vector2=ones(n-2,1);
lower=diag(vector2,-1);
upper=diag(vector2,+1);
A=c*lower+diag(diagonal)+e*upper;
y=inv(A)*B';
newx=[xl x xr];   newy = [yl y' yr];
plot(newx,newy)
```

The output of this program is shown in Figure 6.8.

Problems

1. Apply the program in this project to the following boundary value problems.

   (a) $y'' + 2y' + y = 0,\quad y(0) = 0, y(1) = 1$

   (b) $y'' + y = 0,\quad y(0) = 0, y(2) = 1$

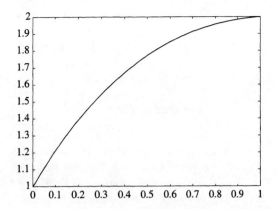

Figure 6.8: The output of the *MATLAB* program.

(c) $2y'' + y' - y = 0$, $y(1) = 1, y(2) = -1$

(d) $y'' + y = \sin 3x$, $y(0) = 0, y(\pi) = \frac{1}{2}$

(e) $y'' - y = \exp(-x)\sin x$, $y(0) = 1, y(0) = -1$

2. Let $f(x,h) = \frac{y(x+h)-y(x-h)}{2h}$.

   (a) Use L'Hôpital's rule and show that

   $$\lim_{h \to 0} f(x,h) = y'(x). \qquad (6.227)$$

   (b) Use the Taylor series expansions of $y(x+h)$ and $y(x-h)$ about $h = 0$ and show that

   $$f(x,h) - y'(x) = \frac{h^2}{6} + \cdots. \qquad (6.228)$$

   Note that the right-hand side of this expression begins with h^2. Hence, $f(x,h)$ is said to be an $O(h^2)$ approximation of $y'(x)$.

3. Let $g(x,h) = \frac{y(x+h)-2y(x)+y(x-h)}{h^2}$

   (a) Use L'Hôpital's rule and show that

   $$\lim_{h \to 0} g(x,h) = y''(x). \qquad (6.229)$$

   (b) Use the Taylor series expansions of $y(x+h)$ and $y(x-h)$ about $h = 0$ and show that

   $$g(x,h) - y''(x) = \frac{h^2}{12} + \cdots. \qquad (6.230)$$

   Hence, $g(x,h)$ is an $O(h^2)$ approximation of $y''(x)$.

4. Modify the *Mathematica* program to allow for a and b to be functions of x. Then:

   (a) Execute the revised program for $a(x) = \alpha = 0.1$, $b(x) = x$, and $f(x) = x$. Plot the approximate solution $y_\alpha(x)$.

(b) Execute this program for $\alpha = 0.01$, 0.001, and 0.0001 while keeping all other parameters fixed. Plot the graphs of $y_\alpha(x)$s on the same screen. How do you expect

$$\lim_{\alpha \to 0} y_\alpha(x) \tag{6.231}$$

would look?

5. Modify the *MATLAB* program so that n, a, and b are taken as input and the plot of the approximate solution is the output. Execute this program for $n = 10, 20, 40, 80, 160,$ and 320 while maintaining a and b as used in the preceding program. Let $y^n(x)$ denote each approximate solution. Do your plots give the impression that

$$\lim_{n \to \infty} y^n(x) \tag{6.232}$$

exists?

6. Use the package

`LinearAlgebra'Tridiagonal'`

of *Mathematica* and simplify the preceding *Mathematica* program by entering the tridiagonal matrix in (6.224) into the command `TridiagonalSolve`.

6.15   Project C—Linear Transformations and Fixed Points

We have seen in the body of this chapter how to solve a linear system such as

$$\mathbf{x} = \mu A\mathbf{x} + \mathbf{f}, \tag{6.233}$$

where A is an $n \times n$ matrix and $\mathbf{f}$ is a given vector in E^n. When $(I - \mu A)^{-1}$ exists, the solution to (6.233) is

$$\mathbf{x} = (I - \mu A)^{-1}\mathbf{f}. \tag{6.234}$$

This solution, given by (6.234), although theoretically easy to obtain, is not often the quickest way to get a handle on $\mathbf{x}$. Of course, the inverse of $I - \mu A$ may not exist, in which case (6.234) is useless; but even in cases where this inverse exists, it may not be prudent to compute it outright. Some of the problems at the end of this section will address this concern.

Instead of going the path of (6.234), we now look at a different way of obtaining $\mathbf{x}$ that uses the concept of "iteration of maps," a concept that has far-reaching applications in other areas of mathematics and physics.

Let T be a linear transformation defined by

$$T(\mathbf{x}) = \mu A\mathbf{x} \tag{6.235}$$

and S the transformation

$$S(\mathbf{x}) = T(\mathbf{x}) + \mathbf{f}. \tag{6.236}$$

Note that S is not linear unless $\mathbf{f}$ vanishes. The relation in (6.233) is equivalent to

$$S(\mathbf{x}) = \mathbf{x}. \tag{6.237}$$

Equation (6.237) states that the vector $\mathbf{x}$ must remain invariant under S. Such vectors are called **fixed points** of the map S. Our goal in this project is to find a convenient scheme of approaching a fixed point of S without resorting to taking inverses of matrices.

The scheme we choose obtains the fixed point of S by taking iterations of this map, starting with an arbitrary vector $\mathbf{x}_0$. Let $\mathbf{x}_1 = S(\mathbf{x}_0)$, $\mathbf{x}_2 = S(\mathbf{x}_1)$, and in general, $\mathbf{x}_n = S(\mathbf{x}_{n-1})$. Suppose, for the moment, that this sequence converges to a vector $\mathbf{x}$. We then have the following string of equalities that show $\mathbf{x}$ must be a fixed point of S:

$$\mathbf{x} = \lim_{n \to \infty} \mathbf{x}_n = \lim_{n \to \infty} S(\mathbf{x}_{n-1}) = S(\lim_{n \to \infty} \mathbf{x}_{n-1}) = S(\mathbf{x}). \tag{6.238}$$

In (6.238) we used the continuity of S to exchange the limiting process with function application. What we present next are the conditions under which the sequence $\mathbf{x}_n$ will converge. First we have a definition and a theorem.

Definition 6.15.1 (Contraction Maps)

A transformation T (linear or nonlinear) is called a contraction map if

$$\|T(\mathbf{x}) - T(\mathbf{y})\| \le \lambda \|\mathbf{x} - \mathbf{y}\| \tag{6.239}$$

for some $\lambda < 1$ and for all $\mathbf{x}$ and $\mathbf{y}$ in a neighborhood of the origin of the vector space. $\|\mathbf{x}\|$ denotes the magnitude of $\mathbf{x}$.

Theorem 6.15.1

Let A in (6.233) be a symmetric $n \times n$ matrix with n distinct eigenvalues $\lambda_1, \lambda_2, \ldots, \lambda_n$. Let

$$\Lambda = \max_{1 \le i \le n} |\lambda_i|. \tag{6.240}$$

Suppose that μ in (6.233) is small enough so that

$$\mu\Lambda < 1. \tag{6.241}$$

Then S is a contraction. Moreover, if a fixed point $\mathbf{x}$ exists, this fixed point is unique and the iteration scheme

$$\mathbf{x}_n = S(\mathbf{x}_{n-1}) \tag{6.242}$$

will converge to this fixed point.

Proof: First we show that S is a contraction. Since A is a symmetric matrix with distinct eigenvalues, there are n mutually orthogonal and normalized eigenvectors

$$\mathbf{e}_1, \mathbf{e}_2, \cdots, \mathbf{e}_n \tag{6.243}$$

that satisfy

$$A\mathbf{e}_1 = \lambda_1 \mathbf{e}_1, \quad A\mathbf{e}_2 = \lambda_2 \mathbf{e}_2, \quad \ldots, \quad A\mathbf{e}_n = \lambda_n \mathbf{e}_n. \tag{6.244}$$

Moreover, these eigenvectors form a basis for E^n. Let $\mathbf{x}$ and $\mathbf{y}$ be any two vectors in E^n. There exist constants $c_1, c_2, \cdots, c_n$ and $d_1, d_2, \cdots, d_n$ such that

$$\mathbf{x} = c_1 \mathbf{e}_1 + c_2 \mathbf{e}_2 + \cdots + c_n \mathbf{e}_n, \quad \mathbf{y} = d_1 \mathbf{e}_1 + d_2 \mathbf{e}_2 + \cdots + d_n \mathbf{e}_n. \tag{6.245}$$

Then

$$S(\mathbf{y}) - S(\mathbf{x}) = \mu \sum_{i=1}^{n} (d_i - c_i) \lambda_i \mathbf{e}_i. \tag{6.246}$$

Since the $\mathbf{e}_i$s are orthonormal, we have

$$\|S(\mathbf{y}) - S(\mathbf{x})\|^2 = (S(\mathbf{y}) - S(\mathbf{x}), S(\mathbf{y}) - S(\mathbf{x})) = \mu^2 \sum_{i=1}^{n} (d_i - c_i)^2 \lambda_i^2. \tag{6.247}$$

We bound the terms λ_i^2 by Λ^2 and note that

$$\sum_{i=1}^{n} (d_i - c_i)^2 = \|\mathbf{y} - \mathbf{x}\|^2 \tag{6.248}$$

to get

$$\|S(\mathbf{y}) - S(\mathbf{x})\| \le \mu \Lambda \|\mathbf{y} - \mathbf{x}\|, \tag{6.249}$$

which shows that S is a contraction as long as $\mu \Lambda < 1$.

Next let $\mathbf{x}$ be a fixed point of S, so that $S(\mathbf{x}) = \mathbf{x}$. Let $\lambda = \mu \Lambda$. Since S is a contraction map, we find that

$$\|\mathbf{x} - \mathbf{x}_n\| = \|S(\mathbf{x}) - S(\mathbf{x}_{n-1})\| \le \lambda \|\mathbf{x} - \mathbf{x}_{n-1}\|. \tag{6.250}$$

In turn, the last term of (6.250) satisfies

$$\|\mathbf{x} - \mathbf{x}_{n-1}\| = \|S(\mathbf{x}) - S(\mathbf{x}_{n-2})\| \le \lambda \|\mathbf{x} - \mathbf{x}_{n-2}\|. \tag{6.251}$$

Inequalities in (6.250) and (6.251) combine to give

$$\|\mathbf{x} - \mathbf{x}_n\| \le \lambda^2 \|\mathbf{x} - \mathbf{x}_{n-2}\|. \tag{6.252}$$

If we continue this process we will end up with

$$\|\mathbf{x} - \mathbf{x}_n\| \le \lambda^n \|\mathbf{x} - \mathbf{x}_0\|. \tag{6.253}$$

Since $\|\mathbf{x} - \mathbf{x}_0\|$ is fixed, the right-side of (6.253) converges to zero as n approaches infinity, so the sequence of approximations $\mathbf{x}_n$ will converge to the fixed point $\mathbf{x}$.

Finally, we show that when a fixed point exists it must be unique. Suppose that both $\mathbf{x}$ and $\mathbf{y}$ are fixed points of S. Then

$$\|\mathbf{y} - \mathbf{x}\| = \|S(\mathbf{y}) - S(\mathbf{x})\| \le \lambda\|\mathbf{y} - \mathbf{x}\|. \tag{6.254}$$

From (6.254) we get $(1 - \lambda)\|\mathbf{y} - \mathbf{x}\| \le 0$. Since $1 - \lambda > 0$, we must have $\|\mathbf{y} - \mathbf{x}\| \le 0$ or

$$\|\mathbf{y} - \mathbf{x}\| = 0, \tag{6.255}$$

which implies $\mathbf{y} = \mathbf{x}$. This completes the proof of the theorem.

Remark 6.15.1: In the statement of this theorem we stated that *if* a fixed point exists, then it satisfies certain properties. As it happens our hypotheses are strong enough to guarantee the existence of such a fixed point. In fact, we can show that the sequence of approximations $\mathbf{x}_n$ form a *Cauchy* sequence; that is, for any $\epsilon > 0$ there is an $N > 0$ such that for $m > N$ and $n > N$ we have

$$\|\mathbf{x}_n - \mathbf{x}_m\| < \epsilon. \tag{6.256}$$

The computation that leads to (6.256) is very similar to the one that led to (6.253). The important fact we have not proved, and whose proof can be found in a text such as W. Rudin's *Real and Complex Analysis*, is that every Cauchy sequence in E^n has a limit in E^n. This abstract result would provide us with the candidate for the fixed point. Instead of developing the techniques that would enable us to prove such a result, we choose to give several computational examples that lend credibility to this claim.

Remark 6.15.2: The conclusions of Theorem 6.15.1 are valid under considerably less stringent hypotheses on S. In particular, one can do without the symmetry of A and without having a complete set of eigenvectors. As the reader may guess, a different proof of this theorem is then needed.

We finish this project by showing how fixed points can be computed in *Mathematica*. The following is such a representative program. Before executing this program the reader should input x, A, and f, the initial guess at the fixed point, the matrix A, and $\mathbf{f}$ of (6.233).

```
fp[mu_, Iterations_]:= Module[{y},
    Do[y = f + mu *A . x; Print[y]; x = y,
        {i, 1, Iterations}]
```

The following is the output of this program for

$$\mathbf{x}_0 = \begin{bmatrix} 0 \\ 0 \end{bmatrix}, \quad A = \begin{bmatrix} 1 & 2 \\ 2 & 1 \end{bmatrix}, \quad \mathbf{f} = \begin{bmatrix} 1 \\ -1 \end{bmatrix}, \tag{6.257}$$

with $\mu = 0.1$ and Iterations = 10.

```
{1, -1}
{0.9, -1.2}
{0.85, -1.21}
{0.843, -1.206}
{0.8431, -1.2049}
{0.84333, -1.2048}
{0.843373, -1.204813}
{0.8433747, -1.2048186}
{0.84337375, -1.20481933}
{0.843373509, -1.204819308}
```

Problems

1. Create a file called `fixed_point.m` that contains the `fp` function defined in the preceding program. Begin a session of *Mathematica*. Input values for A, x_0, and f. Input the file `fixed_point.m` and give the command

   `fp[0.1,10]`

   to get the output listed at the end of this project.

2. Increase the value of μ by increments of 0.1 and repeat Problem 1. How does the output change? What happens when $\mu = 1$? Does it help to increase `Iterations` to 100 or 500? Explain your result in the context of the eigenvalues of A.

3. The Hilbert matrix A is defined by

$$A = [a_{ij}], \quad a_{ij} = \frac{1}{i+j}. \tag{6.258}$$

   (a) Let A be the 3×3 Hilbert matrix. Find the eigenvalues of this matrix and compute Λ. Begin with an f of your own choosing and repeat Problem 1. Decrease the value of μ if you do not reach a fixed point. If you do reach a fixed point, increase the value of μ with increments of 0.1 and repeat the exercise. For what value of μ do you not get convergence, even after 500 iterations? Does this value agree with the value of μ that should make S a contraction?

   (b) Let A be the 4×4 Hilbert matrix and repeat part 3(a) of this problem. Use *Mathematica* and compute the eigenvalues of A. Does the value of μ that gives us a contraction map increase or decrease relative to μ when A is a 3×3 Hilbert matrix? Do you still get "convergence" for a value of μ slightly larger than that allowed by the contraction mapping theorem (that is, do you get a sequence of approximations x_n that seem to repeat themselves after a reasonable number of iterations?)

   (c) Let A be the 8×8 Hilbert matrix. Repeat 3(b) of this problem for this choice of A. What are the difficulties you encounter in determining μ to get convergence?

(d) If you look carefully at the proof of Theorem 6.15.1, you will notice that $\mathbf{f}$ does not enter this proof at all. Does $\mathbf{f}$ enter your experiments with how fast the sequence $\mathbf{x}_n$ seems to converge? Make a few appropriate choices for $\mathbf{f}$, keeping all other parameters fixed, and execute the program to gain an appreciation for the role of $\mathbf{f}$ in this problem.

4. Write a *MATLAB* program that does the computations that lead to a fixed point of the map S. Repeat the previous exercises using this program.

References

[1] Finkbeiner, Daniel T, *Elements of Linear Algebra*, W. H. Freeman, New York, 1972.

[2] Marcus, Marvin, *Matrices and MATLAB*, Prentice Hall, Englewood Cliffs, NJ, 1993.

Chapter 7

Systems of Differential Equations

7.1 Introduction

In the past several chapters we developed various methods for solving ODEs as well as algebraic systems of linear equations. In this chapter we will make use of these methods and obtain solutions to systems of linear differential equations. These systems are characterized by the fact that the evolution of two or more dependent variables are interrelated, thus giving rise to several simultaneous differential equations through which these variables are coupled. Such systems are very common in practice, arising in mechanical systems in which several springs are interconnected or in electrical networks that contain several sources of power as well as multiple capacitors and resistors, to cite two examples. They also appear naturally in many problems in fluid dynamics, ranging from the flow past a cylinder to the convection of fluid particles in a Rayleigh–Bénard flow as well as in the dynamics of rigid bodies such as the motion of a satellite in orbit in space. We will explore several of these examples in detail in this and the subsequent chapters in Volume II.

Our goal in this chapter is to utilize the notation and the terminology of matrices to differential equations. We will apply our understanding of eigenvalues and eigenvectors and reduce the process of solving systems of ODEs to that of solving linear algebraic equations. In addition, eigenvalues and eigenvectors will play a crucial role in classifying the behavior of solutions of such systems; we will identify several so-called "phase spaces" in terms of the eigenvalues of the matrices associated with systems of differential equations. We will also give an introduction to the concept of the stability of equilibrium solutions of linear and nonlinear systems of equations.

All of the computational tools we will develop in this chapter are suited well for implementation in both *Mathematica* and *MATLAB*. In many of the exercises we will take full advantage of `DSolve` and `NDSolve` of *Mathematica* as well as `ode45` of *MATLAB*.

7.2 Examples of Systems of Differential Equations

We will make extensive use of the vector notation and the matrix algebra we developed in Chapter 6. In addition to these tools we need to borrow a few ideas from the calculus of real-valued functions.

A vector-valued function $\mathbf{x}(t)$ is a vector whose entries are functions of t. We often write $\mathbf{x}$ as the column vector

$$\mathbf{x}(t) = \begin{bmatrix} x_1(t) \\ x_2(t) \\ \cdots \\ \cdots \\ x_n(t) \end{bmatrix}, \tag{7.1}$$

where each function x_i is real-valued. When we speak of $\mathbf{x}(t)$ as continuous or differentiable, we mean that each function x_i is continuous or differentiable. The quantity $\mathbf{x}'(t)$ is a vector whose ith component is the derivative of the ith component of $\mathbf{x}$, or

$$\mathbf{x}'(t) = \begin{bmatrix} x_1'(t) \\ x_2'(t) \\ \cdots \\ \cdots \\ x_n'(t) \end{bmatrix}. \tag{7.2}$$

In a similar fashion we extend all of these concepts to matrices. Hence, given an $n \times n$ matrix $A(t)$ with entries $a_{ij}(t)$, by $A'(t)$ we mean the matrix one obtains by differentiating each a_{ij} once. A matrix $A(t)$ is continuous or differentiable according to whether each entry a_{ij} is continuous or differentiable. The algebraic properties of differentiation and integration generalize easily to matrices. For example, by $(A(t)+B(t))'$ and $(A(t)B(t))'$ we mean $A'(t)+B'(t)$ and $A'(t)B(t)+A(t)B'(t)$, respectively. Similarly, $\int A(t)dt$ denotes a matrix, each of whose entries is the integral of the corresponding entry of A. All of these operations are understood by *Mathematica* and *MATLAB*.

We now give several concrete examples of how systems of ODE appear in models that arise in mathematical physics. The systems in the following examples either come about as a result of the direct application of laws of physics (such as the Newton's second law or Kirchhoff's law) or as a reduction of a more complicated model after simplifying assumptions are added to the model. The fluid flow examples that follow all have the latter property.

The systems in the following examples are either all **first order**, meaning that the order of differentiation in each equation is one, or can be reduced to such a state by introducing extra variables into the problem (see Example 7.2.1). These extra variables often have their own significant physical meaning. As a result of this reduction, once we find the solution to the first-order system, we are able to monitor the evolution of the latter extra variables.

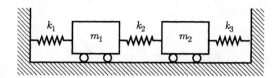

Figure 7.1: A mass-spring system.

Example 7.2.1 (Motion of a Spring System)

Consider a mechanical system in which two masses m_1 and m_2 are connected to each other by three springs as in Figure 7.1. Let us assume that the springs have spring constants k_1, k_2, and k_3, respectively. We define $x_1(t)$ and $x_2(t)$ to be the positions of the centers of mass of m_1 and m_2 away from their equilibrium. Then x_1'' and x_2'' measure the acceleration of each mass. The resultant force acting on each mass is exerted on it by the springs that are attached to it, each force being proportional to the distance the spring is stretched or compressed. For instance, when mass m_1 has moved a distance x_1 to the right of its equilibrium position, the spring to the left of m_1 exerts a restoring force $-k_1 x_1 \mathbf{i}$ on this mass, attempting to return the mass back to its equilibrium position. The vector $\mathbf{i}$ is a unit vector pointing to the right. The spring to the right of m_1 exerts a restoring force $-k_2(x_2 - x_1)\mathbf{i}$ on it; the part $k_2 x_1$ reflects the compression of the middle spring due to the movement of m_1, while $-k_2 x_2$ is due to the movement of m_2 and its influence on the same spring. Following Newton's law of motion, we arrive at the two equations:

$$m_1 x_1'' = -k_1 x_1 + k_2(x_2 - x_1), \quad m_2 x_2'' = -k_2(x_2 - x_1) - k_3 x_2. \tag{7.3}$$

Both of these equations involve x_1 and x_2: The equations in (7.3) are coupled to each other in that one cannot solve one without the knowledge of the other.

We convert each equation in (7.3) to a first-order equation by introducing two new variables y_1 and y_2 representing the velocities of each mass:

$$y_1 = x_1', \quad y_2 = x_2'. \tag{7.4}$$

Using these new dependent variables, we write (7.3) as the following four simultaneous equations in the four unknowns x_1, x_2, y_1, and y_2:

$$\begin{cases} x_1' = y_1, \\ y_1' = \dfrac{-k_1 x_1 + k_2(x_2 - x_1)}{m_1}, \\ x_2' = y_2, \\ y_2' = \dfrac{-k_2(x_2 - x_1) - k_3 x_2}{m_2}. \end{cases} \tag{7.5}$$

It is now clear how to write (7.3) in matrix notation. Let $\mathbf{x}$ be defined as

$$\mathbf{x} = \begin{bmatrix} x_1 \\ y_1 \\ x_2 \\ y_2 \end{bmatrix}. \tag{7.6}$$

The left-hand side of (7.5) is nothing but $\mathbf{x}'$. Define the matrix A to be the matrix of the coefficients on the right-hand side of (7.5), that is,

$$
A = \begin{bmatrix}
0 & 1 & 0 & 0 \\
-\dfrac{k_1 + k_2}{m_1} & 0 & \dfrac{k_2}{m_1} & 0 \\
0 & 0 & 0 & 1 \\
\dfrac{k_2}{m_2} & 0 & -\dfrac{k_1 + k_2}{m_2} & 0
\end{bmatrix}. \tag{7.7}
$$

Hence, (7.5) is equivalent to

$$
\mathbf{x}' = A\mathbf{x}. \tag{7.8}
$$

Each equation in (7.5) is first order. We expect that the process of solving this system will require one integration in time for each equation in this system, leading to four arbitrary constants of integration in the general solution. Typically we know the state of the two springs at a fixed instant of time, say at $t = 0$, in the form

$$
x_1(0) = a, \qquad y_1(0) = b, \qquad x_2(0) = c, \qquad y_2(0) = d, \tag{7.9}
$$

or in vector notation, the vector $\mathbf{x}_0$, given by

$$
\mathbf{x}(0) = \mathbf{x}_0 = \begin{bmatrix} a \\ b \\ c \\ d \end{bmatrix}, \tag{7.10}
$$

is known. These initial conditions will suffice in uniquely determining the solution to (7.5).

Example 7.2.2 (Electrical Network)

Consider the network in Figure 7.2. Let I_1 and I_2 be the currents in the closed loops indicated. Following Kirchhoff's law, that the sum of voltage drop across each closed loop must add up to zero, we have the following two equations for I_1 and I_2:

$$
\begin{aligned}
L_1 \frac{dI_1}{dt} &= -R_1(I_1 - I_2) + E_0, \\
L_2 \frac{dI_2}{dt} &= R_1(I_1 - I_2) - R_2 I_2,
\end{aligned} \tag{7.11}
$$

satisfying the initial conditions

$$
I_1(0) = a, \quad I_2(0) = b. \tag{7.12}
$$

In matrix notation this problem takes the form $\mathbf{x}' = A\mathbf{x} + \mathbf{f}$ with

$$
A = \begin{bmatrix} -\dfrac{R_1}{L_1} & \dfrac{R_1}{L_1} \\ \dfrac{R_1}{L_2} & -\dfrac{R_1 + R_2}{L_2} \end{bmatrix}, \qquad \mathbf{x} = \begin{bmatrix} I_1 \\ I_2 \end{bmatrix}, \quad \text{and} \quad \mathbf{f} = \begin{bmatrix} E_0 \\ 0 \end{bmatrix}, \tag{7.13}
$$

and the initial conditions in (7.12).

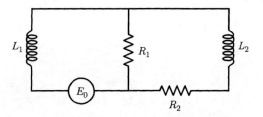

Figure 7.2: An electric Network.

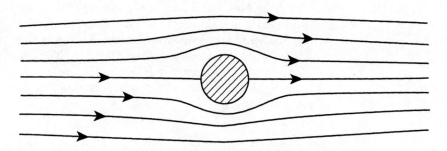

Figure 7.3: The schematic of the flow past a cylinder.

Example 7.2.3 (Flow Past a Cylinder of Radius One)

In a later chapter we will study in detail special fluids and fluid flows, among them the irrotational motion of inviscid and incompressible fluids in several different geometries. The mathematical definitions of these terms will be given later. Here we give an example of such a motion that is described by a system of ODEs.

We consider a two-dimensional steady irrotational fluid flow of an inviscid and incompressible fluid past a cylinder of radius one (see Figure 7.3). The position of each particle of fluid at time t is denoted by $(x(t), y(t))$ and its velocity by $\mathbf{v} = \langle u(t), v(t) \rangle$. The assumptions of incompressibility and irrotationality, as well as the fact that the fluid has reached a steady state, help in determining a function $\psi(x, y)$, known as the **stream function** for this flow, from which the velocity field $\mathbf{v} = \langle u(x, y), v(x, y) \rangle$ is derived from the formulas

$$u = \frac{\partial \psi}{\partial y}, \quad v = -\frac{\partial \psi}{\partial x}. \tag{7.14}$$

In this special case it turns out that ψ has the form

$$\psi(x, y) = y - \frac{y}{x^2 + y^2}. \tag{7.15}$$

Recalling that the position of a fluid particle, $(x(t), y(t))$, and its velocity, $\mathbf{v}$, are related by

$$\frac{dx}{dt} = u(x, y), \quad \frac{dy}{dt} = v(x, y), \tag{7.16}$$

we obtain the following system of nonlinear differential equations for $(x(t), y(t))$ from (7.14):

$$\frac{dx}{dt} = 1 - \frac{x^2 - y^2}{(x^2 + y^2)^2}, \quad \frac{dy}{dt} = -\frac{2xy}{(x^2 + y^2)^2}. \tag{7.17}$$

In a typical setting we are interested in the motion of a particle of fluid whose initial position is (x_0, y_0); that is, the system in (7.17) is supplemented with the initial data

$$x(0) = x_0, \quad y(0) = y_0. \tag{7.18}$$

In the mathematical literature a system of the form (7.14) is called a **Hamiltonian** system and the function ψ its **Hamiltonian**. An important consequence of the special relationship that the hamiltonian ψ enjoys with the dynamics in (7.14) is the invariance of ψ when confined to the path of a particle. To see this, let C be the path of a particle P, located at (x_0, y_0) at time zero, and otherwise moving according to the rules described the system in (7.14). We parametrize C by

$$C = \{\langle \hat{x}(t), \hat{y}(t) \rangle \mid t \geq 0\}.$$

Let f define the value of ψ along C, that is

$$f(t) = \psi(\hat{x}(t), \hat{y}(t)),$$

where $(x(t), y(t))$ satisfies (7.14). Differentiate f with respect to t:

$$f'(t) = \frac{\partial \psi}{\partial x} \frac{d\hat{x}}{dt} + \frac{\partial \psi}{\partial y} \frac{d\hat{y}}{dt}.$$

After substituting the relations (7.14) in the previous expression, we find

$$f'(t) = \frac{\partial \psi}{\partial x} \frac{\partial \psi}{\partial y} - \frac{\partial \psi}{\partial y} \frac{\partial \psi}{\partial x} = 0,$$

from which we deduce that f remains constant along the path of any particle. Thus, the particle paths of the motion in (7.14) are **level curves** or the **contour levels** of the stream function ψ. We will come back to this point repeatedly in the upcoming chapters in Volume II.

In the context of the flow past the cylinder, the previous discussion states that the particle paths of this flow must satisfy

$$\hat{y}(t) - \frac{\hat{y}(t)}{\hat{x}^2(t) + \hat{y}^2(t)} = \text{constant}, \tag{7.19}$$

where the value of the constant is determined by evaluating the left-hand side of (7.19) at $t = 0$. In *Mathematica* the command `ContourPlot[f, domain]` plots the graph of some of the typical contours of the function f in the designated domain. Figure 7.4 shows the output of

```
ContourPlot[y -y/(x^2+y^2), {x, -5, 5}, {y, 0, 3}, PlotPoints ->30]
```

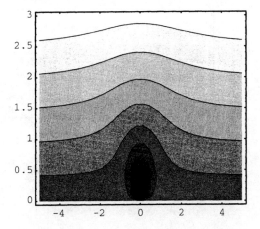

Figure 7.4: Flow past cylinder.

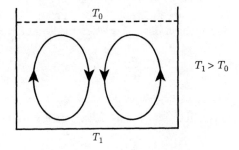

Figure 7.5: The Rayleigh–Bénard flow.

Example 7.2.4 (Rayleigh–Bénard Flow)

Consider a layer of fluid occupying the region

$$D = \{(x, y, z) \mid 0 \leq z \leq 1\}$$

(see Figure 7.5). If we begin to heat the lower plate $z = 0$, the temperature gradient created between the top and the bottom of the region D will eventually force the fluid particles to move. The hotter particles, becoming lighter, move upward. As they approach the plane $z = 1$, they become cooler and will descend. One expects to see a convection of particles due to this exchange of heat. The so-called Rayleigh–Bénard cells exhibit this phenomenon once the temperature difference between the top and bottom reaches a critical value (related to a number known as the Rayleigh number).

We now describe a system of differential equations, derivable from a stream function, whose solutions will exhibit this behavior (see the paper by R. Camassa and S. Wiggins, "Chaotic Advection in a Rayleigh–Bénard Flow," *Physical Review A*, Vol 43, 1991, pp. 774–797, for more motivation. We will study this

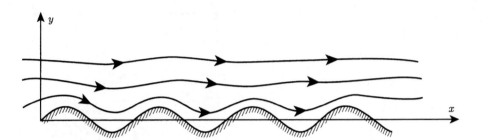

Figure 7.6: Flow past a wavy wall.

problem in more detail later in the text). The stream function that we have in mind is

$$\psi(x, z) = \sin \pi x \sin \pi z. \tag{7.20}$$

Note that ψ is independent of y and t, so the motion is two-dimensional and has reached its steady state. From (7.14) (applied to the x and z variables) we obtain the following system of nonlinear ODEs for the motion of the particles:

$$\frac{dx}{dt} = \pi \sin \pi x \cos \pi z, \qquad \frac{dz}{dt} = -\pi \cos \pi x \sin \pi z, \tag{7.21}$$

with the usual set of initial conditions. Equations (7.21) exhibit the so-called Rayleigh–Bénard cells shown in Figure 7.5.

Example 7.2.5 (Linearized Subsonic Flow)
    The differential equations in this example differ from those in the previous two examples in that they are derived from a potential function rather than a stream function. We say that a velocity field $\mathbf{v} = \langle u(x,y), v(x,y) \rangle$ has a **potential** $V(x,y)$ if the following relations hold between $\mathbf{v}$ and V:

$$u = \frac{\partial V}{\partial x}, \qquad v = \frac{\partial V}{\partial y}. \tag{7.22}$$

An example of such a flow, which gives rise to an interesting set of differential equations, is the motion of particles of gas past a wavy wall. The motion is assumed to be two-dimensional. The equation of the wall past which the gas flows is given by (see Figure 7.6)

$$y = A \cos \frac{2\pi x}{\lambda}, \tag{7.23}$$

with the amplitude A much smaller than the wavelength λ (a model to keep in mind is the wing of an airplane with very small undulations on it). When the flow past this wall is subsonic, a term whose rigorous definition we will give when we cover wave propagation, it can be shown (see the text by Michel Saad, *Compressible Fluid Flow*, Prentice-Hall, Inc., Englewood Cliffs, NJ, 1985, pp. 351–359) that the motion of the particles has the velocity potential

$$V(x,y) = \frac{U_\infty A}{\beta} e^{\frac{-2\pi\beta y}{\lambda}} \sin \frac{2\pi x}{\lambda}. \tag{7.24}$$

In (7.24) β is a positive constant that measures how close the flow is to its transonic limit and U_∞ is the speed of the flow upstream (prior to reaching the wall). After using (7.22), we see that the position of the gas particles satisfy the following system of differential equations:

$$\frac{dx}{dt} = \frac{2\pi U_\infty A}{\lambda \beta} e^{\frac{-2\pi\beta y}{\lambda}} \cos \frac{2\pi x}{\lambda}, \qquad \frac{dy}{dt} = -\frac{2\pi U_\infty A}{\lambda} e^{\frac{-2\pi\beta y}{\lambda}} \sin \frac{2\pi x}{\lambda}. \tag{7.25}$$

We will follow the same terminology we developed in Chapter 3 when we discussed single differential equations. In particular, the concepts of solution, linearity, and homogeneity will be directly generalized to systems.

We define a system of differential equations to be a set of n simultaneous equations involving the n unknowns, $x_1(t), x_2(t), \cdots, x_n(t)$, and their derivatives. Typically systems that we address here can be written in the form

$$\begin{cases} x_1' = f_1(t, x_1, x_2, \cdots, x_n), \\ x_2' = f_2(t, x_1, x_2, \cdots, x_n), \\ \qquad \vdots \\ x_n' = f_n(t, x_1, x_2, \cdots, x_n), \end{cases} \tag{7.26}$$

where the f_is are continuous in all of their arguments. A compact way of writing this system is to introduce the vectors $\mathbf{f} = \langle f_1, f_2, \cdots, f_n \rangle$ and $\mathbf{x} = \langle x_1, x_2, \cdots, x_n \rangle$ and to note that (7.26) is equivalent to

$$\mathbf{x}' = \mathbf{f}(t, \mathbf{x}). \tag{7.27}$$

We say that a system of differential equations (7.27) is **linear** if all of the f_is are linear functions in the x_is; otherwise, the system is called **nonlinear**. A linear system is called **homogeneous** if all the f_is are independent of t and **nonhomogeneous** when at least one f_i depends explicitly on t. A (nonlinear) system is called **autonomous** if all of the f_is in (7.26) are independent of t. When this condition is violated the system is called **nonautonomous**.

Example 7.2.6

The first two systems in this section are examples of linear systems. The first, in Example 7.2.1, is an example of a homogeneous system; while the second system, in Example 7.2.2, is nonhomogeneous. The three systems in Examples 7.2.3 to 7.2.5 are all nonlinear. All three are autonomous.

By a **solution** of system (7.26) we mean a set of functions $(x_1(t), x_2(t), \cdots, x_n(t))$ that satisfy (7.26); that is, when substituted on the left- and right-hand sides of each equation in (7.26), one gets an identity in t. Since there are n

first-order equations in (7.26), we expect that n additional side constraints in the form of initial conditions are needed if we are to determine a unique solution to this system.

Some of the techniques we employed in Chapter 4 on differential equations apply in a straightforward manner to linear systems with constant coefficients. We do not repeat these ideas here but consider them in the exercises. In particular, we emphasize the method of **Laplace transform** as its applies to first order systems. We took up this topic in Chapter 5 and studied in detail how it can be used effectively on systems of differential equations.

As in the case of single differential equations, DSolve and NDSolve of *Mathematica* and dsolve and ode45 of *MATLAB* are powerful tools that apply to systems as well. It is perhaps in the context of systems that the utility of these packages becomes apparent, since even in the case of linear systems with constant coefficients for which we will develop a more than adequate theory for determining the exact solution, the underlying computations become tedious so quickly that one often has to confine oneself to low-dimensional, and less realistic, models. The power of these packages also becomes apparent when we deal with nonlinear system, since, except for very few cases, we are unable to construct their analytic solutions.

DSolve and dsolve primarily apply to linear systems with constant coefficients, while we will use NDSolve and ode45 almost exclusively for nonlinear problems.

Example 7.2.7
Consider the initial value problem

$$x' = 2x - 3y, \quad y' = 3x + 2y, \qquad x(0) = 1, \quad y(0) = -1.$$

The syntax for entering this equation to *Mathematica* is

```
DSolve[{x'[t]  ==  2 x[t]  -  3 y[t],
         y'[t]  ==  3 x[t]  +  2 y[t],
         x[0]  ==  1,  y[0]  ==  -1},
         {x[t], y[t]},  t]
```

The syntax in *MATLAB* is

```
[x y]=dsolve('Dx=2*x-3*y,Dy=3*x+2*y','x(0)=1, y(0)=-1')
```

Problems

1. Show that the function $\mathbf{x}(t)$ is a solution of the system of differential equations. In each case use *Mathematica* or *MATLAB* to check your calculations.

   (a) $\mathbf{x}' = \begin{bmatrix} -1 & 0 \\ 0 & 2 \end{bmatrix} \mathbf{x}, \quad \mathbf{x}(t) = \langle e^{-t}, 2e^{2t} \rangle$

   (b) $\mathbf{x}' = \begin{bmatrix} 0 & 1 \\ -1 & 0 \end{bmatrix} \mathbf{x}, \quad \mathbf{x}(t) = \langle -\sin t, -\cos t \rangle$

(c) $x' = \begin{bmatrix} 0 & 2 \\ -1 & 0 \end{bmatrix} x,$

$x(t) = \langle 2 \sin \sqrt{2}t + 3 \cos \sqrt{2}t, \sqrt{2} \cos \sqrt{2}t - \frac{3}{2}\sqrt{2} \sin \sqrt{2}t \rangle$

(d) $x' = \begin{bmatrix} 1 & 2 \\ 2 & 1 \end{bmatrix} x,$ $x(t) = \langle c_1 e^{3t} + c_2 e^{-t}, c_1 e^{3t} - c_2 e^{-t} \rangle$

(e) $x' = \begin{bmatrix} 0 & 1 & 0 \\ 0 & 0 & 1 \\ 6 & -11 & 6 \end{bmatrix} x,$ $x(t) = \langle -e^t + 2e^{3t}, e^t + 6e^{3t}, e^t + 18e^{3t} \rangle$

(f) $x' = \begin{bmatrix} 0 & \frac{1}{t^2} \\ 1 & 0 \end{bmatrix} x + \begin{bmatrix} 0 \\ 1 - \ln t \end{bmatrix},$ $x(t) = \langle \ln t, t \rangle$

(g) $tx' = \begin{bmatrix} 1 & \frac{1}{t} \\ 0 & 1 \end{bmatrix} x + \begin{bmatrix} -\ln t \\ 0 \end{bmatrix},$ $x(t) = \langle \ln t, t \rangle$

2. Transform the following equations to first-order systems (see Example 7.2.1).

(a) $x'' + 4x = 0$

(b) $y'' + 3y' + 3y = \sin t$

(c) $y''' + y = 0$

(d) $y^{(\text{iv})} + xy''' + x^2 y'' + x^3 y' + x^4 y = 0$

(e) $x'' = \frac{y}{x^2 + y^2},\ y'' = -\frac{x}{x^2 + y^2}$

3. Determine the solution to each initial value problem $x' = Ax + f$, $x(0) = x_0$ using the Laplace transform method.

(a) $A = \begin{bmatrix} -1 & 0 \\ 0 & 2 \end{bmatrix}, f = \begin{bmatrix} 0 \\ 0 \end{bmatrix}, x_0 = \begin{bmatrix} -2 \\ 3 \end{bmatrix}$

(b) $A = \begin{bmatrix} 0 & 1 \\ -1 & 0 \end{bmatrix}, f = \begin{bmatrix} 0 \\ 0 \end{bmatrix}, x_0 = \begin{bmatrix} 1 \\ 1 \end{bmatrix}$

(c) $A = \begin{bmatrix} -1 & 1 \\ 1 & 2 \end{bmatrix}, f = \begin{bmatrix} 0 \\ 0 \end{bmatrix}, x_0 = \begin{bmatrix} 0 \\ 1 \end{bmatrix}$

(d) $A = \begin{bmatrix} 1 & -1 \\ -1 & 1 \end{bmatrix}, f = \begin{bmatrix} 0 \\ 1 \end{bmatrix}, x_0 = \begin{bmatrix} 0 \\ 0 \end{bmatrix}$

(e) $A = \begin{bmatrix} 0 & 1 \\ 1 & 1 \end{bmatrix}, f = \begin{bmatrix} \sin t \\ 0 \end{bmatrix}, x_0 = \begin{bmatrix} 0 \\ 0 \end{bmatrix}$

(f) $A = \begin{bmatrix} -1 & -1 \\ 1 & 0 \end{bmatrix}, f = \begin{bmatrix} \sin 2t \\ \cos t \end{bmatrix}, x_0 = \begin{bmatrix} 0 \\ 0 \end{bmatrix}$

(g) $A = \begin{bmatrix} -1 & 1 \\ -2 & -2 \end{bmatrix}, f = \begin{bmatrix} t \\ 1 - 2t \end{bmatrix}, x_0 = \begin{bmatrix} 0 \\ 0 \end{bmatrix}$

(h) $A = \begin{bmatrix} 1 & -1 \\ 3 & 5 \end{bmatrix}, f = \begin{bmatrix} 2t \\ \sin t \end{bmatrix}, x_0 = \begin{bmatrix} -1 \\ 1 \end{bmatrix}$

(i) $A = \begin{bmatrix} a & b \\ b & a \end{bmatrix}$, $\mathbf{f} = \begin{bmatrix} 0 \\ 0 \end{bmatrix}$, $\mathbf{x}_0 = \begin{bmatrix} -2 \\ 3 \end{bmatrix}$, where a and b are real numbers

(j) $A = \begin{bmatrix} -1 & 0 & 0 \\ 0 & 2 & 0 \\ 0 & 0 & 3 \end{bmatrix}$, $\mathbf{f} = \begin{bmatrix} 0 \\ 0 \\ 0 \end{bmatrix}$, $\mathbf{x}_0 = \begin{bmatrix} -2 \\ 3 \\ -1 \end{bmatrix}$

(k) $A = \begin{bmatrix} -1 & 1 & 0 \\ 0 & 2 & 0 \\ 1 & 0 & 3 \end{bmatrix}$, $\mathbf{f} = \begin{bmatrix} 0 \\ 0 \\ 0 \end{bmatrix}$, $\mathbf{x}_0 = \begin{bmatrix} 0 \\ 0 \\ -1 \end{bmatrix}$

(l) $A = \begin{bmatrix} 0 & 1 & 0 \\ 0 & 0 & 1 \\ 1 & 1 & 1 \end{bmatrix}$, $\mathbf{f} = \begin{bmatrix} 0 \\ 0 \\ 0 \end{bmatrix}$, $\mathbf{x}_0 = \begin{bmatrix} 1 \\ 0 \\ 0 \end{bmatrix}$

(m) $A = \begin{bmatrix} 0 & 0 & 1 \\ 1 & 0 & 0 \\ -3 & -1 & 3 \end{bmatrix}$, $\mathbf{f} = \begin{bmatrix} \sin t \\ 0 \\ 0 \end{bmatrix}$, $\mathbf{x}_0 = \begin{bmatrix} 0 \\ 0 \\ 0 \end{bmatrix}$

(n) $A = \begin{bmatrix} 1 & 0 \\ 0 & 0 & 1 \\ -2 & -3 & -1 \end{bmatrix}$, $\mathbf{f} = \begin{bmatrix} 0 \\ 1-t \\ 0 \end{bmatrix}$, $\mathbf{x}_0 = \begin{bmatrix} 0 \\ 0 \\ 0 \end{bmatrix}$

(o) $A = \begin{bmatrix} 0 & 1 & 0 \\ 0 & 0 & 1 \\ -6a^3 & -11a^2 & -6a \end{bmatrix}$, $\mathbf{f} = \begin{bmatrix} 0 \\ 0 \\ 0 \end{bmatrix}$, $\mathbf{x}_0 = \begin{bmatrix} 0 \\ 0 \\ 1 \end{bmatrix}$, where a is a real number

4. Consider the first-order system $\mathbf{x}' = A\mathbf{x}$, where A is an $n \times n$ matrix with constant coefficients. Show that this system has solutions of the form $\mathbf{x} = e^{\lambda t}\mathbf{c}$, where λ and $\mathbf{c}$ satisfy

$$A\mathbf{c} = \lambda\mathbf{c}, \qquad (7.28)$$

that is, λ is an eigenvalue of A and $\mathbf{c}$ is its eigenvector.

5. Apply the method of Problem 4 to the following first-order systems and obtain as many solutions as possible (note that in some cases the eigenvalues may be complex).

(a) $A = \begin{bmatrix} 1 & 2 \\ -2 & 1 \end{bmatrix}$

(b) $A = \begin{bmatrix} 4 & 1 \\ 2 & -1 \end{bmatrix}$

(c) $A = \begin{bmatrix} 0 & 1 & 0 \\ 0 & 0 & 1 \\ -6 & -11 & -6 \end{bmatrix}$

6. Show that if $\mathbf{x}_1$ and $\mathbf{x}_2$ are solutions of the linear homogeneous system $\mathbf{x}' = A\mathbf{x}$, then so is $\mathbf{y} = c_1\mathbf{x}_1 + c_2\mathbf{x}_2$, for any $c_1, c_2 \in R$.

7. Prove that if $\mathbf{x}_1$ and $\mathbf{x}_2$ are solutions of the linear nonhomogeneous system $\mathbf{x}' = A\mathbf{x} + \mathbf{f}$, then $\mathbf{y} = \mathbf{x}_1 - \mathbf{x}_2$ is a solution of the homogeneous system $\mathbf{x}' = A\mathbf{x}$.

8. Construct the system of differential equations from the given stream functions (see Example 7.2.3). In each case determine the slope to the solution curve at the indicated points (i.e., compute $\frac{dy}{dx}$ at P).

   (a) $\psi(x, y) = x^2 + y^2$; $P = (1, -1)$

   (b) $\psi(x, y) = xy$; $P_1 = (0, 0)$ and $P_2 = (1, 2)$

   (c) $\psi(x, y) = \frac{1}{\sqrt{x^2 + y^2}}$; $P_1 = (-2, 1)$ and $P_2 = (2, 2)$

   (d) $\psi(x, y) = \frac{xy}{\sqrt{x^2 + y^2}}$; $P_1 = (0, -1)$ and $P_2 = (2, -2)$

   (e) $\psi(x, y) = \ln r$, where r is the polar radius; $P_1 = (1, 1)$ and $P_2 = (1, -1)$

   (f) $\psi(x, y) = \theta$, where θ is the polar angle; $P_1 = (1, 0)$ and $P = (2, 0)$

9. Construct the system of differential equations from the given potential functions (see Example 7.2.5). In each case determine the slope to the solution curve at the indicated points (i.e., compute $\frac{dy}{dx}$ at P).

   (a) $V(x, y) = \frac{1}{2}(x^2 + y^2)$; $P = (1, -1)$

   (b) $V(x, y) = xy$; $P_1 = (0, 0)$ and $P_2 = (1, 2)$

   (c) $V(x, y) = \frac{1}{\sqrt{x^2 + y^2}}$; $P_1 = (1, 1)$ and $P_2 = (2, 2)$

   (d) $V(x, y) = \frac{xy}{\sqrt{x^2 + y^2}}$; $P_1 = (1, -1)$ and $P_2 = (2, -2)$

   (e) $V(x, y) = \ln r$, where r is the polar radius; $P_1 = (1, 1)$ and $P_2 = (1, -1)$

   (f) $V(x, y) = \theta$, where θ is the polar angle; $P = (0, 1)$ and $P_2 = (-1, 1 - 1)$

10. Let A be an $n \times n$ matrix with constant entries. Define the $n \times n$ matrix e^A as

$$e^A = I + A + \frac{1}{2!}A^2 + \frac{1}{3!}A^3 + \cdots = \sum_{n=0}^{\infty} \frac{1}{n!}A^n,$$

where $0! = 1$, by definition, and I is the identity matrix. Similarly, define the matrix e^{tA} as

$$e^{tA} = I + tA + \frac{t^2}{2!}A^2 + \frac{t^3}{3!}A^3 + \cdots = \sum_{n=0}^{\infty} \frac{t^n}{n!}A^n, \tag{7.29}$$

where t is a scalar in R. Let $P_N(t)$ be the Nth partial sum of e^{tA}, that is,

$$P_N(t) = \sum_{n=0}^{N} \frac{t^n}{n!}A^n.$$

(a) Show that A and e^{tA} commute, that is $Ae^{tA} = e^{tA}A$.

(b) Show by termwise differentiation that

$$\frac{d}{dt}\left(e^{tA}\right) = Ae^{tA} = e^{tA}A. \tag{7.30}$$

Let $\mathbf{c}$ be a constant $n \times 1$ column vecter. Let

$$\mathbf{x}(t) = e^{tA}\mathbf{c}. \tag{7.31}$$

Show that $\mathbf{x}$ is a solution of the linear system of differential equations

$$\mathbf{x}' = A\mathbf{x}. \tag{7.32}$$

(c) Let

$$A = \begin{bmatrix} 2 & 0 \\ 0 & -1 \end{bmatrix},$$

and consider the system (7.32) with

$$\mathbf{x}(0) = \mathbf{x}_0 = \begin{bmatrix} 3 \\ -2 \end{bmatrix}.$$

i. Use `MatrixExp` of *Mathematica* or `expm` of *MATLAB* and determine e^{tA}. Let $\mathbf{sol} = e^{tA}\mathbf{x}_0$. Compare $\mathbf{sol}$ with the output of `DSolve` or `dsolve` on this system of differential equations.

ii. Compute $A_1 \equiv P_4(t)$ by hand. Let $\mathbf{x}_1 = A_1\mathbf{x}_0$ and $y_1 = \|\mathbf{sol} - \mathbf{x}_1\|$, where $\|\mathbf{a}\|$ stands for the magnitude of the vector $\mathbf{a}$. Draw the graph of y_1 for $t \in (0,3)$.

iii. Next, use *Mathematica* or *MATLAB* to compute $P_{10}(t)$. Call this matrix B. Let $\mathbf{x}_2 = B\mathbf{x}_0$ and $y_2 = \|\mathbf{sol} - \mathbf{x}_2\|$. Draw the graph of y_2 for $t \in (0,3)$.

iv. Repeat the preceding exercise by computing $C = P_{20}(t)$ and construct $\mathbf{x}_3 = C\mathbf{x}_0$. Let $y_3 = \|\mathbf{sol} - \mathbf{x}_3\|$. Draw the graphs of y_1, y_2, and y_3 on the same screen. What can you conclude from this graph about the quality of convergence of $P_N(t)\mathbf{x}_0$ to $e^{tA}\mathbf{x}_0$?

(d) Let

$$A = \begin{bmatrix} 1 & -1 \\ 2 & 3 \end{bmatrix}.$$

Repeat Problem 9(c) with this A.

11. Let L stand for the Laplace transform operator.

(a) Let $\mathbf{x} = \langle x_1, x_2, \cdots, x_n \rangle$. Define

$$L[\mathbf{x}] = \langle L[x_1], L[x_2], \cdots, L[x_n] \rangle.$$

Determine $L[\mathbf{x}]$.

i. $\mathbf{x} = \langle e^{2t}, 1 - t \rangle$

ii. $\mathbf{x} = \langle e^{-t}\cos t, e^{-t}\sin t, t^2 \rangle$

iii. $\mathbf{x} = \langle \sinh 2t, t + \cosh 3t \rangle$

iv. $\mathbf{x} = \langle (2t^2 - 3) \sin t, t \cos t, 1 \rangle$

v. $\mathbf{x} = \langle e^t u(t-1), e^{-t} u(t-3) \rangle$

vi. $\mathbf{x} = \langle \delta(t-1) - \delta(t-2), \delta(t-3) + \delta(t-4) \rangle$

(b) Let $\mathbf{X} = \langle X_1, X_2, \cdots, X_n \rangle$. Let $x_i = L^{-1}[X_i]$. Define $L^{-1}[X] = \langle x_1, x_2, \cdots, x_n \rangle$. Determine $L^{-1}[X]$.

i. $\mathbf{X} = \langle \frac{s}{s+1}, \frac{1-s}{s^2+3} \rangle$

ii. $\mathbf{X} = \langle \frac{2s-3}{s^2+3s+2}, \frac{1}{s^2} \rangle$

iii. $\mathbf{X} = \langle \frac{s}{s+1}, \frac{1-s}{s^2+3} \rangle$

iv. $\mathbf{X} = \langle \frac{e^{-\pi s}}{s^2+1}, \frac{e^{-s}}{s^2-4} \rangle$

v. $\mathbf{X} = \langle e^{-2s} - e^s, \frac{e^{-2s}-e^s}{s} \rangle$

(c) Let $\mathbf{x} = \langle x_1, x_2, \cdots, x_n \rangle$ and A an $n \times n$ matrix.

i. Show that $L[A\mathbf{x}] = AL[\mathbf{x}]$

ii. Show that $L[\mathbf{x}'] = sL[\mathbf{x}] - \mathbf{x}(0)$.

(d) Do Project B at the end of this chapter.

7.3 An Existence and Uniqueness Theorem

In the previous section we presented several examples of systems of differential equations that model physical problems. Some of these systems are linear. In the bulk of the remainder of this chapter we are concerned with finding solutions of such systems and, when possible, relating the information from these solutions to the physical quantities they model. Some of the systems we encountered in Section 7.2 are nonlinear. The techniques we will develop for them have an approximate nature. It is particularly important in the latter case to have the knowledge that solutions exist and are unique under reasonable conditions on the nonlinearities and the initial data. An existence and uniqueness theorem gives us the necessary confidence to seek approximate solutions and the hope that as approximate solutions converge, they converge to the unique solution of the problem. Such a theorem is significant for linear systems as well since it gives us the freedom of seeking solutions by different techniques and feel comfortable that all methods lead to the same solution.

We now present a theorem that, given certain conditions on the functions f_is in (7.26), guarantees the existence of a unique solution to this system. This theorem is the analogue of the existence and uniqueness theorem of Chapter 3 and serves the same purpose for systems as it did for equations.

Theorem 7.3.1 (Existence and Uniqueness)
Consider the initial value problem

$$\mathbf{x}' = \mathbf{f}(t, \mathbf{x}), \quad \mathbf{x}(t_0) = \mathbf{x}_0. \tag{7.33}$$

Suppose that there is a neighborhood of t_0 and $\mathbf{x}_0$ in which $\mathbf{f}$ is continuous in t and continuously differentiable in $\mathbf{x}$ at $\mathbf{x}_0$. Then there is $\delta > 0$ for which there is a unique solution to (7.33) with $t \in (t_0 - \delta, t_0 + \delta)$.

The proof of this theorem can be found in the text by J. Hale referenced in Chapter 3. Its proof follows the line of argument we encountered in the Picard iteration method that was presented in a project in Chapter 3. It is important to note that this theorem guarantees the existence and uniqueness of a solution in an interval of length 2δ about t_0. The magnitude of δ depends primarily on the function $\mathbf{f}$. When $\mathbf{f}$ depends nonlinearly on $\mathbf{x}$, this theorem does not give us a clue as to how large δ could be, only that $\delta > 0$. When $\mathbf{f}$ depends linearly on $\mathbf{x}$, however, the following corollary of this theorem guarantees that the solution to (7.33) exists **globally**; that is, for all t for which $\mathbf{f}$ is continuous.

Corollary 7.3.2 (Existence and Uniqueness for Linear Systems)
Consider the system of differential equations

$$\mathbf{x}' = A(t)\mathbf{x} + \mathbf{f}(t), \quad \mathbf{x}(t_0) = \mathbf{x}_0. \tag{7.34}$$

Suppose that the $n \times n$ matrix $A(t)$ and the $n \times 1$ vector-valued function $\mathbf{f}$ are continuous in the interval (t_0, T). Then there exists a unique solution to (7.34) in the interval (t_0, T).

Example 7.3.1
Consider the system of differential equations in Example 7.2.1. The function $\mathbf{f}$ has four components $\langle f_1, f_2, f_3, f_4 \rangle$, and where $f_1(x_1, y_1, x_2, y_2) = y_2$, $f_2 = \frac{-k_1 x_1 + k_2 (x_2 - x_1)}{m_1}$, for example. Each f_i is continuously differentiable in the x_is and y_is. Hence, the conditions for the existence and uniqueness of a solution are satisfied for any pair $(t_0, \mathbf{x}_0)$. The same analysis applies to the system in Example 7.2.2.

The only point of concern for the system in Example 7.2.3 is the point with coordinates (x_0, y_0), at which the right-hand side of (7.17) blow up. Hence, as long as $\mathbf{x} \neq \mathbf{0}$, the previous theorem guarantees the existence and uniqueness of a solution for the pair $(t_0, \mathbf{x}_0)$.

There is a close relationship between first-order systems and nth-order equations: Any nth-order equation, linear or nonlinear, can be converted to a first-order system. Consider the general equation

$$x^{(n)} = g(t, x, x', x'', \ldots, x^{(n-1)}), \quad x(t_0) = x_0,$$
$$x'(t_0) = x_1, \quad \ldots, \quad x^{(n-1)}(t_0) = x_{n-1}. \tag{7.35}$$

Define n new variables $y_1, y_2, \ldots, y_n$ as

$$y_1 = x, \quad y_2 = x', \quad y_3 = x'', \quad \ldots, \quad y_n = x^{(n-1)}. \tag{7.36}$$

It follows from (7.36) that $y_1' = y_2$ and $y_2' = y_3$. In general $y_i' = y_{i+1}$ for $1 \leq i \leq n-1$. The differential equation for y_n comes to us from (7.35). Similarly, we can read off the initial data for $\mathbf{y}$ from (7.35). Following this procedure we find that (7.35) is equivalent to the first-order system

$$y_1' = y_2, \quad y_2' = y_3, \quad \dots, \quad y_{n-1}' = y_n, \quad y_n' = g(t, y_1, y_2, \dots, y_n), \quad (7.37)$$

will initial data

$$y_1(t_0) = x_0, \quad y_2(t_0) = x_1, \quad \dots, \quad y_n(t_0) = x_{n-1}. \quad (7.38)$$

This system can be written in vector form as

$$\mathbf{y}' = \mathbf{f}(t, \mathbf{y}'), \quad \mathbf{y}(t_0) = \mathbf{y}_0, \quad (7.39)$$

where the vector-valued function $\mathbf{f}$ and $\mathbf{y}_0$ are defined by

$$\mathbf{f} = \begin{bmatrix} y_1 \\ y_2 \\ \cdots \\ \cdots \\ y_{n-1} \\ g(t, y_1, y_2, \dots, y_n) \end{bmatrix}, \quad \mathbf{y}(t_0) = \begin{bmatrix} x_0 \\ x_1 \\ \cdots \\ \cdots \\ x_{n_1} \end{bmatrix}. \quad (7.40)$$

Example 7.3.2
Consider the third-order equation

$$x''' + 2x'' - 3x' + x + x^3 = \sin t, \quad x(0) = 0, \quad x'(0) = -1, \quad x''(0) = 3. \quad (7.41)$$

To convert this equation to a first-order system, we define the vector $\mathbf{y} = \langle y_1, y_2, y_3 \rangle$ where the new dependent variables y_is are related to x by

$$y_1 = x, \quad y_2 = x', \quad y_3 = x''. \quad (7.42)$$

After differentiating each equation in (7.42) once with respect to t and using the relations in (7.42) and (7.41), we have

$$y_1' = y_2, \quad y_2' = y_3, \quad y_3' = -2y_3 + 3y_2 - y_1 - y_1^3 + \sin t, \quad (7.43)$$

with the initial vector

$$y_1(0) = 0, \quad y_2(0) = -1, \quad y_3(0) = 3. \quad (7.44)$$

Example 7.3.3
Consider the linear fourth-order equation

$$x^{(4)} + 0.01x'' + 16x = \gamma \sin \omega t. \quad (7.45)$$

To convert this equation to a system, let $y_1 = x$, $y_2 = x'$, $y_3 = x''$, and $y_4 = x'''$. It is clear that

$$y_1' = y_2, \quad y_2' = y_3, \quad y_3' = y_4, \quad y_4' = -0.01y_3 - 16y_1 + \gamma \sin \omega t, \qquad (7.46)$$

which can be written in matrix notation as

$$\mathbf{y}' = \begin{bmatrix} 0 & 1 & 0 & 0 \\ 0 & 0 & 1 & 0 \\ 0 & 0 & 0 & 1 \\ -16 & 0 & -0.01 & 0 \end{bmatrix} \mathbf{y} + \begin{bmatrix} 0 \\ 0 \\ 0 \\ \gamma \sin \omega t \end{bmatrix}. \qquad (7.47)$$

Example 7.3.4

Consider the nonlinear second-order equation

$$x'' + x'^2 + \sin x = 1 - \cos t, \quad x(0) = 0, \quad x'(0) = -1.$$

With $y_1 = x$ and $y_2 = x'$, we have

$$y_1' = y_2, \quad y_2' = -y_2^2 - \sin y_1 + 1 - \cos t.$$

Thus, $\mathbf{y} = \langle y_1, y_2 \rangle$ satisfies the system $\mathbf{y}' = \mathbf{f}(t, \mathbf{y})$, where

$$f_1(t, \mathbf{y}) = y_2, \quad f_2(t, \mathbf{y}) = -y_2^2 - \sin y_1 + 1 - \cos t,$$

and $\mathbf{y}_0 = \langle 0, -1 \rangle$.

The converse of this procedure, converting first-order systems to single higher order equations, is possible when the system is linear. The procedure is called the **method of elimination**, which we illustrate in the following examples.

Example 7.3.5 (Method of Elimination)

Consider the system of differential equations

$$x' = 2x - 3y + \sin t, \quad y' = -3x + 4y - \cos t. \qquad (7.48)$$

Rewrite this system using the differential operator D:

$$(D - 2)x + 3y = \sin t, \quad 3x + (D - 4)y = -\cos t.$$

Define operator L_1 to L_4 by

$$L_1[x] = (D - 2)x, \quad L_2[y] = 3y, \quad L_3[x] = 3x, \quad L_4[y] = (D - 4)y. \qquad (7.49)$$

We can rewrite (7.48) as

$$L_1[x] + L_2[y] = \sin t, \quad L_3[x] + L_4[y] = -\cos t. \qquad (7.50)$$

Eliminate y from (7.50) by applying L_4 to the first equation in this system, L_2 to the second equation, and subtracting the resulting expressions:

$$L_4[L_1[x]] - L_2[L_3[x]] = L_4[\sin t] - L_2[-\cos t], \tag{7.51}$$

which is equivalent to

$$(D - 4)(D - 2)x - 9x = (D - 4)(\sin t) + \cos t$$

or

$$x'' - 6x' - x = -4\sin t + 2\cos t.$$

We can obtain an equation for y in a similar fashion.

Implementing the method of elimination in *Mathematica* or *MATLAB* is straightforward. The syntax of a program in *Mathematica* as it applies to the system in (7.48) is

```
f1 = Sin[t]; f2 = -Cos[t];
L[1][x_] := D[x, t] - 2*x;
L[2][y_] :=  3*y;
L[3][x_] := 3*x;
L[4][y_] := D[y, t] - 4*y;
eqn[1] = L[1][x[t]] + L[2][y[t]] - f1;
eqn[2] = L[3][x[t]] + L[4][y[t]] - f2;
ans = Simplify[L[4][eqn[1]] - L[2][eqn[2]]]
```

A similar strategy works for higher dimensional and higher order systems, as the exercises will demonstrate.

Problems

1. Verify the hypotheses of the existence and uniqueness theorem for the following systems. Determine all initial data $\mathbf{x}_0$ and t_0 that violate these hypotheses.

   (a) $x' = 2ty + 3x,$ $\qquad y' = (\sin t)x + 2y$

   (b) $tx' = t^2x + 3y,$ $\qquad y' = (\tan t)x + 2y$

   (c) $(t - 2)x' = x + 3(\sin t)y,$ $\qquad y' = x + y$

   (d) $x' = \dfrac{x}{\sqrt{x^2+y^2}},$ $\qquad y' = \dfrac{y}{\sqrt{x^2+y^2}}$ (Hint: Plot the graph of the right-hand side of these equations to get a sense of the behavior of these functions at the origin.)

   (e) $x' = x + \sqrt{y},$ $\qquad y' = x$

2. Convert the following equations to first-order systems. Identify each constant coefficient system and use DSolve of *Mathematica* or dsolve of *MATLAB* to obtain the general solution. If the system is nonlinear use NDSolve or ode45 with appropriate initial data of your choice and plot the graph of the first component of the solution.

(a) $x'' + 2x' - 3x = \sin t$

(b) $x'' - 3x' - a^2 x = 0$, where a is a constant

(c) $y''' + 3y'' - 3y' = e^x$

(d) $y''' + 2y' - 7y = 0$

(e) $x'''' + 4x'' - 16x = 0$

(f) $x'' + 0.1x' + x^2 = 0$

(g) $x'' + (1 - x^2)x' + x = 0$

(h) $y'' + 0.5y' + \sin y = 0$

3. Write the analogue of the program in Example 7.3.4 in *MATLAB*.

4. Convert each of the following systems to a single equation in terms of one of the unknowns in the problem. Use the *Mathematica* or *MATLAB* program of this section and verify the computations in the method of elimination. When the system has order higher than 2, or has nonconstant coefficients, alter the computer programs appropriately to accommodate these cases.

(a) $x' = y, \quad y' = x - y$

(b) $x' = x + y, \quad y' = 2x - y$

(c) $x' = y, \quad y' = x - y$

(d) $x' = tx - y, \quad y' = x + ty$

(e) $x' = \sin t \quad x - t^2 \quad y, \quad y' = \cos t \quad x + ty$

(f) $x' = x - y + z, \quad y' = x + y - 2z, \quad z' = x - y$

(g) $x' = y - z, \quad y' = z - x, \quad z' = y - x$

(h) $x' = y + z, \quad y' = z + x, \quad z' = y + x$

(i) $x' = y, \quad y' = z, \quad z' = -2x - 4y - z$.

5. The equations of motion of two bodies with masses m_1 and m_2 connected in a series to three springs, as described in Example 7.2.1, are given by the system of two second-order equations

$$m_1 x_1'' = -k_1 x_1 + k_2(x_2 - x_1), \quad m_2 x_2'' = -k_2(x_2 - x_1) - k_3 x_2. \quad (7.52)$$

Here k_1, k_2, and k_3 are the spring constants.

(a) Convert the preceding second-order system to a first-order system of order 4. With $m_1 = 1$, $m_2 = 2$, $k_1 = 1$, $k_2 = 2$, and $k_3 = 1$, enter this system to *Mathematica*'s NDSolve or *MATLAB*'s ode45. Use the appropriate number of initial conditions and a terminal time of your choice. Plot the graph of x_1.

(b) Give the second-order system (7.52) directly to *Mathematica*. First use DSolve and the initial conditions you prescribed in the previous part. Plot the graph of x_1 over the same interval of t. Next use NDSolve with the second-order equations and plot x_1. Do you see any differences in the graphs of x_1 in the three cases?

(c) Convert the second-order system in (7.52) to a single fourth-order equation in x_1. Solve this equation with the equivalent initial conditions used in the previous parts using DSolve of *Mathematica* or **dsolve** of *MATLAB*. Plot the graph of x_1 and compare with the previous outputs.

7.4 Wronskian and Fundamental Matrix Solutions

In the next few sections we will concentrate on the properties of solutions of the linear system

$$\mathbf{x}' = A(t)\mathbf{x} + \mathbf{b}(t), \quad \mathbf{x}(t_0) = \mathbf{x}_0, \tag{7.53}$$

where A is an $n \times n$ matrix with continuous entries and $\mathbf{b}$ is a continuous vector in E^n. The existence and uniqueness theorem of the previous section guarantees that this initial value problem has a unique solution in a neighborhood of $(t_0, \mathbf{x}_0)$ as long as the matrix $A(t)$ and the vector function $\mathbf{b}(t)$ are continuous functions in some neighborhood of t_0. Our goal is to develop methods, similar to those developed in Chapters 3 and 4 for equations, that would aid us in constructing this unique solution, both analytically in terms of elementary functions of calculus and numerically when obtaining a solution in closed form might be prohibitive.

As was the case in the analysis we presented for equations, the fact that the right-hand side of (7.53) depends linearly on $\mathbf{x}$ is quite useful in constructing solutions of this equation. First, let us consider the **homogeneous** case

$$\mathbf{x}' = A(t)\mathbf{x}. \tag{7.54}$$

Note that if $\mathbf{x}_1$ and $\mathbf{x}_2$ are solutions of (7.54), so is $c_1\mathbf{x}_1 + c_2\mathbf{x}_2$ when c_1 and c_2 are scalars, since

$$(c_1\mathbf{x}_1 + c_2\mathbf{x}_2)' = c_1\mathbf{x}_1' + c_2\mathbf{x}_2' = c_1 A(t)\mathbf{x}_1 + c_2 A(t)\mathbf{x}_2 = A(t)(c_1\mathbf{x}_1 + c_2\mathbf{x}_2). \tag{7.55}$$

Similarly, given m solutions $\{\mathbf{x}_1, \mathbf{x}_2, \cdots, \mathbf{x}_m\}$ of (7.54), the linear combination

$$c_1\mathbf{x}_1(t) + c_2\mathbf{x}_2(t) + \cdots + c_m\mathbf{x}_m(t),$$

is also a solution of this equation. The above combination is called the **linear superposition** of the functions $\mathbf{x}_1, \mathbf{x}_2, \cdots, \mathbf{x}_n$.

We define the concept of linear independence of vector-valued functions in the same way that linear independence was defined for real-valued functions. Let $\{\mathbf{x}_1, \mathbf{x}_2, \cdots, \mathbf{x}_m\}$ be a set of m vector-valued functions. We say these functions are **linearly independent** if the only way the identity

$$c_1\mathbf{x}_1(t) + c_2\mathbf{x}_2(t) + \cdots + c_m\mathbf{x}_m(t) = \mathbf{0}, \tag{7.56}$$

holds for all t in an interval (a, b) is with $c_1 = c_2 = \cdots = c_m = 0$. Otherwise, this set of functions is called **linearly dependent**. As in the case of real-valued functions, the concept of linear independence emphasizes the point that none of the functions $\mathbf{x}_i$ can be written as a linear combination of the remaining functions in that set.

The concept of **wronskian** of a set of vector-valued functions is the same as its counterpart for real-valued functions. Let $\{\mathbf{x}_1, \mathbf{x}_2, \cdots, \mathbf{x}_n\}$ be a set of functions each having n components. The wronskian of this set is the determinant of the $n \times n$ matrix we construct out of this set by making each function $\mathbf{x}_i$ a column of the matrix

$$
w(\mathbf{x}_1, \mathbf{x}_2, \cdots, \mathbf{x}_n) = \det
\begin{bmatrix}
x_{11}(t) & x_{12}(t) & \cdots & x_{1n}(t) \\
x_{21}(t) & x_{22}(t) & \cdots & x_{2n}(t) \\
\cdots & \cdots & \cdots & \cdots \\
\cdots & \cdots & \cdots & \cdots \\
x_{n1}(t) & x_{n2}(t) & \cdots & x_{nn}(t)
\end{bmatrix}.
\tag{7.57}
$$

Remark 7.4.1: Note that in (7.57) we used the notation x_{ij} to denote the ith component of the jth function $\mathbf{x}_j$.

We noted in Chapter 4 that the wronskian of a set of real-valued functions determined the linear independence of those functions. This concept has the same role for vector-valued functions, as the following theorem shows.

Theorem 7.4.1 (Wronskian and Linear Independence)

Let $\{\mathbf{x}_1(t), \mathbf{x}_2(t), \cdots, \mathbf{x}_n(t)\}$ be n solutions of the linear system (7.54). Let $w(\mathbf{x}_1, \mathbf{x}_2, \cdots, \mathbf{x}_n)$ be the wronskian of these solutions. If the solutions $\mathbf{x}_1, \mathbf{x}_2, \cdots, \mathbf{x}_n$ are linearly independent in an interval (a, b), then the wronskian w is nonzero for every t in (a, b). If, on the other hand, these solutions are linearly dependent, then the wronskian w is identically zero.

The proof of this result is identical with the proof given for the analogous theorem in Chapter 4 and we do not repeat it here.

Example 7.4.1
Let

$$
A(t) = \begin{bmatrix} 0 & 1 \\ -1 & 0 \end{bmatrix}.
$$

The functions

$$
\mathbf{x}_1(t) = \begin{bmatrix} \sin t \\ \cos t \end{bmatrix}, \quad
\mathbf{x}_2(t) = \begin{bmatrix} \cos t \\ -\sin t \end{bmatrix}
$$

are solutions of (7.54) with the previous $A(t)$. The wronskian of these solutions is

$$
w = \det \begin{bmatrix} \sin t & \cos t \\ \cos t & -\sin t \end{bmatrix} = -1.
$$

Hence, by Theorem 7.4.1, the two solutions $\mathbf{x}_1$ and $\mathbf{x}_2$ are linearly independent. On the other hand, the two solutions $\{\mathbf{x}_1, \mathbf{x}_2\}$ with $\mathbf{x}_1$ as defined and

$$\mathbf{x}_2(t) = 2\mathbf{x}_1(t)$$

are linearly dependent since their wronskian is identically zero.

Example 7.4.2

Consider the system (7.54) with

$$A = \begin{bmatrix} 0 & 1 & 0 \\ 0 & 0 & 1 \\ -6 & -11 & -6 \end{bmatrix}.$$

The reader can check that the functions $\mathbf{x}_1$, $\mathbf{x}_2$, and $\mathbf{x}_3$ defined by

$$\mathbf{x}_1(t) = e^{-3t} \begin{bmatrix} 1 \\ -3 \\ 9 \end{bmatrix}, \quad \mathbf{x}_2(t) = e^{-2t} \begin{bmatrix} 1 \\ -2 \\ 4 \end{bmatrix}, \quad \mathbf{x}_3(t) = e^{-t} \begin{bmatrix} 1 \\ -1 \\ 1 \end{bmatrix}$$

are solutions of this linear system and that their wronskian is

$$w = \det \begin{bmatrix} e^{-3t} & e^{-2t} & e^{-t} \\ -3e^{-3t} & -2e^{-2t} & -e^{-t} \\ 9e^{-3t} & 4e^{-2t} & e^{-t} \end{bmatrix} = 2e^{-6t}. \tag{7.58}$$

The functions $\mathbf{x}_1, \mathbf{x}_2$, and $\mathbf{x}_3$ are therefore linearly independent.

Remark 7.4.2: It is worth emphasizing the difference between linear independence of vector-valued functions and solutions of linear systems. Consider the two vector-valued functions

$$\mathbf{x}_1(t) = \begin{bmatrix} t \\ 0 \end{bmatrix}, \quad \mathbf{x}_2(t) = \begin{bmatrix} 1+t \\ 0 \end{bmatrix}. \tag{7.59}$$

Using the definition of linear independence, we can verify that $\mathbf{x}_1$ and $\mathbf{x}_2$ are linearly independent. Their wronskian, however, is zero. This does not contradict the conclusion of Theorem 7.4.1 since $\mathbf{x}_1$ and $\mathbf{x}_2$ cannot be solutions to a system (7.54).

In general, any set of n linearly independent solutions of (7.54) is called a **fundamental system** for this equation. When we have such a set, we can construct an $n \times n$ matrix $X(t)$, each column of which is an element from the fundamental system. Such a matrix is called a **fundamental matrix solution** of (7.54). Since the columns of $X(t)$ are linearly independent, we must have that $\det(X(t)) \neq 0$. We call the matrix X a **principal matrix solution** if X reduces to the identity matrix at the initial time t_0. We prove the following statements about fundamental matrix solutions.

Lemma 7.4.1

1. If $X(t)$ is a fundamental matrix solution of (7.54), then the general solution of this equation is

$$\mathbf{x}(t) = X(t)\mathbf{c}, \tag{7.60}$$

where $\mathbf{c}$ is an arbitrary vector.

2. Let X be a principal matrix solution of (7.54). Then the unique solution to the initial value problem (7.54) with initial data $\mathbf{x}(t_0) = \mathbf{x}_0$ is

$$\mathbf{x}(t) = X(t)\mathbf{x}_0. \tag{7.61}$$

Proof: The proofs of both statements rely on the uniqueness theorem. To prove the first part, let X be a fundamental solution and $\mathbf{y}$ any solution of (7.54). Let t_0 be in the domain of $\mathbf{y}$. Define $\mathbf{c}$ by

$$\mathbf{c} = X^{-1}(t_0)\mathbf{y}(t_0). \tag{7.62}$$

Define $\mathbf{x}(t) = X(t)\mathbf{c}$. Since both $\mathbf{x}$ and $\mathbf{y}$ satisfy the same differential equation and the same initial condition, they must be the same solutions by the uniqueness theorem. The proof of the second part is similar.

The following theorem, known as **Liouville** or **Abel**'s Theorem, gives a closed-form formula for the determinant of a matrix solution of (7.54) in terms of the entries of A. It is a direct consequence of this theorem that if X is a fundamental matrix solution, then its determinant is either never zero or identically zero.

Theorem 7.4.2
Let X be a fundamental solution of (7.54). Then

$$\det(X(t)) = \det(X(t_0))e^{\int_{t_0}^{t} \operatorname{tr}(A(s))\, ds}, \tag{7.63}$$

where tr stands for the trace of A, that is, $\operatorname{tr}(A) = a_{11} + a_{22} + \cdots + a_{nn}$.

Proof: We present the proof of this result only for the case when A is 2×2. The general case follows the same steps. In the 2×2 case X is of the form

$$X(t) = \begin{bmatrix} x_{11} & x_{12} \\ x_{21} & x_{22} \end{bmatrix}$$

and system (7.54) becomes

$$\begin{bmatrix} x'_{11} & x'_{12} \\ x'_{21} & x'_{22} \end{bmatrix} = \begin{bmatrix} a_{11} & a_{12} \\ a_{21} & a_{22} \end{bmatrix} \begin{bmatrix} x_{11} & x_{12} \\ x_{21} & x_{22} \end{bmatrix}. \tag{7.64}$$

It is not difficult to verify that

$$(\det X(t))' = \det \begin{bmatrix} x'_{11} & x'_{12} \\ x_{21} & x_{22} \end{bmatrix} + \det \begin{bmatrix} x_{11} & x_{12} \\ x'_{21} & x'_{22} \end{bmatrix}. \tag{7.65}$$

Thus, the derivative of the determinant of a 2×2 matrix is the sum of the determinants of the matrices one obtains from X by replacing its ith row by the derivative of that row (this statement generalizes to $n \times n$ matrices). Let us consider the first determinant. It follows from (7.64) that $x'_{11} = a_{11}x_{11} + a_{12}x_{21}$ and $x'_{12} = a_{11}x_{12} + a_{12}x_{22}$. We substitute these expressions into the first determinant in (7.65), carry out the necessary algebra, and reach the conclusion that this determinant reduces to

$$a_{11}\det X(t).$$

Similarly the second determinant in (7.65) is equivalent to

$$a_{22}\det X(t).$$

We have therefore shown that $y(t) = \det(X(t))$ satisfies the differential equation

$$y' = (a_{11}(t) + a_{22}(t))y,$$

whose solution is (7.63). This completes the proof of the theorem.

In the next theorem we develop the analogue of the **variation of parameters formula** for systems. But first we need the following lemma.

Lemma 7.4.2 (The Adjoint Equation)
Let $X(t)$ be a fundamental matrix solution of (7.54). Then $X^{-1}(t)$ is a fundamental matrix solution of the **Adjoint Equation**

$$\mathbf{x}' = -\mathbf{x}A(t). \tag{7.66}$$

Proof: Differentiate the identity $X^{-1}(t)X(t) = I$ with respect to t to get

$$(X^{-1})'X + X^{-1}X' = 0. \tag{7.67}$$

Since the matrix X is a solution of (7.54), we can replace X' in (7.67) with AX and obtain

$$(X^{-1})'X = -X^{-1}AX.$$

Multiplying this equation on the right by X^{-1} leads to the conclusion in (7.66).

Theorem 7.4.3 (Variation of Parameters Formula)
Let X be a fundamental matrix solution of the homogeneous linear system (7.54). Then the unique solution to (7.53) is

$$\mathbf{x}(t) = X(t)X^{-1}(t_0)\mathbf{x}_0 + X(t)\int_{t_0}^{t} X^{-1}(s)\mathbf{b}(s)\,ds. \tag{7.68}$$

Proof: Multiply both sides of (7.53) by X^{-1} on the left and use the result of Lemma 7.4.2 to get

$$(X^{-1}(t)\mathbf{x}(t))' = X^{-1}(t)\mathbf{b}(t).$$

The proof of the theorem follows by integrating the previous expression with respect to t from t_0 to t.

We will revisit the variation of parameters formula in more detail in Section 7.7.

Problems

Unless otherwise specified, let $t_0 = 0$.

1. Apply the definition of linear independence of vector-valued functions and determine whether the following collections of functions are linearly independent or not. Verify each answer in *Mathematica* or *MATLAB*.

   (a) $\mathbf{x}_1(t) = \begin{bmatrix} t \\ t^2 \end{bmatrix}$, $\mathbf{x}_2(t) = \begin{bmatrix} 2t \\ 1 \end{bmatrix}$

   (b) $\mathbf{x}_1(t) = \begin{bmatrix} \sin t \\ \cos t \end{bmatrix}$, $\mathbf{x}_2(t) = \begin{bmatrix} \cos t \\ \sin t \end{bmatrix}$

   (c) $\mathbf{x}_1(t) = \begin{bmatrix} e^t \\ e^{2t} \end{bmatrix}$, $\mathbf{x}_2(t) = \begin{bmatrix} te^t \\ 1 \end{bmatrix}$

   (d) $\mathbf{x}_1(t) = \begin{bmatrix} e^t \\ -t \end{bmatrix}$, $\mathbf{x}_2(t) = \begin{bmatrix} -e^t \\ t \end{bmatrix}$

   (e) $\mathbf{x}_1(t) = \begin{bmatrix} t \\ t^2 \\ t^3 \end{bmatrix}$, $\mathbf{x}_2(t) = \begin{bmatrix} 2t \\ 1 \\ 0 \end{bmatrix}$, $\mathbf{x}_3(t) = \begin{bmatrix} 0 \\ 0 \\ 1 \end{bmatrix}$

   (f) $\mathbf{x}_1(t) = \begin{bmatrix} e^t \\ e^{2t} \\ e^{3t} \end{bmatrix}$, $\mathbf{x}_2(t) = \begin{bmatrix} 2e^t \\ 3e^{2t} \\ -e^{3t} \end{bmatrix}$, $\mathbf{x}_3(t) = \begin{bmatrix} e^t \\ 2e^{2t} \\ -2e^{3t} \end{bmatrix}$

2. In each of the following problems show that X satisfies (7.54) with the corresponding A. Use *Mathematica* or *MATLAB* to determine which of the Xs is a fundamental matrix solution.

   (a) $A = \begin{bmatrix} 0 & 1 \\ -1 & 0 \end{bmatrix}$, $X = \begin{bmatrix} \sin t & \cos t \\ \cos t & -\sin t \end{bmatrix}$

   (b) $A = \begin{bmatrix} \frac{2}{t} & -\frac{2}{t^2} \\ 1 & 0 \end{bmatrix}$, $X = \begin{bmatrix} 1 & 2t \\ t & t^2 \end{bmatrix}$

   (c) $A = \begin{bmatrix} 0 & 1 & 0 \\ 0 & 0 & 1 \\ -2 & 1 & 2 \end{bmatrix}$, $X = \begin{bmatrix} e^{-t} & e^t & e^{2t} \\ -e^{-t} & e^t & 2e^{2t} \\ e^t & e^t & e^{4t} \end{bmatrix}$

3. Use Theorem 7.4.3, *Mathematica* or *MATLAB*, and the fundamental matrix solutions of the previous problem and find the solution to (7.53) with the following nonhomogeneities.

   (a) With A and X as in Problem 2(a) and $\mathbf{b} = \begin{bmatrix} t \\ 1 \end{bmatrix}$, $\quad \mathbf{x}_0 = \begin{bmatrix} 0 \\ 0 \end{bmatrix}$

   (b) With A and X as in Problem 2(b) and $\mathbf{b} = \begin{bmatrix} \sin t \\ 0 \end{bmatrix}$, $\quad \mathbf{x}_0 = \begin{bmatrix} 0 \\ 0 \end{bmatrix}$.
   Draw the graph of x_1 in the interval $(0, 6\pi)$.

   (c) With A and X as in Problem 2(c) and $\mathbf{b} = \begin{bmatrix} 0 \\ 0 \\ e^t \end{bmatrix}$, $\quad \mathbf{x}_0 = \begin{bmatrix} 0 \\ 0 \\ 0 \end{bmatrix}$.

4. Show that the two vector-valued functions $\mathbf{x}_1$ and $\mathbf{x}_2$ defined in (7.59) cannot be solutions of a 2×2 system $\mathbf{x}' = A(t)\mathbf{x}$.

5. Let X be a matrix of solutions of (7.54) with A continuous. Show that if $\det X(t_0) \neq 0$ for some t_0, then $\det X(t) \neq 0$ for all t in the domain of X.

6. Let X be a fundamental matrix solution of (7.54). Let B be any nonsingular $n \times n$ matrix with constant entries. Show that $X(t)B$ is also a fundamental matrix solution of (7.54).

7. Use Theorem 7.4.2 to show that the wronskian w of an nth-order differential equation

$$\frac{d^n y}{dt^n} + a_{n-1}(t)\frac{d^{(n-1)}y}{dt^{n-1}} + \cdots + a_1(t)y = 0$$

satisfies

$$w(t) = w(t_0)e^{-\int_{t_0}^{t} a_{n-1}(s)\, ds}.$$

7.5 Eigenvalue Method for Linear Systems

In Chapter 4 we developed an algorithm for determining the solution to an ODE with constant coefficients. We saw that the general solution of such an equation can be described in terms of exponential functions whose exponents were related to the roots of the characteristic polynomial of the differential equation. We now develop an analogue of this algorithm for linear systems with constant coefficients. As we will see shortly, this algorithm reduces to determining the eigenvalues and eigenvectors of the matrix of coefficients.

We start with a linear system of equations of the form

$$\mathbf{x}' = A\mathbf{x}, \tag{7.69}$$

where we assume that the matrix A in (7.69) has constant real entries. Following the steps of the process that was developed for linear equations, we seek solutions to (7.69) of the form

$$\mathbf{x}(t) = e^{\lambda t}\mathbf{e}, \tag{7.70}$$

where λ and $\mathbf{e}$ are as yet unknown. Substitute (7.70) into (7.69), divide by $e^{\lambda t}$, and obtain the algebraic equation

$$A\mathbf{e} = \lambda\mathbf{e}, \tag{7.71}$$

for λ and $\mathbf{e}$. This relation is equivalent to the system

$$(A - \lambda I)\mathbf{e} = \mathbf{0}, \tag{7.72}$$

where I is the $n \times n$ identity matrix. This system has a nontrivial solution $\mathbf{e}$ only if the matrix $A - \lambda I$ is *not* invertible. We recall from Chapter 6 that an $n \times n$ matrix is not invertible (which is equivalent to saying that the matrix is singular) if and only if its determinant is zero. Hence, λ must be a solution to the characteristic equation of A

$$\det(A - \lambda I) = 0. \tag{7.73}$$

Thus, λ must be an eigenvalue of A and, in view of the relation between $\mathbf{e}$ and λ in (7.71), the vector $\mathbf{e}$ must be an eigenvector of A associated with λ. We summarize these findings in the following theorem.

Theorem 7.5.1
The linear system (7.69) has a solution of the form (7.70) if $(\lambda, \mathbf{e})$ is an eigenvalue-eigenvector pair of A.

Example 7.5.1
Consider the linear system

$$\mathbf{x}' = \begin{bmatrix} 2 & 1 \\ 1 & 2 \end{bmatrix}. \tag{7.74}$$

To find the eigenvalues of A, we construct the characteristic polynomial of A:

$$\det(A - \lambda I) = \det\begin{bmatrix} 2 - \lambda & 1 \\ 1 & 2 - \lambda \end{bmatrix} = \lambda^2 - 4\lambda - 3 = 0, \tag{7.75}$$

which has the two roots $\lambda_1 = 1$ and $\lambda_2 = 3$. We next compute the eigenvectors corresponding to each eigenvalue. When $\lambda = 1$, the linear system of algebraic equations (7.72) becomes

$$\begin{cases} c_1 + c_2 = 0, \\ c_1 + c_2 = 0. \end{cases} \tag{7.76}$$

As expected, the two equations in (7.76) are linearly dependent. Setting $c_2 = 1$ in the first equation, we see that

$$\mathbf{e}_1 = \begin{bmatrix} -1 \\ 1 \end{bmatrix} \tag{7.77}$$

is an eigenvector for $\lambda_1 = 1$. Hence, the function $\mathbf{x}_1$ given by

$$\mathbf{x}_1(t) = e^t \begin{bmatrix} -1 \\ 1 \end{bmatrix} \tag{7.78}$$

is a solution of (7.74). Following the same steps with the second eigenvalue $\lambda_2 = 3$, we find that

$$\mathbf{e}_2 = \begin{bmatrix} 1 \\ 1 \end{bmatrix} \tag{7.79}$$

is an eigenvector associated with λ_2 and that

$$\mathbf{x}_2(t) = e^{3t} \begin{bmatrix} 1 \\ 1 \end{bmatrix} \tag{7.80}$$

is a second solution of (7.74). These solutions are linearly independent, as can be seen from their wronskian, or from the fact that the eigenvectors are linearly independent. By putting (7.79) and (7.80) together, we determine a fundamental matrix solution of (7.74):

$$X(t) = \begin{bmatrix} -e^t & e^{3t} \\ e^t & e^{3t} \end{bmatrix}. \tag{7.81}$$

The following program in *Mathematica* produces the same result by generating the general solution of (7.74):

```
A = {{2, 1}, {1, 2}};
eig = Eigenvalues[A];
eigv = Eigenvectors[A];
b = {c1, c2} * Exp[eig t];
sol[t_] = eigv . b
```

Remark 7.5.1: Note that the expression `Exp[eig t]` in this program is exponentiating a vector (`eig t`) and returning a vector. Also note that we used both `*` and `.` as multiplication operations. The first one, `*`, when applied to vectors, creates a vector of appropriate dimensions whose ith entry is the product of the ith entries of the two vectors. The second multiplication, `.`, is the standard matrix multiplication.

Example 7.5.1 is typical of linear systems that have real and distinct eigenvalues. For such systems the eigenvalue method always leads to n linearly independent solutions from which the general solution is constructed.

Example 7.5.2
Consider the linear system

$$\mathbf{x}' = \begin{bmatrix} 0 & 1 \\ -1 & 0 \end{bmatrix} \mathbf{x}. \tag{7.82}$$

The reader can check readily that $\lambda_1 = i$ and $\lambda_2 = -i$ are the eigenvalues of A, while

$$\mathbf{e}_1 = \begin{bmatrix} -i \\ 1 \end{bmatrix}, \qquad \mathbf{e}_2 = \begin{bmatrix} i \\ 1 \end{bmatrix} \tag{7.83}$$

are the associated eigenvectors. Hence, the general solution of this system is

$$\mathbf{x}(t) = c_1 e^{it} \begin{bmatrix} -i \\ 1 \end{bmatrix} + c_2 e^{-it} \begin{bmatrix} i \\ 1 \end{bmatrix}, \tag{7.84}$$

where c_1 and c_2 are complex numbers. As in the case of equations, the real and imaginary parts of (7.84) are also solutions of (7.82). After taking the real part of (7.84) and setting $k_1 = c_1 + c_2$ and $k_2 = i(c_1 - c_2)$, we obtain the following real-valued general solution to (7.82):

$$\mathbf{x}(t) = k_1 \begin{bmatrix} \sin t \\ \cos t \end{bmatrix} + k_2 \begin{bmatrix} \cos t \\ -\sin t \end{bmatrix}. \tag{7.85}$$

It is often desirable to determine the real and imaginary parts of a complex-valued function when we are seeking solutions to systems of differential equations, such as (7.82), that have complex eigenvalues. The reader may find the following program in *Mathematica* helpful in checking over the results. In this program we find the real and imaginary parts of the function $e^{(2+3i)t}$. The main task is to make sure that the software understands that the variable t is real:

```
<<Algebra'ReIm';
condition = {Im[t] -> 0, Re[t] -> t};
f[t_] = Exp[(2 + 3 I) t];
realpart[t_] = Re[f[t]] /. condition;
imagpart[t_] = Im[f[t]] /. condition;
```

The circumstances of the previous example are typical when the matrix A has real entries and its eigenvalues are distinct and complex. In such cases complex eigenvalues must occur in complex conjugate pairs, from which we are able to construct two real-valued solutions. Instead of writing down the cumbersome formulas to formalize this comment (see Problem 2), it is preferable to remember the procedure that leads to a solution such as (7.85). Given a complex eigenvalue λ and its associated complex eigenvector $\mathbf{e}$, first construct one solution from this pair, namely,

$$\mathbf{x}(t) = e^{\lambda t}\mathbf{e}. \tag{7.86}$$

As a second step construct two real-valued solutions of (7.69) by determining the real and the imaginary parts of (7.86), calling them $\mathbf{x}_1$ and $\mathbf{x}_2$. Any real linear combination of $\mathbf{x}_1$ and $\mathbf{x}_2$ is part of the general solution to (7.69), that is,

$$k_1\mathbf{x}_1(t) + k_2\mathbf{x}_2(t) \tag{7.87}$$

is the contribution of the complex eigenvalue (and its complex conjugate) to the general solution.

So far we have considered examples of linear systems that possess distinct eigenvalues, either real or complex. In these cases we have a procedure for constructing the general solution of the differential equation (7.69). This solution always has the form

$$\mathbf{x}(t) = e^{\lambda_1 t}\mathbf{e}_1 + e^{\lambda_2 t}\mathbf{e}_2 + \cdots + e^{\lambda_n t}\mathbf{e}_n, \qquad (7.88)$$

where the $\mathbf{e}_i$s are the associated eigenvectors of A, with appropriate modifications previously when an eigenvalue is complex. The situation is a little more delicate when an eigenvalue is repeated (as was the case for equations as well). It is possible that a matrix A may have repeated eigenvalues and have a full set of n linearly independent eigenvectors. For example, the identity matrix is such a matrix. It has $\lambda = 1$ with multiplicity n and n linearly independent eigenvectors corresponding to the basis vectors in E^n. On the other hand, it is possible that A may have repeated eigenvalues but a deficient number of linearly independent eigenvectors corresponding to this eigenvalue. The matrix

$$A = \begin{bmatrix} 1 & 1 \\ 0 & 1 \end{bmatrix} \qquad (7.89)$$

is an example of such a matrix: Here $\lambda = 1$ is the eigenvalue with multiplicity 2, but there is only one eigenvector, namely

$$\mathbf{e}_1 = \begin{bmatrix} 1 \\ 0 \end{bmatrix} \qquad (7.90)$$

associated with this eigenvalue. The reader recalls that in Chapter 4, where we encountered repeated roots of the characteristic polynomial in the case of equations, we sought solutions that were of the form $te^{\lambda t}$ in order to create a full set of linearly independent solutions. We will follow a modification of the same idea here.

Example 7.5.3
Consider the system of equations

$$\mathbf{x}' = \begin{bmatrix} 1 & 1 \\ 0 & 1 \end{bmatrix} \mathbf{x}. \qquad (7.91)$$

The matrix A in (7.91) has $\lambda = 1$ for an eigenvalue with multiplicity 2. The vector $\mathbf{e}_1$ defined in (7.90) is the only eigenvector associated with this eigenvalue from which we get one solution $\mathbf{x}_1$ to (7.91) given by

$$\mathbf{x}_1(t) = e^t \mathbf{e}_1 = e^t \begin{bmatrix} 1 \\ 0 \end{bmatrix}. \qquad (7.92)$$

To find a second linearly independent solution we could try

$$\mathbf{x}_2(t) = te^t \mathbf{e}_1. \qquad (7.93)$$

We leave it to the reader to substitute (7.93) into (7.91) and observe that there is no such solution (see Problem 3). Instead we try a second solution of (7.91) of the form

$$\mathbf{x}_2(t) = te^t\mathbf{e}_1 + e^t\mathbf{e}_2, \tag{7.94}$$

where $\mathbf{e}_2$ is as yet unknown. Substitute (7.94) in the system of differential equations and simplify to get

$$e^t\mathbf{e}_1 + te^t\mathbf{e}_1 + e^t\mathbf{e}_2 = A(te^t\mathbf{e}_1 + e^t\mathbf{e}_2) = \mathbf{x}_2' = A\mathbf{x}_2 == te^t A\mathbf{e}_1 + e^t A\mathbf{e}_2. \tag{7.95}$$

Note that $A\mathbf{e}_1 = \mathbf{e}_1$. Hence, the terms having te^t cancel out and we end up having to solve the linear set of equations

$$(A - I)\mathbf{e}_2 = \mathbf{e}_1. \tag{7.96}$$

Let the unknown vector $\mathbf{e}_2$ have components v_1 and v_2, in terms of which system (7.96) becomes

$$\begin{bmatrix} 0 & 1 \\ 0 & 0 \end{bmatrix} \begin{bmatrix} v_1 \\ v_2 \end{bmatrix} = \begin{bmatrix} 1 \\ 0 \end{bmatrix}. \tag{7.97}$$

It is easy to verify that

$$\mathbf{e}_2 = \begin{bmatrix} 0 \\ 1 \end{bmatrix}$$

is a solution of (7.97). Thus, returning to (7.94), we see that a second solution to (7.91) is given by

$$\mathbf{x}_2(t) = te^t \begin{bmatrix} 1 \\ 0 \end{bmatrix} + e^t \begin{bmatrix} 0 \\ 1 \end{bmatrix}. \tag{7.98}$$

The two solutions $\mathbf{x}_1$ (given by (7.92)) and $\mathbf{x}_2$ are linearly independent. A linear combination of these solutions is therefore the general solution to (7.91).

Remark 7.5.2: We cannot obtain the solution $\mathbf{e}_2$ to (7.96) by computing the inverse of $A - I$; the inverse of this matrix does not exist since $\lambda = 1$ is an eigenvalue of A. What (7.96) requires is that $\mathbf{e}_1$ be in the *range* of $A - I$, which is verified by the analysis that follows this equation.

The procedure we just went through is typical for systems that have repeated roots and do not possess a full set of associated eigenvectors. Such eigenvalues are called **deficient**. The vector $\mathbf{e}_2$ in Example 7.5.3 is called a **generalized** eigenvector of the matrix A.

Definition 7.5.1 (Deficiency and Generalized Eigenvectors)
*Let A be an $n \times n$ matrix. We say λ is a **deficient** eigenvalue of A if it has multiplicity m and fewer than m eigenvectors associated with it. If there are $k < m$ linearly independent eigenvectors associated with λ, then the integer $r = m - k$ is called the **degree of deficiency** of λ. A vector $\mathbf{e}$ is called a **generalized** eigenvector of A associated with λ if there is an integer $l > 0$ such*

that

$$(A - \lambda)^l \mathbf{e} = \mathbf{0}, \tag{7.99}$$

but

$$(A - \lambda)^{l-1} \mathbf{e} \neq \mathbf{0}. \tag{7.100}$$

According to this definition the vector $\mathbf{e}_2$ in Example 7.5.3 is a generalized eigenvector with $l = 2$; the vector $\mathbf{e}_2$ satisfies (see (7.96)) $(A-I)^2 \mathbf{e}_2 = (A-I)\mathbf{e}_1 = \mathbf{0}$, since $\mathbf{e}_1$ is an eigenvector of A, while $(A - I)\mathbf{e}_2 = \mathbf{e}_1 \neq \mathbf{0}$.

The ideas in Example 7.5.3 were presented in the simple case where the deficiency of the eigenvalue was one, that is, the eigenvalue repeated and there was a single (ordinary) eigenvector associated with the eigenvalue. The strategy described there generalizes to eigenvalues with higher multiplicity and deficiencies. In general, given a matrix A with an eigenvalue λ that has a degree of deficiency r, we construct a set of generalized eigenvectors $\{\mathbf{e}_1, \mathbf{e}_2, \cdots, \mathbf{e}_r\}$ such that

$$(A - \lambda I)\mathbf{e}_r = \mathbf{e}_{r-1}, \quad (A - \lambda I)\mathbf{e}_{r-1} = \mathbf{e}_{r-2}, \quad \cdots, (A - \lambda I)\mathbf{e}_2 = \mathbf{e}_1, \quad (7.101)$$

and $\mathbf{e}_1$ an eigenvector of λ. The set of generalized eigenvectors, in turn, generates the following set of linearly independent solutions of (7.69):

$$\mathbf{x}_1(t) = e^{\lambda t}\mathbf{e}_1, \quad \mathbf{x}_2(t) = (t\mathbf{e}_1 + \mathbf{e}_2)e^{\lambda t}, \quad \mathbf{x}_3(t) = \left(\frac{t^2}{2}\mathbf{e}_1 + t\mathbf{e}_2 + \mathbf{e}_3\right)e^{\lambda t}, \quad \ldots$$
$$(7.102)$$

The reader can guess from the structure of these expressions that the coefficient of each $e^{\lambda t}$ is the antiderivative of the coefficient of the previous term and that the constant of integration is the generalized eigenvector. For instance, the coefficient of $e^{\lambda t}$ in $\mathbf{x}_3$ is the antiderivative of $t\mathbf{e}_1 + \mathbf{e}_2$ with $\mathbf{e}_3$ being the constant of integration. With this in mind, one can write down each $\mathbf{x}_i$.

It turns out that one of the main theorems of linear algebra guarantees that any $n \times n$ matrix has n linearly independent generalized eigenvectors. This theorem enhances considerably the significance of the preceding algorithm since we are assured that at least we are able to get the general solution to (7.69) even though this procedure may be quite tedious.

Example 7.5.4
Consider the system of equations

$$\mathbf{x}' = \begin{bmatrix} 0 & 1 & 0 \\ 0 & 0 & 1 \\ 1 & -3 & 3 \end{bmatrix} \mathbf{x}. \tag{7.103}$$

It is not difficult to check that the matrix A in (7.103) has $\lambda = 1$ for its only eigenvalue and that this eigenvalue has multiplicity 3. When we row reduce $A - I$ we end up with the matrix

$$\begin{bmatrix} -1 & 1 & 0 \\ 0 & -1 & 1 \\ 0 & 0 & 0 \end{bmatrix}, \tag{7.104}$$

from which we deduce that $\lambda = 1$ has

$$\mathbf{e}_1 = \begin{bmatrix} 1 \\ 1 \\ 1 \end{bmatrix} \tag{7.105}$$

as its only eigenvector. Thus, we need to construct two generalized eigenvectors. We mention in passing that the latter information about the eigenvalue, eigenvector, and (7.104) can be obtained from *Mathematica* by entering the following lines:

```
a = {{0, 1, 0}, {0, 0, 1}, {1, -3, 3}};
Eigenvalues[a];
Eigenvectors[a];
RowReduce[a - IdentityMatrix[3]];
```

So far we have one solution of (7.103) in hand, namely,

$$\mathbf{x}_1(t) = e^t \begin{bmatrix} 1 \\ 1 \\ 1 \end{bmatrix}. \tag{7.106}$$

To construct $\mathbf{x}_2$ and $\mathbf{x}_3$, we follow the procedure in (7.101). We seek $\mathbf{e}_2$ such that $(A - I)\mathbf{e}_2 = \mathbf{e}_1$. The latter algebraic system is equivalent to the augmented matrix

$$\begin{bmatrix} -1 & 1 & 0 & 1 \\ 0 & -1 & 1 & 1 \\ 1 & -3 & 2 & 1 \end{bmatrix}. \tag{7.107}$$

The row reduction of (7.107) leads to

$$\begin{bmatrix} -1 & 1 & 0 & 1 \\ 0 & -1 & 1 & 1 \\ 0 & 0 & 0 & 0 \end{bmatrix}. \tag{7.108}$$

Let $\mathbf{e}_2 = \begin{bmatrix} x & y & z \end{bmatrix}^T$. Following (7.108), x, y and z satisfy the simultaneous equations

$$\begin{cases} -x + y &= 1, \\ -y + z &= 1. \end{cases} \tag{7.109}$$

Let $z = 0$ in (7.109), which implies that $x = -2$ and $y = -1$. Hence, the first generalized eigenvector is

$$\mathbf{e}_2 = \begin{bmatrix} -2 \\ -1 \\ 0 \end{bmatrix}$$

and

$$\mathbf{x}_2 = (t\mathbf{e}_1 + \mathbf{e}_2)e^t = \left(t \begin{bmatrix} 1 \\ 1 \\ 1 \end{bmatrix} + \begin{bmatrix} -2 \\ -1 \\ 0 \end{bmatrix} \right) e^t \tag{7.110}$$

is a second linearly independent solution of (7.103). We could have obtained the same information from *Mathematica* by row reducing (7.107) or by using the Solve command with the three equations that result from $(A - I)\mathbf{e}_2 = \mathbf{e}_1$.

Now that we have $\mathbf{e}_2$ we look for the second generalized eigenvector $\mathbf{e}_3$. We start with the formula $(A - I)\mathbf{e}_3 = \mathbf{e}_2$, which leads to the augmented matrix

$$
\begin{bmatrix}
-1 & 1 & 0 & -2 \\
0 & -1 & 1 & -1 \\
1 & -3 & 2 & 0
\end{bmatrix}.
\tag{7.111}
$$

We leave it to the reader to show that

$$
\mathbf{e}_3 = \begin{bmatrix} 3 \\ 1 \\ 0 \end{bmatrix}
$$

which in turn implies that the third linearly independent solution to (7.103) is

$$
\mathbf{x}_3(t) = \left(\frac{t^2}{2} \begin{bmatrix} 1 \\ 1 \\ 1 \end{bmatrix} + t \begin{bmatrix} -2 \\ -1 \\ 0 \end{bmatrix} + \begin{bmatrix} 3 \\ 1 \\ 0 \end{bmatrix} \right) e^t.
\tag{7.112}
$$

The general solution to the system (7.103) is a linear combination of (7.106), (7.110), and (7.112).

Problems

1. Find the general solution to the linear systems (7.69) whose matrix A is given. In each case use *Mathematica* and *MATLAB* to check your answer.

   (a) $A = \begin{bmatrix} 1 & 3 \\ -3 & 7 \end{bmatrix}$

   (b) $A = \begin{bmatrix} 2 & 5 \\ 1 & -2 \end{bmatrix}$

   (c) $A = \begin{bmatrix} -1 & -5 \\ -3 & 1 \end{bmatrix}$

   (d) $A = \begin{bmatrix} 2 & 4 \\ -4 & 7 \end{bmatrix}$

   (e) $A = \begin{bmatrix} -1 & 1 \\ 3 & -1 \end{bmatrix}$

   (f) $A = \begin{bmatrix} 3 & -1 & 3 \\ 0 & 0 & 1 \\ 1 & 0 & 0 \end{bmatrix}$

   (g) $A = \begin{bmatrix} -3 & 0 & 1 \\ 1 & -1 & 0 \\ -4 & 0 & 1 \end{bmatrix}$

(h) $A = \begin{bmatrix} 3 & 0 & 0 & 1 \\ 7 & 1 & 0 & 3 \\ -14 & 0 & 1 & -6 \\ -4 & 0 & 0 & -1 \end{bmatrix}$

(i) $A = \begin{bmatrix} 0 & 1 & 0 & 0 \\ 0 & 0 & 1 & 0 \\ 0 & 0 & 0 & 1 \\ -16 & 32 & -24 & 8 \end{bmatrix}$

2. Suppose that $\lambda = a + bi$ is a complex eigenvalue of the linear system (7.69), where A is a constant matrix with real entries. Let e be an eigenvector associated with λ.

   (a) Show that $\bar{\lambda}$ must also be an eigenvalue with $\bar{e}$ as its eigenvector.

   (b) Let $e = e_1 + ie_2$, where e_1 and e_2 are the real and imaginary parts of e. Let $x(t) = e^{\lambda t}e$. Show that

   $$\begin{aligned} x_1(t) &= \mathrm{Re}(x(t)) = e^{at}(\cos bt\, e_1 - \sin bt\, e_2), \\ x_2(t) &= \mathrm{Im}(x(t)) = e^{at}(\cos bt\, e_1 + \sin bt\, e_2) \end{aligned} \qquad (7.113)$$

   are (real-valued) solutions of (7.69). Show that x_1 and x_2 are linearly independent.

3. Show by direct substitution that $x(t) = te^t e$ cannot be a solution of (7.91) for any $e \neq 0$.

4. Find the solution to the following initial value problems.

   (a) $x' = \begin{bmatrix} 1 & 1 \\ 0 & 1 \end{bmatrix} x$, and $x(0) = \begin{bmatrix} 1 \\ -1 \end{bmatrix}$

   (b) $x' = \begin{bmatrix} -1 & 1 \\ 1 & 1 \end{bmatrix} x$, and $x(0) = \begin{bmatrix} -1 \\ 1 \end{bmatrix}$

   (c) $x' = \begin{bmatrix} 2 & 1 \\ -1 & 2 \end{bmatrix} x$, and $x(0) = \begin{bmatrix} 2 \\ 1 \end{bmatrix}$

   (d) $x' = \begin{bmatrix} 0 & 1 \\ -1 & 0 \end{bmatrix} x$, and $x(0) = \begin{bmatrix} 0 \\ 1 \end{bmatrix}$

   (e) $x' = \begin{bmatrix} a & a \\ 0 & a \end{bmatrix} x$, and $x(0) = \begin{bmatrix} 1 \\ -1 \end{bmatrix}$ where a is a real number.

5. Solve each initial value problem $x' = Ax$, $x(0) = x_0$.

   (a) $A = \begin{bmatrix} 0 & 1 & 0 \\ 0 & 0 & 1 \\ 1 & -3 & 3 \end{bmatrix}$, and $x(0) = \begin{bmatrix} 0 \\ 0 \\ 1 \end{bmatrix}$

   (b) $A = \begin{bmatrix} 0 & 1 & 0 \\ 0 & 0 & 1 \\ 3 & -7 & 5 \end{bmatrix}$, and $x(0) = \begin{bmatrix} 0 \\ 0 \\ 1 \end{bmatrix}$

(c) $A = \begin{bmatrix} 0 & 1 & 0 \\ 0 & 0 & 1 \\ 8 & -12 & 6 \end{bmatrix}$, and $\mathbf{x}(0) = \begin{bmatrix} 0 \\ 0 \\ 1 \end{bmatrix}$

(d) $A = \begin{bmatrix} 0 & 0 & 1 \\ 0 & 1 & 0 \\ 8 & -12 & 6 \end{bmatrix}$, and $\mathbf{x}(0) = \begin{bmatrix} 0 \\ 0 \\ 1 \end{bmatrix}$

6. Consider the system of differential equations

$$\mathbf{x}' = \begin{bmatrix} 1 & 1 \\ \epsilon & 1 \end{bmatrix} \mathbf{x}, \qquad \mathbf{x}(0) = \begin{bmatrix} 1 \\ -1 \end{bmatrix}.$$

(a) Determine the solution to this system with $\epsilon > 0$ and $\epsilon = 0$.

(b) Let $\mathbf{x}(t, \epsilon)$ be the solution to the above initial value problem. Find $\lim_{\epsilon \to 0} \mathbf{x}(t, \epsilon)$. Is this limit the same at the solution to the initial value problem when $\epsilon = 0$?

(c) Plot the graphs of the solution when $\epsilon = 0.01$ and $\epsilon = 0$. Are these solutions close to each other over all values of t?

7. Consider the set of matrices

$$A(\epsilon) = \begin{bmatrix} 0 & 0 & 1 \\ 0 & 1 & 0 \\ 8 & -12 & 6+\epsilon \end{bmatrix}.$$

(a) Determine the eigenvalues and eigenvectors of $A(0)$. Does this matrix have a full set of eigenvectors?

(b) Determine the eigenvalues and eigenvectors of $A(0.0001)$. Does this matrix have a full set of eigenvectors?

(c) Execute the following commands in *Mathematica*:

```
A = {{0,1,0}, {0,0,1}, {8,-12,6+eps}};
f[x_] := Max[Abs[Eigenvalues[A /. eps -> x]]];
Plot[f[x], {x, 0, 1}]
```

What is f measuring?

7.6 Diagonalization and Linear Systems

One of the important results we encountered in Chapter 6 showed how, for certain special matrices, one could use information about eigenvalues of the matrix and introduce a transformation that diagonalizes it. Specifically, associated with a matrix A with a full set of eigenvectors is a matrix Q, whose columns consist of these eigenvectors, such that

$$Q^{-1}AQ = \operatorname{diag}(\lambda_1, \lambda_2, \ldots, \lambda_n). \tag{7.114}$$

The notation diag($\mathbf{u}$) stands for a matrix that has the vector $\mathbf{u}$ on its diagonal and is zero otherwise. For instance, we saw in Chapter 6 that a matrix that has a full set of *distinct* eigenvalues is diagonalizable.

In this section we will explore the implications of this result in the context of linear systems of differential equations

$$\mathbf{x}' = A\mathbf{x}, \quad \mathbf{x}(0) = \mathbf{x}_0, \tag{7.115}$$

where A has constant coefficients and satisfies (7.114). As we will see shortly, we are able to develop an alternative method of determining the solution of such systems. The key idea of this method is to use Q, change variables in (7.115), and reduce this system to one consisting only of uncoupled equations, that is, a system of n equations where each equation depends only on a single unknown.

We will assume for the remainder of this section that the matrix A has a complete set of eigenvectors. Let Q be its matrix of eigenvectors. We introduce a new dependent variable $\mathbf{y}$ through the change of variables

$$\mathbf{x} = Q\mathbf{y}. \tag{7.116}$$

Since A has constant entries, the matrix Q is constant as well and $\mathbf{x}' = Q\mathbf{y}'$. Moreover, $A\mathbf{x} = AQ\mathbf{y}$, so (7.115) is transformed to

$$\mathbf{y}' = Q^{-1}AQ\mathbf{y}. \tag{7.117}$$

It now follows from (7.114) that (7.117) is equivalent to

$$\mathbf{y}' = \text{diag}(\lambda_1, \ldots, \lambda_n)\mathbf{y} \tag{7.118}$$

or, in component form,

$$y_1'(t) = \lambda_1 y(t), \quad y_2'(t) = \lambda_2 y_2(t), \quad \ldots, \quad y_n'(t) = \lambda_n y_n(t). \tag{7.119}$$

System (7.119) is uncoupled; each differential equation in (7.119) depends on a single dependent variable. Since each equation in (7.119) is first order, we easily obtain its solution by an integration:

$$y_1(t) = c_1 e^{\lambda_1 t}, \quad y_2(t) = c_2 e^{\lambda_2 t}, \quad \ldots, \quad y_n(t) = c_n e^{\lambda_n t}, \tag{7.120}$$

where the constants c_is need to be computed from the initial condition in (7.115). Since $\mathbf{x}(0) = Q\mathbf{y}(0) = Q\mathbf{c}$, where $\mathbf{c} = (c_1, c_2, \ldots, c_n)$, it follows from (7.115) that

$$\mathbf{c} = Q^{-1}\mathbf{x}_0. \tag{7.121}$$

Thus, we see that the solution to (7.115) is $\mathbf{x}(t) = Q\mathbf{y}(t)$, where $\mathbf{y}$ and $\mathbf{c}$ satisfy (7.120) and (7.121), respectively. We have proved the following theorem.

Theorem 7.6.1

Suppose that the matrix A in (7.115) is daigonalizable. Let Q be the matrix of eigenvectors of A. Then the solution to the initial value problem (7.115) is

$$\mathbf{x}(t) = Q\mathbf{y}(t), \tag{7.122}$$

where $\mathbf{y}$ is given by (7.120) and $\mathbf{c}$ by (7.121).

Example 7.6.1

Consider the system of differential equations

$$\mathbf{x}' = \begin{bmatrix} 1 & 2 \\ 2 & 1 \end{bmatrix} \mathbf{x}, \quad \mathbf{x}_0 = \begin{bmatrix} 1 \\ -1 \end{bmatrix}. \tag{7.123}$$

In this example the eigenvalues of A are 3 and -1, and Q is

$$Q = \begin{bmatrix} 1 & 1 \\ 1 & -1 \end{bmatrix}. \tag{7.124}$$

We define the new variable $\mathbf{y}$ by

$$\mathbf{x} = Q\mathbf{y}. \tag{7.125}$$

According to (7.118) $\mathbf{y}$ satisfies

$$y_1' = 3y_1, \quad y_2' = -y_2. \tag{7.126}$$

The general solution to this system is

$$y_1(t) = c_1 e^{3t}, \quad y_2(t) = c_2 e^{-t}. \tag{7.127}$$

Since $\mathbf{x} = Q\mathbf{y}$, we have

$$x_1(t) = c_1 e^{3t} + c_2 e^{-t}, \quad x_2(t) = -c_1 e^{3t} + c_2 e^{-t}. \tag{7.128}$$

The constants c_1 and c_2 can now be computed from the initial conditions.

Example 7.6.2

Consider the system of differential equations

$$\frac{dx}{dt} = y, \quad \frac{dy}{dt} = -5x - 2y. \tag{7.129}$$

Here the matrix A is $\begin{bmatrix} 0 & 1 \\ -5 & -2 \end{bmatrix}$, whose eigenvalues are $-1 \pm 2i$ with the corresponding eigenvectors

$$\begin{bmatrix} 1 \\ -1 + 2i \end{bmatrix}, \quad \begin{bmatrix} 1 \\ -1 - 2i \end{bmatrix}. \tag{7.130}$$

Put the eigenvectors together to form the matrix Q:

$$Q = \begin{bmatrix} 1 & 1 \\ -1 + 2i & -1 - 2i \end{bmatrix}. \tag{7.131}$$

Following our procedure of the previous example, we introduce a new vector of variables $\mathbf{w}$ such that

$$\mathbf{x} = \begin{bmatrix} x \\ y \end{bmatrix} = Q\mathbf{w}. \tag{7.132}$$

As this theory predicts, $\mathbf{w}$ satisfies the uncoupled system of differential equations

$$\frac{dw_1}{dt} = (-1 + 2i)w_1, \qquad \frac{dw_2}{dt} = (-1 - 2i)w_2, \qquad (7.133)$$

whose general solution is given by

$$\begin{aligned} w_1(t) &= c_1 e^{-t}(\cos 2t + i\sin 2t), \\ w_2(t) &= c_2 e^{-t}(\cos 2t - i\sin 2t). \end{aligned} \qquad (7.134)$$

We are now able to obtain $x(t)$ and $y(t)$ from (7.132)

$$\begin{aligned} \begin{bmatrix} x(t) \\ y(t) \end{bmatrix} &= Q\mathbf{w} \\ &= e^{-t}\left\{ (c_1 + c_2)\begin{bmatrix} \cos 2t \\ -\cos 2t - 2\sin 2t \end{bmatrix} + i(c_1 - c_2)\begin{bmatrix} \sin 2t \\ 2\cos(2t - \sin(2t) \end{bmatrix} \right\}. \end{aligned}$$
$$(7.135)$$

With k_1 and k_2 defined by

$$k_1 = c_1 + c_2, \quad k_2 = i(c_1 - c_2), \qquad (7.136)$$

the general solution to (7.129) takes the form

$$\begin{bmatrix} x(t) \\ y(t) \end{bmatrix} = e^{-t}\left\{ k_1\begin{bmatrix} \cos 2t \\ -\cos 2t - 2\sin 2t \end{bmatrix} + k_2\begin{bmatrix} \sin 2t \\ \cos 2t - \sin 2t \end{bmatrix} \right\}. \qquad (7.137)$$

Example 7.6.3

In this example we will find the displacements of two masses, m_1 and m_2, that are connected to each other and to two rigid walls by three springs with spring constants k_1, k_2, and k_3. As we saw in Example 7.2.1, the equations of motion of the two masses are derived from Newton's second law. They are

$$m_1\frac{d^2 x}{dt^2} = -(k_1 + k_2)x + k_2 y, \qquad m_2\frac{d^2 y}{dt^2} = -(k_2 + k_3)y + k_2 x. \qquad (7.138)$$

Here x and y measure the displacement (deviation) of the two masses from their equilibrium positions. These equations are supplemented with initial data of the form

$$x(0) = x_0, \quad y(0) = y_0, \quad \left.\frac{dx}{dt}\right|_{t=0} = x_1, \quad \left.\frac{dy}{dt}\right|_{t=0} = y_1. \qquad (7.139)$$

In order to apply the technique of this section to this system of second-order equations, we first need to transform it to a first-order system. To this end, we define a vector $\mathbf{z}$ by

$$z_1 = x, \quad z_2 = \frac{dx}{dt}, \quad z_3 = y, \quad z_4 = \frac{dy}{dt} \qquad (7.140)$$

and note that $\mathbf{z}$ satisfies

$$\frac{dz_1}{dt} = z_2,$$

$$\frac{dz_2}{dt} = 1/m_1[-(k_1 - k_2)z_1 + k_2 z_3],$$

$$\frac{dz_3}{dt} = z_4,$$

$$\frac{dz_4}{dt} = 1/m_2[-(k_2 - k_3)z_3 - k_2 z_1].$$

$$(7.141)$$

The function $\mathbf{z}$ also satisfies the initial conditions

$$z_1(0) = x_0, z_2(0) = x_1, z_3(0) = y_0, z_4(0) = y_1. \qquad (7.142)$$

System (7.141) has the form $\mathbf{z}' = A\mathbf{z}$, where A is

$$A = \begin{bmatrix} 0 & 1 & 0 & 0 \\ -\dfrac{k_1 + k_2}{m_1} & 0 & \dfrac{k_2}{m_1} & 0 \\ 0 & 0 & 0 & 1 \\ \dfrac{k_2}{m_2} & 0 & -\dfrac{k_2 + k_3}{m_2} & 0 \end{bmatrix}. \qquad (7.143)$$

We now proceed to solve (7.141), (7.142) for a specific set of parameters. Although it is possible to determine the eigenvalues of A by hand, this computation is rather tedious. On the other hand, these types of calculations are quite simple in *Mathematica* and *MATLAB*. We will illustrate the syntax in *Mathematica* for carrying out the computations that lead to the matrix of eigenvectors Q, and eventually to the solution $\mathbf{z}$. Consider the concrete example

$$m_1 = 1, \quad m_2 = 2, \quad k_1 = k_2 = k_3 = 0.1, \qquad (7.144)$$

with initial data

$$x_0 = 0.8, \quad x_1 = y_0 = y_1 = 0. \qquad (7.145)$$

The matrix A then takes the form

$$A = \begin{bmatrix} 0 & 1 & 0 & 0 \\ -0.2 & 0 & 0.1 & 0 \\ 0 & 0 & 0 & 1 \\ 0.05 & 0 & -0.1 & 0 \end{bmatrix}. \qquad (7.146)$$

The following program in *Mathematica* that solves for $x(t)$ and $y(t)$ and plots their graphs:

```
A={{0,1,0,0}, {-0.2,0,0.1,0}, {0,0,0,1}, {0.05,0,-0.1,0}};
b=Eigenvalues[A];
c=Eigenvectors[A];
Q=Transpose[c];
```

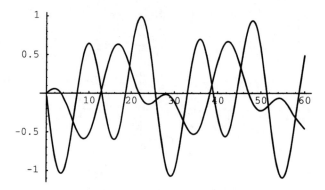

Figure 7.7: The graphs of $x(t)$ and $y(t)$.

```
initial={0.8,0,0,0};
d=Inverse[Q].initial;
solution=Q.(Exp[t*b]*d);
solution=Chop[FullSimplify[solution]];
Plot[{solution[[1]], solution[[3]]}, {t, 0, 50}]
```

The expression solution contains all of the information about the motion of
the two masses. x and y are the first and third entries of this vector. Figure 7.7
shows their evolution.

Problems

1. Determine which of the following systems of differential equations has a
   diagnolizable matrix of coefficients. For those that do, solve for the solution
   of the initial value problem. Draw the graphs of all of the components on
   the same screen. Verify your answer in *Mathematica* or *MATLAB*.

   (a) $A = \begin{bmatrix} 2 & 3 \\ 3 & 2 \end{bmatrix}, \qquad x_0 = \begin{bmatrix} -1 \\ 1 \end{bmatrix}$

   (b) $A = \begin{bmatrix} 1 & 3 \\ 3 & 1 \end{bmatrix}, \qquad x_0 = \begin{bmatrix} 2 \\ -1 \end{bmatrix}$

   (c) $A = \begin{bmatrix} 0 & -1 \\ 1 & 0 \end{bmatrix}, \qquad x_0 = \begin{bmatrix} -2 \\ 2 \end{bmatrix}$

   (d) $A = \begin{bmatrix} 0 & 1 & 0 \\ 0 & 0 & 1 \\ -6 & 11 & -6 \end{bmatrix}, \qquad x_0 = \begin{bmatrix} 0 \\ 0 \\ 1 \end{bmatrix}$

   (e) $A = \begin{bmatrix} 0 & 0 & 1 \\ 1 & 0 & 0 \\ 0 & 1 & 0 \end{bmatrix}, \qquad x_0 = \begin{bmatrix} 0 \\ 2 \\ 0 \end{bmatrix}$

   (f) $A = \begin{bmatrix} 0 & 1 & 0 \\ 0 & 0 & 1 \\ -1 & 0 & 0 \end{bmatrix}, \qquad x_0 = \begin{bmatrix} 1 \\ 1 \\ 1 \end{bmatrix}$

2. Convert each equation to a first-order system and apply the eigenvalue technique, when applicable, and obtain the solution to the initial value problem. Verify your answer in *Mathematica* or *MATLAB*.

(a) $x'' + 3x' + 2x = 0$, $\quad x(0) = 1$, $x'(0) = 0$

(b) $x'' + 4x = 0$, $\quad x(0) = 0$, $x'(0) = 1$

(c) $y'' + ay = 0$, $\quad y(0) = 1$, $y'(0) = -1$, and a is a positive number

(d) $x''' + 6x'' - 11x' + 6x = 0$, $\quad x(0) = 1$, $x'(0) = x''(0) = 0$

(e) $y^{(iv)} + 64y = 0$, $\quad y(0) = y'(0) = y''(0) = 0$, $y'''(0) = 1$

3. Find the solution to $\mathbf{x}' = A\mathbf{x}$ where

$$A = \begin{bmatrix} 2 & 0 & 0 \\ 0 & 0 & -1 \\ 0 & 1 & 0 \end{bmatrix} \tag{7.147}$$

and the initial condition is given by

$$\mathbf{x}_0 = \begin{bmatrix} 1 \\ 1 \\ 1 \end{bmatrix}. \tag{7.148}$$

Use *Mathematica* or *MATLAB* and compute

$$\int_0^t \|\mathbf{x}(s)\|^2 ds \quad \text{and} \quad \int_0^t \|\mathbf{x}'(s)\|^2 ds.$$

Draw the graphs of these functions on the same screen for $t \in (0, 10)$.

4. Create a file that contains the program in Example 7.6.3. Input this file to *Mathematica* and draw the graphs of $x(t)$ and $y(t)$.

5. Solve (7.142), (7.143) with parameter values (7.144) and the following two sets of initial conditions:

$$\mathbf{x}_0 = \begin{bmatrix} 1 \\ 0 \\ 0 \\ 0 \end{bmatrix} \quad \text{and} \quad \mathbf{x}_0 = \begin{bmatrix} 0 \\ 0 \\ 1 \\ 0 \end{bmatrix}.$$

What do these initial conditions represent physically? Draw the graphs of $x(t)$ in each case.

6. Write a *MATLAB* program that begins with A and the parameter values m_is, k_is, and the initial data $\mathbf{x}_0$ and returns the four components of $\mathbf{x}$, each in a quarter of the screen.

7. Use the preceding *MATLAB* program and do Problem 5.

8. Alter the *Mathematica* program in this section appropriately and find the solution to

$$\frac{dx}{dt} = -x + y,$$
$$\frac{dy}{dt} = -x - y,$$

(7.149)

with initial data

$$x(0) = 0.1, \quad y(0) = -0.2.$$

(7.150)

Use DSolve with (7.149) and compare the solutions obtained by the two methods.

7.7   Nonhomogeneous Linear Systems

In Chapter 4 we discussed two methods, the method of underdetermined coefficients and the variation of parameter formula, for finding the particular solution of differential equations. In this section we will describe briefly the natural generalizations of these methods to the linear systems

$$\mathbf{x}' = A\mathbf{x} + \mathbf{f}(t).$$

(7.151)

We recall that once we have found a particular solution $\mathbf{x}_p$ of (7.151), the general solution of this equation is the sum of $\mathbf{x}_p$ and the complementary solution $\mathbf{x}_c$ of the homogeneous system

$$\mathbf{x}' = A\mathbf{x}.$$

(7.152)

7.7.1   Method of Undetermined Coefficients

This method is successful when the matrix A in (7.151) is constant and $\mathbf{f}$ is a linear combination of sines and cosines, exponentials and polynomials. For such problems the appropriate choice for the particular solution is a general linear combination of vectors in the form of the functions that appear in $\mathbf{f}$ together with all their derivatives. For instance, if $\mathbf{f}$ is

$$\mathbf{f} = \begin{bmatrix} -2\sin 3t \\ 4\cos 2t \end{bmatrix},$$

(7.153)

then one chooses $\mathbf{x}_p$ in the form

$$\mathbf{x}_p = \mathbf{a}\sin 3t + \mathbf{b}\cos 2t.$$

(7.154)

Similarly, if

$$\mathbf{f} = \begin{bmatrix} -2e^{-2t} \\ e^{4t} \end{bmatrix},$$

(7.155)

then the appropriate choice for x_p is

$$\mathbf{x}_p = \mathbf{a}e^{-2t} + \mathbf{b}e^{4t}.$$

(7.156)

So the main difference between linear systems and linear equations is that in the former case we will use *vectors* of undetermined coefficients. The general strategy is the same. We substitute the choice for $\mathbf{x}_\mathrm{p}$ into the linear system and determine the unknown vectors $\mathbf{a}$ and $\mathbf{b}$.

In much the same way as we discussed in Chapter 4, the choice of $\mathbf{x}_\mathrm{p}$ depends crucially on whether any part of the input vector $\mathbf{f}$ is already a solution of the complementary solution. When that is the case, our guess of $\mathbf{x}_\mathrm{p}$ must be multiplied by an appropriate power of t, as the ensuing examples show.

Example 7.7.1
Consider the nonhomogeneous linear system

$$\mathbf{x}' = \begin{bmatrix} 0 & 1 \\ -1 & 0 \end{bmatrix} \mathbf{x} + \begin{bmatrix} -4e^{-3t} \\ e^{-2t} \end{bmatrix}. \tag{7.157}$$

Note that

$$A = \begin{bmatrix} 0 & 1 \\ -1 & 0 \end{bmatrix} \tag{7.158}$$

and that $\mathbf{f}$ is equivalent to

$$\mathbf{f} = -4 \begin{bmatrix} 1 \\ 0 \end{bmatrix} e^{-3t} + \begin{bmatrix} 0 \\ 1 \end{bmatrix} e^{-2t} = -4\mathbf{e}_1 e^{-3t} + \mathbf{e}_2 e^{-2t} \tag{7.159}$$

with obvious definitions for $\mathbf{e}_1$ and $\mathbf{e}_2$. Here the complementary solution is

$$\mathbf{x}_\mathrm{c}(t) = c_1 \begin{bmatrix} \cos t \\ -\sin t \end{bmatrix} + c_2 \begin{bmatrix} \sin t \\ \cos t \end{bmatrix} = c_1 \sin t \begin{bmatrix} 1 \\ -1 \end{bmatrix} + c_2 \cos t \begin{bmatrix} 1 \\ 1 \end{bmatrix}. \tag{7.160}$$

Note that $\mathbf{f}$ and the complementary solution do not have any part in common. We therefore choose $\mathbf{x}_\mathrm{p}$ in the form

$$\mathbf{x}_\mathrm{p} = \mathbf{a}e^{-3t} + \mathbf{b}e^{-2t}. \tag{7.161}$$

Substitute $\mathbf{x}_\mathrm{p}$ into the linear system (7.157) and write the resulting equation as

$$\mathbf{0} = (3\mathbf{a} + A\mathbf{a} + 4\mathbf{e}_1)e^{-3t} + (2\mathbf{b} + A\mathbf{b} + \mathbf{e}_2)e^{-2t}. \tag{7.162}$$

Since the functions e^{-3t} and e^{-2t} are linearly independent, their coefficients must be zero, from which we obtain two equations in terms of $\mathbf{a}$ and $\mathbf{b}$

$$(A + 3I)\mathbf{a} = -4\mathbf{e}_1, \quad (A + 2I)\mathbf{b} = -\mathbf{e}_2. \tag{7.163}$$

The matrix coefficients of $\mathbf{a}$ and $\mathbf{b}$ are invertible. (Why?) Hence,

$$\mathbf{a} = -(A + 3I)^{-1}(4\mathbf{e}_1) = -\frac{1}{5} \begin{bmatrix} 6 \\ 2 \end{bmatrix},$$
$$\mathbf{b} = -(A + 2I)^{-1}(\mathbf{e}_2) = \frac{1}{5} \begin{bmatrix} 1 \\ -2 \end{bmatrix}. \tag{7.164}$$

After substituting the values of $\mathbf{a}$ and $\mathbf{b}$ in (7.161), we find a particular solution of (7.157):

$$\mathbf{x}_p = - \begin{bmatrix} \frac{6}{5}e^{-3t} - \frac{1}{5}e^{-2t} \\ \frac{2}{5}e^{-3t} + \frac{2}{5}e^{-2t} \end{bmatrix}. \tag{7.165}$$

This procedure contains the types of computations that are suitable for the symbolic capabilities of *Mathematica* and *MATLAB*. We present the program in *Mathematica* that delivers (7.165) and leave to the reader to write the analogous program in *MATLAB*.

```
A={{0,1}, {-1, 0}};
f={4 Exp[-3 t], Exp[-2 t]};
a={a1, a2};
b={b1, b2};
L[x_] := D[x, t] - A . x - f;
x[t_] = a*Exp[-3 t] + b*Exp[-2 t];
eqn = L[x[t]];
eqn1 = Coefficient[eqn, Exp[-3 t]];
eqn2 = Coefficient[eqn, Exp[-2 t]];
aa = Solve[eqn1 == 0, a];
bb = Solve[eqn2 == 0, b];
xp[t_]=a*Exp[-3 t] + b*Exp[-2 t] /. aa /. bb
```

Example 7.7.2

Consider the system of differential equations

$$x' = y + \sin t, \quad y' = -x \tag{7.166}$$

or

$$\mathbf{x}' = A\mathbf{x} + \mathbf{e}_1 \sin t, \tag{7.167}$$

with A defined in (7.158). The vectors $\mathbf{e}_1$ and $\mathbf{e}_2$ are the natural basis for E^2. The homogeneous part of this system is the same as that in the preceding example. Note that the forcing term

$$\mathbf{f} = \sin t \, \mathbf{e}_1 \tag{7.168}$$

has the $\sin t$ in common with with $\mathbf{x}_c$ (see (7.160)). Hence, our guess for a particular solution is

$$\mathbf{x}_p = \mathbf{a} \sin t + \mathbf{b} \cos t + \mathbf{c} \, t \sin t + \mathbf{d} \, t \cos t. \tag{7.169}$$

Note, in particular, that, unlike the case of equations, we included in $\mathbf{x}_p$ undetermined vectors for $\sin t$ and $\cos t$ as well. Substitute $\mathbf{x}_p$ into (7.167); use the linear independence of $\sin t$, $\cos t$, $t \sin t$, and $t \cos t$; and arrive at the following four matrix equations in terms of $\mathbf{a}$, $\mathbf{b}$, $\mathbf{c}$, and $\mathbf{d}$:

$$-\mathbf{b} + \mathbf{c} = A\mathbf{a} + \mathbf{e}_1, \quad \mathbf{a} + \mathbf{d} = A\mathbf{b}, \quad -\mathbf{d} = A\mathbf{c}, \quad \mathbf{c} = A\mathbf{d}. \tag{7.170}$$

These equations constitute a system of eight simultaneous equations in the eight unknowns $\{a_1, a_2\}$, $\{b_1, b_2\}$, $\{c_1, c_2\}$, and $\{d_1, d_2\}$. We studied several ways of solving such a system, Gaussian elimination being one of the prime candidates. Here, however, we pursue a more ad hoc method and try to reduce the complexity of the problem by eliminating some of the unknowns. We begin by recognizing that

$$A^2 = -I. \tag{7.171}$$

Next, multiply the last two equations in (7.170) by A and use the identity in (7.171). It follows that $\mathbf{c}$ and $\mathbf{d}$ are eigenvectors of A^2 associated with eigenvalue -1. Apply the matrix A to both sides of the first equation in (7.170) and use the second and third equations and (7.171) to get

$$\mathbf{d} = \frac{1}{2}\mathbf{e}_2. \tag{7.172}$$

But $\mathbf{c} = A\mathbf{d}$. Therefore,

$$\mathbf{c} = \frac{1}{2}\mathbf{e}_1. \tag{7.173}$$

The first two equations in (7.170) now reduce to

$$A\mathbf{a} + \mathbf{b} = -\frac{1}{2}\mathbf{e}_1, \quad \mathbf{a} - A\mathbf{b} = -\frac{1}{2}\mathbf{e}_2. \tag{7.174}$$

These equations are dependent (to see this apply A to the first equation and obtain the negative of the second equation). Therefore, for simplicity, let

$$\mathbf{a} = \mathbf{0}. \tag{7.175}$$

Consequently,

$$\mathbf{b} = -\frac{1}{2}\mathbf{e}_1. \tag{7.176}$$

Substituting (7.172), (7.173), (7.175), and (7.176) into (7.169) leads to the following particular solution of (7.166):

$$\mathbf{x}_\mathrm{p} = \frac{1}{2} \begin{bmatrix} -\cos t + t\sin t \\ t\cos t \end{bmatrix}. \tag{7.177}$$

Example 7.7.3
Consider the linear system

$$\mathbf{x}' = \begin{bmatrix} 1 & 3 \\ 3 & 1 \end{bmatrix} \mathbf{x} + \begin{bmatrix} e^{-2t} \\ 0 \end{bmatrix}. \tag{7.178}$$

The eigenvalues of the matrix A in this case are $\lambda_1 = 4$ and $\lambda_2 = -2$. Therefore, the function e^{-2t} is a part of the complementary solution of (7.178). Hence, the trial particular solution $\mathbf{x}_\mathrm{p} = \mathbf{a}e^{-2t}$ will not get anywhere. It is equally fruitless to look for a solution of the form $\mathbf{x}_\mathrm{p} = te^{-2t}\mathbf{a}$ (this point is not so obvious until you try to substitute $\mathbf{x}_\mathrm{p}$ into (7.178)). The right choice is

$$\mathbf{x}_\mathrm{p} = te^{-2t}\mathbf{a} + e^{-2t}\mathbf{b}. \tag{7.179}$$

We leave it to the reader to substitute (7.179) into (7.178) and determine the unknowns $\mathbf{a}$ and $\mathbf{b}$ (see Problem 4).

7.7.2   Method of Variation of Parameters

The method of undetermined coefficients is effective when the matrix A in (7.151) is constant and $\mathbf{f}$ has a special form. The method of variation of parameters, on the other hand, works, at least theoretically, even when A depends on t and f belongs to a much larger class of vector-valued functions that includes the special class of exponentials, polynomials, and sines and cosines. The starting point in the method of variation of parameters requires that we have at hand a full set of linearly independent complementary solutions of (7.151). One then proceeds with the construction of a particular solution by allowing the constants in the linear combination of the complementary solution to vary with t.

Before proceeding to derive the main formula, we recall the concept of the fundamental matrix solution of (7.152) (see Section 7.4). Let $\{\mathbf{x}_1(t), \mathbf{x}_2(t), \ldots, \mathbf{x}_n(t)\}$ be a sequence of n linearly independent solutions of (7.152). The fundamental matrix solution of (7.152), denoted by $X(t)$, is defined as the $n \times n$ matrix

$$X(t) = [\ \mathbf{x}_1, \quad \mathbf{x}_2, \quad \ldots, \quad \mathbf{x}_n\ ]. \tag{7.180}$$

The fundamental matrix solution has two important properties. Since each of its columns is a solution of (7.152), the entire matrix X satisfies that equation, that is,

$$X'(t) = A(t)X(t). \tag{7.181}$$

Second, the columns of X are linearly independent. So the general solution of

$$\mathbf{x}' = A\mathbf{x}$$

is a linear combination of these columns, or in other words

$$\mathbf{x}(t) = X(t)\mathbf{c}, \tag{7.182}$$

where $\mathbf{c}$ is an arbitrary constant n-vector. The method of variation of paramters seeks a particular solution to

$$\mathbf{x}' = A\mathbf{x} + \mathbf{f}$$

in the form (7.182) by allowing $\mathbf{c}$ to depend on t:

$$\mathbf{x}_\mathrm{p}(t) = X(t)\mathbf{c}(t). \tag{7.183}$$

We begin the search for $\mathbf{c}(t)$ by substituting (7.183) into the nonhomogeneous system (7.151). We get

$$X'\mathbf{c} + X\mathbf{c}' = AX\mathbf{c} + \mathbf{f}. \tag{7.184}$$

But, as (7.181) shows, $X' = AX$, so $X'\mathbf{c} = AX\mathbf{c}$. This fact reduces (7.184) to

$$X\mathbf{c}' = \mathbf{f}. \tag{7.185}$$

The matrix X, being the fundamental matrix solution, is invertible. Multiply both sides of (7.185) by X^{-1} to yield the following simple differential equation for $\mathbf{c}'$:

$$\mathbf{c}'(t) = X^{-1}(t)\mathbf{f}(t). \tag{7.186}$$

Integrate this equation once to find that $\mathbf{c}$ is given by

$$\mathbf{c}(t) = \int^t X^{-1}(s)\mathbf{f}(s)\,ds. \tag{7.187}$$

Combine this information with the original form of $\mathbf{x}_\mathrm{p}$. We thus obtain the main formula of this section, that a particular solution of (7.151) is of the form

$$\mathbf{x}_\mathrm{p}(t) = X(t)\int^t X^{-1}(s)\mathbf{f}(s)\,ds. \tag{7.188}$$

The major obstacle in determining a particular solution using the formula in (7.188) is in carrying out the integration. However, having access to symbolic manipulators such as *Mathematica* or *MATLAB* considerably reduces the painstaking and tedious calculations one often has to go through to find an explicit formula for $\mathbf{x}_\mathrm{p}$.

An equation such as (7.151) is generally complemented by the initial data

$$\mathbf{x}(0) = \mathbf{x}_0. \tag{7.189}$$

We can now determine the unique solution to this problem by noting that we are free to choose the lower limit of the integration in (7.188) to be whatever we wish. The appropriate choice is 0, the value of t in the initial data. In that case the solution to (7.151), (7.188) is

$$\mathbf{x}(t) = X(t)X^{-1}(0)\mathbf{x}_0 + X(t)\int_0^t X^{-1}(s)\mathbf{f}(s)\,ds. \tag{7.190}$$

This result is important enough that we summarize it in the following theorem.

Theorem 7.7.1
The unique solution to the initial value problem

$$\mathbf{x}' = A(t)\mathbf{x} + \mathbf{f}(t), \quad \mathbf{x}(0) = \mathbf{x}_0$$

is given by (7.190), where $X(t)$ is a fundamental matrix solution of the homogeneous system.

Example 7.7.4
Returning to the system of differential equation (7.178), we look for a particular solution of this problem by the variation of parameter method. The general

solution of the homogeneous system is

$$\mathbf{x}_c(t) = c_1 \begin{bmatrix} 1 \\ -1 \end{bmatrix} e^{-2t} + c_2 \begin{bmatrix} 1 \\ 1 \end{bmatrix} e^{4t}. \tag{7.191}$$

The two solutions multiplying c_1 and c_2 are linearly independent. Therefore, the matrix

$$X(t) = \begin{bmatrix} e^{-2t} & e^{4t} \\ -e^{-2t} & e^{4t} \end{bmatrix} \tag{7.192}$$

is a fundamental matrix solution of (7.178). Although the rest of the calculations that leads to a particular solution $\mathbf{x}_p$ is relatively simple in this case, we write a short program in *Mathematica* to carry them out in preparation for more complicated settings. The ensuing program takes full advantage of the symbolic resourcefulness of *Mathematica* and its ability to integrate rather complicated vector-valued functions and invert abstract matrices.

```
X[t_] = {{Exp[-2 t], Exp[4 t]}, {-Exp[-2 t], Exp[4 t]}};
f[t_] = {Exp[-2 t], 0};
invX[t_] = Inverse[X[t]];
xp[t_] = Simplify[X[t].Integrate[invX[s].f[s], {s, 0, t}]]
```

Mathematica's output is

```
        6 t                   6 t
 -1 + E      + 6 t  -1 + E     - 6 t
{---------------- , ----------------}
                         
        2 t                   2 t
     12 E                  12 E
```

Problems

1. Input the program in Example 7.7.1 to *Mathematica* and compare its output with the particular solution in that example.

2. Write the analogue of the program in Example 7.7.1 in *MATLAB* and use it to derive the particular solution of the linear system in that example.

3. Write a *Mathematica* or a *MATLAB* program whose output is the particular solution (7.177).

4. Complete the analysis in Example 7.7.3.

5. Apply the method of undetermined coefficients and obtain a particular solution of the linear systems. Check your results using the *Mathematica* or *MATLAB* programs written in the preceding problems. Alternatively, use DSolve or dsolve to check your answers.

   (a) $\mathbf{x}' = \begin{bmatrix} 0 & 1 \\ -1 & 0 \end{bmatrix} \mathbf{x} + \begin{bmatrix} \sin 2t \\ \cos 3t \end{bmatrix}$

   (b) $\mathbf{x}' = \begin{bmatrix} 1 & 1 \\ -1 & 1 \end{bmatrix} \mathbf{x} + \begin{bmatrix} e^t \\ 0 \end{bmatrix}$

(c) $\mathbf{x}' = \begin{bmatrix} 4 & 2 \\ 3 & -1 \end{bmatrix} \mathbf{x} + \begin{bmatrix} e^{5t} \\ e^{5t} \end{bmatrix}$

(d) $\mathbf{x}' = \begin{bmatrix} 0 & 1 \\ -1 & 0 \end{bmatrix} \mathbf{x} + \begin{bmatrix} t \\ 0 \end{bmatrix}$

(e) $\mathbf{x}' = \begin{bmatrix} 0 & 1 \\ -1 & 0 \end{bmatrix} \mathbf{x} + \begin{bmatrix} t \\ t+1 \end{bmatrix}$.

6. Apply the method of variation of parameters to determine the particular solution to the following systems. Plot the graph of the first component of the solution in each problem.

(a) $\mathbf{x}' = \begin{bmatrix} 0 & 1 \\ -1 & 0 \end{bmatrix} \mathbf{x} + \begin{bmatrix} t^2 \\ 1 - 2t + t^3 \end{bmatrix}$

(b) $\mathbf{x}' = \begin{bmatrix} -1 & 2 \\ -1 & 2 \end{bmatrix} \mathbf{x} + \begin{bmatrix} \sin t \\ 0 \end{bmatrix}$

(c) $\mathbf{y}' = \begin{bmatrix} 2 & 1 \\ 1 & 2 \end{bmatrix} \mathbf{y} + \begin{bmatrix} u(t-2) \\ u(t-3) \end{bmatrix}$, where $u(t-c)$ is the Heaviside function

(d) $\mathbf{x}' = \begin{bmatrix} -1 & 1 \\ 1 & 3 \end{bmatrix} \mathbf{x} + \begin{bmatrix} \delta(t-2) \\ \delta(t-3) \end{bmatrix}$

7. Consider the system of differntial equations $\mathbf{z}' = \begin{bmatrix} a & b \\ b & a \end{bmatrix} \mathbf{z} + \begin{bmatrix} 1 \\ 1 \end{bmatrix}$, where a and b are real numbers. Determine a particular solution to this problem using

(a) the method of undetermined coefficients,

(b) the method of variation of parameters.

8. Write the analogue of the *Mathematica* program in Example 7.7.1 to apply to the following higher dimensional linear systems.

(a) $\mathbf{x}' = \begin{bmatrix} 0 & 1 & 0 \\ 0 & 0 & 1 \\ -2 & 1 & -3 \end{bmatrix} \mathbf{x} + \begin{bmatrix} t^2 \\ 0 \\ 0 \end{bmatrix}$

(b) $\mathbf{x}' = \begin{bmatrix} -1 & 1 & 1 \\ 1 & 0 & 1 \\ 2 & -1 & 3 \end{bmatrix} \mathbf{x} + \begin{bmatrix} 0 \\ 0 \\ e^t \end{bmatrix}$

9. Write the analogue of the program in Example 7.7.4 in *MATLAB*.

10. Use the method of variation of parameters to find the particular solution of each of the linear systems in Problem 5.

11. Modify the program in Example 7.7.4 and apply them to the linear systems in Problem 8.

12. Describe how one could use DSolve or dsolve to construct a fundamental matrix solution of the linear system

$$\mathbf{x}' = \begin{bmatrix} 1 & 3 \\ 3 & 1 \end{bmatrix} \mathbf{x}.$$

13. Consider the system

$$\mathbf{x}' = \begin{bmatrix} 0 & 1 \\ a & 1 \end{bmatrix} \mathbf{x} + \begin{bmatrix} 1 \\ 1 \end{bmatrix}.$$

Find all values of a for which the particular solution to this equation approaches zero as t approaches infinity.

14. Find a particular solution for each of the following systems. Use *Mathematica* or *MATLAB* to determine the eigenvalues of the matrix of coefficients. Compare your final answer with the output of DSolve or dsolve.

(a) $\mathbf{x}' = \begin{bmatrix} 5 & 1 & 1 \\ -3 & 1 & 2 \\ 1 & 0 & 3 \end{bmatrix} \mathbf{x} + \begin{bmatrix} 1 \\ 0 \\ 0 \end{bmatrix}$

(b) $\mathbf{x}' = \begin{bmatrix} 8 & -4 & 1 \\ 0 & 8 & -1 \\ -2 & 10 & 2 \end{bmatrix} \mathbf{x} + \begin{bmatrix} 0 \\ 1 \\ 0 \end{bmatrix}$

(c) $\mathbf{x}' = \begin{bmatrix} 5 & 1 & 0 & 0 \\ 1 & 5 & 0 & 0 \\ 1 & 0 & 3 & -1 \\ 0 & 1 & -1 & 3 \end{bmatrix} \mathbf{x} + \begin{bmatrix} 1 \\ 1 \\ 1 \\ 1 \end{bmatrix}$

(d) $\mathbf{x}' = \begin{bmatrix} 5 & 1 & 1 & 0 \\ 1 & 5 & 0 & 1 \\ 0 & 0 & 3 & -1 \\ 0 & 0 & -1 & 3 \end{bmatrix} \mathbf{x} + \begin{bmatrix} 1 \\ 1 \\ 1 \\ 1 \end{bmatrix}$

7.8 Numerical Methods

It is fair to say that most differential equations that one encounters in practice are too difficult or impossible to solve exactly and explicitly. Even if the physical model leads to a linear system of equations, the source of complexity may be the shear size of the system in cases where a large number of variables is required to describe adequately the motion or the deformation in the problem. In many problems, however, the physics behind the model leads to a system of nonlinear differential equations that does not lend itself to explicit integration. In such cases we often resort to numerical discretization and approximation of the problem at hand.

Numerical analysis is an old and rich subject in mathematics. Very early in the development of calculus the founders of modern mathematics recognized the necessity for this branch of science and laid the foundation for future generations. In the past two hundred years a far reaching theory of numerical methods has been developed, allowing students and researchers in a variety of branches of science to investigate finer and more complex physical structures. In the past few decades rapid advances in the development of algorithms coupled with the unparalleled progress in computer technology have created an environment in which a large number of very powerful algorithms is available to us through software packages such as *Mathematica* and *MATLAB*. Among these algorithms are some of the most efficient methods for solving system of differential equations. We now give an introduction to this subarea of numerical methods and study finite difference methods and the role they play in differential equations.

We will touch upon three standard finite difference methods: a) **Euler**'s, b) **Modified Euler**'s, and c) **Runge–Kutta** methods, as representative of numerous techniques available for solving differential equations. We will also discuss the internal routines `NDSolve` of *Mathematica* and **ode45** of *MATLAB* as examples of sophisticated and powerful routines and compare their nearly flawless performance with the discrete methods we develop here.

The goal of the finite difference methods is to create a table of values of the dependent variable(s) $x(t)$ corresponding to the known values of the independent variable t in an interval (t_0, T). These methods all have their roots in the Taylor approximation of one function or other, although finite difference algorithms are considerably different from the series approximations we studied in Chapter 4. One of the aspects of the former methods is that they apply to linear and nonlinear problems with nearly the same degree of difficulty in implementation. This property has made finite difference algorithms one of the favorites of the practitioners of numerical methods.

7.8.1   Euler's Method for First-Order Equations

Let us begin with a first-order initial-value problem

$$x'(t) = f(t, x), \quad x(t_0) = x_0. \tag{7.193}$$

According to this information, the point $P = (t_0, x_0)$ is on the solution curve of (7.193). Additionally, the slope of the tangent line to this curve at P is known since

$$x'(t_0) = f(t_0, x_0)$$

from the differential equation. Euler's method approximates the value of the exact solution $x(t)$ at a t near t_0 by using the equation for the tangent line:

$$\bar{x}(t) = x_0 + f(t_0, x_0)(t - t_0). \tag{7.194}$$

We expect that the closer t is to t_0 the better $\bar{x}(t)$ approximates the exact value of $x(t)$.

To develop an iterative scheme based on approximation by the tangent line, we first choose a fixed **step size** h and represent the time interval (t_0, T) by

$$t_0, \quad t_1 = t_0 + h, \quad t_2 = t_0 + 2h, \quad \cdots, \quad t_n = t_0 + nh = T. \tag{7.195}$$

Note that $t_n = t_{n-1} + h$. We denote by

$$x_1, \, x_2, \cdots, \, x_n \tag{7.196}$$

the approximate values corresponding to the exact values $x(t_1)$, $x(t_2)$, $\cdots$, $x(t_n)$. It follows from (7.194) that

$$x_1 = x_0 + hf(t_0, x_0), \quad x_2 = x_1 + hf(t_1, x_1), \quad \cdots, \quad x_n = x_{n-1} + hf(t_{n-1}, x_{n-1}).$$
$$\text{(7.197)}$$

We define the abbreviation

$$f_i = f(t_i, x_i) \tag{7.198}$$

and note that Euler's method can be written compactly as

$$x_{i+1} = x_i + hf_i, \quad i = 0, 1, \cdots, n, \cdots. \tag{7.199}$$

Equation (7.199) is called the **difference equation** associated with Euler's method.

Algorithm 7.8.1 (Euler's Method)

The difference equation of Euler's method corresponding to the initial-value problem

$$x' = f(t, x), \quad x(t_0) = x_0$$

is

$$x_{n+1} = x_n + hf(t_n, x_n), \quad t_n = t_0 + nh, \tag{7.200}$$

where $n = 0, 1, \ldots$.

Example 7.8.1

Let x be the solution to the initial value problem

$$x' = x + \sin t, \quad x(0) = 1. \tag{7.201}$$

The exact solution of this equation is

$$x(t) = \frac{3}{2}e^t - \frac{1}{2}(\cos t + \sin t), \tag{7.202}$$

as is easily seen by entering

```
DSolve[{x'[t] == x[t] + Sin[t], x[0] == 1}, x[t], t]}
```

to *Mathematica*. Let $h = 0.1$. The approximation from the iteration formula (7.199) takes the form

$$x_{i+1} = x_i + h(x_i + \sin t_i), \quad i = 0, 1, 2, \cdots, \tag{7.203}$$

with $t_0 = 0$, $t_1 = 0.1$, $t_2 = 0.2$, $\cdots$, $t_n = 0.1n$. For instance, x_1, the approximation to $x(t_1) = x(t_0 + h) = x(0.1)$, is determined by substituting $i = 0$ in (7.203):

$$x_1 = x_0 + hf_0 = x_0 + hf(t_0, x_0) = 1 + 0.1(1 + \sin 0) = 1.1.$$

Similarly,

$$x_2 = x_1 + hf_1 = 1.1 + 0.1(1.1 + \sin 0.1) = 1.21998$$

and

$$x_3 = x_2 + hf_2 = 1.21998 + 0.1(1.21998 + \sin 0.2) = 1.36185.$$

Table 7.1 tabulates the first 10 iterations of this scheme and compares the result with the exact solution.

n	t_n	x_n	$x(t_n)$	Error	Rel Err
0	0	1	1	0	0
1	0.1	1.1	1.11034	0.0103376	0.00931031
2	0.2	1.21998	1.24274	0.0227528	0.0183087
3	0.3	1.36185	1.39936	0.0375113	0.026806
4	0.4	1.52759	1.5825	0.0549119	0.0346995
5	0.5	1.71929	1.79458	0.075292	0.0419553
6	0.6	1.93916	2.03819	0.0990321	0.0485883
7	0.7	2.18954	2.3161	0.126562	0.0546445
8	0.8	2.47291	2.63128	0.158368	0.0601865
9	0.9	2.79194	2.98694	0.194997	0.0652833
10	1.	3.14947	3.38654	0.23707	0.0700037

Table 7.1: Approximations of $x(t)$ for the differential equation (7.201) with step size $h = 0.1$.

In the preceding table the column labeled "Error" refers to the difference between the exact and approximate solutions, while that labeled by "Rel Err" refers to the relative error, that is, the ratio of the error to the exact solution:

$$\text{Error} = x(t_n) - x_n, \quad \text{Rel Err} = \frac{x(t_n) - x_n}{x(t_n)}. \tag{7.204}$$

We expect that these approximations get better if we make h smaller. Table 7.2 shows the iterations one obtains with $h = 0.01$. Every 10 iterations are tabulated to save space and still allow comparison with the first table.

n	t_n	x_n	$x(t_n)$	Error	Rel Err
0	0	1	1	0	0
10	0.1	1.10924	1.11034	0.00109676	0.00098777
20	0.2	1.24032	1.24274	0.0024183	0.00194595
30	0.3	1.39537	1.39936	0.00399429	0.00285437
40	0.4	1.57664	1.5825	0.00585822	0.00370188
50	0.5	1.78653	1.79458	0.00804801	0.00448463
60	0.6	2.02758	2.03819	0.0106066	0.00520391
70	0.7	2.30252	2.3161	0.0135824	0.00586435
80	0.8	2.61425	2.63128	0.0170306	0.00647237
90	0.9	2.96592	2.98694	0.0210135	0.00703512
100	1.	3.36093	3.38654	0.0256015	0.0075598

Table 7.2: Approximations of $x(t)$ for the differential equation (7.201) with step size $h = 0.01$.

The numbers in Table 7.1 were produced using the following program in *Mathematica*:,

```
Clear[n,h,t,x,y];
h =0.1; t = 0; x = 1; n = 10;
f[t_, x_] := x + Sin[t]; sol[t_] := 3/2 Exp[t] -
                 1/2(Cos[t] + Sin[t]);
Print[t," ",x," ",sol[t]];
Do[y = x  + h*f[t, x]; t = t+h; exact = sol[t]; error = exact - y;
   rel=error/exact; Print[i," ",t," ",y," ", exact," ",error,
           " ", rel];
   x = y, {i, 1, n}]
```

Remark 7.8.1: Replace the two instances of := with = and execute this program. What is the source of error?

Table 7.2 is basically the same as Table 7.1 with the only changes coming from the values of h and n and the fact that we wish to print every 10 values of x_n. The latter is accomplished by using If and Mod commands of *Mathematica*. The line beginning with Do in the previous program is replaced by

```
Do[y = x  + h*f[t, x]; t = t + h; exact = sol[t]; error =
     exact - y;
   rel=error/exact;
   If[Mod[i,10] == 0, Print[i," ",t," ",y," ", exact," ",error,"
          ", rel]];
   x = y, {i, 1, n}]
```

The preceding programs have a similar flavor in *MATLAB*. The following two M-files produce the analogue of Table 7.1:

```
Clear[n,h,t,x,y];
n = 10; h = 0.1; t = 0; x = 1;
%
output = [0 0 1 1 0 0];
for i=1:n
    y = x + h*yprime(t,x);
    t = t + h;
    exact = 3/2*exp(t) - 0.5*(cos(t) + sin(t));
    error = exact - y;
    rel = error/exact;
    output=[output;i t y exact error rel];
    x=y;
end
output
```

The function **yprime** called in this program is listed in the following M-file. It contains the definition of the right-hand side of the differential equation (7.201).

```
function y = yprime(t,x);
%
y = x + sin(t);
```

To obtain the contents of Table 7.2 in *MATLAB* we introduce a counting device
in the main program that keeps track of the increase in the index i. The variable
every_so_often specifies the programer's wish as to when the output should be
written to the matrix output. The altered main program reads as

```
n = 100; h = 0.01; t = 0; x = 1;
%
count = 2; every_so_often = 10;
output = [0 0 1 1 0 0];
for i=1:n
    y = x + h*yprime(t,x);
    t=t+h;
    exact = 3/2*exp(t) - 0.5*(cos(t)+sin(t));
    error = exact - y;
    rel = error/exact;
    if count > every_so_often
       output=[output;i t y exact error rel];
       count = count - every_so_often;
    end
    x=y;
count = count + 1;
end
output
```

7.8.2   Euler's Method for Systems

The Euler's method we described was in the context of a single first-order equation. It turns out that this method generalizes in a straightforward manner to
first-order systems of ODEs. Because higher order equations and systems can
always be reduced to first-order systems, Euler's algorithm can be applied to
essentially any ODE we will see in applications.

Let $\mathbf{x}(t)$ be a solution of the first-order system of equations

$$\mathbf{x}'(t) = \mathbf{f}(t, \mathbf{x}), \quad \mathbf{x}(t_0) = \mathbf{x}_0. \tag{7.205}$$

We approximate $\mathbf{x}(t_n)$ by $\mathbf{x}_n$ where $\mathbf{x}_n$ satisfies the iterative equation

$$\mathbf{x}_{n+1} = \mathbf{x}_n + h\mathbf{f}(t_n, \mathbf{x}_n). \tag{7.206}$$

Compare this equation with its counterpart (7.200). Although we no longer
can appeal to the geometric interpretation that motivated (7.200), nevertheless
we think of (7.206) as the natural generalization of Euler's method to higher
order equations and systems. The next example shows the necessary steps in
implementing this method in the context of a nonlinear system of equations.

Example 7.8.2

Consider the system of differential equations

$$\frac{dx}{dt} = 1 - \frac{x^2 - y^2}{(x^2 + y^2)^2}, \quad \frac{dy}{dt} = -\frac{2xy}{(x^2 + y^2)^2}, \tag{7.207}$$

with initial data

$$x(0) = -3, \quad y(0) = 0.1. \tag{7.208}$$

We discussed this system in Section 7.2 as a set of differential equations that models the flow of an incompressible fluid past a cylinder of radius one. The solution $(x(t), y(t))$ describes the path of a particle located initially at $(-3, 0.1)$.

Because these differential equations are nonlinear, we are unable to find an exact solution easily. On the other hand, knowing the physical interpretation of the equations helps in imagining how the solutions must look. We will use Euler's method to obtain a solution of this problem and plot its trajectory for various values of h in order to get a sense of how well or poorly this method applies to (7.207). Following (7.206) (with $\mathbf{x} = (x, y)$), we find that the ordered pair (x_n, y_n) satisfies the finite difference equations

$$x_{n+1} = x_n + h\left(1 - \frac{x_n^2 - y_n^2}{(x_n^2 + y_n^2)^2}\right), \quad y_{n+1} = y_n - h\left(\frac{2x_n y_n}{(x_n^2 + y_n^2)^2}\right) \tag{7.209}$$

and

$$x_0 = -3, \quad y_0 = 0.1. \tag{7.210}$$

We can now proceed as we did in Example 7.8.1 and determine the first few x_is and y_is by hand. Instead of carrying this step out, we consider the changes we need to make to the computer programs in the previous example in order to accommodate the complexity of (7.209). The *Mathematica* program that generates the iterations in (7.209) is

```
Clear[h,n,den,x,y,f,ff,g]
h =0.01; den = (x^2 + y^2)^2;
(* The right-side of the system *)
f[t_, {x_, y_}] =  {1 - (x^2-y^2)/den, -2 x y/den};
(* Initial data *)
t = 0; x = {-3, 0.1}; Print[t,"    ",x];
solution = {};
Do[y = x  + h*f[t, x]; t = t + h; If[Mod[i,10] ==0,
      Print[i, "  ",t, "   ",y]];
      solution = Append[solution, y];
      x = y,  {i, 1, 1000}];
output1 = ListPlot[solution, PlotJoined->True,
               PlotStyle->Dashing[{0.05, 0.05}]]
(* Solution obtained using NDSolve *)
Clear[x,y,t];
ff[x_, y_] = 1 - (x^2-y^2)/den;
```

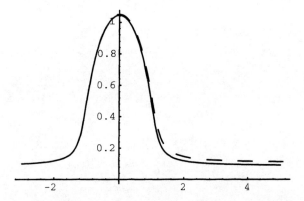

Figure 7.8: The output of Euler's method on *Mathematica* for the flow past the cylinder.

```
g[x_, y_] = -2 x y/den;
sol = NDSolve[{x'[t] == ff[x[t], y[t]],
               y'[t] == g[x[t], y[t]],
               x[0] == -3, y[0] == 0.1},
               {x, y}, {t, 0, 10}];
output2 = ParametricPlot[Evaluate[{x[t], y[t]}/. sol],
                 {t, 0, 10}];
output3 = Show[output1, output2]
```

Figure 7.8 shows the output of this program together with the approximate solution we obtain by applying the internal function NDSolve of *Mathematica* to the system of differential equations (7.207)–(7.208). The solution obtained via Euler's method agrees well with the solution from NDSolve for small values of t, but deviates from it considerably for large values of t. We get a better approximation from Euler's method if we let h be 0.001 in the preceding program.

The logic behind the syntax of the *MATLAB* program that yields Figure 7.9 is very similar to the previous *Mathematica* program. We need to make a few changes to the first *MATLAB* program we listed since we are now dealing with a system of equations rather than a single equation. We will also use the internal package ode45 of *MATLAB* to generate the solution to the initial value problem for comparison. The listing of the M-files follows. The main program is

```
n = 1000; h = 0.01; t = 0; x = [-3 0.1];
%
count = 2; every_so_often = 10;
output = [0 0 x];
for i=1:n
    y = x +h*yprimes(t,x)';
```

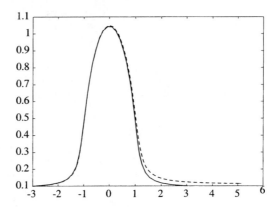

Figure 7.9: The output of Euler's method on *MATLAB* for the flow past the cylinder.

```
    t=t+h;
    if count > every_so_often
        output=[output;i t y];
        count = count - every_so_often;
    end
    x=y;
count = count + 1;
end
plot(output(:,3),output(:,4),'--')
hold
[s,w]=ode45('yprimes',[0 10],[-3 0.1]);
plot(w(:,1),w(:,2))
```

The M-file that defines the function **yprimes** called by the previous program is

```
function y = yprimes(t,x);
%
den =   (x(1)^2 + x(2)^2)^2;
y =[1-(x(1)^2-x(2)^2)/den;   -2*x(1)*x(2)/den];
```

Figure 7.9 shows the output of this program.

As these examples show, the approximation we obtain from Euler's method deteriorates as n becomes large and t is not close to the initial time t_0. One source of error is in the discretization process where we replace the exact solution $x(t_n)$ by the approximate value $x_n = x_{n-1} + hf(t_{n-1}, x_{n-1})$. We note that even if we used the exact value of the solution at x_{n-1} (namely, $x(t_{n-1})$) in the latter formula, we would still incur an error by computing x_n since we are approximating the exact graph of the solution by its tangent line. So, not only does Euler's method approximate by using the tangent line, it uses the tangent line at approximate values, the accumulative effect being that the quality of the approximation deteriorates with increasing n. The terms **local error** and **global error** are often used to express these sources of error. Local error is the error one introduces at each step of the discretization assuming that exact values of the solution were obtained in the previous steps. Global error measures the error at step n not assuming that the exact solution was obtained in the previous steps. An important aspect of the analysis of finite difference schemes concerns obtaining estimates for the local and global errors in terms of the given data of the problem, such as bounds on the nonlinearity $\mathbf{f}$ in (7.205) and the length of the interval of the existence of the solution and its approximation. Estimates on local errors are often found from the properties of the Taylor series expansion of $x(t_n + h)$, its relation to $x(t_n)$, while utilizing the remainder formula for Taylor series. Estimates of global errors generally require an in-depth study of the differential equation itself. We quote an example of the type of theorem that has been proved for Euler's method.

Theorem 7.8.1
Let each component of the function $\mathbf{f}$ *in (7.193) be continuously differentiable in each variable in its domain. Suppose further that the second derivative of the solution* $\mathbf{x}$ *is bounded in the domain* (t_0, T)*. Then there is a constant* C *such that for all* $t_i \in (t_0, T)$ *we have the etimate*

$$||\mathbf{x}(t_i) - \mathbf{x}_i|| \leq Ch. \tag{7.211}$$

The reader may consult pages 209–211 of Reference [2] for a proof of a similar theorem.

Because of the preceding inequality, Euler's method is called an **order** h scheme. One way to interpret this inequality is as follows. If everything else was kept fixed while h was allowed to vary, one expects to gain a significant digit more accurately by reducing h by a factor of 10.

Another source of error in our approximations is due to **roundoff**. This error is generated at each step of the algorithm because the computer on which the computation is being carried out is only capable of storing numbers with finitely many significant digits. Usually this error does not manifest itself until a large number of algebraic operations has been carried out. For an algorithm such as Euler's method the roundoff error may be significant when we compare approximations with $h = 0.1$ and $h = 0.00001$, say. On the one hand, the approximate sequence we obtain with the latter value of h will have a smaller global error than the the sequence computed from the former h. On the other hand, with the number of algebraic operations increased dramatically by taking such a small value for h, there is a significant probability that the roundoff error in the case of $h = 0.00001$ calculation is 10,000 times larger than the $h = 0.1$ case.

We used the term probability because there is an element of randomness in the process of rounding off. In general, one does not expect that every instance of rounding off contributes to increasing the error in an algorithm. In fact, it is possible that in a string of algebraic calculations the roundoff residual from one set of operations cancels out the roundoff residual from another set, so on the average the contribution of this error is rather small or negligible. It is a difficult matter to track how errors, in general, and the roundoff error, in particular, propagate in a string of computations. This subject is to this day a vital and ongoing area of research in numerical analysis.

We end this section by pointing out that both *Mathematica* and *MATLAB* are capable of exact arithmetic that eradicates roundoff as a source of error. Unfortunately using this facility is impractical since for most lengthy calculations an extremely large storage capability is needed, which is simply not available on ordinary computers at this point. A few of the ensuing exercises will address this idea.

Problems

1. Create a file that contains either the *Mathematica* or the *MATLAB* programs for the differential equation $x' = x + \sin t$, with $x(0) = 1$. Input this program to the appropriate software and generate the corresponding version of Table 7.1 and Table 7.2.

   (a) Let $h = 0.1$ and $n = 10$. Plot the exact solution and the approximate solution on the same screen.

   (b) Let $h = 0.01$ and $n = 100$ and repeat (a).

2. Repeat the preceding problem for the following differential equations.

   (a) $x' = -x + \sin t$, $x(0) = 1$. First let $h = 0.1$ and $n = 60$. Next let $h = 0.01$ and $n = 600$. In each case compare the results with the exact solution.

   (b) $x' = x^2 - 1$, $x(0) = 0$. Let $h = 0.001$ and $n = 1000$. Compare the graph of the approximate solution from Euler's method to the exact solution. (Hint: You can obtain the exact solution either by separating variables or using the DSolve of *Mathematica* or dsolve of *MATLAB*. Alternatively, both NDSolve and ode45 provide excellent approximations to the exact solution.)

   (c) Repeat (b) for $x' = -x^2 - 1$, $x(0) = 0$.

3. Use either of the two programs in this section and generate the graph of the path of the fluid particle in the flow past the cylinder.

4. Consider the system of differential equations for the fluid flow past the cylinder. Modify the appropriate program in the text and draw the paths of the particles whose initial positions are

$$(-3, 0.1), \quad (-3, 0.2), \quad (-3, 0.3), \quad (-3, 0.4), \quad (-3, 0.5).$$

First use the step size $h = 0.1$ and $n = 50$. Next, generate the same graphs with $h = 0.01$ and $n = 500$.

5. Consider the system of differential equations for the fluid particles moving in a Rayleigh–Bénard flow. Here

$$\frac{dx}{dt} = \pi \sin \pi x \cos \pi z, \qquad \frac{dz}{dt} = -\pi \cos \pi x \sin \pi z.$$

Use Euler's method and find the paths of the particles whose initial positions are

$$(0.9, 0.1), \quad (0.8, 0.2), \quad (0.7, 0.3), \quad (0.6, 0.4).$$

Use $h = 0.01$ and $n = 200$. Compare the results with graphs you get from either NDSolve or ode45.

6. Both *Mathematica* and the *Symbolic Math TOOLBOX* of *MATLAB* are capable of performing the necessary algebraic operations in the programs we listed in this section exactly and without rounding off the outcomes. In the first *Mathematica* program we listed in this section all variables are introduced in such a way that exact arithmetic is performed, except in the statement $h = 0.1$. Replace the latter input statement with

```
h = 1/10;
```

and run the program with $n = 10$. Compare the results of x_{10} with the exact value of the solution, $x(1)$, where x_{10} is generated first with $h = 0.1$ and next with $h = 1/10$. Use the internal function N to get the numerical values of the two x_{10} with 20 significant digits (i.e., N[x, 20]). Do these values differ from each other? Is one of them closer to the exact value?

7. Write an interactive program in *Mathematica* based on the Euler's method. To that end, use Input of *Mathematica* to ask the user to input h, n, and, f, and the program outputs the graph of the approximate solution.

8. (**Higher Order Taylor Methods**) Euler's method is an example of a class of methods known as Taylor methods, whose difference equation is derived from an application of the Taylor series. We note that $x(t+h)$ and $x(t)$ are related by

$$x(t + h) = x(t) + hx'(t) + \frac{h^2}{2!}x''(t) + \frac{h^3}{3!}x'''(t) + \cdots. \qquad (7.212)$$

When x is a solution of the differential equation $x' = f(t, x)$ we can replace the derivatives of x in (7.212) by f and its partial derivatives. For example, when we cut off the series in (7.212) after just two terms and replace x' with f, we end up with Euler's method. We thus say that Euler's method is equivalent to a first-order Taylor method.

(a) Show that the difference equation we get from the second-order Taylor method (i.e., keeping only the first three terms on the right-hand side of (7.212)) is

$$x_{n+1} = x_n + hf_n + \frac{h^2}{2}((f_t)_n + f(f_x)_n). \qquad (7.213)$$

(Hint: Use the chain rule on $x' = f(t, x)$ to show that $x'' = f_t + f_x f$.)

(b) Let $x' = \sin x - \cos t$ with $x(0) = 0$. First use Euler's method with $h = 0.01$ and obtain an approximate solution in the interval $(0, 2\pi)$. Next write a program for the second-order Taylor method and determine an approximate solution in the same interval. Graph these two approximate solutions with the output of either NDSolve or ode45 for comparison.

Remark 7.8.2: It is possible to show that the global error of the second-order Taylor method is proportional to h^2.

(c) Find the difference equation for the third-order Taylor method.

7.9 Runge–Kutta Methods

We saw in the previous section that Euler's method is a first-order algorithm, meaning that its global error is proportional to the first power of the step size h. In this section we seek techniques for developing higher order algorithms (see also Problem 8 of the previous section in this connection).

We first discuss a simple and clever improvement of Euler's method that generates a second-order algorithm. We recall that this method is applied to the initial value problem

$$x' = f(t, x), \quad x(t_0) = x_0$$

by replacing the differential equation with the difference equation

$$x_{n+1} = x_n + hf(t_n, x_n). \tag{7.214}$$

In this expression x_n and x_{n+1} are the approximations to the exact values $x(t_n)$ and $x(t_{n+1})$. Since $h = t_{n+1} - t_n$, the difference equation states that the ordered pairs (t_n, x_n) and (t_{n+1}, x_{n+1}) are linearly related with the slope of the straight line being the slope of the tangent line to the solution curve at (t_n, x_n). Once we have computed an approximate value for x_{n+1} we can determine an approximation to the slope of the tangent line at (t_{n+1}, x_{n+1}) by evaluating f at this point:

$$f(t_{n+1}, x_{n+1}). \tag{7.215}$$

It turns out that we can achieve an improvement in Euler's method if we use the slope at (t_{n+1}, x_{n+1}) to improve the value of the slope at (t_n, x_n). Specifically, we replace $f(t_n, x_n)$ in (7.214) with the average of the slopes at x_n and x_{n+1}. We end up with the difference equation

$$x_{n+1} = x_n + \frac{h}{2}(f(t_n, x_n) + f(t_{n+1}, x_{n+1})). \tag{7.216}$$

This equation does turn out to give us a more accurate scheme than the Euler method, but we do pay a price in its implementation: In the new scheme x_{n+1} appears on both sides of the difference equation, or equivalently, x_{n+1} appears *implicitly* in the difference equation. So, to find x_{n+1} we need to apply a root-finding algorithm in addition to the iterative scheme itself. One way to get around this complication is to use the value of x_{n+1} predicted by Euler's method on the right-hand side of (7.216), that is, to have a scheme that is a hybrid of Euler's method and the given implicit algorithm:

$$x_{n+1} = x_n + \frac{h}{2}(f(t_n, x_n) + f(t_{n+1}, x_n + hf(t_n, x_n))). \qquad (7.217)$$

This difference scheme is called the **modified Euler method**. It is somewhat more complicated than Euler's method, however it should be noted that it is still an explicit iterative scheme whose implementation relies only on function evaluations of f alone (and not on its derivatives; see Problem 8 of Section 7.8). Most importantly, the modified Euler method is a second order scheme.

Algorithm 7.9.1 (Modified Euler Method)
The difference equation for the Modified Euler Method is

$$x_0 = given, \quad x_{n+1} = x_n + \frac{h}{2}(s_1 + s_2) \qquad (7.218)$$

where s_1 and s_2 are the approximate slopes at (t_n, x_n) and $(t_n, x_n + hs_1)$ or

$$s_1 = f(t_n, x_n), \quad s_2 = f(t_{n+1}, x_n + hs_1). \qquad (7.219)$$

Example 7.9.1
Consider the differential equation

$$x' = x^2 - \cos t, \quad x(0) = 1. \qquad (7.220)$$

Here $f(t, x) = -\cos t + x^2$. We choose the step size $h = 0.1$ and compute the first two iterates x_1 and x_2. First, to compute x_1 we need to evaluate its s_1 and s_2:

$$s_1 = f(t_0, x_0) = 0, \quad s_2 = f(0.1, 1) = 1 - \cos 0.1.$$

Thus,

$$x_1 = x_0 + \frac{h}{2}(s_1 + s_2) = 1 + 0.05(1 - \cos(0.1)) = 1.00024979.$$

Next, we compute x_2:

$$s_1 = f(t_1, x_1) = (1.00024979)^2 - \cos 0.1 = .00549548$$

and

$$s_2 = f(0.2, 1.0002498 + 0.1 \times 0.00549548) = 0.0215327.$$

Finally,

$$x_2 = x_1 + \frac{h}{2}(s_1 + s_2) = 1.00024979 + 0.05\,(0.00549548 + 0.0215327) = 1.0016012.$$

The following program in *Mathematica* shows how to implement the modified Euler method for (7.220).

```
h = 0.1; n = 10; f[t_, x_] = x^2 - Cos[t];
t = 0; x = 1; h2 = h/2;
Do[s1 = f[t, x]; t = t + h; pred = x + h*s1; s2 = f[t, pred];
    x = x + h2*(s1 + s2); Print[t,",",x], {i, n}];
```

The output of this program is

```
0,   1.
0.1,   1.00024979
0.2,   1.0016012
0.3,   1.00538357
0.4,   1.01320243
0.5,   1.02700284
0.6,   1.04916547
0.7,   1.08265886
0.8,   1.13128914
0.9,   1.20012147
1.,   1.2962153
```

Analogue of the preceding program in *MATLAB* is

```
t=0; x=1; n=10;
h=0.1; h2=h/2;
ans = [t x];
for i=1:n
    s1=fun(t,x);
    t=t+h;
    pred=x+h*s1;
    s2=fun(t,pred);
    x=x+h2*(s1+s2);
    ans=[ans;t x];
end
ans
```

The M-file fun.m contains

```
function y=fun(t,x);
%
y=-cos(t)+x^2;
```

MATLAB's output is

```
ans =

         0    1.0000
    0.1000    1.0002
    0.2000    1.0016
    0.3000    1.0054
    0.4000    1.0132
```

0.5000	1.0270
0.6000	1.0492
0.7000	1.0827
0.8000	1.1313
0.9000	1.2001
1.0000	1.2962

The modified Euler method is an example of a **predictor-corrector** algorithm. In such an algorithm one uses some means of predicting a first approximation to the output, and then appeals to other indicators to correct this value further. In the preceding method we used one step of the Euler method to predict a value for x_{n+1}, and subsequently corrected this value by taking the average of the slopes at t_n and t_{n+1}.

Before proceeding to the next scheme, we point out an analytic interpretation of the difference equation (7.217). Note that

$$\int_t^{t+h} x'(s)\, ds = x(t+h) - x(t). \tag{7.221}$$

Since $x' = f(t, x)$, we have that $x(t)$ and $x(t+h)$ are related by

$$x(t+h) - x(t) = \int_t^{t+h} f(s, x(s))\, ds. \tag{7.222}$$

In Euler's method we approximate the integrand of (7.222) with the *constant* value $f(t, x(t))$, so that the integral reduces to $hf(t, x(t))$. In (7.216) we approximate the integral in (7.222) by the area of the trapezoid with vertices at $(x(t), 0)$, $(x(t+h), 0)$, $(x(t+h), f(t+h, x(t+h)))$, and $(x(t), f(t, x(t)))$. Finally, to get the modified Euler method we use the Euler method to predict the value of $x(t+h)$.

The modified Euler method is a second-order finite difference scheme that relies only on function evaluations of f. A class of higher order methods known as the **Runge–Kutta** schemes have several properties in common with the latter method, perhaps the most important property they share is that their determination too requires function evaluations only and no derivatives of f need to be evaluated. The fourth-order Runge–Kutta scheme is very much in the same spirit as the modified Euler method in that its finite difference scheme is in the form

$$x_{n+1} = x_n + \frac{h}{6}(s_1 + 2s_2 + 2s_3 + s_4), \tag{7.223}$$

where the term that multiplies h is a weighted average of the slope of the tangent line at $(t+h, f(t+h))$. Here we give the formulas for s_is and refer the reader the pages 219–221 of Reference [2] for the analysis that leads to these formulas. Also see a project at the end of the chapter in connection with this scheme. The term s_1 is the slope at (t_n, x_n), that is,

$$s_1 = f(t_n, x_n). \tag{7.224}$$

while s_2 estimates the slope at the midpoint $t_n + \frac{h}{2}$, which in turn uses the Euler method for $x_{n+\frac{1}{2}}$, that is,

$$s_2 = f\left(t_n + \frac{h}{2}, x_n + \frac{h}{2}s_1\right). \tag{7.225}$$

The term s_3 is a correction of this predicted value using the same formula we obtained from the modified Euler method:

$$s_3 = f\left(x_n + \frac{h}{2}, x_n + \frac{h}{2}s_2\right). \tag{7.226}$$

Finally, s_4 represents a correction of the slope at t_{n+1} using the value of s_3:

$$s_4 = f(t_{n+1}, x_n + hs_3). \tag{7.227}$$

We have not justified why this scheme gives a reasonable difference equation for the problem at hand. A rather lengthy Taylor series expansion of the terms in (7.223) eventually leads to the fact that this scheme is fourth-order. This method is often considerably simpler to implement in practice than a fourth-order Taylor method (see Problem 8 of Section 7.8) since we do not need to take partial derivatives of f, which could be a costly process. It turns out that the preliminary mathematical analysis and the creative planning that went into discovering the Runge–Kutta scheme have paid out a large dividend since at the present time one variation or another of this scheme is prevalent among the software packages in the market.

Algorithm 7.9.2 (Runge–Kutta Method)
The difference equation for the Runge–Kutta is given by (7.223) and the slopes s_is are defined by (7.224)–(7.227).

Example 7.9.2
We consider again the differential equation in the previous example and apply the Runge–Kutta scheme to it with $h = 0.1$ and go through one step of this algorithm. Since $t_0 = 0$ and $x_0 = 1$, we have

$$s_1 = f(0, 1) = 0.$$

Similarly,

$$s_2 = f\left(t_0 + \frac{h}{2}, x_0 + \frac{h}{2}s_1\right) = f(0.05, 1) = 0.00124974,$$

while

$$s_3 = f\left(t_0 + \frac{h}{2}, x_0 + \frac{h}{2}s_2\right) = f(0.05, 1.00006) = 0.00137472.$$

Finally,

$$s_4 = f(t_0 + h, x_0 + hs_3) = f(0.1, 1.00014) = 0.00527079.$$

Substituting these values in (7.223), we arrive at the following estimate for $x(t_1)$:

$$x_1 = 1 + \frac{0.1}{6}(0.00527079 + 2 \times 0.00137472 + 2 \times 0.00124974) = 1.00017533.$$

The following program in *Mathematica* carries out the steps in the Runge–Kutta algorithm.

```
Clear[t,x,f];
h = 0.1; n = 10; f[t_, x_] = x^2 - Cos[t];
t = 0; x = 1; h2 = h/2; Print["The initial data: t =  ",t,",
          x = ",x];
sol = {{t, x}};
Do[s1 = f[t, x]; pred1 = x + h2*s1; s2 = f[t+h2, pred1];
    pred2 = x + h2*s2; s3 = f[t+h2, pred2];
    pred3 = x + h*s3; s4 = f[t+h, pred3];
    x = x + h/6*(s1 + 2*s2 + 2*s3 + s4);t=t+h;
    Print["t = ",t,", x = ", x];
    sol = Append[sol, {t, x}],
{i, n}];
```

Mathematica's output is

```
The initial data: t =  0, x = 1
t =  0.1, x = 1.00018
t =  0.2, x = 1.00148
t =  0.3, x = 1.00524
t =  0.4, x = 1.0131
t =  0.5, x = 1.02703
t =  0.6, x = 1.04944
t =  0.7, x = 1.08336
t =  0.8, x = 1.13268
t =  0.9, x = 1.20263
t =  1., x = 1.30054
```

The modified Euler and the fourth-order Runge–Kutta methods generalize to systems of differential equations in the same fashion that the Euler method was extended to systems. In each case the finite difference equation is applied to individual equations in the system. Some of the ensuing exercises address this process.

Problems

1. Use the *Mathematica* program for the modified Euler method and obtain a table of values for the following differential equations. Use a step size h and a number of iterations n of your own choosing. In each case compare the output with the solution one obtains from NDSolve.

   (a) $x' = -3x + 2t$, $x(0) = 0$
   (b) $x' = \sin 2x + \frac{1}{t^2}$, $x(1) = 0$

(c) $x' = x^2$, $x(0) = 0.1$

(d) $y' = \sin t + \tan y$, $y(0) = \frac{\pi}{4}$

2. Use the *Mathematica* program for the fourth-order Runge–Kutta and get a table of values for the following differential equations. Use a step size h and a number of iterations n of your own choosing. In each case compare the output with the solution one obtains from NDSolve.

   (a) $x' = -x + t$, $x(0) = 0$

   (b) $x' = -\sin x + \frac{1}{t}$, $x(1) = 0$

   (c) $x' = x^2$, $x(0) = 0.1$

   (d) $y' = \tan y$, $y(0) = \frac{\pi}{4}$

3. Develop the analogue of the program in Example 7.9.2 in *MATLAB*. Follow the directions in Problem 2 and compare the output to the solution one obtains from ode45.

4. Develop a *Mathematica* or a *MATLAB* program for the fourth-order Runge-Kutta scheme for the 2×2 system

$$x' = f(t, x, y), \quad y' = g(t, x, y).$$

Apply this program to the following systems. Use h, n, and initial conditions of your own choosing. In each case graphically compare the output with that of either NDSolve or ode45.

   (a) $x' = y$, $y' = -x$

   (b) $x' = 2x - 3y$, $y' = x + y$

   (c) $x' = y$, $y' = -\sin x$

   (d) $x' = y$, $y' = -y - \sin x$

   (e) $x' = 1 - \frac{x^2 - y^2}{(x^2 + y^2)^2}$, $y' = -\frac{2xy}{(x^2 + y^2)^2}$

7.10 NDSolve and ode45

The numerical schemes we studied in the past two sections were based on natural and intuitive arguments that lead to rather good approximations of solutions of differential equations, especially when the domain of approximation is a small neighborhood of the initial data. We have stated several results that relate the global error in a numerical scheme to the step size h in the form

$$|x(t) - x_n| \le Ch^p, \tag{7.228}$$

where the constant C depends on the smoothness of the solution x and the size of the interval of approximation and p is related to the order of the difference scheme. For example, we saw that p is 1 and 2 for the Euler and the modified Euler methods, respectively, while it has the value 4 for the Runge–Kutta scheme.

It seems reasonable to expect from the error estimate (7.228) that we may achieve more accuracy by choosing a smaller step size h. This is, of course, true to a point since reducing h by a factor of 10 reduces the error in (7.228) by a factor of 10^p. The price we pay for smaller h, and one that is not reflected in (7.228), is that more algebraic operations must be carried out to cover the same domain. We expect that reducing h by a factor of 10 increases the number of algebraic operations tenfold. This increase in the number of operations will be reflected in an increase in the roundoff error. If the latter error turns out to outweigh the gain in the global error, we see that the strategy of reducing the step size would not always have the desired outcome. It is not clear what the optimal value of h is in a calculation or how one may get a sense of this optimal value other than employing a trial-and-error strategy. It is fair to say, however, that the price of reducing the step size due to the roundoff becomes steep rather quickly, an indication that we may have reached an "optimal" value for h.

It is important to note that the strategy of reducing the step size uniformly across the interval of approximation may not be wise for a differential equation whose solution behaves nonuniformly in that interval. For example, if a solution remains nearly constant in a large part of the interval (a, b) but varies rather sharply near a and b, then it is not very economical to reduce the step size in the entire interval and increase accuracy across the board, since the smaller subintervals are needed near a and b while away from these points a coarse grid will probably pick out the nearly constant behavior of the solution. We would like to employ an **adaptive** strategy and reduce the step size in regions of high transitions and oscillations in the solution but leave the grid rather coarse otherwise. The price one pays for an adaptive scheme is in the delicate decision process of discovering when and where a solution is going through a region of sharp transition. A rather large amount of theoretical work on adaptive difference schemes for ODEs has resulted in several software packages with this idea as the core of their design. NDSolve of *Mathematica* and ode45 of *MATLAB* are two such applications, both of which are based on an adaptive extension of the fourth-order Runge–Kutta method.

Typically adaptive schemes employ two error tolerances that indicate the maximum and minimum errors we are willing to tolerate when approximating the value of the solution at the nth iteration step. For the sake of argument, suppose that we use any of the aforementioned numerical methods and compute two approximate values at t_n, one using the step size h, which we denote by y_n, and the other with $\frac{h}{2}$ and denoted by z_n, the latter using twice as many operations as the former. We expect that the value of z_n to be a more accurate approximation of the exact solution. The difference between y_n and z_n is an error that we compare to the minimum and maximum tolerance errors. If this difference falls between the tolerance errors, y_n is a reasonable approximation and we proceed with the scheme using h as a step size. If the difference is below the minimum tolerance error, then y_n is a reasonable value but we double the step size for the next iteration. If, on the other hand, the difference in y_n and z_n is above the maximum tolerance error we suspect that the solution may be going through a region of sharp transition, reject both y_n and z_n, go back to the $n - 1$ stage, and recompute the preceding quantities with h replaced with $\frac{h}{2}$.

Figure 7.10: Motion of parcels of fluid past a unit cylinder in *Mathematica*.

We do not attempt here to implement this modification, or many alternative schemes based on this procedure, instead we embark on illustrating how one extracts information from the adaptives schemes NDSolve and ode45 when applied to some of the examples we introduced earlier in this chapter.

Example 7.10.1

In this example we develop a program in *Mathematica* which draws the snapshots of the flow in a two-dimensional system of differential equations. The idea here is to simulate the motion of a parcel of fluid under the action of the flow. We write this program for the flow past the cylinder and leave it to the reader to modify it for other flows.

Let $\{P_0, P_1, \ldots, P_m\}$ be a sequence of points on the boundary of a region in R^2. Typically we start with a region in the shape of a disc and choose P_is on its circular boundary. We follow the evolution of these points under the action of the system of differential equations

$$\frac{dx}{dt} = 1 - \frac{x^2 - y^2}{(x^2 + y^2)^2}, \quad \frac{dy}{dt} = -\frac{2xy}{(x^2 + y^2)^2}, \tag{7.229}$$

which are also derivable from the stream fuction (see Section 7.2)

$$\psi(x, y) = y - \frac{y}{x^2 + y^2}.$$

The following is the *Mathematica* program that produces Figure 7.10. Prior to going through this program the reader may wish to review the commands ListPlot, RGBColor, Map, and Polygon.

```
psi=y-y/(x^2+y^2);   (* The stream function or the hamiltonian *)
f[x_,y_]=D[psi,y];
g[x_,y_]=-D[psi,x];
m=4;                 (*Number of parcels*)
dy = 0.5;            (*Distance between parcels at time 0*)
n=20;                (*Number of points on the boundary
                       of parcels*)
circle =  Table[{Cos[t], Sin[t]}, {t, 0, Pi, Pi/20}];
```

```
GraphOfCircle = ListPlot[circle, PlotJoined->True,
                   DisplayFunction->Identity];
ColoredCircle=Graphics[{RGBColor[0,0,1],
                   Map[Polygon, circle, {0}]}];
diffeqn[tfinal_,a_,b_]:=NDSolve[{x'[t]==f[x[t],y[t]],
                             y'[t]==g[x[t],y[t]],
        x[0]==a, y[0]==b},
        {x,y}, {t,0,tfinal}];
oldsolution=Table[diffeqn[8,-2+0.1*Cos[w],
        0.3+(h-1)*dy+0.1*Sin[w]], {h, 1, m},
        {w, 0, 2*Pi, 2*Pi/n}];
data=Table[{x[t], y[t]} /. oldsolution, {t, 0, 5, 0.5}];
ColoredSnapshots=Table[Graphics[
           {RGBColor[Cos[Mod[i,Pi/2]], Sin[Mod[i,Pi/2]], 0],
            Map[Polygon, Flatten[data[[i]][[j]],1], {0}]}],
            {i, Length[data]},
        {j, 1, m}]
output=Show[ColoredCircle, ColoredSnapshots,
     AspectRatio->Automatic];
```

There is quite a bit of information in Figure 7.10 that will become more accessible to the reader once we study the physics of this flow carefully in later chapters. To summarize a few points from this figure, the flow moves from left to right, where the coloring of the parcels is synchronized with the time at which a snapshot of the four parcels is taken. The parcels are identical at time zero but deform as they move past the cylinder. Since the flow is incompressible, meaning that the area of a parcel of fluid is preserved as it evolves, each parcel must stretch and thin out as it passes the cylinder. The parcels of fluid that are near the cylinder must slow down in response to the increase in the pressure ahead of the cylinder, but they accelerate as they move over the cylinder. The flow downstream approaches a uniform state, as can be observed from the figure, because the shape of the individual parcels remain relatively unchanged once they pass the cylinder. Finally, once a parcel of fluid passes the cylinder, it must slow down and try to match its speed with the downstream speed at $x = +\infty$. A signature of this slowdown can be seen in the shape of the parcel of fluid closest to the cylinder after it has passed this obstacle, noting that the snapshots begin to fatten as time evolves.

Example 7.10.2

We now apply the same arguments that led to the preceding program and develop a similar program in MATLAB. Figure 7.11 shows the output of this program.

```
n=20;tfinal=0.5;m=15;
pts=0:0.01:pi; circle=[cos(pts);sin(pts)]';
plot(circle(:,1),circle(:,2));
```

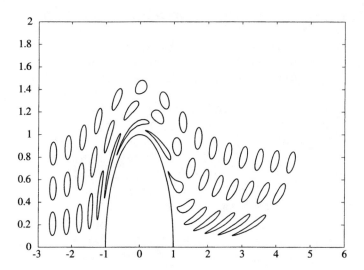

Figure 7.11: Motion of parcels of fluid past a unit cylinder in *MATLAB*.

```
hold on
h=0;
for l=1:3
  data=[];
  for i=1:n
  xcoord=-3+0.1*cos(2*pi*(i-1)/n);
  yccord= 0.2+h+0.1*sin(2*pi*(i-1)/n);
    data=[data;xccord ycoord];
  end
  data=[data;data(1,:)];
  for k=1:m
      sol=[];
      for j=1:n+1
          [t,y]=ode45('fpc',[0 tfinal],data(j,:));
          sol=[sol;y(length(t),:)];}
      end
      plot(sol(:,1),sol(:,2))}
      hold on
    data = sol;
  end
h=h+0.3;
end
```

The M-file `fpc.m` used in the main program is

```
function yprime=fpc(t,y)
%
```

```
term=1/(y(1)^2+y(2)^2)^2;
%
yprime=[1-(y(1)^2-y(2)^2)*term; -2*y(1)*y(2)*term];
```

Problems

1. Use either of the programs in the preceding examples to obtain Figures 7.10 or 7.11.

2. Use the animation capabilities of either *Mathematica* or *MATLAB* to animate the snapshots in Figures 7.10 or 7.11.

3. Modify the programs in Examples 7.10.1 and 7.10.2 appropriately and apply them to the following set of equations.

   (a) $x' = y$, $y' = -x$

   (b) $x' = \dfrac{y}{\sqrt{x^2+y^2}}$, $y' = -\dfrac{x}{\sqrt{x^2+y^2}}$

   (c) $x' = y - x$, $y' = -x + y$

   (d) $x' = \sin y \cos x$, $y' = -\sin x$

   (e) $x' = \ln |y|$, $y' = -x$

   (f) The following set of equations model the flow in a semi-infinite bay occupying the region $D = \{(x,y)| \, x < 0, \quad 0 < y < 1\}$:

   $$x' = -0.01\pi \sinh \pi x \cos \pi y, \quad y' = 0.01\pi \cosh \pi x \sin \pi y. \quad (7.230)$$

   Start with a parcel of fluids contained in a disc of radius 0.1 centered at $(-3, 0.3)$ and draw the snapshots of its deformation at later times.

7.11 Stability of Differential Equations and Systems

Most of the techniques and methods we have developed so far give us *quantitative* information about the solutions of a differential equation or system. For a small class of systems this information is exact since we are able to determine explicitly the solution to an initial value problem. We have seen, however, that most equations of interest are nonlinear and require approximate methods. In this section we seek methods that give us *qualitative* information about a differential system, a kind of information that may not lead to the determination of a solution, exact or approximate, but will add to our understanding of how a solution will behave as time evolves.

We confine most of our discussion to first-order systems consisting of two equations, although the main ideas are readily extended to smaller or larger systems. Let x and y satisfy the system

$$\frac{dx}{dt} = f(t,x,y), \quad \frac{dy}{dt} = g(t,x,y). \quad (7.231)$$

This system is called **autonomuous** if f and g do not depend on t, otherwise it is called **nonautonomous**. The great majority of physical systems we encounter are autonomous since the laws of physics (such as Newton's second law or Kirchhoff's law) are time independent. Nonautonomous systems come about often as a result of time-dependent forcing terms or special manipulations of the equations (such as linearization of the equations about a time-dependent solution). We call (7.231) a T-periodic system when f and g depend periodically on t with period T

By a **trajectory** of the system (7.231) we mean the set of points $(t, x(t), y(t))$ where the pair $(x(t), y(t))$ is a solution pair of (7.231) satisfying the initial conditions (x_0, y_0). By an **orbit** or a **path** of a solution we mean the projection of a trajectory in the xy-plane, that is, the set of points $(x(t), y(t))$. For the most part the illustrations of solutions we have made in the past sections and chapters have been of the orbits of a differential system and not its trajectories. This interpretation agrees with our intuitive outlook on the physical world where a motion occurs in the **physical space** R^2 (for a two-dimensional system) and time is a parameter, and hence graphing orbits is a device designed to track the different frames of this physical space.

The uniqueness theorem of Section 7.3 guarantees that under reasonable hypotheses on f and g different trajectories of (7.231) do not intersect in the (t, x, y) space. A remarkable fact is that this statement remains valid for orbits when we deal with autonomous systems. This is the result of the next lemma.

Lemma 7.11.1

Let f and g in (7.231) be independent of t and continuously differentiable in x and y. Let $(x(t), y(t))$ be a solution pair of the system

$$\frac{dx}{dt} = f(x, y), \quad \frac{dy}{dt} = g(x, y). \tag{7.232}$$

Then the pair $(x(t - c), y(t - c))$ is also a solution pair for any $c \in R$. Moreover, either two orbits of (7.232) coincide in the interval (a, b) or they never intersect.

Proof: Let (x, y) be a solution pair for (7.232). Define $(x_c(t), y_c(t)) \equiv (x(t - c), y(t - c))$. Evidently

$$\frac{dx_c(t)}{dt} = \frac{dx(t - c)}{dt} = f(x(t - c), y(t - c)) = f(x_c(t), y_c(t)),$$

with a similar statement for y_c. Hence, (x_c, y_c) is a solution pair for (7.232). Note that (x_c, y_c) has the same orbit as (x, y).

Now, let $(x_1(t), y_1(t))$ and $(x_2(t), y_2(t))$ be two solution pairs of (7.232). Suppose that their orbits intersect at $\mathbf{z}$. Hence, there are at least two values of $t \in (a, b)$, denoted by t_1 and t_2, such that

$$(x_1(t_1), y_1(t_1)) = (x_2(t_2), y_2(t_2)) = \mathbf{z}. \tag{7.233}$$

Let $c = t_2 - t_1$. Using the previous argument we see that $(u(t), v(t)) \equiv (x_1(t - c), y_1(t - c))$ is a solution of (7.232) and that (u, v) and (x_1, y_1) have the same orbits. But $u(t_2) = x_1(t_2 - c) = x_1(t_1) = x_2(t_2)$, by (7.233), and, similarly, $v(t_2) = x_2(t_2)$. Thus, the pairs (u, v) and (x_2, y_2) satisfy the same equations and initial conditions. Since solutions of (7.232) are unique, these pairs must be identical for all $t \in (a, b)$ and, hence, have the same orbits. Thus, the pairs (x_1, y_1) and (x_2, y_2) must have the same orbits. This completes the proof of the lemma.

Definition 7.11.1

The **phase portrait** *of the autonomous system (7.232) is a collection of its orbits in* R^2.

Definition 7.11.2

We say a point $(a, b) \in R^2$ *is a* **critical point** *of the autonomous system (7.232) if*

$$f(a, b) = 0, \quad g(a, b) = 0. \tag{7.234}$$

In that case the solution $(x(t), y(t)) \equiv (a, b)$ *is called an* **equilibrium** *solution of (7.232).*

Example 7.11.1

Consider the system

$$x' = x - y, \quad y' = x + y. \tag{7.235}$$

Here f and g are the right-hand sides of (7.235). The critical points of this system satisfy the simultaneous system

$$x - y = 0, \quad x + y = 0, \tag{7.236}$$

whose only solution is $(0, 0)$. Thus, there is only one equilibrium solution to (7.235), namely, the trivial solution $(x(t), y(t)) \equiv (0, 0)$.

Example 7.11.2

Consider the second-order equation

$$x'' + \sin x = 0. \tag{7.237}$$

This equation models oscillations in a pendulum. Here $x(t)$ measures the angle that the pendulum makes with its "equilibrium" or rest position at any time t. We begin our analysis by writing (7.237) as a first-order system. Let $y = x'$, so that

$$x' = y, \quad y' = -\sin x. \tag{7.238}$$

Critical points of (7.238) satisfy $y = 0$ and $\sin x = 0$. These are the points $(n\pi, 0) \in R^2$, where n is an integer. Note that when $n = 0$, the equilibrium solution $(x(t), y(t)) = (0, 0)$ corresponds to the pendulum standing still and hanging down, while when $n = 1$ the equilibrium solution corresponds to $(x(t), y(t)) = (\pi, 0)$, which has the physical interpretation of the pendulum standing still but in its upright position.

The previous example and the physical representation of each equilibrium solution brings us to the concept of stability of solutions. Clearly the second equilibrium solution, $(\pi, 0)$, is "unstable" physically, in that a small perturbation of this solution (a gust blowing at the pendulum, for instance) will cause the pendulum to move far away from that position. Whereas, the first equilibrium solution, $(0, 0)$, is "stable," in that small perturbations of it will cause the pendulum to move away from this position but not by an appreciable amount. We make the definition of stability precise in the following statement.

Definition 7.11.3 (Stability)

*Let $\mathbf{x}_0$ be a critical point of the differential system (7.232). The equilibrium solution associated with $\mathbf{x}_0$ is said to be **stable** if for every $\epsilon > 0$ there is $\delta > 0$ such that if*

$$||\mathbf{z}_0 - \mathbf{x}_0|| < \delta, \tag{7.239}$$

then

$$||\mathbf{x}(t) - \mathbf{x}_0|| < \epsilon \tag{7.240}$$

*for all $t > t_0$. Here $\mathbf{x}(t_0) = \mathbf{z}_0$. An equilibrium point is **unstable** if it is not stable.*

One way to interpret (7.240) is as follows. A critical point $\mathbf{x}_0$ is stable if no matter how small a neighborhood of this point we choose (referring to the ϵ inequality), we can find yet another (possibly smaller) neighborhood of $\mathbf{x}_0$ such that if we follow the trajectory of the solution through any initial data in the latter neighborhood, this trajectory will remain in the former neighborhood for all forward time (cf. Figure 7.12). In Figure 7.12 the ϵ neighborhood is denoted by U and the δ neighborhood by D.

An important feature of this definition to keep in mind is that (7.240) must hold for *all t* greater than t_0 and not just a finite interval containing the initial time. In fact the definition of stability is closely related to the definition of continuity of a solution as a function of its initial data. Let $\mathbf{x}(t, t_0, \mathbf{x}_0)$ denote the solution to (7.232), where the dependence of the solution on t_0 and $\mathbf{x}_0$ is explicitly expressed and, in particular, $\mathbf{x}(t_0, t_0, \mathbf{x}_0) = \mathbf{x}_0$. When $\mathbf{x}_0$ is stable, Definition 7.11.3 states considerably more than the fact that $\mathbf{x}$ is continuous as a function of its variables. Inequality (7.240) must hold for all forward time for $\mathbf{x}_0$ to be stable, while the continuity of $\mathbf{x}$ on $\mathbf{x}_0$ requires that (7.240) holds in a neighborhood of t_0. In the language of **dynamical systems**, when $\mathbf{x}$ is continuous as a function of t, t_0, and $\mathbf{x}_0$, the system of differential equations is said to **depend continuously** on its initial data. We note that according to the existence and uniqueness theorem of Section 7.3, under rather mild assumptions on the nonlinearities, a

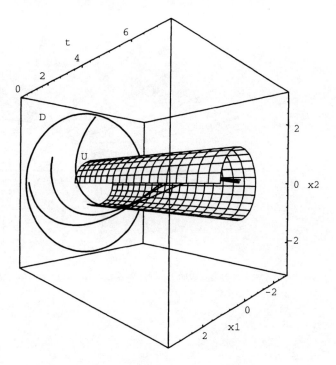

Figure 7.12: The definition of stability.

solution of a system of differential equations depends continuously on its initial data. As we will see later, the concept of stability imposes considerably more restrictive conditions on the nonlinearities in a system.

One way to determine whether an equilibrium point $\mathbf{x}_0$ of an autonomous system is stable or not is to find all solutions of the system and study the left-hand side of (7.233) as an explicit function of t. This approach is often impractical since we do not expect to find the general solution of nonlinear systems or linear systems with nonconstant coefficients. In the following example we will see, however, that elements of Definition 7.11.3 can be verified when the system of equations is linear with constant coefficients.

Example 7.11.3

Consider the system

$$x_1' = -\frac{3}{2}x_1 + \frac{1}{2}x_2 - 2, \quad x_2' = \frac{1}{2}x_1 - \frac{3}{2}x_2 + 2. \tag{7.241}$$

This system has only one equilibrium point

$$\mathbf{x}_0 = \begin{bmatrix} -1 \\ 1 \end{bmatrix}, \tag{7.242}$$

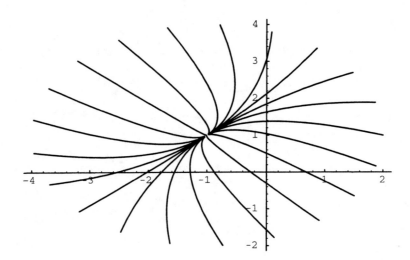

Figure 7.13: Orbits of the system of differential equations $x_1' = -\frac{3}{2}x_1 + \frac{1}{2}x_2 - 2$, $\quad x_2' = \frac{1}{2}x_1 - \frac{3}{2}x_2 + 2$.

while its general solution is

$$x_1(t) = c_1 e^{-t} + c_2 e^{-2t} - 1, \quad x_2(t) = c_1 e^{-t} - c_2 e^{-2t} + 1. \tag{7.243}$$

Let $t_0 = 0$. Let $\mathbf{z}_0 = \begin{bmatrix} z_1 \\ z_2 \end{bmatrix}$ be a typical initial data near $\mathbf{x}_0$. It follows from (7.243) that $c_1 = \frac{1}{2}(z_1 + z_2)$ and $c_2 = \frac{1}{2}(z_1 - z_2) + 1$. Next, we compute the left-hand side of (7.240):

$$||\mathbf{x} - \mathbf{x}_0||^2 = (c_1 e^{-t} + c_2 e^{-2t})^2 + (c_1 e^{-t} - c_2 e^{-2t})^2 = (c_1^2 + c_2^2)(e^{-2t} + e^{-4t}). \tag{7.244}$$

Similarly, the left-hand side of (7.239) is

$$||\mathbf{z}_0 - \mathbf{x}_0||^2 = 2(c_1^2 + c_2)^2. \tag{7.245}$$

Putting (7.244) and (7.245) together, we have

$$||\mathbf{x} - \mathbf{x}_0|| = \frac{1}{\sqrt{2}}||\mathbf{z}_0 - \mathbf{x}_0||\sqrt{e^{-2t} + e^{-4t}}. \tag{7.246}$$

The $\epsilon - \delta$ definition of stability is satisfied if we choose

$$\delta = \epsilon, \tag{7.247}$$

since $e^{-2t} + e^{-4t}$ is less than 2 for $t > 0$. Hence, the equilibrium point $\mathbf{x}_0$ is stable.

Figure 7.13 shows some of the orbits of (7.241). It is clear from this figure that not only is the equilibrium solution $(-1, 1)$ stable, all solutions of this system converge to this point as time approaches infinity. When a system satisfies the property that all solutions in a neighborhood of an equilibrium point converge to that point (in addition to the equilibrium point being stable), we say that the equilibrium point is asymptotically stable. We have the following definition.

Definition 7.11.4 (Asymptotic Stability)

An equilibrium point $\mathbf{x}_0$ *of (7.232) is said to be* **asymptotically stable** *if, in addition to being stable, it satisfies*

$$||\mathbf{x}(t) - \mathbf{x}_0|| \to 0. \tag{7.248}$$

In Example 7.11.3 we were fortunate enough to be able to find the general solution to (7.241) and from it deduce that the equilbrium point $(-1,1)$ is asymptotically stable. In the next section we develop a different technique that allows us to reach the same conclusion without actually determining the exact solution. This method is particularly helpful when it is combined with a general theorem about the stability of nonlinear systems.

Problems

1. Find all equilibrium solutions of the following systems and equations. In each case either use NDSolve of *Mathematica* or ode45 of *Matlab* and draw several trajectories in a small neighborhood of each equilibrium point and determine formally its state of stability.

   (a) $x_1' = 2x_1 - 3x_2, \quad x_2' = 3x_1 - 7x_2$

   (b) $x_1' = 3x_1 + x_2, \quad x_2' = -2x_1 + x_2$

   (c) $x_1' == 0.1x_1 + x_2, \quad x_2' = -x_1 - 0.1x_2$

   (d) $x'' + \sin x = 0$

   (e) $y'' + 0.1y + \sin y = 0$

2. Follow the procedure outlined in Example 7.11.3 and show that the equilibrium solution of the following equations are stable.

   (a) $x_1' = x_2, \quad x_2' = -x_1$

   (b) $x_1' = x_1 - 2x_2, \quad x_2' = 2x_1 - 3x_2$

   (c) $x_1' = x_2, \quad x_2' = -x_1 - x_2$

3. Find all equilibrium solutions of the systems in Examples 7.2.1, 7.2.3, 7.2.4, and 7.2.5. Use NDSolve or ode45 and determine formally which of the equilibria is stable.

7.12 Stability of Linear Systems

In Example 7.11.3 we studied a system of linear equations whose coefficients were constant. We were able to apply the definition of stability directly to this system since we were successful in determining its general solution in closed form. If we look closely at this solution, it becomes apparent that the (asymptotic) stability of the equilibrium solution depends on the growth of the exponential terms in the general solution; in this case, since they are both exponentially decaying, one expects that all solutions eventually converge to the equilibrium solution.

Was it necessary to determine the general solution of Example 7.11.1 in such detail to deduce that all orbits must approach $(-1, 1)$? The answer is no: We know from previous experience that the difference between any solution of (7.241) and the equilibrium solution is a solution to the homogeneous part of this system. Moreover, the general solution of the latter is spanned by exponential functions whose exponents are the eigenvalues of the matrix of coefficients, in this case, the matrix

$$A = \begin{bmatrix} -\frac{3}{2} & \frac{1}{2} \\ \frac{1}{2} & -\frac{3}{2} \end{bmatrix}. \tag{7.249}$$

The eigenvalues of A are -1 and -2, which attest to the fact that the difference between any solution of (7.241) and its equilibrium solution must approach zero as t goes to infinity. This scenario is typical for linear systems, which we summarize after a definition.

Definition 7.12.1
We say a critical point $\mathbf{x}_0$ of the autonomous system (7.232) is **isolated** *if there is a neighborhood of $\mathbf{x}_0$ in which $\mathbf{x}_0$ is the only critical point.*

Example 7.12.1
Consider the linear system

$$\mathbf{x}' = A\mathbf{x}, \tag{7.250}$$

where

$$A = \begin{bmatrix} 0 & 1 \\ -1 & 0 \end{bmatrix}.$$

Since A is nonsingular, the only solution to $A\mathbf{x} = \mathbf{0}$ is $\mathbf{x} = \mathbf{0}$. Thus, the equilibrium solution $\mathbf{x}_0 = \mathbf{0}$ is an isolated critical point of the system. On the other hand, suppose that A is given by

$$A = \begin{bmatrix} 0 & 0 \\ 0 & 1 \end{bmatrix}.$$

Equation $A\mathbf{x} = \mathbf{0}$ has infinitely many solutions given by $\mathbf{x} = \begin{bmatrix} c \\ 0 \end{bmatrix}$, so $\mathbf{x}_0 = \mathbf{0}$ is not isolated.

Theorem 7.12.1

Let $\mathbf{x}_0$ be an isolated critical point of

$$\mathbf{x}' = A\mathbf{x} + \mathbf{f} \qquad (7.251)$$

(i.e., $\mathbf{x}_0 = -A^{-1}\mathbf{f}$). Then one of the following cases occurs (note that because A is invertible, zero is not an eigenvalue of A).

1. *A has two negative real eigenvalues. In that case $\mathbf{x}_0$ is an asymptotically stable equilibrium point and is called an **attracting node** or a **sink**.*

2. *Eigenvalues of A are complex conjugates of each other with negative real part. Then $\mathbf{x}_0$ is asymptotically stable and is called an **attracting spiral**. If the real part is positive, then $\mathbf{x}_0$ is unstable.*

3. *A has purely imaginary roots. Then $\mathbf{x}_0$ is stable and is called a **center**.*

4. *A has multiple eigenvalues λ. If $\lambda > 0$, then $\mathbf{x}_0$ is unstable. If $\lambda < 0$, then $\mathbf{x}_0$ is asymptotically stable. The nature of this stability depends on how many eigenvectors A has. If A has two eigenvectors associated with λ, then $\mathbf{x}_0$ is an **attracting node**. If A has only a single eigenvalue associated with λ, then $\mathbf{x}_0$ is called an **improper node**.*

5. *If A has a positive eigenvalue, then $\mathbf{x}_0$ is unstable. If one eigenvalue is positive and the other negative, the equilibrium solution is called a **saddle**. If both eigenvalues are positive, the equilibrium solution is called a **source**.*

Proof: Without loss of generality, we can assume that the equilibrium solution $\mathbf{x}_0$ is the zero solution and that $\mathbf{f} = \mathbf{0}$. To see this, note that $\mathbf{x}_0 = -A^{-1}\mathbf{f}$ and so $\mathbf{u} = \mathbf{x} - \mathbf{x}_0$ satisfies

$$\mathbf{u}' = \mathbf{x}' = A\mathbf{x} + \mathbf{f} = A\mathbf{u} + A\mathbf{x}_0 + \mathbf{f} = A\mathbf{u}. \qquad (7.252)$$

Thus, the difference between any solution $\mathbf{x}$ and $\mathbf{x}_0$ satisfies the homogeneous system. The behavior of the general solution of the latter system depends entirely on A and its eigenvalues. We leave it to the reader to extract the separate cases of this theorem by considering all of the different possibilities that the eigenvalues of an invertible matrix can take.

Example 7.12.2

Consider the system (7.251) with $\mathbf{f} = \mathbf{0}$ and A given by

$$A = \begin{bmatrix} -2 & 1 \\ 1 & -2 \end{bmatrix}. \qquad (7.253)$$

The eigenvalues of A are $\lambda = -1$ and -3. The equilibrium solution of the system $\mathbf{x}' = A\mathbf{x}$, which is $\mathbf{x}_0 = \mathbf{0}$, is therefore asympotically stable (cf. Figure 7.14).

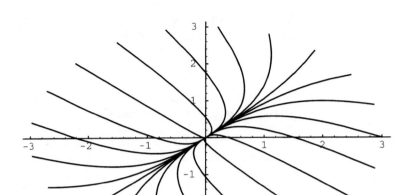

Figure 7.14: Phase portrait of $x_1' = -2x_1 + x_2, \quad x_2' = x_1 - 2x_2$.

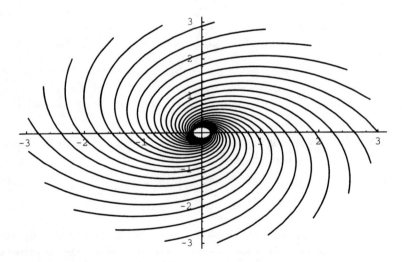

Figure 7.15: Phase portrait of $x_1' = -x_1 - 2x_2, \quad x_2' = 2x_1 - x_2$.

Next, let A be defined by

$$A = \begin{bmatrix} -1 & -2 \\ 2 & -1 \end{bmatrix}. \tag{7.254}$$

Its eigenvalues are $-1 \pm 2i$. The zero equilibrium solution is then asymptotically stable and an attracting spiral (cf. Figure 7.15).

We will see examples of other types of stability and instability in the ensuing problems. In the next section we consider the effect of adding nonlinearities to linear systems with the stability behavior listed in Theorem 7.11.1.

Problems

1. Identify the stability type of the equilibrium solution of the following linear systems. Use NDSolve or ode45 and draw some of the representative orbits of each system in a neighborhood of the equilibrium solution.

   (a) $x' = y, \quad y' = -2y - x$

   (b) $x_1' = 4x_1 - x_2, \quad x_2' = x_1 + x_2$

   (c) $x' = y, \quad y' = y - x$

   (d) $x' = y - 1, \quad y' = 2y + x - 2$

   (e) $x_1' = x_2 + 1, \quad x_2' = -x_1 - 1$

2. Consider the system $x_1' = x_2$, $x_2' = ax_1 + bx_2$. Find all values of a and b for which $\mathbf{x}_0 = \mathbf{0}$ is a:

   (a) sink

   (b) source

   (c) center

   (d) saddle

   (e) an improper node

   (f) an attracting spiral.

3. Consider the system $x' = y$, $y' = -x - \epsilon y$.

   (a) Show that when $\epsilon = 0$ the equilibrium solution $\mathbf{x}_0 = \mathbf{0}$ is a center, and draw the graphs of a few representative orbits of this system in a neighborhood of the origin.

   (b) Show that when $0 < \epsilon < 2$ the equilibrium solution $\mathbf{x}_0 = \mathbf{0}$ is an attracting spiral. Draw the graphs of a few representative orbits of this system in a neighborhood of the origin.

   (c) Show that when $-2 < \epsilon < 0$ the equilibrium solution $\mathbf{x}_0 = \mathbf{0}$ is an unstable spiral center, and draw the graphs of a few representative orbits of this system in a neighborhood of the origin.

7.13 Stability of Nonlinear Systems

Some of the stability results we listed in the previous section generalize to nonlinear systems as long as the nonlinearities are "small" compared with the linear trems. Small perturbations of linear equations may come about because there is a small parameter in the model that had been ignored in the first analysis of the model or because the circumstances of external forces have changed. Forces due to friction in a mass-spring model or the viscosity of the fluid in a hydrodynamic problem are examples of internal constraints that are often ignored at the first stage of analysis, but brought to bear at a later stage of a model's study. Adding small time-periodic forces to the Rayleigh–Bénard flow, to model certain external disturbances on the experimental apparatus, is an example of including external forces that may have been ignored in the first go around of the analysis.

Before proceeding with the general discussion of stability of nonlinear systems, let us note that the eigenvalues of a 2×2 matrix A depend continuously on the coefficients of A. This fact can be easily verified by the quadratic formula. Let $A = \begin{bmatrix} a & b \\ c & d \end{bmatrix}$. Its eigenvalues are determined from the coefficients of A through the relations

$$\lambda = \frac{a + d \pm \sqrt{(a + d)^2 - 4(ad - bc)}}{2}. \tag{7.255}$$

Thus, each λ is a continuous function of a, b, c, and d; and small perturbations of these coefficients will induce small variations in λ. An important consequence of this continuity is that eigenvalues that are either positive or negative will remain positive or negative under small enough perturbations. Similarly, eigenvalues that have negative or positive real parts retain these properties after small perturbations of A. It turns out that these considerations help us deduce the stability properties of a nonlinear system by investigating the sign of the eigenvalues of the *linearization* of the system of differential equations about an equilibrium solution.

Let

$$x'(t) = f(x, y), \quad y'(t) = g(x, y) \tag{7.256}$$

be a system of differential equation. Let $\mathbf{x}_0 = \begin{bmatrix} x_0 \\ y_0 \end{bmatrix}$ be an equilibrium solution of (7.256). We assume that f and g are smooth enough at $\mathbf{x}_0$ so that we can expand these functions about this point to any order we wish. We use the Taylor series of f and g about $\mathbf{x}_0$ and write (7.256) in the form

$$\begin{bmatrix} u' \\ v' \end{bmatrix} = A \begin{bmatrix} u \\ v \end{bmatrix} + \text{h.o.t.}, \tag{7.257}$$

where

$$A = \begin{bmatrix} \dfrac{\partial f}{\partial x} & \dfrac{\partial f}{\partial y} \\[2mm] \dfrac{\partial g}{\partial x} & \dfrac{\partial g}{\partial y} \end{bmatrix}, \tag{7.258}$$

evaluated at the equilibrium point $\mathbf{x}_0$, $u = x - x_0$ and $v = y - y_0$, and h.o.t. stands for higher order terms in u and v. The following theorem classifies the conditions on f and g under which the stability properties of the equilibrium solution $\mathbf{x}_0$ can be deduced from the stability properties of the linear system

$$\mathbf{x}' = A\mathbf{x}. \tag{7.259}$$

We call (7.259) the **linearization** of (7.256).

Theorem 7.13.1 (Liapunov's Theorem)

Let $\mathbf{x}_0$ be an equilibrium solution of (7.256). Let $\mathbf{x} = \mathbf{0}$ be an isolated critical point of (7.259).

1. *Suppose that the eigenvalues of A have negative real parts. Then $\mathbf{x}_0$ is an asymptotically stable solution of the nonlinear system (7.256).*

2. *Suppose that A has at least one eigenvalue with positive real part. Then $\mathbf{x}_0$ is an unstable solution of (7.256).*

3. *Suppose that A has only eigenvalues with zero real parts. Then the stability of $\mathbf{x}_0$ cannot be determined from the linear system (7.259).*

The proof of this theorem can be found in Reference [1].

Example 7.13.1
Consider the equation

$$x'' + 0.1x' + \sin x = 0, \tag{7.260}$$

which models the oscillations of a damped pendulum. First, we transform this equation to a first-order system by letting $x' = y$:

$$x' = y, \quad y' = -0.1y - \sin x. \tag{7.261}$$

Two of (7.261)'s equilibria are $\begin{bmatrix} 0 \\ 0 \end{bmatrix}$ and $\begin{bmatrix} \pi \\ 0 \end{bmatrix}$. The linearization of (7.261) about the first equilibrium point is

$$x' = y, \quad y' = -0.1y - x \tag{7.262}$$

with

$$A = \begin{bmatrix} 0 & 1 \\ -1 & -0.1 \end{bmatrix}, \tag{7.263}$$

whose eigenvalues are $-0.05 \pm 0.998749i$. Therefore, the linear system associated with the first equilibrium solution has eigenvalues with negative real parts, which implies that the zero solution of the nonlinear system is asymptotically stable.

Next, we linearize (7.261) about the second equilibrium solution $\mathbf{x}_0 = \begin{bmatrix} \pi \\ 0 \end{bmatrix}$.
Here A is given by

$$A = \begin{bmatrix} 0 & 1 \\ 1 & -0.1 \end{bmatrix}, \tag{7.264}$$

whose eigenvalues are -1.05125 and 0.951249. Since one of the eigenvalues of the linearization is positive, the linear system as well as the nonlinear system are unstable at this equilibrium point.

In the previous example the two equilibria have concrete physical interpretations: the first equilibrium solution describes the pendulum at rest pointing downward, while the second equilibrium corresponds to the pendulum at rest but pointing upward. Our analysis shows that the first equilibrium solution is asymptotically stable, in that a small perturbation of the pendulum away from this particular rest position is eventually damped out and the pendulum returns to rest as time approaches infinity. The situation for the second equilibrium solution is quite different in that a small perturbation causes the pendulum to move far away from its rest position and, perhaps, converge to the first (stable) equilibrium solution.

We note that Theorem 7.13.1 only addresses systems whose linearizations have eigenvalues with real parts that are either positive or negative. This theorem does not give any information about systems of differential equations whose linearizations have centers about the equilibrium solution. In general, a small perturbation of a center may lead to an attracting spiral or an unstable spiral, depending on how the nonlinearities affect the evolution of systems. Some of the most interesting systems in nature have mathematical models that fall in this uncertain category and the behavior of their solutions depend crucially on the sign of the physical parameters in these systems. We will give examples of such systems in a project in the next chapter when we touch on the subject of **bifurcation** in vector fields.

Problems

1. Find all equilibrium solutions of the following systems and test for their stability. Use *Mathematica* or *MATLAB* in the calculation of the equilibria, the linearization about these points, and the computation of their eigenvalues. Draw the graphs of a few orbits with initial data near the equilibria and compare the overall behavior of these graphs with the qualitative behavior predicted by the stability of the equilibria.

   (a) $x' = y \quad y' = -x - y^3$

   (b) $x' = x - y - 1, \quad y' = 3x - y - x^3 - y^3$

   (c) $x' = y - xy(x^2 + y^2), \quad y' = -x - xy(x^2 + y^2)$

   (d) $x' = x + y(1 - x), \quad y' = -y + xy$

2. Find all equilibria of the fluid flow past the cylinder defined in Example 7.2.3. What is the stability state of each equilibrium point?

3. Consider the system of differential equations

$$x' = ax(1 - by), \quad y' = cy(1 - dx) \qquad (7.265)$$

   where a, b, c, and d are real numbers.

   (a) Determine all equilibrium points of this system.

   (b) Determine the state of stability of each equilibrium point in terms of the parameters in the system.

   (c) Let $a = b = d = 1$, and $c = -1$. Use NDSolve or ode45 to construct the phase portrait of this system in the first quadrant. Are there any closed curves in this part of the plane? Assuming that x and y are populations of two competing species (predators and preys), what is the physical significance of a closed curve in the phase portrait?

(d) Let $a = c = 1$. Find all values of b and d for which the second equilibrium solution is stable.

4. Consider the Van der Pol equation

$$x'' + \epsilon(x^2 - 1)x' + x = 0$$

where ϵ is a non-negative parameter.

(a) Write this equation as a first-order system and classify the stability of all of its equilibrium points in terms of ϵ.

(b) Let $\epsilon = 0$. Plot the phase portrait of this equation with initial conditions near the origin as well as far away from it. How would you classify the behavior of the solutions that start out far from the origin?

(c) Let $\epsilon = 1$. Plot the phase portrait of this equation with initial conditions near the origin and far away from it. How would you classify the behavior of the solutions that start out near the origin? Far from the origin?

5. Consider the Duffing equation

$$x'' + x' + kx + \alpha x^3 = 0$$

where k and β are real numbers.

(a) Identify the equilibrium points of this system and classify their stability in terms of k and α.

(b) Plot the phase portrait of the Duffing equation with $k = -1$ and $\alpha = 1$. How do the solutions with initial conditions near the origin behave? How about the ones that are initially far from the origin?

6. Consider the system of differential equations

$$x' = x(a_1 - a_2 x - a_2 y), \quad y' = y(b_1 - b_2 x - b_3 y)$$

where the a_is and b_is are real numbers.

(a) Determine all equilibrium points of this system.

(b) With $a_i > 0$ and $b_i > 0$, classify the stability of each equilibrium point.

(c) Let $a_1 = 17$, $a_2 = 3$, $a_3 = 5$, $b_1 = 33$, $b_2 = 4$, and $b_3 = 3$. Classify the state of stability of all equilibrium points. Plot the phase portrait of this system in a large enough region that includes all equilibria.

7.14 Project A—NDSolve and Phase Portraits

In this project we develop an interactive program in *Mathematica* that draws the phase portrait of a first-order autonomous system of differential equations in two variable. The main ingredients of the program are the NDSolve and the Input commands of *Mathematica*. The program is as follows:

```
eqns = Input["\n\n Input the differential equations\n
for example {x'[t] == x[t] - y[t], y'[t] == x[t] + 2 y[t]}\n\n"];
data = Input["\n\n Input the initial points \n\n
  for example {{1.1, 1.2}} for a single point,
  {{1, 1}, {1, -1}} for a list of points,
  or Table[{-1 + 3 Cos[t], 1 + 3 Sin[t]},
  {t, 0, 2 Pi, 0.3}] for a table of points.\n\n"];
tfinal = Input["\n\n Input the value of t_final
      (for example 3 or -5)\n"];
sol = Table[NDSolve[Flatten[{eqns, x[0] == data[[i,1]],
      y[0] == data[[i,2]]}],
  {x, y}, {t, 0, tfinal}], {i, 1, Length[data]}];
solution = Flatten[sol, 1];
output = ParametricPlot[Evaluate[{x[t], y[t]} /. solution],
      {t, 0, tfinal}]
```

Figure 7.16 shows the output of this program for the Van der Pol equation

$$x'' + (x^2 - 1)x' + x = 0$$

with initial data

$$\{(0.1, 0.1), (0.2, 0.2), (1, 1), (2, 2), (3, 3)\}$$

and the tfinal value of 10.

1. Create a file called **phase.m** containing the above program. Then input the program to *Mathematica* and test it on the following systems.

   (a) $x' = y$, $y' = -0.1y - \sin x$

   (b) $x' = x + 2y + 2x^2 - 3y^2$, $y' = x - y + x^2 + y^2$

   (c) $x' = -y + x(1 - x^2 - y^2)$, $x + y(1 - x^2 - y^2)$

2. Add appropriate commands to the program to alert the user that the output is stored in **output** and is available for printing.

3. Add an appropriate command that allows the user to add a label to the output.

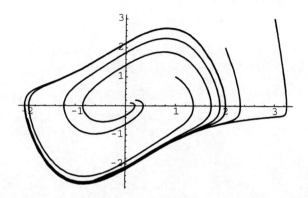

Figure 7.16: The phase portrait of the Van der Pol equation.

7.15 Project B—Laplace Transform and Systems

We saw in Chapter 5 that the Laplace transform is an effective tool in solving linear ODEs with constant coefficients, especially when combined with *Mathematica*. In this project we pursue this topic further in the context of first-order systems.

Consider the initial-value problem

$$\mathbf{x}' = A\mathbf{x} + \mathbf{b}, \quad \mathbf{x}|_{t=0} = \mathbf{x}_0. \tag{7.266}$$

We begin by taking the Laplace transform of both sides of (7.266), recalling that this transform is a linear operator:

$$L[\mathbf{x}'] = AL[\mathbf{x}] + L[\mathbf{b}], \tag{7.267}$$

where by $L[\mathbf{x}]$ we mean the vector whose ith entry is the Laplace transform of the ith component of $\mathbf{x}$. We recall that $L[f'] = sL[f] - f(0)$. Therefore, (7.267) reduces to

$$(sI - A)L[\mathbf{x}] = \mathbf{x}_0 + L[\mathbf{b}]. \tag{7.268}$$

Assuming that $(sI - A)$ is invertible, we end up with

$$L[\mathbf{x}] = (sI - A)^{-1}\mathbf{c}, \tag{7.269}$$

where $\mathbf{c}$ is the right-hand side of (7.268). The solution $\mathbf{x}$ to the original system of equations is then obtained by determining the inverse Laplace transform of (7.269).

This procedure is often too cumbersome to carry out by hand but is well suited for a program like *Mathematica*. One implements this algorithm for the specific system

$$x' = y, \quad y' = -x - y, \quad x(0) = 1, \quad y(0) = -1 \tag{7.270}$$

via the following program. Start a new session of *Mathematica* and enter

```
<<Calculus'LaplaceTransform';
```

The following commands take the necessary parameters from (7.270) and return the solution to this equation.

```
A = {{0, 1}, {-1, -1}};   x0 = {1, -1};
F[s_] = Inverse[s IdentityMatrix[2] - A] . x0;
x[t_] = InverseLaplaceTransform[F[s], s, t]
```

1. Test this program on the 2×2 systems $\mathbf{x}' = A\mathbf{x}$ whose matrix of coefficients and initial data are given by

   (a) $A = \begin{bmatrix} 1 & -3 \\ 3 & 7 \end{bmatrix}$, $\mathbf{x}_0 = \begin{bmatrix} 1 \\ -1 \end{bmatrix}$

   (b) $A = \begin{bmatrix} 2 & 5 \\ 1 & -2 \end{bmatrix}$, $\mathbf{x}_0 = \begin{bmatrix} 1 \\ 0 \end{bmatrix}$

   (c) $A = \begin{bmatrix} -1 & -5 \\ -3 & 1 \end{bmatrix}$, $\mathbf{x}_0 = \begin{bmatrix} 0 \\ 1 \end{bmatrix}$

2. Extend the program appropriately to include forcing terms $\mathbf{b}$. Apply the new program to the systems $\mathbf{x}' = A\mathbf{x} + \mathbf{b}$ where

   (a) $A = \begin{bmatrix} -1 & 3 \\ 2 & 1 \end{bmatrix}$, $\mathbf{x}_0 = \begin{bmatrix} 1 \\ -1 \end{bmatrix}$, $\mathbf{b} = \begin{bmatrix} 1 \\ 1 \end{bmatrix}$

   (b) $A = \begin{bmatrix} 2 & -5 \\ 0 & -1 \end{bmatrix}$, $\mathbf{x}_0 = \begin{bmatrix} 1 \\ 0 \end{bmatrix}$, $\mathbf{b} = \begin{bmatrix} 1 \\ -3 \end{bmatrix}$

   (c) $A = \begin{bmatrix} -1 & -5 \\ -3 & 1 \end{bmatrix}$, $\mathbf{x}_0 = \begin{bmatrix} 0 \\ 0 \end{bmatrix}$, $\mathbf{b} = \begin{bmatrix} \sin t \\ \cos t \end{bmatrix}$

3. Test the limitations of this algorithm on the following 3×3 and 4×4 systems.

   (a) $A = \begin{bmatrix} 3 & -1 & 3 \\ 0 & 0 & 1 \\ 1 & 0 & 0 \end{bmatrix}$, $\mathbf{x}_0 = \begin{bmatrix} 1 \\ 0 \\ -1 \end{bmatrix}$

   (b) $A = \begin{bmatrix} 0 & 1 & 0 & 0 \\ 0 & 0 & 1 & 0 \\ 0 & 0 & 0 & 1 \\ -16 & 32 & -24 & 8 \end{bmatrix}$, $\mathbf{x}_0 = \begin{bmatrix} 1 \\ 0 \\ 0 \\ 1 \end{bmatrix}$

4. Write an interactive program in *Mathematica* that asks the user to input the matrix A and the initial conditions and returns the solution $\mathbf{x}$.

7.16 Project C—Coupled Mass-Spring Motion

In this project we study the motion of a set of bodies attached to springs, obtain the solution of the underlying system of differential equations, understand the connection between the physical parameters and the natural frequencies of the system, and ultimately animate the motion.

We concentrate here on a system of two bodies with masses M_1 and M_2 connected to three springs with constants k_1, k_2, and k_3, as described in Example 7.2.1, although the analytic and numerical computations are easily generalized to more complicated systems. The system of differential equations that govern the motion of the two masses is

$$m_1 x_1'' = -k_1 x_1 + k_2(x_2 - x_1), \quad m_2 x_2'' = -k_2(x_2 - x_1) - k_3 x_2. \qquad (7.271)$$

1. Consider the special case $m_1 = m_2 = m$ and $k_1 = k_2 = k_3 = k$. Show that (7.271) reduce to

$$mx_1'' + k(2x_1 - x_2) = 0, \quad mx_2'' + k(2x_2 - x_1) = 0. \qquad (7.272)$$

Write this system as a first-order system. Show that the eigenvalues of the system are $\pm \omega_1 i$ and $\pm \omega_2 i$, where

$$\omega_1 = \sqrt{\frac{3k}{m}}, \quad \omega_2 = \sqrt{\frac{k}{m}}. \qquad (7.273)$$

Moreover, the four eigenvectors associated with these eigenvalues are

$$\mathbf{e}_1 \begin{bmatrix} \frac{i}{\sqrt{a}} \\ 1 \\ \frac{i}{\sqrt{a}} \\ 1 \end{bmatrix}, \quad \begin{bmatrix} \frac{-i}{\sqrt{a}} \\ 1 \\ \frac{-i}{\sqrt{a}} \\ 1 \end{bmatrix}, \quad \begin{bmatrix} \frac{-i}{\sqrt{3a}} \\ -1 \\ \frac{i}{\sqrt{3a}} \\ 1 \end{bmatrix}, \quad \begin{bmatrix} \frac{i}{\sqrt{3a}} \\ -1 \\ \frac{-i}{\sqrt{3a}} \\ 1 \end{bmatrix}, \qquad (7.274)$$

where $a = \sqrt{\frac{k}{m}}$. Use *Mathematica* or *MATLAB* and verify all of these computations.

2. Show that the solution to the initial value problem (7.272) with

$$x_1(0) = A, \quad x_1'(0) = 0, \quad x_2(0) = A, \quad x_2'(0) = 0 \qquad (7.275)$$

is

$$x_1(t) = A \cos \omega_2 t, \quad x_2(t) = A \cos \omega_2 t. \qquad (7.276)$$

How do the initial data in (7.275) relate to the eigenvectors in (7.274)? What are the physical and geometric interpretations of this initial data? What are the physical and geometric interpretations of the solution (7.276)? This mode of oscillation is called the **symmetric mode**.

3. A good way to understand the geometric interpretation of the solution in (7.276) is to draw the graphs of x_1 and x_2. A better way is to animate the positions of the masses as time evolves. Here is the *Mathematica* program that accomplishes this goal for $A = 1$ and $\omega_2 = 4$. The natural length of each spring is taken to be 5.

```
A = 1; omega2 = 4; spring = 5;
x1[t_] = A Cos[omega2 t]; x2[t_] = A Cos[omega2 t];
solution[t_]={spring+x1[t], 2*spring + x2[t]};
out1 = Table[{{solution[t][[1]], 0}, {solution[t][[2]], 0}},
             {t, 0, 2Pi/omega2, 0.1}];
out2 = Table[ListPlot[out1[[i]],
        PlotRange->{{spring-2*A, 2*spring+2*A}, {-1, 1}},
        Prolog -> AbsolutePointSize[30]],
        {i, Length[out1]}];
```

Study the syntax and the logic of this program carefully. Create a file called **system.m** that contains this program and execute it into *Mathematica*.

4. In (7.275) we took the initial velocities to be zero. Prove the more general result that if

$$x_1(0) = x_2(0) \quad \text{and} \quad x_1'(0) = x_2'(0), \tag{7.277}$$

then $x_1(t) = x_2(t)$ for all t. First prove this statement analytically. Next, enter (7.272) and (7.277) into DSolve of *Mathematica*. Use the arbitrary initial data $x_1(0) = x_2(0) = A$ and $x_1'(0) = x_2'(0) = B$. Construct $x_1(t) - x_2(t)$ and observe that *Mathematica* returns the value zero.

5. The **antisymmetric mode** of oscillation is defined by the initial data

$$x_1(0) = A, \quad x_1'(0) = 0, \quad x_2(0) = -A, \quad x_2'(0) = 0. \tag{7.278}$$

Show that the solution to the initial value problem (7.272) and (7.278) is given by

$$x_1(t) = A \cos \omega_1 t, \quad x_2(t) = -A \cos \omega_1 t. \tag{7.279}$$

Is there a physical reason why the frequency of the antisymmetric mode is larger than the symmetric mode's?

6. Formulate and prove the analogue of Problem 4 for the antisymmetric case.

7. Modify the preceding *Mathematica* program and animate the solution of the antisymmetric mode.

8. Solve the initial value problem (7.272) with the initial data

$$x_1(0) = 1, \quad x_1'(0) = -1, \quad x_2(0) = -1.2, \quad x_2'(0) = 1. \tag{7.280}$$

Is the solution periodic? What is the frequency of vibration? Animate this solution. Does the motion resemble either of the symmetric or antisymmetric modes?

9. A somewhat better *Mathematica* program than the one we presented previously is a program that uses DSolve to determine the exact solution of (7.272) before animating its relevant components. Write such a program,

keeping in mind that the following line is a good way of accessing the solution (sol is written for the initial data in (7.275)). Call this new program system1.m and execute it with the parameter values of the two animation problems discussed previously.

```
sol= DSolve[{x[1]'[t] == y[1][t],
              y[1]'[t] == -2*a x[1][t] + a*x[2][t],
              x[2]'[t] == y[2][t],
              y[2]'[t] == -2*a x[2][t] + a*x[1][t],
              x[1][0] == A, y[1][0] == 0,
              x[2][0] == A, y[2][0] == 0},
              {x[1][t], y[1][t], x[2][t], y[2][t]}, t]
```

Here a is defined as before, namely, $a = \sqrt{\frac{k}{m}}$.

7.17 Project D—Systems and Resonance

We have already encountered the concept of resonance for single equations. When a mass-spring system, say, is periodically forced, resulting in a governing differential equation of the form

$$mx'' + kx = A \cos \omega t, \tag{7.281}$$

the system responds with a growing amplitude of oscillation when the input frequency ω matches the natural frequency $\sqrt{\frac{k}{m}}$ of the homogeneous system. In this project we will study the concept of resonance in the context of mass-spring systems.

We return to the mass-spring system we considered in Project C, with the important addition of the following special periodic forcing terms:

$$mx_1'' + k(2x_1 - x_2) = f_1 \cos \omega t, \quad mx_2'' + k(2x_2 - x_1) = f_2 \cos \omega t. \tag{7.282}$$

Our goal is to understand for which values of the input frequency ω the amplitude of the response of the system grows without bound.

1. Show that when f_1 and f_2 are zero, then all solutions of (7.282) are bounded, that is, $\|\mathbf{x}\| < \infty$ where $\mathbf{x} = \langle x_1, x_1', x_2, x_2' \rangle$. (Hint: Show that all solutions of this system consist of sines and cosines.)

2. We can determine the solution of an initial value problem associated with (7.282) using either the method of undertermined coefficients or the variation of parameters formula. Instead we will use DSolve of *Mathematica* (or equivalently, dsolve of *MATLAB*) to arrive at the same conclusions had we used these analytic methods. We concentrate on the special case

$$m = 1, \quad k = 4, \quad f_1 = f_2 = 1 \tag{7.283}$$

and the initial data

$$x_1(0) = 0, \quad x_2(0) = 0, \quad x_1'(0) = 0, \quad x_2'(0) = 0, \tag{7.284}$$

thus emphasizing the role of the forcing frequency ω.

3. Use `DSolve` of *Mathematica* or `dsolve` of *MATLAB* and determine the solution of (7.282)–(7.284). Plot the graph of $x_1(t)$ for $t \in (0, 20)$ as ω varies between 1 and 6 at increments of 0.3. Are there values of ω for which x_1 grows substantially more than other values? How are these values related to the natural frequencies of this system? Once you have located the resonant frequencies approximately, choose a small interval of ω about these values and a small increment and try to pinpoint what the exact value of the resonant frequencies are? How do *Mathematica* or *MATLAB* respond when you attempt to pass through the exact value of the natural frequencies? Finally, use `DSolve` or `dsolve` with ω equal to the natural frequency and plot the graph of x_1 in each case.

4. The following program in *Mathematica* finds the solution to (7.282)–(7.284) and plots $x_1(t)$ for various values of ω. Execute this program and compare its output to Figure 7.17.

```
sol=DSolve[{x[1]'[t] == y[1][t],
             y[1]'[t] == - 8 x[1][t] + 4 x[2][t] + Sin[w t],
             x[2]'[t] == y[2][t],
             y[2]'[t] == - 8 x[2][t] + 4 x[1][t] + Sin[w t],
             x[1][0] == 0, y[1][0] == 0, x[2][0] == 0,
                  y[2][0] == 0},
             {x[1][t], y[1][t], x[2][t], y[2][t]}, t];
graph = {};
x[1][t_]=Simplify[ComplexExpand[x[1][t]/.sol]];
Do[llabel=StringJoin["omega = ", ToString[w]];
   graph=Append[graph, Plot[x[1][t], {t, 0, 20},
                            PlotLabel -> llabel]],
     {w, 1.3, 2.55, 0.25}]
```

5. Apply the appropriate modification of the preceding program to the following problems and determine in each case that the approximate values of the resonant frequencies match the natural frequencies of the system.

   (a) $m = 1$, $k = 4$, $f_1 = \omega^2$, $f_2 = \omega^2$
   (b) $m = 50$, $k = 2500$, $f_1 = 0$, $f_2 = 50$
   (c) $m = 1$, $k = 100$, $f_1=1$, $f_2 = 0$

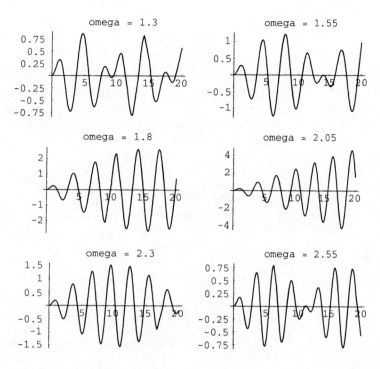

Figure 7.17: The graph of $x_1(t)$ for several values of the forcing frequency ω.

7.18 Project E—The Method of Lines in *MATLAB*

The method of lines is an effective numerical technique for solving a large number of PDEs. The main strategy in this method is to replace the spatial derivatives in a PDE by finite differences while leaving the temporal derivatives intact. This process converts such a PDE to a large system of simultaneous ODEs, which can then be addressed by the techniques we have been studying in this chapter. In this project we will discuss the details of the method of lines for the heat equation

$$u_t = \lambda u_{xx} \tag{7.285}$$

and present a program in *MATLAB* for solving the resulting system of ODEs.

The function $u(x,t)$ in (7.285) is the temperature in a heat-conducting rod at the point x and at time t. The constant λ is a lumped parameter that contains the density, specific heat, and the thermal conductivity of the material. Equation (7.285) is supplemented with the boundary conditions

$$u(0,t) = u(l,t) = 0, \tag{7.286}$$

which state that the two ends of the rod of length l are kept at temperature zero (ice), and initial data

$$u(x, 0) = u_0(x). \tag{7.287}$$

We will see in Volume II how to construct the exact solution of (7.285)–(7.287). Here, however, we are concerned with constructing a good approximation to $u(x, t)$ for $x \in (0, l)$ and $t \in (0, T)$. We begin by subdividing the interval $(0, l)$ into n equal subintervals. Let $\{x_1, x_2, \ldots, x_{n-1}\}$ be the set of $n - 1$ equidistant points in the interval $(0, l)$, with

$$h = x_i - x_{i-1}, \tag{7.288}$$

the step size. Note that $h = \frac{l}{n}$.

   We recall the formula

$$u_{xx}(x, t) = \lim_{h \to 0} \frac{u(x + h, t) - 2u(x, t) + u(x - h, t)}{h^2}. \tag{7.289}$$

Using the above center-difference approximation of u_{xx}, we approximate $u_{xx}(x_i, t)$ by

$$\frac{1}{h^2} \left(u(x_{i+1}, t) - 2u(x_i, t) + 2u_{(x_{i-1}, t)} \right). \tag{7.290}$$

Returning to the heat equation, we now replace the exact expression

$$u_t(x_i, t) = \lambda u_{xx}(x_i, t),$$

by

$$u_t(x_i, t) = \frac{1}{h^2} \left(u(x_{i+1}, t) - 2u(x_i, t) + u(x_{i-1}, t) \right). \tag{7.291}$$

where we have used (7.290) in place of $u_{xx}(x_i, t)$. We have succeeded in constructing a linear system of ODEs to approximate the original PDE (7.285). To illustrate this point better, set $n = 4$ and and let us write out the differential equations in (7.291). In this example, $h = \frac{l}{4}$ and

$$x_1 = \frac{l}{4}, \quad x_2 = \frac{l}{2}, \quad x_3 = \frac{3l}{4}. \tag{7.292}$$

Let $u_i(t) \equiv u(x_i, t)$. Then (7.291) is equivalent to

$$u_1' = \frac{16}{l^2} (u_2 - 2u_1), \quad u_2' = \frac{16}{l^2} (u_3 - 2u_2 + u_1), \quad u_3' = \frac{16}{l^2} (-2u_3 + u_2), \tag{7.293}$$

where we used the boundary conditions (7.286) to set u_0 and u_4 equal to zero. The initial data for the linear system in (7.293) is calculated from (7.287):

$$u_1(0) = u_0 \left(\frac{l}{4} \right), \quad u_2(0) = u_0 \left(\frac{l}{2} \right), \quad u_3(0) = u_0 \left(\frac{3l}{4} \right). \tag{7.294}$$

In practice we need to consider more points than just three in order for (7.291) to represent a good approximation to (7.285). It should be clear, however, that the larger n is the larger the dimension of the linear system in (7.291). We now present a program in *MATLAB* that uses **ode45** and solves this system of equations. This program is written with the following set of parameters:

$$l = 1, \quad n = 40, \quad u_0(x) = \sin \pi x, \quad \lambda = 0.01. \tag{7.295}$$

The exact solution $u(x, t)$ of (7.285) with the initial data given in (7.295) is

$$u(x, t) = e^{-\lambda \pi^2 t} \sin \pi x. \tag{7.296}$$

```
global n h matrix lambda;
nographs=5; lambda = 0.1;
n=40;h=1/n;
%
x=1/n:1/n:1-1/n;
vector1=ones(size(1:n-2));
vector2=ones(size(1:n-1));
matrix1=diag(vector1,-1);
matrix2=diag(vector1,1);
matrix3=diag(vector2);
matrix=matrix1+matrix2-2*matrix3;
%
u0=sin(pi*x);
x=[0 x 1];
%
for i=1:nographs
[t,u]=ode45('oneDheat',0.1*(i-1),0.1*i,u0,10^(-7));
approximate=[0 u(length(t),:) 0];
subplot(211)
plot(x,approximate)
title(['1D Heat Equation, Method of Line, n=',num2str(n)]);
hold on
subplot(212)
exact=exp(-lambda*pi*pi*t(length(t)))*sin(pi*x);
plot(x,exact-approximate)
xlabel('Error')
hold on
u0 = u(length(t),:);
end
hold off
```

This M-file calls on another M-file, identified as **oneDheat.m** in the program, which contains the definition of the right-hand side of the equivalent of (7.293) when $n = 40$. The listing of **oneDheat.m** is

```
function uprime=oneDheat(t,u);
%
global n h matrix lambda;
uprime=(lambda/h^2)*matrix*u;
```

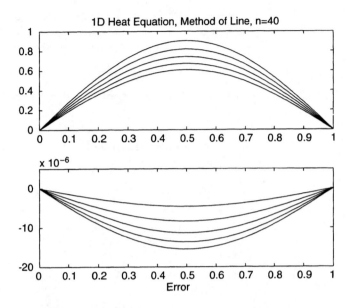

Figure 7.18: The graph of $u(x,t)$ with $u(x,0) = \sin \pi x$.

1. Create the two M-files mol1d.m and oneDheat.m containing the preceding listings. Study carefully the syntax of this program, especially the role played by matrix in mol1d.m, which is then used in oneDheat.m. Execute mol1d.m in *MATLAB*. Compare your output with Figure 7.18.

2. Change n to 10 and execute mol1d.m. How significantly does the output differ from Figure 7.18? Change n to 100 and repeat this part.

3. The exact solution of the initial boundary value problem (7.285)–(7.286) with initial data

$$u(x,0) = \sin \pi x + \sin 2\pi x + \sin 3\pi x \qquad (7.297)$$

is

$$u(x,t) = \sum_{n=1}^{3} e^{-n^2\pi^2 t} \sin n\pi x. \qquad (7.298)$$

Alter mol1d.m appropriately to allow for this initial data and execute this program in *MATLAB*. Compare the output with Figure 7.19.

4. Consider the initial-boundary value problem

$$u_t = u_{xx} + u, \quad u(0,t) = u(1,t) = 0$$

and

$$u(x,0) = \sin \pi x.$$

Make appropriate changes to mol1d.m to accommodate the addition of the linear term to the heat equation. Obtain the analogue of Figure 7.18a.

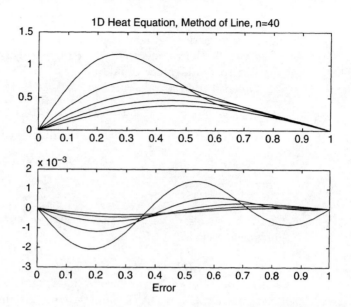

Figure 7.19: The graph of $u(x,t)$ with $u(x,0) = \sum_{n=1}^{3} \sin n\pi x$.

5. Consider the initial-boundary value problem

$$u_t = u_{xx} - 2u, \quad u(0,t) = u(1,t) = 0$$

and

$$u(x,0) = \sin \pi x.$$

Make appropriate changes to `mol1d.m` to accommodate the term $-2u$. Obtain the analogue of Figure 7.18a.

6. Consider the nonlinear initial-boundary value problem

$$u_t = u_{xx} + u^2, \quad u(0,t) = u(1,t) = 0$$

and

$$u(x,0) = \sin \pi x.$$

Make appropriate changes to `mol1d.m` to accommodate the addition of the u^2 term to the heat equation. Obtain the analogue of Figure 7.18a.

7.19 Project F—Designing Numerical Schemes in *Mathematica*

We introduced the Modified Euler and Runge-Kutta methods in this chapter without elaborating much as to how one discovers such schemes. As mentioned in Section 7.9, these schemes are improvements over Euler's method because they are higher order schemes, meaning the truncation error is proportional to

h^p with $p > 1$. In this project we will use *Mathematica* and explore how one may design a numerical scheme such as the Modified Euler method. Our starting point will be Euler's method followed by certain natural extensions of the ideas that led to its difference equation.

Given the initial value problem

$$x' = f(t, x), \quad x(t_0) = x_0, \tag{7.299}$$

Euler's method converts (7.299) to the difference equation

$$x_{i+1} = x_i + hf(t_i, x_i) \tag{7.300}$$

where

$$t_i = t_0 + ih$$

and h is the step size. We seek a class of numerical schemes defined by the difference scheme

$$x_{i+1} = x_i + h[af(t_i, x_i) + bf(t_i + ch, x_i + dh)]. \tag{7.301}$$

Euler's method is a special case of this difference scheme when $a = 1$ and $b = 0$. In comparing (7.301) to (7.300), we note that in (7.301) we have replaced the slope of the tangent line at (t_i, x_i) by a linear combination of the slopes at (t_i, x_i) and $(t_i + ch, x_i + dh)$. At this point the parameters a, b, c, and d are arbitrary. We will choose these parameters in such a way so as to reach a desired truncation error in (7.301).

Our strategy will be to compute x_{i+1} in two ways, once using (7.301), and the other using the differential equation in (7.299). Assuming that the function x could be derived from (7.299), we can obtain an approximation to $x(t_i + h)$ by computing the Taylor series of x as follows:

$$x(t_i + h) = x(t_i) + hx'(t_i) + \frac{h^2}{2!}x''(t_i) + \frac{h^3}{3!}x'''(t_i) + O(h^4). \tag{7.302}$$

We may replace the first two terms in (7.302) by x_{i+1} and x_i. Moreover, all derivatives of x at t_i can be determined from the differential equation $x' = f(t, x)$; for instance,

$$x'(t_i) = f(t_i, x_i), \quad x''(t_i) = \frac{d}{dt}\left(f(t, x(t))\right)|_{t=t_i} = f_t(t_i, x_i) + f_x(t_i, x_i)f(t_i, x_i).$$

Generally speaking, determining derivatives of the function f could be computationally costly. We would therefore like to design schemes that have the accuracy of (7.302), but rely only on function evaluations of f and not on its derivatives. Hence, the motivation behind the scheme (7.301).

We need to determine the Taylor series of the right-hand side of (7.301) in order to compare it with the right-hand side of (7.302). These calculations depend primarily on the Taylor series expansion in h of the various terms involving f. Since t_i and x_i are kept constant in these computations, we replace them with t and x for convenience. Thus the right-hand side of (7.301) becomes

$$x + h[af(t, x) + bf(t + ch, x + dh)]. \tag{7.303}$$

We carry out the remainder of the computations in *Mathematica*. These calculations are designed for a second order scheme. First we define (7.303):

```
rhs = x + h*(a*f[t,x] + b*f[t+c*h,x+d*h)];
```

Next we compute the Taylor series of rhs:

```
series1 = Series[rhs, {h, 0, 3}]
```

Mathematica returns

```
x + (a f[t, x] + b f[t, x]) h +

        (0,1)              (1,0)         2
 b (d f      [t, x] + c f       [t, x]) h  +

      2 (0,2)
     d  f      [t, x]           (1,1)
 b (--------------- + c d f       [t, x] +
           2


      2 (2,0)
     c  f      [t, x]   3         4
 --------------------) h  + O[h]
           2
```

Next, we compute the right-hand side of (7.302). In preparation for this calculation, we introduce the differential equation to *Mathematica*:

```
z[t_] = f[t, x[t]]
```

Now, the right-side of (7.302) is evaluated via

```
series2 = Sum[D[z[t],{t,i}]*h^i/i!, {i, 3}];
series2 = series2 /. x[t] -> x
```

which results in

```
x + h f[t, x] +

    2          (0,1)          (1,0)
   h  (f[t, x] f      [t, x] + f       [t, x])
 -------------------------------------------- +
                       2


   3   (0,1)                    (0,1)
 (h  (f      [t, x] (f[t, x] f      [t, x] +

          (1,0)                    (1,1)
         f      [t, x]) + f[t, x] f      [t, x] +

                       (0,2)          (1,1)
     f[t, x] (f[t, x] f     [t, x] + f      [t, x]) +

          (2,0)
         f      [t, x])) / 6
```

We would like `series1` and `series2` to match, at least up to h^2 terms. Hence, we look at their difference:

`difference = series2 - series1`

which is
```
\begin{verbatim}
(f[t, x] - a f[t, x] - b f[t, x]) h +

          (0,1)             (1,0)
  f[t, x] f     [t, x] + f      [t, x]
  (----------------------------------- -
                   2

          (0,1)               (1,0)         2
     b (d f     [t, x] + c f       [t, x])) h  +

      (0,1)                 (0,1)
  ((f      [t, x] (f[t, x] f      [t, x] +

            (1,0)                     (1,1)
         f       [t, x]) + f[t, x] f       [t, x] +

                        (0,2)
       f[t, x] (f[t, x] f      [t, x] +

              (1,1)             (2,0)
           f       [t, x]) + f       [t, x]) / 6 -

        2  (0,2)
       d  f     [t, x]              (1,1)
     b (--------------- + c d f       [t, x] +
              2

        2  (2,0)
       c  f     [t, x]    3          4
       ----------------)) h  + O[h]
              2
```

We now access the coefficients of h and h^2. First

`Factor[Coefficient[difference, h]]`

gives us

`(1 - a - b) f[t, x]`

Thus, a and b should be chosen so that

$$a + b = 1.$$ (7.304)

Similarly,

`Factor[Coefficient[difference, h^2]]`

leads to

```
          (0,1)                    (0,1)
(-2 b d f      [t, x] + f[t, x] f       [t, x] +

    (1,0)                    (1,0)
  f      [t, x] - 2 b c f       [t, x]) / 2
```

Therefore, a, b, c, and d must satisfy

$$2b(df_y + cf_x) = ff_y + f_x \qquad (7.305)$$

where each function is evaluated at (t_i, x_i).

1. Show that a, b, c, and d must satisfy

$$a + b = 1, \quad 2bd = f(t_i, x_i), \quad 2bc = 1, \qquad (7.306)$$

    in order to have a second order scheme that does not require any function evaluation of the derivatives of f.

2. Show that we would obtain the Modified Euler scheme if we let

$$a = b = \frac{1}{2}, \quad c = 1, \quad d = f(t_i, x_i). \qquad (7.307)$$

3. Consider the initial value problem

$$x' = -x + \sin t, \quad x(0) = -2, \qquad (7.308)$$

    Apply `DSolve` to this problem to obtain the exact solution.

4. Let $a = 0.01$. Determine b, c, and d from (7.306). Apply this scheme (which we will refer to as scheme E) to the initial value problem (7.308) with $h = 0.01$ and $n = 1000$. Generate a table of (t, x) values where every 50 points of data are displayed.

5. Compare the scheme in Problem 4 with the Modified Euler method. Which scheme is more accurate?

6. Draw the graphs of the exact solution, the Modified Euler solution, and the solution you obtained from scheme E. Compare your graph with Figure 7.20.

7. Compare the result of applying scheme E to (7.308) to the case where $a = 0.1$ in (7.306). Is there any appreciable change in the amount of error induced by the new scheme? What if $a = 0.9$? How about $a = 1.2$?

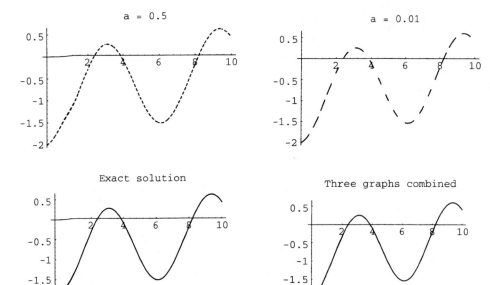

Figure 7.20: The results of a second-order scheme for the initial value problem $x' = -0.1x + \sin t$, $x(0) = -2$.

8. To obtain a second-order scheme we equated coefficients of h up to order 2 in the expressions (7.301) and (7.302). We now seek a fourth-order scheme of the form

$$x_{i+1} = x_i + h[a_1 f(t_i, x_i) + a_2 f(t_i + a_3(h), x_i + a_4(h)) + \qquad (7.309)$$

$$a_5 f(t_i + a_6(h), x_i + a_7(h)) + a_8 f(t_i + a_9(h), x_i + a_{10}(h))].$$

Determine the series expansion of this expression about $h = 0$, keeping terms of up to $O(h^5)$. Equate the coefficients of h^i in this expression with those in the expansion of (7.302). Show that one possible set of parameter values for the a_is that leads to the fourth-order Runge–Kutta scheme

$$x_{i+1} = x_i + \frac{1}{6}(s_1 + 2s_2 + 2s_3 + s_4), \qquad (7.310)$$

is

$$s_1 = hf(t_i, x_i), \qquad (7.311)$$

$$s_2 = hf(t_i + \frac{h}{2}, x_i + \frac{s_1}{2})), \qquad (7.312)$$

$$s_3 = hf(t_i + \frac{h}{2}, x_i + \frac{s_2}{2})), \qquad (7.313)$$

$$s_4 = hf(t_{i+1}, x_i + s_3). \qquad (7.314)$$

Write a program to implement (7.310) and (7.314) for a typical first-order initial value problem. Apply this program to the following equations with n and h of your own choosing. When possible, compare the approximate solution to the exact solution.

(a) $x' = t + x$, $x(0) = -2$

(b) $x' = t + x^2$, $x(0) = 1$

(c) $x' = \sin t + \cos x$, $x(0) = \pi$

(d) $x' = \frac{1}{t+x}$, $x(0) = -1$

References

[1] Hale, J. K., *Ordinary Differential Equations*, John Wiley and Sons, New York, NY, 1974.

[2] Burden, R. L. & J. D. Faires, *Numerical Analysis*, 3rd Edition, Wiley & Sons, New York, NY, 1985.

Answers to selected problems

Chapter 1

Section 1.2
2(a). `Solve[a x^ 2 + b x + c == 0, x]` **2(c).** `NSolve[x^ 2 + 1 == 0, x]`
2(e). `Solve[x^ (1/3) - x + 1 == 0, x]` **2(g).** `Solve[Sin[x]^2 - 1/3 == 0, x]`
2(i). `FindRoot[Sin[x] - x == 0, {x, 1}]`
3(a). `FindRoot[Tan[x] - 3 x + 1 == 0, {x, 0.5}]`
3(b)ii. `Solve[a x - y == 0, x + a y == 1, {x, y}]`
3(b)iv. `Solve[{x^3 - y^3 == 1, x^2 - 3 x y + y^2 == 8}, {x, y}]` or the same expression
with Solve replaced by `NSolve`
3(b)vi. `Solve[{3 x^2 - 4 y^2 + 3 z == 1, x + y + z == 0, z^3 - x^2 y == -1}, {x, y,
z}]`, or `FindRoot[{3 x^2 - 4 y^2 + 3 z == 1, x + y + z == 0, z^3 - x^2 y == -1}, {x,
0.1}, {y, 0.1}, {z, 0.1}]`

Section 1.4
1(a). `Simplify[Sin[2 x] - 2 Sin[x] Cos[x]]`
1(c). `Simplify[Sin[3 x] - 4 (Sin[x])^3 - 3 Sin[x]]`
2(a). Look up the syntax for `TrigExpand`. **2(c).** Use `TrigExpand` with `Cos[2x]` first.
3(a). `TrigExpand[Cos[x+y]+Cos[x-y]]` **3(c).** `TrigExpand[Cos[x+y]+Sin[x+y]]`
3(e). `Simplify[(Cos[a x])^2- (Sin[a x])^2]`

Section 1.5
1(a). `f[x_] = (1-x)/(1+x); f[0]` **1(b).** `f[x_] = Log[x + Sqrt[1-x^2]];`
3. `f[x_, t_, omega_] = Sin[x - omega t]; FindRoot[f[1, t, 4] == 0, {t, 0}]`

Section 1.6
1(b). `D[Sin[4 x]^3 Cos[7*(x^2 - 2x +1)]^5, x]` **3(b).** `Integrate[x Sin[x], x]`
3(e). `Integrate[Exp[x] Sin[x], x]` **3(h).** `Integrate[t Exp[t^2], t]`
5. `f[x_] = Integrate[1/(x+t^5), {t, 0, x}]; a = D[f[x], x] /. x->1/2; N[a]`
7. `Integrate[D[f[t],t], {t, a, b}]`

Section 1.7
1(a). `Plot[Sin[5 x], {x, 0, 2 Pi}]` **1(c).** `Plot[Sin[2 x] + Sin[Sqrt[2] x], {x, 0, 10
Pi}, PlotPoints->100]`  **3(a).** `Plot[Exp[-x^2], {x, -5, 5}, PlotRange->All]`
4(a). `ParametricPlot[{Sin[t]^2, Cos[t]^2}, {t, 0, 2Pi}]`
4(e). `ParametricPlot[{Sin[t]^5, Cos[1+2t]}, {t, 0, 2 Pi}]`

Section 1.8
1(a). `ParametricPlot3D[{t, t, t}, {t, 0, 1}]`

1(c). `ParametricPlot3D[{Exp[-t/4] Sin[3 t], Exp[-t/4] Cos[3 t], t/12}, {t, 0, 4Pi}, PlotPoints->300]`

1(e). `ParametricPlot3D[{Sinh[t/6], Sin[4 t], Cosh[t/6]}, {t, 0, 4Pi}]`

2(a). `ParametricPlot3D[{2 Cos[t], 2 Sin[t], 0}, {t, 0, 2 Pi}]`

2(c). `ParametricPlot3D[{3 Cos[t], 2 Sin[t], 0}, {t, 0, 2 Pi}]`

3(a). `Plot3D[x^2 + y^2, {x, -3, 3}, {y, -3, 3}]`

3(c). `Plot3D[3 x^2 + 4 y^2, {x, -3, 3}, {y, -3, 3}]`

3(e). `Plot3D[Sin[Sqrt[x^2 + y^2]], {x, -Pi, Pi}, {y, -Pi, Pi}]`

3(g). `Plot3D[Sin[x^2 + y^2] Cos[y], {x, -Pi, Pi}, {y, -Pi, Pi}]`

4. `ParametricPlot3D[{-1 + Cos[u]Sin[v], 1 + Sin[u]Sin[v], Cos[v]}, {u, 0, 2Pi}, {v, 0, Pi}]`

8. `ParametricPlot3D[{u Cos[v], u Sin[v], 3(1-u)}, {u, 0, 1}, {v, 0, 2 Pi}]`

9(a). `Table[Plot[Sin[3 x] Cos[t], {x, 0, Pi}, PlotRange->{{0, 2Pi}, {-1, 1}}], {t, 0, 2 Pi, Pi/10}].` Then highlight the cells containing all graphs and select **Animate Selected Graphics** from **Cell** in the Notebook menu.

Section 1.9

1(a). `DSolve[x'[t] + 3 x[t] == 0, x[t], t]`

1(c). `DSolve[x'[t] + x[t]^3 == 0, x[t], t]`

1(e). `DSolve[x'''[t] + x'[t] + x[t] == 0, x[t], t]`

1(g). `DSolve[x''[t] + x[t] == Sin[2 t], x[t], t]`

2(a). `NDSolve[{x''[t] + x[t] == 0, x[0] == 0, x'[0] == 1}, x, {t, 0, 3}]`

2(c). `NDSolve[{y''[t] + 0.1 y'[t] + Sin[y[t]] == 0, y[0] == 0, y'[0] == 3}, y, {t, 0, 5}]`

3(a). `sol[a_, b_] := DSolve[{x'[t] == y[t], y'[t] == -x[t], x[0] == a, y[0] == b}, x[t], y[t], t];`
`solution1 = sol[1,1]; solution2 = sol[2,2];`
`graph1 = ParametricPlot[Evaluate[{x[t], y[t]} /. solution1], {t, 0, 5}];`
`graph2 = ParametricPlot[Evaluate[{x[t], y[t]} /. solution2], {t, 0, 5}];`
`Show[graph1, graph2]`

3(d). `sol[a_, b_] := NDSolve[{x'[t] == y[t], y'[t] == -x[t] - y[t]^2, x[0] == a, y[0] == b}, {x, y}, {t, 0, 5}];`
The rest follows the pattern in 3(a).

Section 1.10

3. `A = {{a, -a, b}, {-a, b, a}, {b, a, 2a}};`
`determinant = Det[A]; Solve[determinant == 0, a]`
Or alternatively, try `N[Solve[determinant /. b->3, a]]`

4(a). `A = {{a, b}, {b, a}}; MatrixPower[A, 5]`   **5.** `A = {{-1, 1}, {1, 3}};`
`s[n_] := Sum[IdentityMatrix[2] + Sum[1/i! MatrixPower[A,i], {i, 0, n}]`

7(a). See Figure 1

8. See Figure 2

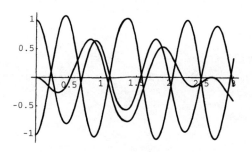

Figure 1: Problem 7(a).

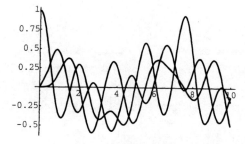

Figure 2: Problem 8.

Section 1.11

2(a). Solve[x^2 - 4 == 0, x]   **2(c).** D[Sin[t], t] /. t-> 3

Section 1.12

1(a). Sum[N[i], {i, 0, 20}]

1(c). Sum[1/i^2, {i, 1, 100}]; Sum[N[1/i^2]], {i, 1, 100}]   **2(a).** 0.947724

2(c). The sequence oscillates between -1.895 and 1.895 **2(e).** 1.98704   **2(g).** Undefined

3. f[y_]:=Integrate[Exp[-t] y, {t, 0, x}]; y = Sin[t]; Do[z=f[y];z=z/.x->t;Print[{y, z}];y=z, {i, 10}]

Chapter 2

Section 2.2

2(a). solve('a*x^2+b*x+c','x')   **2(c).** solve('x^2+1','x') or roots([1 0 1])

2(e). solve('x^1/3-x+1','x')   **2(g).** solve('sin(x)^2-1/3', 'x')

2(i). solve('sin(x)-x','x')   **3(b)i.** [x,y]=solve('3*x-2*y=2', 'x+y=7')

3(b)iv. [x,y]=solve('x^3-y^3=1','x^2-3*x*y+y^2=8')

3(b)vi. [x,y,z]=solve('3*x^2-4*y^2+3*z=1','x+y+z=0', 'z^3-x^2*y=-1')

Section 2.4

1(a). x=sym('x'); expand(sin(2*x)-2*sin(x)*cos(x))

1(c). x=sym('x'); expand(sin(3*x)- 3*sin(x)+4*sin(x)^3)

2(a). x=sym('x'); y=sym('y'); expand(cot(x+y))

2(c). y=sym('y'); expr=expand(cos(2*y)); subs(expr,x/2,y)

3(a). x=sym('x'); y=sym('y'); expand(cos(x+y)+cos(x-y))

3(d). x=sym('x'); expand(cos(2*x)^2-sin(2*x)^2)

Section 2.5

1(a). x=sym('x'); f = (1-x)/(1+x); subs(f,0,x); subs(f,0.5,x); subs(f,pi,x);

1(c). syms x t omega; f = sin(x-omega*t); solve(subs(subs(f,1,x),4,omega),t)

2(c). x=sym('x'); f=(x^2-1)/(x-2)+x; solve(f,x)

Section 2.6

2. `syms a b c d; A=det([a b;c d]); solve(A)`

3. `syms a b; A=det([a -a b; -a b a; b a 2*a]); solve(A)`

4. `syms a b; A=[a b; b a]; simplify(A^5)`   5(a). `x=0:0.01:2*pi; plot(x,sin(x))`

5(c). `x=0:0.01:2*pi; plot(x,sin(5*x)-sin(4*x))`

5(f). `x=0:0.01:2*pi; plot(x,log(1+abs(sin(x))))`

Section 2.7

1(a). `diff('ln(x/(x+1))','x')`   1(c). `diff('x^(x-1)', 'x')`

2. First plot this function via `x=0:0.01;2*pi; plot(x,x.*sin(x).^2)`. Next, look up the syntax for fzero. Try `fzero(diff('x*sin(x)^2','x'), 3)` to find a zero of the derivative of f near 3.

3(b). `int('x*sin(x)', 'x')`   3(d). `int('x^10*sin(x)', 'x')`   3(g). `int('sin(x^2)','x')`

4(a). `int('1/(1+x^2)', 'x', -inf, inf)`   4(c). `int('e^(-a*t^2)','t',0,inf)`

5. `f=int('1/(x+t^5)','t',0,'x'); fp=diff(f,'x'); numeric(subs(fp,'x',0.5))`

6. `int(diff('f(t)', 't'), 't','a','b')`

7. `f=int('sin(x*t^3)','t',0,1); g=diff(f,'x'); numeric(subs(g,'x',1/2))`

Section 2.8

1(b). `x=0:0.01:2*pi; plot(x,sin(2*x)+3*sin(3*x))`; period is 2π.

2.

```
x=0:0.01:2;
for i=1:10
  plot(x,x.^i)
  hold on
end
```

4(a). `t=0:0.01:2*pi; plot(sin(t).^2, cos(t).^2)`

4(c). `t=0:0.01:2*pi; plot(sin(t).^5, cos(t).^5)`

Section 2.9

1(b). `t=0:0.01:3; plot3(t.^2, t, t.^2)`   1(d). `t=0:0.01:3; plot3(sinh(t), t, cosh(t))`

2(b). `t=0:0.01:2*pi; plot3(2*cos(t), 2*sin(t), ones(size(t)))`

3(a). `[X, Y] = meshgrid(-3:0.1:3, -3:0.1:3); mesh(X.^2 + Y.^2);`

3(d). `[X, Y] = meshgrid(-pi:0.1:pi, -pi:0.1:pi); mesh(sin(X.^2 + Y.^2));`

Section 2.10

1(a). `dsolve('D2y + y=0')`   1(c). `dsolve('D2y+y=0','y(0)=1','Dy(0)=-1')`

1(e). `dsolve('D2y+y=sin(t)')`

2(a). First convert the equation to system: $x' = y,\ y' = -x$. Next, create the M-file `prob2a.m` as

```
function yprime=prob2a(t,y)
yprime=[y(2); -y(1)];
```

Finally, execute `[t,y]=ode45('prob2a',[0 5],[0 1])`.

2(c). First convert the equation to system: $y' = x$, $x' = -0.1x - \sin y$. Next, create the M-file prob2c.m as

```
function yprime=prob2c(t,y)
yprime=[y(2); -0.1*y(1)-sin(y(2))];
```

Finally, execute [t,y]=ode45('prob2c',[0 5],[0 3]).

Section 2.11
1(a). 9.7656   **1(c).** $1.5597e + 21$ **1(e).** 7.9961

Chapter 3

Section 3.1
1(a). ODE. In *Mathematica* try y = 3 Exp[-2 x]; Simplify[D[y,x] + 2 y]
1(c). ODE. In *Mathematica* try y = -7Cos[2 x]; Simplify[D[y, x,2] + 4y]
1(f). PDE. In *MATLAB* try

```
syms u x t a
u = 'exp(-a^2*t)*sin(a*x)';
simplify(diff(u,t)-diff(u,2,x))
```

1(g). PDE. In *Mathematica* try

```
u = Exp[-9 a^2 t]Cos[a x];
Simplify[D[u,t] - 9 D[u, {x, 2}]]
```

2(a). In *MATLAB* try

```
syms x y
y=exp(-2*x);
diff(y,2,x)+3*diff(y,x)+2*y
```

In *Mathematica* try

```
y=Exp[-2*x];
D[y,{x,2}]+3*D[y,x]+2*y
```

2(c). In *Mathematica* try

```
x=1/2 Sin[t];
D[x,{t,2}]+3*x
```

2(g). In *MATLAB* try

```
syms x y
y=x^(1/2)*log(x);
4*x^2*diff(y,2,x)+y
```

In *Mathematica* try

```
y=Sqrt[x] Log[x];
4 x^2 D[y,{x,2}] + y
```

2(h). In *Mathematica* try

```
Plot[Evaluate[Table[Cos[2 t]Sin[x], {t, 0, 0.5, 0.1}]], {x, 0, Pi}]
```

In *MATLAB* try

```
x=0:0.01:pi;
for i=1:6
    t=(i-1)*0.1;
    plot(x,cos(2*t)*sin(x))
    hold on
end
```

3(a). In *Mathematica* try Solve[c Exp[-6 t] == -1/. t->1, c]

3(c). In *MATLAB* try

```
syms x t c1 c2 eqn1 eqn2
x = c1*sin(2*t)+c2*cos(2*t);
eqn1=subs(x,t,0);
dx=diff(x,t);
eqn2=subs(dx,t,0)-1;
[c1 c2]=solve(eqn1,eqn2-1)
```

In *Mathematica* try

```
x = c1 Sin[2 t] + c2 Cos[2 t];
eqn1 = x /. t -> 0;
eqn2 = D[x, t] /. t -> 0;
Solve[{eqn1 == 0, eqn2 == 1}, {c1, c2}]
```

3(e). $c_1 = -\frac{1}{5}$, $c_2 = \frac{1}{2}$

4(a). In *Mathematica* try

```
Plot[Table[Evaluate[c Exp[x], {c, -1, 1, 0.1}]], {x, 0, 1}]
```

4(e). In *MATLAB* try

```
x=-1:0.01:1;
for i=1:21
    c=-1+(i-1)*0.1;
    plot(x,c*exp(-x.*x))
    hold on
end
```

5(a). $y'(0) = -1$, $y''(0) = 1$ **5(c).** $x'(0) = 0$, $x''(0) = 3$ **5(e).** $y''(1) = -5$
5(g). $y''(\pi) = a - \pi$, $y'''(\pi) = 2a - 1$ **6.** 6.389 s; velocity at impact is 62.61 m/s.
7. 6.7478 s; velocity at impact is 56.13 m/s.   **8.** $T' = k(100 - T)$

Section 3.2
1(a). Linear   **1(b).** Nonlinear   **1(c).** Linear   **1(d).** Linear   **1(e).** Linear   **1(f).** Nonlinear
2(a). Linear   **2(b).** Nonlinear   **2(c).** Linear   **2(d).** Nonlinear   **2(e).** Linear   **4(a).** $c = \pm 3$
4(b). FindRoot[c^3 + 3 c^2 + 1 == 0, {c, 1, 2}]

Section 3.3
3(a). See Figure 3

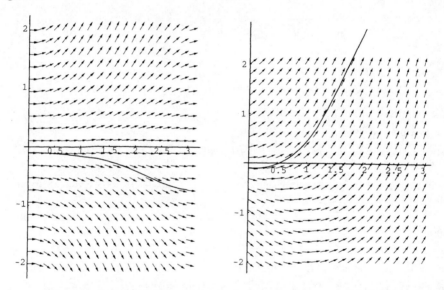

Figure 3: Problem 3(a).   Figure 4: Problem 3(c).

3(c). See Figure 4
4(e). See Figure 5
5(g). See Figure 6
6(a). Yes   **6(b).** Yes   **6(c).** No   **6(d).** No   **6(e).** Yes   **6(f).** No   **6(g).** Yes

Section 3.4
3(a). $y(x) = ce^{2x}$ **3(c).** $y(x) = 1/(c - \ln(1 + x))$ **3(e).** $y(x) = c_1 e^{-x} + c_2$ **3(g).** $y(x) = c\sec(x)$
3(i). $x(t) = c\exp\left(-(1 + t)\exp(-t)\right)$ **3(k).** $w(t) = \text{Arcsin}(c + \frac{t^2}{2})$ **3(m).** $x(t) = \frac{c}{c-t}$
3(o). $y(x) = 1/4(-4 + x^2 - 2xc_1 + c_1)$ **4(c).** $x(t) = \sqrt{\frac{\pi}{2}}\exp(-t^2/2)\text{erfi}(t/\sqrt{2})$
5. The temperature reaches 65° in 17.91 minutes, 69° in 34.01 minutes.
7. In *Mathematica* try DSolve[{m v'[t] == m g - k v[t], v[0] == 0}, v[t], t]
12(b)i. $x(t) = \frac{1}{2}\left(t + \sqrt{3}t\tan(\frac{\sqrt{3}}{2}(c + \ln t))\right)$

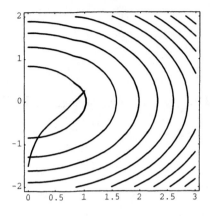

Figure 5: Problem 4(e).

Figure 6: Problem 5(g).

Section 3.5

4(a). $P(t) = \exp(2t/3)$ **4(c).** $P(t) = e^{2t}$ **4(e).** $P(t) = \exp\left(\int^t \exp(-s^3)\,ds\right)$ **5(c).** $P(x) = x$

5(d). $P(x) = \frac{1}{x}$ **8(a).** $x(t) = 2te^{-t}$ **8(c).** $x(t) = \frac{1}{5}\left(e^{-2t} - \cos t + 2\sin t\right)$

10. Terminal velocity is $\frac{mg}{k}$

11. Terminal velocity is 137.2 m/s. It takes 9.704 seconds for the person to reach half of his terminal velocity.

12. The governing initial-value problem is $x'(t) = 6 - 3x/200$, $x(0) = 0$. Eventually, there will be 400 grams of salt in the tank.

Section 3.6

2(a). $F(t, x) = t^2/2 + x^2/2$ **2(c).** $F(x, y) = -x^3/3 + y^3/3$ **2(e).** $F(x, y) = -x^2y + y^3/3$

2(g). $F(t, x) = 3t^2/2 - \cos x + 1/4e^{-4t}\sin x$ **3(a).** $\rho(t) = t^{-3/2}$ **3(c).** $\rho(t) = \exp(1/t)$

8(a). `ContourPlot[x^2+t, {t,0,1}, {x,0,1}]`   **8(c).** `Solve[Sin[x]-x==0,x]`   **9(a).** None

9(c). `[t,x]=meshgrid(0:0.1:1;-1:0.1:1)`

Chapter 4

Section 4.1

2(a). Linear. Second order. In *Mathematica* try

```
y=Exp[-3 t];
D[y, {t,2}] + 4 D[y, t] + 3 y
```

In *MATLAB* try

```
syms t y
y=exp(-3*t);
diff(y,2,t) + 4*diff(y, t) + 3*y
```

2(b). Nonlinear. Second order. In *Mathematica* try

```
x1=Log[1-t]; x2=Log[t-1] ;
L[x_]:= D[x, {t,2}] + D[x, t]^2;
L[x1]
L[x2]
```

In *MATLAB* try

```
syms t
x1=log(1-t);x2=log(t-1);
l=diff('y(t)',2,t) + diff('y(t)', t)^2
subs(l,x1,'y(t)')
subs(l,x2,'y(t)')
```

2(d). $\omega = 2$
2(e). In *MATLAB* try

```
syms x
y='cos(log(x))';
l=x^2*diff('y(x)',2,x)+x*diff('y(x)',x)+'y(x)'
answer=subs(l,y,'y(x)')
simplify(answer)
```

2(f). In *Mathematica* try

```
y = 1/x;
L=x^2 D[y, {x,2}] + x D[y, x] - y
```

3(a). $a = 9$
3(c). $r = -1 \pm \sqrt{2}$
3(e). `NSolve[lambda^3 + 2 lambda^2 - 3 lambda + 17 == 0, lambda]` in *Mathematica* returns

```
{{lambda -> -3.89262}, {lambda -> 0.946309 - 1.86326 I},
```

```
   {lambda -> 0.946309 + 1.86326 I}}
```

4(a). Let $z = y'$. Then $z = \sqrt{c + 2y - y^2}$, from which we get $\int 1/\sqrt{c + 2y - y^2}\, dy = \pm x + k$.
Mathematica returns the implicit solution

$$- \arctan\left(\frac{(-1 + y)\, \sqrt{c + 2\,y - y^2}}{-c - 2\,y + y^2} \right) = \pm x + k.$$

4(b). $\int 1/\sqrt{c + 2\cos y}\, dy = \pm x + k$ **4(e).** $\int 1/\sqrt{c - 2\ln y}\, dy = \pm x + k$
4(g). $\int 1/\sqrt{c - y^3/3}\, dy = \pm x + k$ **4(i).** $y = \ln(x + k)$ **4(l).** $\int 1/\sqrt{2y^3 + 4y + c}\, dy = \pm x + k$
6(a). See Figure 7
6(b). See Figure 8
6(c). See Figure 9

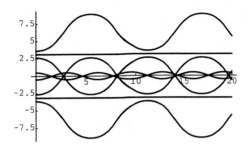

Figure 7: Problem 6(a).

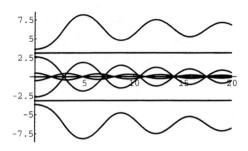

Figure 8: Problem 6(b).

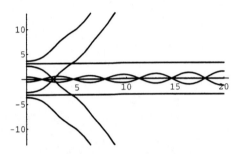

Figure 9: Problem 6(c).

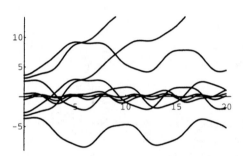

Figure 10: Problem 6(d).

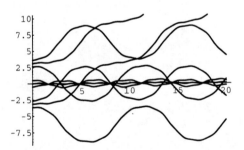

Figure 11: Problem 6(e).

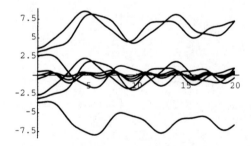

Figure 12: Problem 6(f).

6(d). See Figure 10
6(e). See Figure 11
6(f). See Figure 12

Section 4.2

1(a). $W(1 - t - t^2, t^3 + 1) = 1 + 2t + 3t^2 - 2t^3 - t^4$ **1(c).** $W(1/(1 - t^2), 1 - t^2) = \frac{4t}{-1+t^2}$

1(e). $W(\cos x, \cos 2x, \cos 3x) = -4 \left(9 \cos(x) + \cos(3x)\right) \sin(x)^3$

1(g). $W(\exp(x), \exp(2x), \exp(3x)) = 2 \exp(6x)$ **1(i).** $W(\exp(at), t \exp(at)) = \exp(2at)$

2(a). Linearly independent   **2(c).** Linearly independent   **2(e).** Linearly independent
2(g). Linearly independent   **4(a).** $x(t) = (-2 - 3t)\exp(-3t)$ **4(c).** $x(t) = -2\cos(t)\exp(-t)$
4(e). $x(t) = 1/10\left(e^{3(t-\pi)} + \cos t + 23\sin t\right)$ **5.** $a \neq 0$

Section 4.3
1(a). $\cos 2 + i\sin 2$ **1(c).** $e^{-2}\cos 3 + ie^{-2}\sin 3$ **1(e).** $\cos bt + i\sin bt$

1(g). $\exp(t/2)\left(\cos\frac{\sqrt{3}}{2}t + i\sin\frac{\sqrt{3}}{2}t\right)$

2(b). In *Mathematica* try DSolve[x''[t] + 4 x'[t] - x[t]==0, x[t], t]. It responds with
$x(t) = c_1 e^{(-2-\sqrt{5})t} + c_2 e^{(-2+\sqrt{5})t}$

2(d). $x(t) = e^{-2t}(c_1 + c_2 t)$ **3(a).** $x(t) = \cos(2t) - \sin(2t)$

3(c). $y(t) = e^{\frac{-3t}{2}}\left(\cos(\frac{\sqrt{7}t}{2}) + \sin(\frac{\sqrt{7}t}{2})/\sqrt{7}\right)$ **3(g).** $x(t) = \frac{a+2at}{e^{2t}}$

4(a). In *Mathematica* try

```
sol=DSolve[{x''[t]+x'[t]+x[t]==0, x[0]==2, x'[0]==-1}, x[t], t];
solution = ComplexExpand[Simplify[First[x[t]/. sol]]];
energy=Integrate[solution^2+D[solution,t]^2, {t, 0, 1}]
```

Mathematica returns $2\exp(t/2)\cos\sqrt{3}/2t$ for **solution** and $5 - \frac{4+\cos(\sqrt{3})}{e}$ for **energy**.
4(c). $x(t) = -e^{-t}\cos(\sqrt{3}t)$; $\int_0^1 x^2 + x'^2\,dt = \frac{1}{16e^2}(25e^2 - (20 + 5\cos(2\sqrt{3}) + 3\sqrt{3}\sin(2\sqrt{3})))$
5(a). In *Mathematica* try

```
data={-2, -1, 0, 1, 2};
eqn = {2 x''[t] + 3 x'[t] + x[t] == 0};
sol=Table[DSolve[Flatten[{eqn, x[0]==data[[i]],x'[0]==0}], x[t], t],
                {i, Length[data]}];
Plot[Evaluate[Table[x[t] /. sol[[i]], {i, Length[sol]}]], {t, 0, 3}]
```

The program's output is shown in Figure 13.

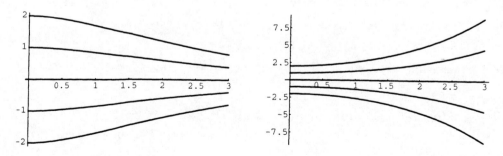

Figure 13: Problem 5(a).   Figure 14: Problem 5(c).

5(c). See Figure 14
6. With DSolve[{y''[t]+y'[t]+4 y[t]==0, y[0]==1, y[2]==3}, y[t], t], *Mathematica*

returns

```
              Sqrt[15] t
{{y[t] -> (Cos[----------] -
                   2

                                        Sqrt[15] t
    (Cot[Sqrt[15]] - 3 E Csc[Sqrt[15]]) Sin[----------]) /
                                                 2

        t/2
      E   }}
```

7(a). $a \geq 4$ **11(a).** $y(t) = \cos(\sqrt{3}\,t) - \left(-2 + \cos(\sqrt{3}\,\pi)\right)\csc(\sqrt{3}\,\pi)\sin(\sqrt{3}\,t)$

11(c). $x(t) = \cos(2t) + \sin(2t)\tan 2$

11(e). Let $A = \sqrt{a^2 - 4a}$.
$x(t) = 1/\left(-1 + \exp(2A)\right)\left[\exp\left(-\sqrt{a}/2\left(\sqrt{a-4} + \sqrt{a}\right)(t-2)\right)\left(-1 + \exp(At)\right)\right]$

Section 4.4

2(a). The roots of the characteristic polynomial are $\lambda_1 = 0.532089$, $\lambda_2 = -0.652704$, $\lambda_3 = -2.87939$. The general solution is $x(t) = c_1 e^{\lambda_1 t} + c_2 e^{\lambda_2 t} + c_3 e^{\lambda_3 t}$.

2(c). The roots of the characteristic polynomial are $\lambda_1 = \sqrt{2 - \sqrt{3}}$, $\lambda_2 = \sqrt{2 - \sqrt{3}}$, $\lambda_3 = -\sqrt{2 + \sqrt{3}}$, $\lambda_4 = \sqrt{2 + \sqrt{3}}$. The general solution is $x(t) = c_1 e^{\lambda_1 t} + c_2 e^{\lambda_2 t} + c_3 e^{\lambda_3 t} + c_4 e^{\lambda_4 t}$.

2(e). The roots of the characteristic polynomial are $\lambda_1 = -1$, $\lambda_2 = 1$, $\lambda_3 = -i$, $\lambda_4 = i$. The general solution is $x(t) = c_1 e^{-t} + c_2 e^{t} + c_3 \cos t + c_4 \sin t$.

2(g). The roots of the characteristic polynomial are $\lambda_1 = -2$, $\lambda_2 = \lambda_3 = \lambda_4 = 2$. The general solution is $x(t) = c_1 e^{-2t} + c_2 e^{2t} + c_3 t e^{2t} + c_4 t^2 e^{2t}$.

3(a). $y(t) = e^{-t}/6\left(-2 + \left(1 + \sqrt{3}\right)e^{t+(2-\sqrt{3})t} + \left(1 - \sqrt{3}\right)e^{t+(2+\sqrt{3})t}\right)$

3(c). $x(t) = \frac{1}{3}\left(e^{-t} - e^{t/2}\cos\left(\sqrt{3}t/2\right) + \sqrt{3}e^{t/2}\sin\left(\sqrt{3}t/2\right)\right)$. This solution is obtained in *Mathematica* via

```
sol=DSolve[{x'''[t]+ x[t]==0, x[0]==0, x'[0]==0, x''[0]==1},x[t], t];
Simplify[ComplexExpand[First[x[t]/.sol]]]
```

4(a). In *MATLAB* try dsolve('D3x-x=0','x(0)=0','Dx(0)=1','x(1)=3')

4(b). In *MATLAB* try

```
dsolve('D4x+16*x=0', 'x(0)=0', 'D2x(0)=0', 'x(pi)=1', 'D2x(pi)=0')
```

Section 4.5

2(a). $D^2 + 4$ **2(c).** $D^4 - 24$ **2(e).** $(D+3)^3$ **2(g).** $D^2 - 2D + 5$ **2(i).** $(D-1)^4(D^2+4)$

3(a). $x(t) = c_1 \cos t + c_2 \sin t + 1/2\exp(t)$ **3(c).** $x(t) = 1/2\cos t + 1/2t\sin t + c_1 \cos t + c_2 \sin t$

3(e). $y(t) = e^{-3t}\left(t^2 + c_1 + t c_2\right) - \frac{3\cos t}{50} + \frac{2\sin t}{25}$

3(g). $x(t) = -2 + t + c_1 e^{\lambda_1 t} + c_2 e^{\lambda_2 t} + c_3 e^{\lambda_3 t}$ where λ_i are the roots of $\lambda^3 - 3\lambda^2 + 3\lambda + 1 = 0$.

4(a). $x(t) = \frac{\sin t}{3} - \frac{\sin(2t)}{6}$ **4(d).** $y(t) = \frac{-1}{5}e^{-2t} + \frac{1}{2}e^{-t} + \frac{-3\cos t + \sin t}{10}$ **4(e).** $z(t) = -2 + e^{-t} + t$

4(g). $y(t) = 1/136\left(-\left(-31 + 9\sqrt{2}\right)e^{t-\sqrt{2}t}\right) + 1/136\left(\left(31 + 9\sqrt{2}\right)e^{t+\sqrt{2}t}\right) +$
$1/68\left(-34\cos t + 3\cos(3t) - 34\sin t - 5\sin(3t)\right)$

5(b). $y(t) = -4e^{-t} + 1/2\left(8 - 8t + 4t^2 - t^3\right)$

5(d). $z(t) = -\sqrt{2}/4\left(\exp(t/\sqrt{2}) - \exp(-t/\sqrt{2})\right)\cos(t/\sqrt{2}) + 1/2\sin t$

5(f). $y(t) = -1/4e^{-t}\left(t\left(2 + t\right)\right) + 1/4\left(-t\cos t + 3\sin t - t\sin t\right)$ **6(a).** $y(t) = t$

6(c). $y(t) = t + 1/2\left(-1 + t\right)\sin t$

6(e). Execute the following program in *Mathematica*.

```
sol=DSolve[{y'''[t]-y[t]==1, y[0]==0, y'[0]==0, y[1]==1}, y[t],t];
solution=Chop[Simplify[N[ComplexExpand[First[y[t]/.sol]]]]]
```

It results in

$$1. + 0.878842\,e^{1.\,t} + e^{-0.5\,t} - 0.121158\,\cos(0.866025\,t) - e^{-0.5\,t} - 0.944849\,\sin(0.866025\,t).$$

Section 4.6

2(b). $x(t) = \left(\frac{-t}{4} + c_1\right)\cos(2t) - c_2\sin(2t)$

2(d). $y(t) = c_1\cos t + c_2\sin t - \cos t\sin t + \cos t\left(\log(\cos(\frac{t}{2}) - \sin(\frac{t}{2})) - \log(\cos(\frac{t}{2}) + \sin(\frac{t}{2}))\right) + \sin t$

2(f). $x(t) = c_1 x + c_2/x - 1$ **2(h).** $x(t) = c_1\sinh t + c_2\cosh t + 1/2\left(t\cosh(t) - \sinh(t)\right)$

5(a). In *Mathematica* execute

```
change[a_,c_,j_]:=Block[{},b=a;Table[b=ReplacePart[b,c[[i]],{i,j}],
         {i,Length[c]}];b]
x1=Exp[t]; x2=t Exp[t]; x3=t^2 Exp[t];
a={x1,x2,x3};A={a, D[a,t], D[a, {t,2}]};
w=Det[A];
a1={0,0,Exp[-4t]};
w1=Det[change[A,a1,1]];
w2=Det[change[A,a1,2]];
w3=Det[change[A,a1,3]];
c1=Integrate[w1/w,t];c2=Integrate[w2/w,t];c3=Integrate[w3/w,t];
xp=c1*x1+c2*x2+c3*x3
```

The program returns $-1/125e^{-4t}$.

5(d). Apply the appropriate modification of the previous *Mathematica* program. It results in $1/54\left(\cos 3t + \sin 3t\right)$.

Section 4.7

4(a). $-\cos t$ **4(c).** $1 + t/2\log(-1 + t) - t/2\log(1 + t)$ **4(e).** $t^2 + 1$

7(a). $r_1 = 4.94883$, $r_2 = 1.21718$, $r_3 = -0.166013$. The general solution is $x(t) = c_1 t^{r_1} + c_2 t^{r_2} + c_3 t^{r_3}$.

7(c). $r = 12.3813$, $r_2 = 0.794817$, $r_3 = -0.176116$ **7(e).** $r_1 = 1.86081$, $r_2 = 0.254102$, $r_3 - 2.11491$

Section 4.8

1(a). $x(t) = 2 + 6t + 9t^2 + 9t^3 + \frac{27t^4}{4} + \frac{81t^5}{20} + \frac{81t^6}{40}$ **1(c).** $x(t) = 2 + t^2 + \frac{t^4}{4} + \frac{t^6}{24}$

1(e). $y(t) = 2 + 6t + \frac{19t^2}{2} + \frac{19t^3}{2} + \frac{85t^4}{12} + \frac{17t^5}{4} + \frac{1531t^6}{720} + \frac{1531t^7}{1680} + \frac{6889t^8}{20160} + \frac{6889t^9}{60480} + \frac{124003t^{10}}{3628800} + \frac{11273t^{11}}{1209600} + \frac{558013t^{12}}{239500800}$

1(g). $y(t) = 2 + 3t + 2t^2 + \frac{5t^3}{6} + \frac{t^4}{4} + \frac{7t^5}{120} + \frac{t^6}{90}$ **1(i).** $y(t) = 2 - 2t + t^2$

2(a). $y(t) = 1 - \frac{t^2}{2} + \frac{t^4}{24} - \frac{t^6}{720}$ **2(c).** $y(t) = 1 - t + 2t^2 - \frac{13t^3}{6} + \frac{43t^4}{24} - \frac{71t^5}{60} + \frac{469t^6}{720}$

2(e). $y(t) = \frac{t^3}{6} - \frac{t^5}{60} + \frac{t^7}{1680} - \frac{t^9}{90720} + \frac{t^{11}}{7983360}$

2(g). $y(t) =$
$\frac{t^3}{6} - 0.00416667\,t^4 - 0.0165833\,t^5 + 0.000415278\,t^6 + 0.000587321\,t^7 - 0.0000147572\,t^8 - 0.000010749\,t^9 + 2.71459\,10^{-7}\,t^{10} + 1.20303\,10^{-7}\,t^{11} - 3.05903\,10^{-9}\,t^{12} - 9.0823\,10^{-10}\,t^{13} + 2.32952\,10^{-11}\,t^{14} + 4.93432\,10^{-12}\,t^{15} - 1.27903\,10^{-13}\,t^{16} - 2.02\,10^{-14}\,t^{17} + 5.30205\,10^{-16}\,t^{18} + 6.44943\,10^{-17}\,t^{19} - 1.71775\,10^{-18}\,t^{20} - 1.64951\,10^{-19}\,t^{21} + 4.46785\,10^{-21}\,t^{22} + 3.45246\,10^{-22}\,t^{23} - 9.53245\,10^{-24}\,t^{24}$

2(i). $y(t) = \frac{t^3}{6} + \frac{t^4}{24} - \frac{t^5}{120} - \frac{t^6}{360} + \frac{t^8}{20160} + \frac{t^9}{362880} - \frac{t^{10}}{3628800} - \frac{t^{11}}{39916800}$

2(k). $y(t) = \frac{t^3}{6} - 0.00416667\,t^4 - 0.00825\,t^5 + 0.000276389\,t^6 + 0.00019248\,t^7 - 7.34152\,10^{-6}\,t^8 - 2.59176\,10^{-6}\,t^9 + 1.0749\,10^{-7}\,t^{10} + 2.25843\,10^{-8}\,t^{11} - 1.00252\,10^{-9}\,t^{12}$

3(a). See Figure 15

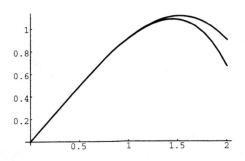

Figure 15: Problem 3(a).

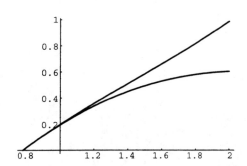

Figure 16: Problem 3(e).

3(e). See Figure 16

Section 4.9

2(a). See Figure 17.

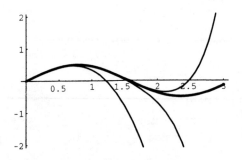

Figure 17: Problem 2(a).

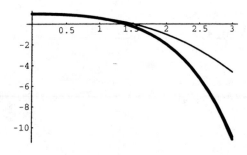

Figure 18: Problem 2(c).

2(c). See Figure 18.
4(a). See Figure 19.
4(b). See Figure 20.
4(c). See Figure 21.

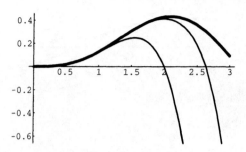

Figure 19: Problem 4(a). The exact solution is displayed together with its series approximate solutions with $n = 5$, 10, and 15.

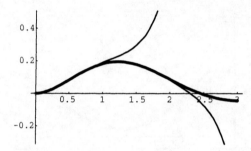

Figure 20: Problem 4(b). The exact solution is displayed together with its series approximate solutions with $n = 5$, 10, and 15.

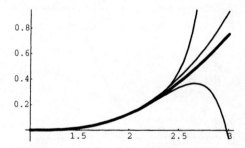

Figure 21: Problem 4(c). The exact solution is displayed together with its series approximate solutions with $n = 5$, 10, and 15.

5(a). See Figure 22.
6(a). See Figure 23.
6(c). See Figure 24.

Section 4.10
7(a). In *Mathematica* try

```
f = BesselJ[0, t];
a=Integrate[f, {t, 0, 1}]
```

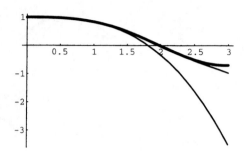

Figure 22: Problem 5(a). The exact solution is displayed together with its series approximate solutions with $n = 5$, 10, and 15.

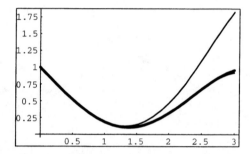

Figure 23: Problem 6(a). The exact solution is displayed together with its series approximate solutions with $n = 5$, 10, and 15.

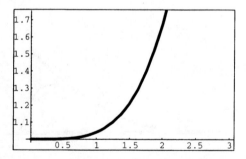

Figure 24: Problem 6(c). The exact solution is displayed together with its series approximate solutions with $n = 5$, 10, and 15.

returns

$$\text{HypergeometricPFQ}[\{-\}, \{1, -\frac{3}{2}\}, -(-\frac{1}{4})]$$

Applying the operator N to a results in 0.91973. Similarly

```
NIntegrate[f, {t, 0, 1}]
```

returns 0.91973.

7(b). 1.06701

7(c). To get this result in *Mathematica* try

`Integrate[BesselJ[0,t], {t, 0, Infinity}]`

Interestingly,

`NIntegrate[BesselJ[0,t], {t, 0, Infinity}]`

returns an incorrect answer with ample warning. To gain more insight into this problem, try

```
h = Integrate[BesselJ[0,t], t];
Plot[h, {t, 0, 30}]
```

Figure 25 shows the output.

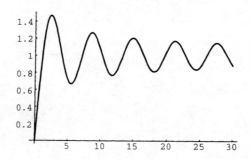

Figure 25: Problem 7(c). The graph of $\int_0^x J_0(t)\,dt$.

7(e). In *Mathematica* the command

`ff=Integrate[(BesselJ[0,x])^2, {x, 0, Infinity}]`

returns

```
EulerGamma + Log[8]
--------------------
        Pi
```

Chapter 5

Section 5.2

1(a). $1/(s+a)$; $s > -a$ **1(c).** $1/(s+a)^2$ **1(e).** $n!/s^{n+1}$ **1(g).** $b/((s-a)^2 + b^2)$

1(i). $2/((s+3)^2 + 4)$ **2(a).** $1/s(1 - \exp(-2s))$ **2(c).** $(1 - e^{-2\pi s})/(1 + s^2)$ **4(a).** $s/(s^2 - a^2)$

4(c). $2s/(1 + s^2)^2$ **4(e).** $2/(4s + s^3)$ **6(a).** $\Gamma(3/2)/(s^{3/2})$ **6(c).** $\Gamma(5/2)/(s^{5/2})$

7(a). $2/5\exp(t) + 8/5\exp(-4t)$ **7(c).** $-1/3\exp(t) + 2/3\exp(-t/2)\left(2\cos\frac{\sqrt{3}}{2}t + \sqrt{3}\sin\frac{\sqrt{3}}{2}t\right)$

8(a). $1/3\exp(-t) - 2/3\exp(t/2)\cos(\pi/3 + \sqrt{3}/2t)$

8(c). $1/5\exp(-t) - 2/5\exp\left((1 + \sqrt{5})t/4\right)\cos\left(\pi/5 + \sqrt{5 - \sqrt{5}}/2, t/2\right) -$

$-2/5\exp\left((1 - \sqrt{5})t/4\right)\cos\left(\pi/5 + \sqrt{5 + \sqrt{5}}/2, t/2\right)$ **11(c)i.** $1 - \exp(-t)$

11(c)ii. $1/a^2 - (a+1)/(2a^2) \exp(-at) + (a-1)/(2a^2) \exp(at)$ **12(c)i.** $1/(s-a)^2$
12(c)iii. $(s^2 - 4)/(s^2 + 4)^2$ **12(c)v.** $(6s^2 + 2)/(s^2 - 1)^3$

Section 5.3
3(a). $x(t) = 6/5 \exp(-2t) + 1/5 (-\cos t + 2\sin t)$
3(c). $x(t) = 33/29 \exp(-7t/3) + 1/29 (-4\cos t + 19\sin t)$
3(e). $y(t) = \frac{(-a+a^2 y_0 + b^2 y_0) \exp(-at) + a\cos bt + b\sin bt}{a^2 + b^2}$ **3(g).** $y(t) = \frac{t^2}{2} - \frac{t^3}{6}$
4(a). $y(t) = c_1 \cos 2t + c_2 \sin 2t$ **4(c).** $x(t) = \frac{3+5t}{50\,e^{3t}} + \frac{-3\cos t + 4\sin t}{50}$ **4(e).** $x(t) = -\frac{t\cos\omega t}{2\omega} + \frac{\sin\omega t}{2\omega^2}$
5(a). $x(t) = c_1 e^{-t} + c_2 e^{t/2} \cos \frac{\sqrt{3}t}{2} + c_3 e^{t/2} \sin \frac{\sqrt{3}t}{2}$
5(c). $y(t) = -2/3 e^{-t} + t^2 + 2/3 e^{t/2} \cos \frac{\sqrt{3}t}{2} - 2/\sqrt{3} e^{t/2} \sin \frac{\sqrt{3}t}{2}$
5(e). $y(t) = c_1 e^{rt} + c_2 e^{at} \cos bt + c_3 e^{at} \sin bt$, where $r = -0.682328$, $a = 0.341164$, and $b = -1.16154$

Section 5.4
1(a). $(x(t), y(t)) = (-\sin t, \cos t - \sin t)$
1(c). $(x(t), y(t)) = (1/30 \left(\left(-5 - \sqrt{5} - 5e^{\sqrt{5}t} + \sqrt{5}e^{\sqrt{5}t}\right) e^{(3-\sqrt{5})t/2} + 10\cos t + 10\sin t\right),$
$- 1/30 \left(-5 - 3\sqrt{5} - 5e^{\sqrt{5}t} + 3\sqrt{5}e^{\sqrt{5}t} + 10e^{(-3+\sqrt{5})t/2} \cos t\right) e^{(-3+\sqrt{5})t/2})$
1(e). $(x(t), y(t)) = \left(-e^t + 1/2 e^{(1-\sqrt{5})t/2} - \sqrt{5}/2 e^{(1-\sqrt{5})t/2} + 1/2 e^{(1+\sqrt{5})t/2}\right.$
$\left. +\sqrt{5}2 e^{(1+\sqrt{5})t/2}, -e^t + 1/2 e^{(1-\sqrt{5})t/2} - 3\sqrt{5}/2 e^{(1-\sqrt{5})t/2} + 1/2 e^{(1+\sqrt{5})t/2} +3\sqrt{5}2 e^{(1+\sqrt{5})t/2}\right)$
2(a). See Figure 26.

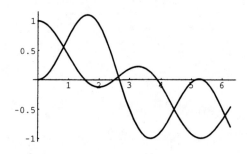

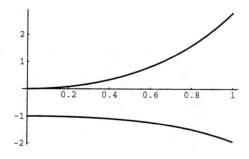

Figure 26: Problem 2(a).   Figure 27: Problem 2(c).

2(c). See Figure 27.
2(e). See Figure 28.
4(a). See Figure 29.
4(c). See Figure 30.
4(e). See Figure 31.

Section 5.5
1(a). See Figure 32.
1(c). See Figure 33.

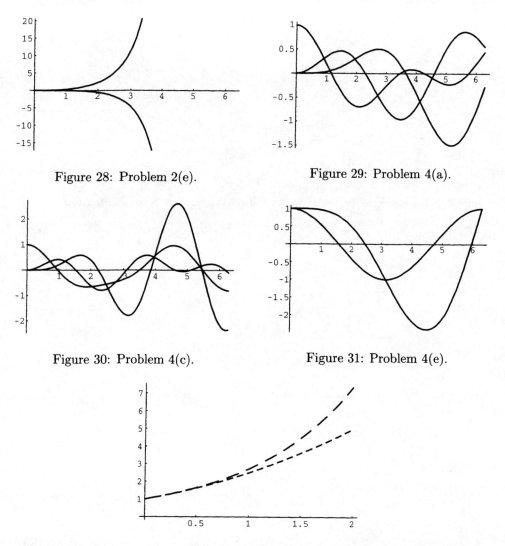

Figure 28: Problem 2(e).

Figure 29: Problem 4(a).

Figure 30: Problem 4(c).

Figure 31: Problem 4(e).

Figure 32: Problem 1(a). A three-term Taylor polynomial approximation is used.

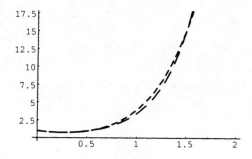

Figure 33: Problem 1(c). A five-term Taylor polynomial approximation is used.

1(e). See Figure 34.

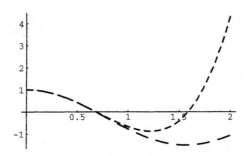

Figure 34: Problem 1(e). A five-term Taylor polynomial approximation is used.

2(a). See Figure 35.

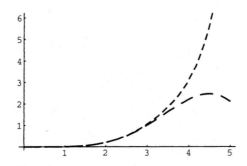

Figure 35: Problem 2(a). An eleven-term Taylor polynomial approximation is used.

2(c). See Figure 36.

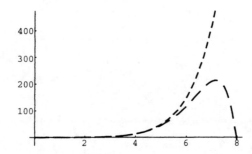

Figure 36: Problem 2(c). An eleven-term Taylor polynomial approximation is used.

3(a). See Figure 37.
3(c). See Figure 38.
4(a). See Figure 39.
4(c). See Figure 40.

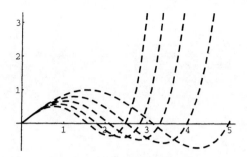

Figure 37: Problem 3(a). An eleven-term Taylor polynomial approximation is used.

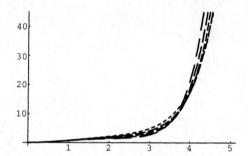

Figure 38: Problem 3(c). An eleven-term Taylor polynomial approximation is used.

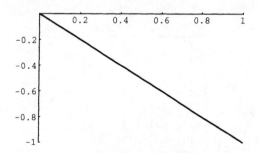

Figure 39: Problem 4(a). The exact solution is shown in solid line. A fifteen-term Taylor Polynomial approximation is used.

Section 5.6

2(a). See Figure 41.

2(c). See Figure 42.

2(e). See Figure 43.

3(a). $\frac{3}{e^{7s}s} + \frac{2}{e^{4s}s} + \frac{1}{e^s s}$ **3(c).** $\frac{1}{e^{2s}(1+s^2)}$

3(e). $\frac{1}{e^{10s}s} - \frac{1}{e^{9s}s} + \frac{1}{e^{8s}s} - \frac{1}{e^{7s}s} + \frac{1}{e^{6s}s} - \frac{1}{e^{5s}s} + \frac{1}{e^{4s}s} - \frac{1}{e^{3s}s} + \frac{1}{e^{2s}s} - \frac{1}{e^s s}$

4(a). $-\left(\sin(3-t)\,u(-3+t)\right)$ **4(c).** $\left(\frac{e^{1-t}}{3} - \frac{2\,e^{\frac{-1+t}{2}}\cos(\frac{\pi}{3}+\frac{\sqrt{3}(-1+t)}{2})}{3}\right)u(-1+t)$

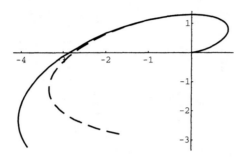

Figure 40: Problem 4(c). The exact solution, shown in solid line, was obtained using `NDSolve`. The approximate solution uses a fifteen-term Taylor polynomial.

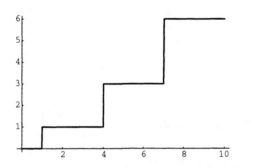

Figure 41: Problem 2(a).

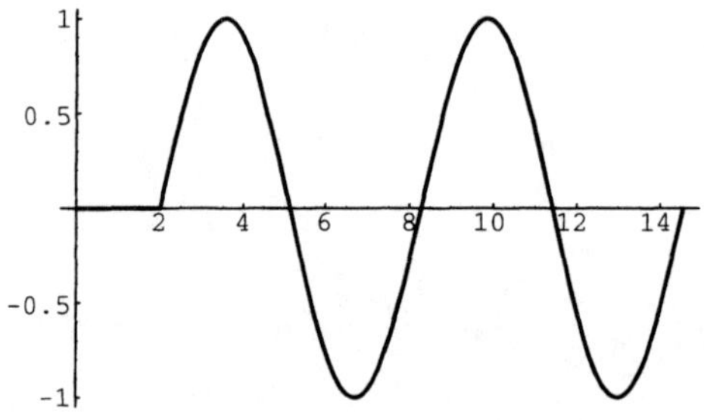

Figure 42: Problem 2(c).

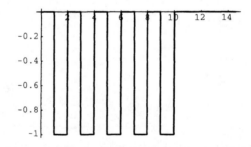

Figure 43: Problem 2(e).

5(a). $y(t) = \frac{u(-2+t) - \cos(4-2t)u(-2+t)}{4}$ **6(a).** $e^t \sin t$ **6(c).** $e^t \cos t - e^t \sin t$ **8.** $\frac{1-(s+1)e^{-s}}{s^2(1-e^{-s})}$

12(a). See Figure 44.

Section 5.7

1(a). $x(t) = 2/\sqrt{3} \exp\left(\frac{1-t}{2}\right) \sin \frac{\sqrt{3}}{2} (t-1) u(t-1)$

1(c). $y(t) = -(\sin(2-t)u(-2+t)) - \sin(1-t)u(-1+t)$ **2.** $x(t) = e^{\pi-x} \sin xu(\pi - x)$

7(a). See Figure 45.

7(c). See Figure 46.

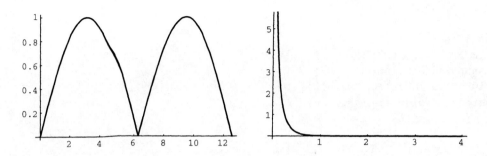

Figure 44: Problem 12(a).

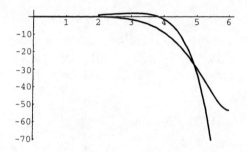

Figure 45: Problem 7(a).

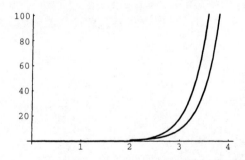

Figure 46: Problem 7(c).

Section 5.8

1(a). $\frac{t^3}{6}$ **1(c).** $t - \sin t$ **1(e).** $-1 + e^t - t$ **1(g).** $\frac{e^{2t}}{5} + \frac{-\cos(t) - 2\sin(t)}{5}$ **2(a).** $t - \sin t$

2(c). $\frac{-1}{4e^t} + \frac{e^t}{4} - \frac{\sin(t)}{2}$ **7(a).** $x(t) = \frac{1}{2} + \frac{1}{2e^{2t}}$ **7(c).** $x(t) = \frac{3\cos(\sqrt{2}\,t)}{2} + \frac{2 + 2t - 2t^2 + \sqrt{2}\sin(\sqrt{2}\,t)}{4}$

Chapter 6

Section 6.2

1(a). $\langle 2, 1, 3\rangle$; $\sqrt{14}$ **1(c).** $\langle -5/4, 3/4 - \sqrt{2}, 3\rangle$; $\sqrt{105/8 - 3\sqrt{2}/2}$ **2(a).** $\langle 3\cos 30, 3\sin 30\rangle$

3(a). $6/\sqrt{13}\,(1/2\mathbf{i} + 1/3\mathbf{j})$ **3(c).** $1/\sqrt{2}\langle 1, -1\rangle$ **3(e).** $ab/\sqrt{a^2 + b^2}\langle 1/a, -1/b, 0\rangle$

4(d). In *Mathematica* try

```
norm2[a_]:=a.a;
a={a1, a2}; b={b1, b2};
leftside=norm2[a+b]+norm2[a-b];
rightside=2*(norm2[a]+norm2[b]);
expression=leftside-rightside
Simplify[expression]
```

7(a).

```
<<Calculus'VectorAnalysis'
a={12,-3,0}; b={1, -1, 0};
CrossProduct[a, b]
```

8. $c = -0.682328, 0.341164 - 1.16154i, 0.341164 + 1.16154i$ **9.** $c = 1$
14. $a = 0.483752, -1.24188 - 2.15853i, -1.24188 + 2.15853i$ **15(a).** $\langle 0,0 \rangle$ **15(c).** $1/9\langle -1, 2, -1 \rangle$
15(e). $1/3\langle x + y + z, x + y + z, x + y + z \rangle$ **16.** $\langle a, b, 2a - 3b \rangle$

Section 6.3
2. No.   **4.** $a = 1$ and $a = 2$ **6(a).** The xy-plane

Section 6.4
1(a). Yes   **1(c).** No   **1(e).** Yes   **1(g).** Yes if $a(-b^2 + b - 1)$ does not vanish
3. No; The plane $4x - 3y - z = 0$ **5(a).** $a = 1$ and $a = -1$ **6.** All a and b
14(b). $\{\cos x, \cos 3x, \cos 5x\}$; 3   **15(b).** $\{1, t, t^2, \cdots, t^n\}$; $n + 1$ **17.** $\langle 2, -1 \rangle = 3/2\langle 1, -1 \rangle + 1/6\langle 3, 3 \rangle$

Section 6.5
5. See Figure 47.

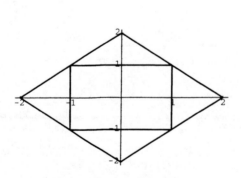

Figure 47: Problem 5.

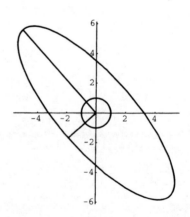

Figure 48: Problem 7.

7. $t_1 = 0.735564$; $t_2 = 3.87716$; See Figure 48.

Section 6.6

$$\textbf{2(b). } \sum_{i=1}^{10} 1/i! A^i = \begin{bmatrix} -\frac{7073}{8960} & \frac{3523}{22400} & 0 & 0 \\ \frac{3523}{22400} & -\frac{7073}{8960} & 0 & 0 \\ \frac{1715}{20736} & \frac{326359}{1814400} & -\frac{7073}{8960} & \frac{3523}{22400} \\ \frac{10249}{453600} & \frac{1715}{20736} & \frac{3523}{22400} & -\frac{7073}{8960} \end{bmatrix}$$

6(a). Symmetric part $= \begin{bmatrix} 1 & \frac{5}{2} \\ \frac{5}{2} & 4 \end{bmatrix}$; Anti-symmetric part $= \begin{bmatrix} 0 & -\frac{1}{2} \\ \frac{1}{2} & 0 \end{bmatrix}$

6(d). Symmetric part $= \begin{bmatrix} 1 & \frac{1}{2} & ,2 \\ \frac{1}{2} & 0 & 0 \\ 2 & 0 & 2 \end{bmatrix}$; Anti-symmetric part $= \begin{bmatrix} 0 & \frac{3}{2} & 1 \\ -\frac{3}{2} & 0 & 1 \\ -1 & -1 & 0 \end{bmatrix}$

6(h). Symmetric part $= \begin{bmatrix} 1 & t & t^2 \\ t & t^2 & \frac{-t^2+t^3}{2} \\ t^2 & \frac{-t^2+t^3}{2} & t^4 \end{bmatrix}$; Anti-symmetric

part $= \begin{bmatrix} 0 & 0 & 0 \\ 0 & 0 & \frac{t^2+t^3}{2} \\ 0 & \frac{-t^2-t^3}{2} & 0 \end{bmatrix}$

13. 2   **14.** Basis: $\left\{ \begin{bmatrix} 1 & 1 \\ 0 & 0 \end{bmatrix}, \begin{bmatrix} 0 & 1 \\ 0 & 1 \end{bmatrix} \right\}$; Dimension $= 2$

Section 6.7

2(a). $\begin{bmatrix} 1 & -1 \\ 1 & 1 \end{bmatrix}$

5(a). Each vector in the domain is mapped to one with the same direction but twice as long.

5(f). each vector is rotated 90 degrees counterclockwise.   **6.** A^2 rotates each vector by 2θ

Section 6.8

3(a). $x = 39/5$, $y = 6/5$ **3(c).** $x = -13/70$, $y = -6/35$ **3(e).** $x = 19$, $y = 13$

8. $1/(-ceg + bfg + cdh - afh - bdi + aei) \begin{bmatrix} -fh + ei & ch - bi & -ce + bf \\ fg - di & -cg + ai & cd - af \\ -eg + dh & bg - ah & -bd + ae \end{bmatrix}$

Section 6.9

1(a). -1; $\begin{bmatrix} 0 & 0 & 1 \\ 0 & 1 & 0 \\ 1 & 0 & 0 \end{bmatrix}$ **1(c).** $2a^2$; $1/(2a) \begin{bmatrix} 1 & -1 \\ 1 & 1 \end{bmatrix}$ **1(g).** 209; $1/(209) \begin{bmatrix} 56 & 15 & 4 & 1 \\ 15 & 60 & 16 & 4 \\ 4 & 16 & 60 & 15 \\ 1 & 4 & 15 & 56 \end{bmatrix}$

Section 6.10

1(a). Unique solution; $x = 5/2$, $y = -1/2$, $z = 3/2$ **1(c).** No solution

1(e). In *Mathematica* try

```
vars=Table[x[i], {i, 5}];
Table[eqn[j]=Sum[1/(i+j) x[i], {i, 5}]-1/j, {j, 5}];
Solve[Table[eqn[i]==0, {i, 5}], vars]
```

Mathematica responds with

```
{{x[1] -> 30,x[2] -> -210,x[3] -> 560,x[4] -> -630,
   x[5] -> 252}}
```

Alternatively, you may try

```
M=Table[1/(i+j), {i, 5}, {j, 5}];
F=Table[1/j, {j, 5}];
LinearSolve[M,F]
```

1(g). Two parameter family of solutions; $x = -1/2 - 2w$, $y = -1 - 3w + z$; $u = -1/2 + w$ **2(a).** 4
2(b). $b \neq 1$ **4(a).** 2 **4(c).** 1 **4(e).** 3 **4(g).** 3 if $a \neq 0$; 0 if $a = 0$ **4(i).** 4

Section 6.11
1(a). $\mathbf{x} = 1/5(2\mathbf{i} - \mathbf{j}) + 6/5(1/2\mathbf{i} + \mathbf{j})$ **1(c).** $\mathbf{x} = 2/5(\mathbf{i} + 3\mathbf{j}) + 3/10(2\mathbf{i} - 2/3\mathbf{j})$
2(a). $\mathbf{x} = 3/5(2\mathbf{i} - \mathbf{k}) + 2\mathbf{j} - 1/5(\mathbf{i} + 2\mathbf{k})$
2(c). $\mathbf{x} = -1/2(\mathbf{i} - \mathbf{j}) + 9/17(2\mathbf{i} + 2\mathbf{j} - 3\mathbf{k}) + 5/34(3\mathbf{i} + 3\mathbf{j} + 4\mathbf{k})$
4. $\langle 3, -2, 3 \rangle = 1/3 \langle 1, 2, 3 \rangle - 8/3 \langle -1, 1, -1 \rangle$

Section 6.12
1(a). $\frac{1-\sqrt{5}}{2}$ and $\frac{1+\sqrt{5}}{2}$; $\langle \frac{1-\sqrt{5}}{2}, 1 \rangle\}$ and $\langle \frac{1+\sqrt{5}}{2}, 1 \rangle$
1(c). $1 - \sqrt{5}$ and $1 + \sqrt{5}$; $\langle -2 - \sqrt{5}, 1 \rangle$ and $\langle -2 + \sqrt{5}, 1 \rangle$
1(e). $(1 - i)a$ and $(1 + i)a$; $\langle i, 1 \rangle$ and $\langle -i, 1 \rangle$ **1(g).** a, a, and a; $\langle 1, a, a^2 \rangle$
1(i). -1, 1, and 1; $\langle -1, 1, 0 \rangle$, $\langle 0, 0, 1 \rangle$, and $\langle 1, 1, 0 \rangle$
1(l). $1/2(3 - \sqrt{5})$, $1/2(5 - \sqrt{5})$, $1/2(3 + \sqrt{5})$, and $1/2(5 + \sqrt{5})$; $\langle 1, 1/2(1 + \sqrt{5}), 1/2(1 + \sqrt{5}), 1 \rangle$,
$\langle -1, 1/2(1 - \sqrt{5}), 1/2(-1 + \sqrt{5}), 1 \rangle$, $\langle 1, 1/2(1 - \sqrt{5}), 1/2(1 - \sqrt{5}), 1 \rangle$, $\langle 0, 0, 1 \rangle$, and $\langle 1, 1, 0 \rangle$

2. $Q = 1/\sqrt{2} \begin{bmatrix} -1 & 1 \\ 1 & 1 \end{bmatrix}$ **3.** $Q = \begin{bmatrix} 0.0455793 & -0.826926 & -0.56046 \\ -0.239134 & -0.553763 & 0.797597 \\ 0.969916 & -0.0976712 & 0.222986 \end{bmatrix}$

Chapter 7

Section 7.2
2(a). $x' = y$, $y' = -4x$ **2(c).** $y' = x$, $x' = z$, $z' = -y$
2(e). $x' = z$, $z' = y/(x^2 + y^2)$, $y' = w$, $w' = -x/(x^2 + y^2)$ **3(a).** $\langle x(t), y(t) \rangle = \langle -2e^t, 3e^{2t} \rangle$
3(c). $\langle x(t), y(t) \rangle =$
$\langle 1/\sqrt{13} e^{\frac{t-\sqrt{13}t}{2}} \left(-1 + e^{\sqrt{13}t} \right), 1/26 e^{\frac{t-\sqrt{13}t}{2}} \left(13 - 3\sqrt{13} + 13 e^{\sqrt{13}t} + 3\sqrt{13} e^{\sqrt{13}t} \right) \rangle$

3(e). $\langle x(t), y(t) \rangle = \langle 1/10 \left(-e^{\frac{t-\sqrt{5}t}{2}} \left(-3 - \sqrt{5} - 3 e^{\sqrt{5}t} + \sqrt{5} e^{\sqrt{5}t} \right) - 6 \cos t + 2 \sin t \right),$
$-1/10 e^{-t} \left(-e^{-1/2(-3+\sqrt{5})t} \left(-1 - \sqrt{5} - e^{\sqrt{5}t} + \sqrt{5} e^{\sqrt{5}t} \right) - 2e^t Cos[t] + 4e^t Sin[t] \right) \rangle$

3(g). $\langle x(t), y(t)\rangle =$
$\langle \frac{1}{2} - 1/2 e^{-3t/2} \cos(\frac{\sqrt{7}t}{2} - 3/2\sqrt{7}e^{-3t/2} \sin(\frac{\sqrt{7}t}{2}), \frac{1}{2} - t - 1/2 e^{-3t/2} \cos\left(\frac{\sqrt{7}t}{2}\right) + 5/14 e^{-3t/2} \sin\left(\frac{\sqrt{7}t}{2}\right)\rangle$

3(i). $\langle x(t), y(t)\rangle = \langle \frac{-5 e^{(a-b)t}}{2} + \frac{e^{(a+b)t}}{2}, \frac{5 e^{(a-b)t}}{2} + \frac{e^{(a+b)t}}{2}\rangle$ **3(k).** $\langle x(t), y(t)\rangle = \langle 0, 0, -e^{3t}\rangle$

3(m). $\langle x(t), y(t)\rangle = \langle 0.166667 e^{-0.259921\,t} - 0.5\,\cos t + 0.333333\,e^{1.62996\,t}\cos(1.09112\,t) - 0.5\,\sin t,$
$-0.64122 e^{-0.259921\,t} + 0.5\,\cos t + 0.0112598\,e^{1.62996\,t}\cos(1.09112\,t) - 0.5\,\sin t -$
$0.169568\,e^{1.62996\,t}\sin(1.09112\,t), -0.0433202 e^{-0.259921\,t} - 0.5\,\cos t - 0.58664\,e^{1.62996\,t}\cos(1.09112\,t) -$
$0.5\,\sin t - 0.288675\,e^{1.62996\,t}\sin(1.09112\,t)\rangle$

3(o). $\langle x(t), y(t)\rangle = \langle \frac{1}{2a^2 e^{3at}} - \frac{1}{a^2 e^{2at}} + \frac{1}{2a^2 e^{at}}, \frac{-3}{2a e^{3at}} + \frac{2}{a e^{2at}} - \frac{1}{2a e^{at}}, \frac{9}{2 e^{3at}} - \frac{4}{e^{2at}} + \frac{1}{2 e^{at}}\rangle$

5(a). $\langle e^{(1-2i)t}\begin{bmatrix} i \\ 1 \end{bmatrix}, e^{(1+2i)t}\begin{bmatrix} -i \\ 1 \end{bmatrix}\rangle$ **5(c).** $\langle e^{-t}\begin{bmatrix} 1 \\ -1 \\ 1 \end{bmatrix}, e^{-2t}\begin{bmatrix} 1 \\ -2 \\ 4 \end{bmatrix}, e^{-3t}\begin{bmatrix} 1 \\ -3 \\ 9 \end{bmatrix}\rangle$

8(a). $x' = 2y,\ y' = -2x;\ \frac{dy}{dx}\big|_{(1,-1)} = 1$ **8(c).** $x' = -\frac{y}{(x^2+y^2)^{\frac{3}{2}}},\ y' = \frac{x}{(x^2+y^2)^{\frac{3}{2}}};\ \frac{dy}{dx}\big|_{2,2} = -1$

8(e). $x' = \frac{y}{x^2+y^2},\ y' = -\frac{x}{x^2+y^2};\ \frac{dy}{dx}\big|_{1,1} = -1,\ \frac{dy}{dx}\big|_{1,-1} = 1$ **9(a).** $x' = x,\ y' = y;\ \frac{dy}{dx}\big|_{1,1} = -1$

9(c). $x' = -\frac{x}{(x^2+y^2)^{\frac{3}{2}}},\ y' = -\frac{y}{(x^2+y^2)^{\frac{3}{2}}};\ \frac{dy}{dx}\big|_{1,1} = 1,\ \frac{dy}{dx}\big|_{2,2} = 1$

9(e). $x' = \frac{x}{x^2+y^2},\ y' = \frac{y}{x^2+y^2};\ \frac{dy}{dx}\big|_{1,1} = 1,\ \frac{dy}{dx}\big|_{1,-1} = -1$

10(d). $e^{tA} = \begin{bmatrix} e^{2t}\,(\cos t - \sin t) & -e^{2t}\,\sin t \\ 2\,e^{2t}\,\sin t & e^{2t}\,(\cos t + \sin t) \end{bmatrix}$. In *Mathematica* try

```
A={{1,-1}, {2,3}};
Simplify[ComplexExpand[MatrixExp[t A]]]
```

Section 7.3
1(a). $t_0 \in R,\ \mathbf{x}_0 \in E^2$ **1(c).** $t_0 \neq 2$ **1(e).** $y_0 > 0$ **2(a).** $x' = y,\ y' = 3x - 2y + \sin t$
2(c). $y' = z,\ z' = w,\ w' = 3z - 3w + e^x$. Using `NDSolve` with initial data $y(0) = 1,\ y'(0) = -1,$
$y''(0) = 2$ leads to Figure 49.

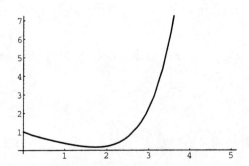

Figure 49: Problem 2(c).

2(e). $x' = y,\ y' = z,\ z' = w,\ w' = -4z + 16x$. See Figure 50 where $x(0) = 1,\ x'(0) = -1,\ x''(0) = 3,$
and $x'''(0) = 4$ are used.
2(g). $x' = y,\ y' = -(1 - x^2)y - x$. See Figure 51 where $x(0) = 1$ and $x'(0) = -1$ are used.
4(a). $x'' + x' - x = 0$ **4(c).** $x'' + x' - x = 0$

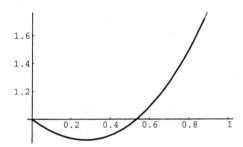

Figure 50: Problem 2(e).

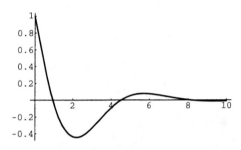

Figure 51: Problem 2(g).

4(e). First solve the first equation for ty in terms of the remaining terms. Next, differentiate the first equation with respect to t, replace y' and ty using the second equation and the result of the first step, respectively, to obtain

$$x'' = (\sin t + t + 2/t)x' + \left(\cos t - t\sin t - t^2\cos t - 2\frac{\sin t}{t}\right)$$

4(g). $x''' - x' = 0$ **4(i).** $x''' + x'' + 4x' + 2x = 0$ **5(c).** $x_1^{(iv)} + 9/2x_1'' + 5/2x_1 = 0$

Section 7.4
1(a). Linearly independent   **1(c).** Linearly independent   **1(e).** Linearly independent

3(a). $\langle 2 - 2\cos t, -t + 2\sin t\rangle$

3(c). $\langle\frac{-1 - 3e^{2t} + 4e^{3t} - 6e^{2t}t}{12e^t}, \frac{1}{12e^t} - \frac{3e^t}{4} + \frac{2e^{2t}}{3} - \frac{e^t t}{2}, \frac{-1}{12e^t} - \frac{5e^t}{4} + \frac{4e^{2t}}{3} - \frac{e^t t}{2}\rangle$

Section 7.5
1(a). $\langle(e^{4t} - 3e^{4t}t)\,c_1 + 3e^{4t}t\,c_2, -3e^{4t}t\,c_1 + (e^{4t} + 3e^{4t}t)\,c_2\rangle$ **1(c).** $\langle\frac{5c_1}{e^{4t}} - e^{4t}\,c_2, \frac{3c_1}{e^{4t}} + e^{4t}\,c_2\rangle$

1(e). $\langle -\frac{e^{(-1-\sqrt{3})t}\,c_1}{\sqrt{3}} + \frac{e^{(-1+\sqrt{3})t}\,c_2}{\sqrt{3}}, e^{(-1-\sqrt{3})t}\,c_1 + e^{(-1+\sqrt{3})t}\,c_2\rangle$

1(g). $\langle\left(e^{-t} - \frac{2t}{e^t}\right)c_1 + \frac{tc_3}{e^t}, \left(\frac{t}{e^t} - \frac{t^2}{e^t}\right)c_1 + \frac{c_2}{e^t} + \frac{t^2c_3}{2e^t}, \frac{-4tc_1}{e^t} + \left(e^{-t} + \frac{2t}{e^t}\right)c_3\rangle$

1(i). $x(t) = -1/3e^{2t}\left(2\left(-3 + 6t - 6t^2 + 4t^3\right)c_1 - 6t\left(1 - 2t + 2t^2\right)c_2 + t\left(3c_3 - 6tc_3 + tc_4\right)\right)$,
$y(t) = 1/6e^{2t}\left(c_2 + 6t\left(-2c_2 + c_3\right) + 2t^3\left(-8c_1 + 12c_2 - 6c_3 + c_4\right) + 3t^2\left(4c_2 - 4c_3 + c_4\right)\right)$,
$z(t) = -1/3e^{2t}\left(-3c_3 + t\left(6c_3 - 3c_4\right) + 6t^2\left(4c_1 - 8c_2 + 5c_3 - c_4\right) + 2t^3\left(8c_1 - 12c_2 + 6c_3 - c_4\right)\right)$, $w(t) = -2te^{2t}\left(8c_1 - 16c_2 + 12c_3 - 3c_4 + 6t^2\left(16c_1 - 28c_2 + 16c_3 - 3c_4\right) + 4t^3\left(8c_1 - 12c_2 + 6c_3 - c_4\right) - 3c_4\right)$ **4(a).** $\langle -\left(e^t\left(-1 + t\right)\right), -e^t\rangle$

4(c). $\langle e^{2t}(\sin t - 2\cos t), e^{2t}(\cos t - 2\sin t)\rangle$ **4(e).** $-\langle e^{at}(-1+at), e^{at}\rangle$

5(a). $\langle \frac{e^t t^2}{2}, \frac{e^t t(2+t)}{2}, \frac{e^t(2+4t+t^2)}{2}\rangle$

5(d). $\langle \frac{1}{2\sqrt{17}}\left(-1 + e^{2\sqrt{17}t}e^{(-3+\sqrt{17})t}\right), 0, \left(\frac{1}{2} - \frac{3}{2\sqrt{17}}\right)e^{(-3+\sqrt{17})t} + \frac{1}{34}(17+3\sqrt{17})e^{(3+\sqrt{17})t}\rangle$

Section 7.6

1(a). Diagonalizable. $Q = \begin{bmatrix} -1 & 1 \\ 1 & 1 \end{bmatrix}$. $\mathbf{x}(t) = \langle -e^{-t}, e^{-t}\rangle$

1(c). Diagonalizable. $Q = \begin{bmatrix} -i & i \\ 1 & 1 \end{bmatrix}$. $\mathbf{x}(t) = \langle -\cos t - \sin t, \cos t - \sin t\rangle$

1(e). Diagonalizable. $Q = \begin{bmatrix} -\frac{1}{2}+\frac{i\sqrt{3}}{2} & -\frac{1}{2}-\frac{i\sqrt{3}}{2} & 1 \\ 1 & 1 & 1 \\ -\frac{1}{2}-\frac{i\sqrt{3}}{2} & -\frac{1}{2}+\frac{i\sqrt{3}}{2} & 1 \end{bmatrix}$

$\langle x(t), y(t), z(t)\rangle = \langle \frac{2e^t}{3} - \frac{2\sqrt{3}}{3}e^{-t/2}\left(\sin\frac{\sqrt{3}t}{2} + \cos\frac{\sqrt{3}t}{2}\right), \frac{2e^t}{3} + \frac{4\sqrt{3}}{3}\cos\frac{\sqrt{3}t}{2},$

$\frac{2e^t}{3} + \frac{2\sqrt{3}}{3}e^{-t/2}\left(\sin\frac{\sqrt{3}t}{2} - \cos\frac{\sqrt{3}t}{2}\right)\rangle$

2(a). $x' = y,\ y' = -2x - 3y;\ A = \begin{bmatrix} 0 & 1 \\ -2 & -3 \end{bmatrix},\ Q = \begin{bmatrix} -1 & -1 \\ 2 & 1 \end{bmatrix}$

2(c). $y' = z,\ z' = -ay;\ A = \begin{bmatrix} 0 & 1 \\ -a & 0 \end{bmatrix},\ Q = \begin{bmatrix} i/\sqrt{a} & -i/\sqrt{a} \\ 1 & 1 \end{bmatrix}$

2(e). $y' = x,\ x' = z,\ z' = w,\ w' = -64y;\ A = \begin{bmatrix} 0 & 1 & 0 & 0 \\ 0 & 0 & 1 & 0 \\ 0 & 0 & 0 & 1 \\ -64 & 0 & 0 & 0 \end{bmatrix}$,

$Q = \begin{bmatrix} 1+i & 1-i & -1+i & -1-i \\ -4i & 4i & 4i & -4i \\ -8+8i & -8-8i & 8+8i & 8-8i \\ 32 & 32 & 32 & 32 \end{bmatrix}$

3. $\mathbf{x}(t) = \langle e^{2t}, \cos t - \sin t, \cos t + \sin t\rangle;\ \int_0^1 ||\mathbf{x}(s)||^2\,ds = \frac{e^4+7}{4};\ \int_0^1 ||\mathbf{x}'(s)||^2\,ds = e^4 + 1$

Section 7.7

5(a). $\mathbf{x}_p = \langle -\frac{1}{8}\cos 3t - \frac{2}{3}\cos 2t, \frac{1}{3}\sin 2t + \frac{3}{8}\sin 3t\rangle$

5(c). Let $\mathbf{e} = \begin{bmatrix} 1 \\ 1 \end{bmatrix}$. Look for a solution of the type $\mathbf{x}_p = e^{5t}\mathbf{a} + te^{5t}\mathbf{b}$. Show that $\mathbf{a}$ and $\mathbf{b}$ satisfy the equations $(A - 5I)\mathbf{a} = \mathbf{b} + \mathbf{e}$ and $A\mathbf{b} = 5\mathbf{b}$. Conclude that $\mathbf{b}$ is an eigebvector of A associated with the eigenvalue 5 and $(A - 5I)^2\mathbf{a} = (A - 5I)\mathbf{e}$. Finally, show that $\mathbf{a} = \begin{bmatrix} 13/7 \\ 1 \end{bmatrix}$ and $\mathbf{b} = \begin{bmatrix} \frac{8}{7} \\ \frac{4}{7} \end{bmatrix}$.

5(e). $\langle 2+t, 1-t\rangle$ **6(a).** $\langle 1 - 6t + t^3, 2(t^2 - 3)\rangle$

6(c). $\langle \frac{(2e^9+e^{3t}-3e^{6+t})u(-3+t)+e^3(-4e^6+e^{3t}+3e^{4+t})u(-2+t)}{6e^9},$

$\frac{(-4e^9+e^{3t}+3e^{6+t})u(-3+t)+e^3(e^2-e^t)^2(2e^2+e^t)u(-2+t)}{6e^9}\rangle$

10(a). $\langle \frac{-2\cos(2t)}{3} - \frac{\cos(3t)}{8}, \frac{\sin(2t)}{3} + \frac{3\sin(3t)}{8} \rangle$ **10(c).** $\langle \frac{e^{5t}(-1+56t)}{49}, \frac{e^{5t}(3+28t)}{49} \rangle$

10(e). $\langle 2+t, 1-t \rangle$ **13.** $\mathbf{x_p}(t) = \begin{bmatrix} 0 \\ -1 \end{bmatrix}$ is a particular solution for any a. **14(a).** $\frac{1}{25}\langle -3, -11, 1 \rangle$

14(c). $-\frac{1}{12}\langle 2, 2, 5, 5 \rangle$

Section 7.8
2(b). See Figure 52

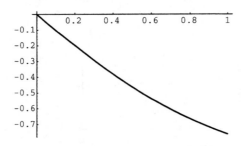

Figure 52: Problem 2(b). The approximate and the exact solutions are shown.

2(c). See Figure 53

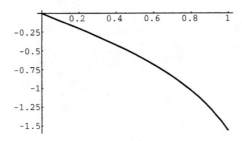

Figure 53: Problem 2(c). The approximate and the exact solutions are shown.

5. See Figure 54

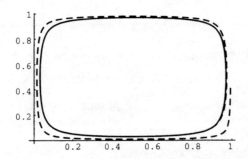

Figure 54: Problem 5. The approximate and the exact solutions through the point $(0.9, 0.1)$ are shown.

Section 7.9

0	0
0.1	0.01
0.2	0.03445
0.3	0.0696653
0.4	0.112901
0.5	0.162111
0.6	0.215773
0.7	0.272751
0.8	0.332199
0.9	0.393488
1.	0.456149

1(a). Let $h = 0.1$ and $n = 10$.

0	0.1
0.1	0.10101
0.2	0.102041
0.3	0.103093
0.4	0.104166
0.5	0.105263
0.6	0.106383
0.7	0.107526
0.8	0.108695
0.9	0.10989
1.	0.11111

1(c). Let $h = 0.1$ and $n = 10$.

0	0
0.1	0.0048375
0.2	0.0187309
0.3	0.0408184
0.4	0.0703203
0.5	0.106531
0.6	0.148812
0.7	0.196586
0.8	0.249329
0.9	0.30657
1.	0.36788

2(a). Let $h = 0.1$ and $n = 10$.

0	0.1
0.1	0.10101
0.2	0.102041
0.3	0.103093
0.4	0.104167
0.5	0.105263
0.6	0.106383
0.7	0.107527
0.8	0.108696
0.9	0.10989
1.	0.111111

2(c). Let $h = 0.1$ and $n = 10$.

4(a). Let $h = 0.1$, $n = 10$, $x_0 = 1$, and $y_0 = 2$. The fourth-order Runge–Kutta yields

0	1	2
0.1	1.19467	1.89017
0.2	1.3774	1.76146
0.3	1.54638	1.61515
0.4	1.6999	1.4527
0.5	1.83643	1.27574
0.6	1.95462	1.08603
0.7	2.05328	0.885468
0.8	2.13142	0.676059
0.9	2.18826	0.459895
1.	2.22324	0.239135

4(c). Let $h = 0.1$, $n = 10$, $x_0 = 1$, and $y_0 = 2$.

0	1	2
0.1	1.19563	1.91109
0.2	1.38198	1.81515
0.3	1.55854	1.71578
0.4	1.72513	1.61615
0.5	1.88186	1.51897
0.6	2.02908	1.42639
0.7	2.16734	1.34008
0.8	2.29735	1.26127
0.9	2.41988	1.19082
1.	2.53581	1.12929

0	−3	0.1
0.1	−2.91141	0.100776
0.2	−2.82353	0.101633
0.3	−2.73642	0.102581
0.4	−2.65015	0.103635
0.5	−2.56479	0.104807
0.6	−2.48043	0.106117
0.7	−2.39715	0.107586
0.8	−2.31506	0.109238
0.9	−2.23426	0.111103
1.	−2.15487	0.113216

4(e). Let $h = 0.1$, $n = 10$, $x_0 = 1$, and $y_0 = 2$.

Section 7.10

3(a). See Figure 55.

Figure 55: Problem 3(a).

3(c). See Figure 56.

Figure 56: Problem 3(c).

3(e). See Figure 57.

Figure 57: Problem 3(e).

Section 7.11

1(a). $(0,0)$; unstable   **1(c).** $(0,0)$; stable

1(e). $y' = z$, $z' = -0.1z - \sin y$; Equilibrium points are $(0,0)$ and $(\pi, 0)$; $(0,0)$ is stable; $(0,\pi)$ is unstable.

3(a). Example 7.2.1; There is only one equilibrium point, namely, $(0,0,0,0)$, which is stable. Example 7.2.3: There are two equilibrium points at $(-1,0)$ and $(1,0)$, both of which are unstable. Example 7.2.4: There are equilbrium points at $(\pm n, \pm m)$ where m and n are integers, all of which are unstable. Example 7.2.5: There are no equilibrium points.

Section 7.12

1(a). $(0,0)$ is asymptotically stable and an improper node.   **1(c).** $(0,0)$ is unstable.

1(e). $(-1,-1)$ is a stable center.   **2(a).** $b < 0$ and $-b^2/4 < a < 0$ **2(c).** $b = 0$ and $a < 0$

2(e). $b < 0$ and $b^2 = -4a$

Section 7.13

1(a). $(0,0)$ is a center. See Figure 58.

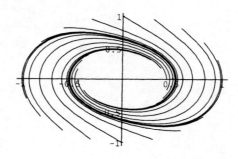

Figure 58: Problem 1(a).

1(c). $(0,0)$ is a center. See Figure 59.

2. Both equilibrium points $(1,0)$ and $(-1,0)$ are unstable saddle points.

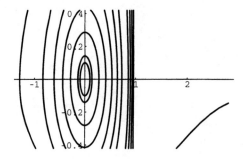

Figure 59: Problem 1(c).

3(a). $(0,0)$, $(0,1/d)$, $(1/b,0)$, $(1/b,1/d)$ are the equilibrium points.

3(c). See Figure 60

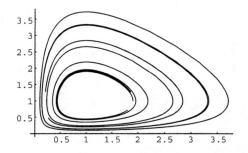

Figure 60: Problem 3(c).

Index

Three easy ways to receive more information on MATLAB:

Fax this form to (508) 647-7101

Mail this form to The MathWorks, Inc., 24 Prime Park Way, Natick, MA 01760

@ Send e-mail to *info@mathworks.com* and request kit KP108

Send me a *free* copy of the *MATLAB® Product Catalog.*

This catalog provides information on MATLAB, Toolboxes, SIMULINK®, Blocksets, and more.

I am currently a MATLAB user: ☐ Yes ☐ No

Computer Platform: ☐ PC or Macintosh   ☐ UNIX workstation

For the fastest response, fax to (508) 647-7101 or send e-mail to *info@mathworks.com* and request KP108.

NAME

E-MAIL

TITLE

COMPANY/UNIVERSITY

DEPT. OR M/S

ADDRESS

CITY/STATE/COUNTRY/ZIP

PHONE

FAX

GRADUATION DATE IF APPLICABLE

R-BK-JOH/411v0/KP108